AUTOMATION AND CONTROL ENGINEERING

ADVANCES IN MISSILE GUIDANCE, CONTROL, AND ESTIMATION

T0225492

AUTOMATION AND CONTROL ENGINEERING

ADVANCES IN MISSILE GUIDANCE, CONTROL, AND ESTIMATION

S. N. BALAKRISHNAN, A. TSOURDOS, AND B. A. WHITE

CRC Press
Taylor & Francis Group
Boca Raton London New York

CRC Press is an imprint of the
Taylor & Francis Group, an informa business

CRC Press
Taylor & Francis Group
6000 Broken Sound Parkway NW, Suite 300
Boca Raton, FL 33487-2742

First issued in paperback 2017

© 2013 by Taylor & Francis Group, LLC
CRC Press is an imprint of Taylor & Francis Group, an Informa business

No claim to original U.S. Government works

Version Date: 20120719

ISBN 13: 978-1-138-07295-4 (pbk)
ISBN 13: 978-1-4200-8313-2 (hbk)

Library of Congress Cataloging-in-Publication Data

Advances in missile guidance, control, and estimation / editors, S.N. Balakrishnan, A. Tsourdos, B.A. White.
      p. cm. -- (Control series)
   Includes bibliographical references and index.
   ISBN 978-1-4200-8313-2 (hardback)
   1. Guided missiles--Guidance systems. I. Balakrishnan, S. N. II. Tsourdos, A. III. White, B. A.

UG1310.A38 2012
623.4'519--dc23   2012024643

Visit the Taylor & Francis Web site at
http://www.taylorandfrancis.com

and the CRC Press Web site at
http://www.crcpress.com

We would like to thank the contributors for their patience

over the period of the production of this book.

We would also like to dedicate this work to the following people:

My wife Roja and sons Vijay, Karthik, and Vasanth for the joy they bring

and for showing me that real life is about more than just engineering reasoning,

and my beloved Midwest for its disarming splendor and for the fairness,

warmth, and generosity of its people.—Sivasubramanya N. Balakrishnan.

My wife Liz, who has always kept my feet on the ground and

my head out of the clouds, for providing the support to keep our

growing family happy and together.—Brian A. White.

My parents for their unconditional support and love

all these years.—Antonios Tsourdos

Contents

Preface

This book is based on the knowledge, experience, and findings of the authors gained over decades of research, development, and teaching. To our knowledge, this book is unique in bringing together all aspects of missile system development—guidance, control, navigation/estimation, and implementation. The richness of the contents of this book is due to having contributors in guidance and control, and estimation theories and implementation by authors who are themselves leaders in their respective fields. The chapter contributors are at the forefront of modern developments in missile systems and are the originators of the concepts and algorithms described in each chapter, or have put them into practice in current missile systems. Many chapters cover the basic theory in a tutorial form that makes this book almost a text book on the latest topics in guidance theory, nonlinear control, or estimation.

We thank all contributors for taking the time to make valuable contributions to this book. We are especially grateful to the contributors from industry, for whom this book was a long and arduous endeavor, undertaken in addition to their regular job demands.

Contributors

Daniel Alazard
SUPAERO
Toulouse, France

S. N. Balakrishnan
Department of Mechanical and
  Aerospace Engineering
Missouri University of Science and
  Technology
Rolla, Missouri

L. Bruyere
School of Engineering
Cranfield University
Swindon, United Kingdom

Hangju Cho
The Agency for Defense
  Development
Daejeon, Republic of Korea

Michael Dancer
IST-Rolla
Rolla, Missouri

Pini Gurfil
Faculty of Aerospace Engineering
Technion–Israel Institute of
  Technology
Haifa, Israel

Nathan Harl
Integrated Military Systems
  Development Center
Sandia National Laboratories
Albuquerque, New Mexico

G. Hexner
Rafael Advanced Defense Systems
Haifa, Israel

M. Lauzon
Defence R&D Canada–Valcartier
Quebec City, Quebec, Canada

N. Léchevin
Defence R&D Canada–Valcartier
Quebec City, Quebec, Canada

Ernest J. Ohlmeyer
Aero Science Applications
King George, Virginia

C. Phillips
Naval Surface Warfare Center
  Dahlgren Division
Dahlgren, Virginia

C. A. Rabbath
Defence R&D Canada–Valcartier
Quebec City, Quebec, Canada

Karthikeyan Rajagopal
Department of Mechanical and
  Aerospace Engineering
Missouri University of Science and
  Technology
Rolla, Missouri

Chang-Kyung Ryoo
Department of Aerospace
  Engineering
Inha University
Incheon, Republic of Korea

David Salmond
QinetiQ
Farnborough, United Kingdom

Tal Shima
Faculty of Aerospace Engineering
Technion-Israel Institute of
    Technology
Haifa, Israel

Josef Shinar
Faculty of Aerospace Engineering
Technion-Israel Institute of
    Technology
Haifa, Israel

Yuri B. Shtessel
Department of Electrical and
    Computer Engineering
University of Alabama in
    Huntsville
Huntsville, Alabama

Min-Jea Tahk
Department of Aerospace
    Engineering
KAIST
Daejeon, Republic of Korea

Christian H. Tournes
Davidson Technologies
Huntsville, Alabama

Antonios Tsourdos
School of Engineering
Cranfield University
Swindon, United Kingdom

H. Weiss
Rafael Advanced Defense Systems
Haifa, Israel

Brian A. White
School of Engineering
Cranfield University
Swindon, United Kingdom

Kevin A. Wise
Senior Technical Fellow
The Boeing Company
St. Louis, Missouri

Introduction

Over the last 30 years, there have been many studies in the area of missile guidance. The result has been a great deal of progress, and several approaches to the problem have emerged. The basic problem is to intercept a target with great accuracy in an environment that is uncertain and noisy.

One of the earliest forms of missile guidance is that of command to line-of-sight guidance. This involves establishing a line of sight between the tracking sensor and the target, as shown in Figure I.1.

There are four positions to be identified in order that the geometry be defined. These are the tracking radar, the missile position, the target position, and the *impact point*. The missile is commanded onto or along the line of sight by use of a command link. Hence, the missile is fairly simple and only needs a communication link back to the guidance computer, which is usually located with the tracking sensor. The guidance computer and the tracking sensor can be located on a stationary platform, such as a ground station for area defense, or on a mobile platform, such as an aircraft or helicopter. This form of guidance is usually short range as a line of sight has to be established between the tracking sensor and the target and maintained over the whole of the engagement. It also requires an accurate tracking sensor for both the target and the missile and a means to discriminate between the two.

The other form of guidance is homing guidance, which has become the predominant mode of guidance in the recent past. The geometry of the guidance of homing missiles is shown in Figure I.2.

There are three reference points defining the engagement geometry: (1) the missile plus homing head position, (2) the target position, and (3) the impact point. If the tracking sensor is placed in the nose of the missile, the sensor will get closer to the target as the engagement progresses. This makes the tracking of the target easier, but the main disadvantage is that the sensor must fit into the diameter of the missile and consume power that is stored on board. Hence, the range and accuracy of the tracking sensor are much less than those of the line-of-sight guidance. The guidance computer also has to be carried on board; hence, its size and power will also be reduced in comparison. All of these facts make the homing missile more complex and costly than the line-of-sight missile. The main advantage of the homing missile is the fact that when launched, it does not require a separate tracking sensor, which could become vulnerable to counterattack. It also means that several homing missiles can be launched at the same time. There has been a lot of interest in the development of homing guidance techniques in the literature, mostly on two themes: that of *proportional navigation* and the use of more modern control techniques such as *optimal control*. Variants on the proportional navigation algorithm that improve the performance against

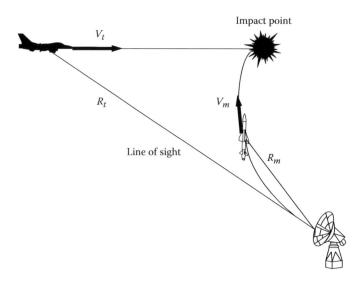

FIGURE I.1
Command to line-of-sight guidance.

maneuvering targets have been proposed. These rely on information about the target motion, usually the velocity and acceleration vectors, and so are more difficult to realize than the basic proportional navigation algorithm.

All of these rely on accurate measurements that can position the missile and the target, as well as estimate the target motion in terms of a velocity and acceleration vector. The main way of doing this is by means of the tracking sensor, which measures properties of the sight line between the missile and the target. Most guidance algorithms rely on the line-of-sight rotation rate and target relative range, with extra information coming from the range rate

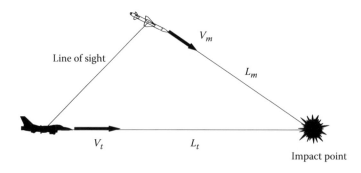

FIGURE I.2
Homing geometry.

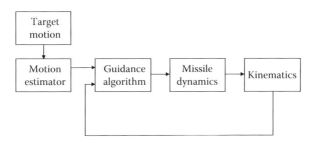

FIGURE I.3
Homing guidance loop.

if available. Modern trends are to also obtain information from third-party sensors with a networked capable system, but this is usually for the mid-course guidance phase where the tracking sensor is not active. The research work detailed in this book deals only with the terminal guidance phase, between any midcourse guidance and the warhead fusing phase.

In order to develop algorithms to accurately guide the missile onto the target, a good understanding of the kinematics and dynamics of the engagement geometry is required. The kinematics describes the motion of the missile and the target in inertial space. This is necessary to define the effect of the guidance commands on the geometry of the engagement. The dynamics are associated with the motion of the missile airframe in response to the changes in fin or thrust vectoring actuation of the missile. This mechanism produces the required maneuver of the missile derived from the guidance geometry. Hence, the guidance algorithm measures the engagement geometry, estimates the target dynamics, and produces commands that change the missile dynamics in a closed-loop form, shown in Figure I.3.

To help put the following chapters in context, a short description of the kinematics of the engagement and the dynamics of the target and missile is given here. The description is not meant to be exhaustive but is sufficient to inform the reader about where each chapter fits into the engagement system.

Homing Guidance Kinematics

Figure I.4 is drawn for the case of a target that is flying in a straight line at constant velocity V_t. The missile is flying at velocity V_m, also in a straight line. Both trajectories are assumed to intercept at the impact point at point I. The target and missile centers of gravity (cg) together with the impact point form a triangle, which will be called the *impact triangle*. Two sides of the triangle form the predicted straight line trajectories of the missile and the target, while the third is formed by connecting the missile and the target. If an observer was positioned at the missile cg, this line would establish a line of sight from the observer to the missile and is thus labeled the *line of sight*. In the real system, the observer is replaced by a homing head and the *line of sight* established between the missile homing head (usually sited in the missile nose, rather than the missile cg) and the cg of the target. For this simple analysis, the *line of sight* will be defined between the two cg's.

To establish the conditions for impact, consider a time T such that the target has traveled in a straight line and at constant velocity from its initial position in Figure 3 to the impact point. The length of this trajectory L_t will be

$$L_t = V_t T. \tag{1}$$

In order for the missile to arrive at the impact point at the same time as the target, it must travel a distance L_m in the same time T, that is,

$$L_m = V_m T. \tag{2}$$

The ratio of the trajectory lengths is then given by

$$\frac{L_m}{L_t} = \frac{V_m T}{V_t T}$$

$$= \frac{V_m}{V_t}. \tag{3}$$

The time of impact is a useful measure of the engagement progression and is sometimes used explicitly in the guidance algorithm. As such, it is labeled *time to go* or T_{go}.

Equation 3 shows that in order to impact on a target flying at constant velocity in a straight line, the missile must maneuver until the trajectory lengths of the *impact triangle* are in the same ratio as the target and missile velocities. As the target velocity and heading are either unknown or estimated, and targets can maneuver, there must be an active control system to acquire and maintain this impact geometry: this is the role of the guidance algorithm. The geometry is not fixed, however, as only the trajectory lengths need matching with the missile and target velocities.

Figure I.4 shows the locus of possible impact triangles, where the missile position lies on a circle of radius L_m (the *impact circle*), and the missile velocity vector V_m lies along the radius of the impact circle. From this figure, it can be seen that an impact triangle can be produced to give a head-on collision (point A) or a tail chase (point B) and any variant in between.

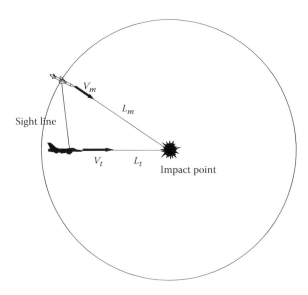

FIGURE I.4
Range of impact triangle locus.

Line of Sight Kinematics

In order to form an impact triangle, the missile and target velocity vector must both point directly at the impact point. If the impact triangle is not established, the missile must maneuver until it is established. To investigate this maneuver, consider the geometry of the target and missile connected by the *line of sight*. Figure I.5 shows the geometry of a two-dimensional or planar engagement with a set of axes centered on the missile. The axes are defined such that the x axis is pointing up the sight line, the y axis is normal to the sight line in the plane of the engagement, and the z axis forms a right-handed set and points directly out of the engagement plane. The basis vectors i, j define these directions. The *line of sight* is also rotating at ω_s rad/s around the z axis as the target moves relative to the missile.

The target position d_{tr} relative to the missile, in sight line axes is given by

$$d_{tr} = Ri$$

$$= \begin{pmatrix} R \\ 0 \end{pmatrix} \tag{4}$$

where R is the *closing range* of the target. The velocity and acceleration of the target in sight line axes can be obtained by differentiation; hence, the target's velocity v_{tr} relative to the sight line axes is given by

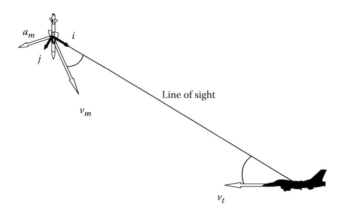

FIGURE I.5
Homing engagement dynamic geometry.

$$v_{tr} = \frac{\mathrm{d}d_{tr}}{\mathrm{d}t}$$

$$= \dot{d}_{tr} + \omega_s \times d_{tr} \tag{5}$$

$$= \dot{d}_{tr} + \Omega_s d_{tr}$$

where

$$\Omega_s = \begin{pmatrix} 0 & -\omega_s \\ \omega_s & 0 \end{pmatrix}. \tag{6}$$

Hence

$$v_{tr} = \begin{pmatrix} \dot{R} \\ \omega_s R \end{pmatrix}. \tag{7}$$

The target acceleration a_{tr} relative to the *line of sight* axes is obtained by further differentiation and is given by

$$a_{tr} = \frac{\mathrm{d}v_{tr}}{\mathrm{d}t}$$

$$= \dot{v}_{tr} + \omega_s \times v_{tr}$$

$$= \dot{v}_{tr} + \Omega_s v_{tr} \tag{8}$$

$$= \begin{pmatrix} \ddot{R} - \omega_s^2 R \\ \dot{\omega}_s R + 2\omega_s \dot{R} \end{pmatrix}.$$

Inspection of Equations 7 and 8 shows that if the missile matches the target velocity normal to the sight line (the y axis), the sight line angular velocity and angular acceleration must be zero. Zero angular velocity follows from Equation 7, as to get the y axis velocity to zero for any *closing range* R, we must have

$$\omega_s = 0 \tag{9}$$

Using this information in Equation 8 shows that for zero relative acceleration in the y direction, and any closing velocity V_c, where

$$V_c = -\dot{R}, \tag{10}$$

the *line of sight* angular acceleration $\dot{\omega}_s$ must be zero, that is,

$$\dot{\omega}_s = 0. \tag{11}$$

For the zero conditions in Equations 10 and 11, Equations 7 and 8 become

$$v_{tr} = \begin{pmatrix} -V_c \\ 0 \end{pmatrix} \quad a_{tr} = \begin{pmatrix} -\dot{V}_c \\ 0 \end{pmatrix}. \tag{12}$$

This implies that the *line of sight* maintains its orientation in space and shrinks in size until impact at a velocity of $-V_c$ and an acceleration of $-\dot{V}_c$, as shown in Figure I.6.

If the target maneuvers or the missile is not on a collision course, then the *line of sight* will rotate and

$$v_{tr} = \begin{pmatrix} v_{trx} \\ v_{try} \end{pmatrix}$$

$$= \begin{pmatrix} -V_c \\ \omega_s R \end{pmatrix} \tag{13}$$

$$a_{tr} = \begin{pmatrix} a_{trx} \\ a_{try} \end{pmatrix}$$

$$= \begin{pmatrix} -\dot{V}_c - \omega^2 R \\ \dot{\omega}_s R - 2\omega_s V_c \end{pmatrix} \tag{14}$$

$$= \begin{pmatrix} -\dot{V}_c - \omega_s^2 R \\ \dot{v}_{try} - \omega_s V_c \end{pmatrix}.$$

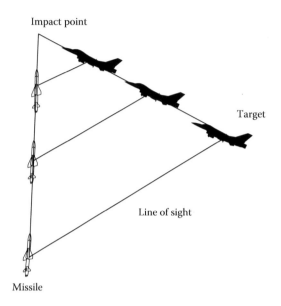

FIGURE I.6
Shrinking impact triangle.

Referring to Figure I.5, the missile velocity vector subtends an angle of θ_m to the *line of sight*, and the target submits an angle θ_t; hence

$$
v_{tr} = \begin{pmatrix} v_{trx} \\ v_{try} \end{pmatrix}
$$

$$
= \begin{pmatrix} V_t \cos(\theta_t) \\ V_t \sin(\theta_t) \end{pmatrix} - \begin{pmatrix} V_m \cos(\theta_m) \\ V_m \sin(\theta_m) \end{pmatrix} \tag{15}
$$

$$
= \begin{pmatrix} V_t \cos(\theta_t) - V_m \cos(\theta_m) \\ V_t \sin(\theta_t) - V_m \sin(\theta_m) \end{pmatrix}.
$$

Similarly, the target relative acceleration a_{tr} is given by

$$
a_{tr} = \begin{pmatrix} a_{trx} \\ a_{try} \end{pmatrix}
$$

$$
= \begin{pmatrix} a_{tp} - a_m \sin(\theta_m) \\ a_{tn} - a_m \cos(\theta_m) \end{pmatrix} \tag{16}
$$

where a_{tp} is the target acceleration parallel to the *line of sight*, a_{tn} is the target acceleration normal to the sight line, and a_m is the missile latax. Note that Equation 16 implies that the target can accelerate both longitudinally and laterally, while the missile accelerates only laterally. Hence, the acceleration vector of the missile is normal to the velocity vector, as shown in Figure I.5.

Any guidance algorithm must determine the missile normal acceleration in response to any mismatch in geometry caused by initial heading errors or by target maneuver. The geometry is controlled much more effectively if the target accelerations in Equation 16 are known. This required some form of an estimator as the sensor measurements will not measure acceleration directly.

Airframe Equations of Motion

In the study of the control of missiles, the understanding of the dynamics of the airframe plays an important role. The dynamic behavior of the airframe dictates the way the missile responds to the control surfaces, how it reacts to disturbances, and how it flies in steady state or trim conditions. It is also important to understand the dynamics of the target, and the dynamic equations developed in this section are relevant to the target as well as the missile. The difference is that the aerodynamics and dynamics of the target airframe are not known, and so an approximation to these dynamics equation is necessary.

General free body motion will be considered, which is subject to generalized forces and moments. The equations of motion of the generalized free body are obtained by considering the six degrees of freedom (DOFs) of motion. Figure I.7 shows the DOFs of the body of a missile airframe.

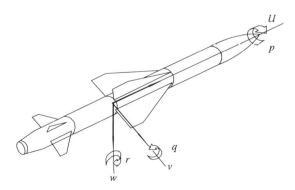

FIGURE I.7
Airframe coordinate system.

The motion can be categorized into two forms of motion: translational motion and rotational motion. There are three DOFs associated with each category, and these are summarized in Table I.1.

These DOFs will be used to describe the motion of the free body in space, and the description of the motion will be a function of the application. If an autopilot is to be designed, it will be necessary to control the rotational velocities and the translational velocities by suitable measurement and actuation. If the control system is required to perform some form of guidance (inertial guidance/navigation, for instance), then the added control of the translational displacements is required. In either case, the dynamic behavior of the airframe in response to the movement of control surfaces is required.

Free body motion describes how the airframe responds to the application of translational and rotational forces (moments). The forces and moments produce translational and rotational accelerations, which cause changes in the velocity and displacement of the airframe. Figure I.7 shows velocities measured in axes that are normally fixed in the airframe. As any axes set fixed in the airframe body will be subject to both rotational and translational accelerations, the dynamic description of the airframe will be complex as the equations of motion generated by Newton's laws of motion assume a nonaccelerating set of axes. Another complication is apparent in most real control systems. This relates to the definition of the aerodynamic forces acting on the body. Both actuators and sensors will be fixed in body axes and so will measure quantities relative to those axes. Inertia calculations are also easier in body axes if they are aligned with the principal axes of the body. Aerodynamic forces and moments, however, are usually referred to as "wind axes." These are axes with the *x*-axis aligned with the direction of the airflow arriving at the airframe. These coincide only when the airframe is flying at zero incidence. For small incidence, the difference between the two axis sets will be small and can be ignored. For highly maneuverable missiles that pull significant incidence ($\geq 15°$ for example), this approximation might have to be reviewed.

Consider first the motion of the missile airframe expressed in body axes. These axes are used mainly because the actuators and instruments are fixed

TABLE I.1

Six DOFs of a Free Body

Motion	Direction	Displacement	Velocity
Translational	Forward	x	u
	Sideslip	y	v
	Vertical	z	w
Rotational	Roll	ϕ	p
	Pitch	θ	q
	Yaw	ψ	r

in the airframe body. The translational and rotational velocities can be written in vector form as

$$v = \begin{pmatrix} u \\ v \\ w \end{pmatrix} = ui_b + vj_b + wk_b = (i_b \quad j_b \quad k_b) \begin{pmatrix} u \\ v \\ w \end{pmatrix} \qquad (17)$$

and

$$\omega = \begin{pmatrix} p \\ q \\ r \end{pmatrix} = pi_b + qj_b + rk_b = (i_b \quad j_b \quad k_b) \begin{pmatrix} p \\ q \\ r \end{pmatrix} \qquad (18)$$

where the components i_b, j_b, and k_b are unit vectors along the airframe body axes as shown in Figure I.8.

The translational and rotational free body equations are obtained by use of Newton's second law of motion. This states that for any mass, the rate of change of momentum is equal to the applied force. If the translational and angular momentum are given by h_t and h_a, respectively, then

$$\frac{dh_t}{dt} = F + G \qquad (19)$$

$$\frac{dh_a}{dt} = T \qquad (20)$$

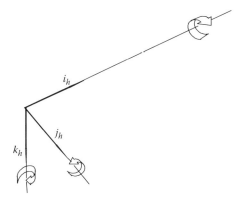

FIGURE I.8
Airframe defining vectors.

where F is a vector of forces applied along the airframe axes such that

$$F = \begin{pmatrix} X \\ Y \\ Z \end{pmatrix}. \tag{21}$$

G is the gravitational force given by

$$G = \begin{pmatrix} G_x \\ G_y \\ G_z \end{pmatrix} \tag{22}$$

and T is a vector of torques or moments such that

$$T = \begin{pmatrix} L \\ M \\ N \end{pmatrix}. \tag{23}$$

Translational dynamics

Consider first the translational momentum equation 19. The translational momentum h_t is given by

$$h_t = mv. \tag{24}$$

Substituting for h_t, and noting that m is a scalar variable that will be considered to be constant, yields

$$m\frac{dv}{dt} = F + G. \tag{25}$$

As the axes are fixed in the body, the defining orthonormal basis vectors i_b, j_b, and k_b will have angular velocity and hence will possess derivatives. In vector form, we have

$$\frac{dv}{dt} = v + \omega \times v \tag{26}$$

where $\times$ is the vector cross product. This can also be expressed in component form as

$$\frac{dv}{dt} = (\dot{u}i_b + \dot{v}j_b + \dot{w}k_b) + \left(u\frac{di_b}{dt} + v\frac{dj_b}{dt} + w\frac{dk_b}{dt} \right). \tag{27}$$

Comparing equations gives

$$\omega \times v = \left(u\frac{di_b}{dt} + v\frac{dj_b}{dt} + w\frac{dk_b}{dt} \right). \tag{28}$$

Figure I.9 shows the derivation of the rate of change of the unit vectors. Hence

$$\frac{di_b}{dt} = (rj_b - qk_b)$$

$$\frac{dj_b}{dt} = (pk_b - ri_b)$$

$$\frac{dk_b}{dt} = (qi_b - pj_b).$$

Collecting terms yields

$$\omega \times v = \Omega v \tag{29}$$

where

$$\Omega = \begin{pmatrix} 0 & -r & q \\ r & 0 & -p \\ -q & p & 0 \end{pmatrix}. \tag{30}$$

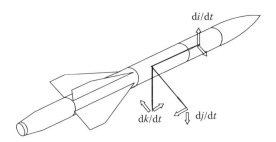

FIGURE I.9
Rate of change of unit vectors.

The translational acceleration of the free body is then obtained by combining Equations 25, 26, and 29 to give, in matrix form,

$$mv + m\Omega v = (F + G). \tag{31}$$

Rearranging and multiplying both sides by m^{-1} yields

$$v = -\Omega v + m^{-1}(F + G). \tag{32}$$

Equation 32 defines the translational dynamics of the free body.

Rotational dynamics

The rotational or angular momentum is more complex than the translational momentum. It is written in a similar manner as

$$h_a = J\omega \tag{33}$$

where J is the inertia matrix given by

$$J = \begin{pmatrix} J_{xx} & -J_{xy} & -J_{xz} \\ -J_{xy} & J_{yy} & -J_{yz} \\ -J_{xz} & -J_{yz} & J_{zz} \end{pmatrix}. \tag{34}$$

The diagonal terms are moments of inertia given by

$$J_{xx} = \int x^2 \, dm$$
$$J_{yy} = \int y^2 \, dm \tag{35}$$
$$J_{zz} = \int z^2 \, dm$$

and the off-diagonal elements are products of inertia given by

$$J_{xy} = \int xy \, dm$$
$$J_{xz} = \int xz \, dm \tag{36}$$
$$J_{yz} = \int yz \, dm.$$

Substituting for the angular momentum h_a in Equation 20 yields

$$\frac{d(J\omega)}{dt} = T. \tag{37}$$

Again, as the axes are rotating, and noting that the inertia terms are treated as constant, we have

$$\frac{d(J\omega)}{dt} = J\omega + \omega \times (J\omega). \tag{38}$$

The cross product can also be written as

$$\omega \times (J\omega) = \Omega J\omega. \tag{39}$$

Hence, the rotational dynamics can be written as

$$J\omega + \Omega J\omega = T. \tag{40}$$

Multiplying by J^{-1} and rearranging gives

$$\omega = -J^{-1}\Omega J\omega + J^{-1}T. \tag{41}$$

Equation 41 defines the rotational dynamics of the free body airframe.

Inertial/body axis transformation

Airframe motion is required in terms of a fixed inertial frame that is aligned with the earth when used in guidance geometries. As the dynamic equations for the airframe are expressed in airframe body axes, a translation between these frames is necessary. The basis for the translation is a coordinate transformation. This transformation takes a vector quantity such as airframe velocity or acceleration (translational or rotational) expressed in one set of axes and transforms it into another set of axes. Consider a vector v_b expressed in body axes as

$$v_b = \begin{pmatrix} v_{b_x} \\ v_{b_y} \\ v_{b_z} \end{pmatrix} = v_{b_x} i_b + v_{b_y} j_b + v_{b_z} k_b \tag{42}$$

and the same vector expressed in a fixed inertial axis set v_e as

$$v_e = \begin{pmatrix} v_{e_x} \\ v_{e_y} \\ v_{e_z} \end{pmatrix} = v_{e_x} i_e + v_{e_y} j_e + v_{e_z} k_e \tag{43}$$

where both axis sets share a common origin, as shown in Figure I.10.

Note that the subscripts b and e on vector v denote the description of the vector in different axes and do not denote different vectors. The relationship between the components v_{b_x}, v_{b_y}, and v_{b_z} in the body axes and v_{e_x}, v_{e_y}, and v_{e_z} in the earth axes can be explored by noting that v_{b_x} is the projection of vector v onto the x-axis of the body axis. This can be calculated by the inner product of the unit vector i_b, defining the body x-axis, and vector v. Hence

$$v_{b_x} = i_b \cdot v$$

$$v_{b_y} = j_b \cdot v \tag{44}$$

$$v_{b_z} = k_b \cdot v.$$

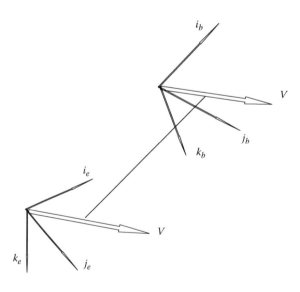

FIGURE I.10
Inertial and body axes.

Similarly, the components of v in the earth axes can be calculated by

$$v_{e_x} = i_{e'} v$$
$$v_{e_y} = j_{e'} v$$
$$v_{e_z} = k_{e'} v.$$

(45)

Substituting for vector v in Equation 43 into Equation 44 gives

$$v_{b_x} = i_{b'}(v_{e_x} i_e + v_{e_y} j_e + v_{e_z} k_e)$$
$$v_{b_y} = j_{b'}(v_{e_x} i_e + v_{e_y} j_e + v_{e_z} k_e)$$
$$v_{b_z} = k_{b'}(v_{e_x} i_e + v_{e_y} j_e + v_{e_z} k_e).$$

(46)

This can be rearranged to give

$$\begin{pmatrix} v_{b_x} \\ v_{b_y} \\ v_{b_z} \end{pmatrix} = \begin{pmatrix} i_{b'} i_e & i_{b'} j_e & i_{b'} k_e \\ j_{b'} i_e & j_{b'} j_e & j_{b'} k_e \\ k_{b'} i_e & k_{b'} j_e & k_{b'} k_e \end{pmatrix} \begin{pmatrix} v_{e_x} \\ v_{e_y} \\ v_{e_z} \end{pmatrix}.$$

(47)

If the earth components are calculated in the same manner by substituting for vector v in Equation 42 in Equation 45, this gives

$$\begin{pmatrix} v_{e_x} \\ v_{e_y} \\ v_{e_z} \end{pmatrix} = \begin{pmatrix} i_{e'} i_b & i_{e'} j_b & i_{e'} k_b \\ j_{e'} i_b & j_{e'} j_b & j_{e'} k_b \\ k_{e'} i_b & k_{e'} j_b & k_{e'} k_b \end{pmatrix} \begin{pmatrix} v_{b_x} \\ v_{b_y} \\ v_{b_z} \end{pmatrix}.$$

(48)

Equation 48 can be written in compact form as

$$v_e = Rv_b$$

(49)

where

$$R = \begin{pmatrix} i_{e'} i_b & i_{e'} j_b & i_{e'} k_b \\ j_{e'} i_b & j_{e'} j_b & j_{e'} k_b \\ k_{e'} i_b & k_{e'} j_b & k_{e'} k_b \end{pmatrix}$$

(50)

with the inverse transformation in Equation 47 as

$$v_b = R^{-1}v_e. \tag{51}$$

Noting that for inner products, the order of the vectors is not significant, so that

$$i'j = j'i. \tag{52}$$

Inspection of Equations 47 and 48 yields

$$R^{-1} = R'. \tag{53}$$

This implies that the transformation matrix R is unitary. Note also that the terms in the R matrix represent the inner product of the three unit vectors in body axes with each of the unit vectors in earth axes. Hence, for any product such as $i'_b i_e$,

$$i'_b i_e = \cos(\theta_{be}) \tag{54}$$

where θ_{be} is the angle between the two vectors.

Three methods are mainly used to generate the transformation matrix R from body axes to earth axes: Euler angles, direction cosines, and quaternions. Analysis of these transformation representations is beyond the scope of this book. Suffice it to say that the rotation matrix can be represented in direction cosine form by

$$v_e = Rv_b$$

$$= \begin{pmatrix} l_1 & m_1 & n_1 \\ l_2 & m_2 & n_2 \\ l_3 & m_3 & n_3 \end{pmatrix} v_b \tag{55}$$

where l_i, m_i, and n_i are called the direction cosines and are summarized in Table I.2.

TABLE I.2

Direction Cosines

Direction Cosine	Value
l_1	$\cos(\psi)\cos(\theta)$
l_2	$\sin(\psi)\cos(\theta)$
l_3	$-\sin(\theta)$
m_1	$\cos(\psi)\sin(\theta)\sin(\phi) - \sin(\psi)\cos(\phi)$
m_2	$\sin(\psi)\sin(\theta)\sin(\phi) + \cos(\psi)\cos(\phi)$
m_3	$\cos(\theta)\sin(\phi)$
n_1	$\cos(\psi)\sin(\theta)\cos(\phi) + \sin(\psi)\sin(\phi)$
n_2	$\sin(\psi)\sin(\theta)\cos(\phi) - \cos(\psi)\sin(\phi)$
n_3	$\cos(\theta)\cos(\phi)$

Autopilot algorithm development

The forces and moments in Equation 25 are usually functions of the translational and rotational vector components:

$$\mathbf{F} = \mathbf{F}(u, v, w, p, q, r)$$
$$\mathbf{T} = \mathbf{T}(u, v, w, p, q, r). \tag{56}$$

These can then be used to formulate the equations of motion for autopilot design. The equations governing the translational dynamics can also be written in terms of incidence angles (u, β, γ) rather than velocity vector components (u, v, w), where

$$\beta = \frac{w}{u}$$
$$\gamma = \frac{v}{u}. \tag{57}$$

This does not alter the kinematics but requires the aerodynamics to be described in terms of the forward speed and the incidence angles:

$$\mathbf{F} = \mathbf{F}(u, \beta, \gamma, p, q, r)$$
$$\mathbf{T} = \mathbf{T}(u, \beta, \gamma, p, q, r). \tag{58}$$

Guidance algorithm development

In most cases, the dynamics of the missile are not considered in the design of guidance algorithms, and the kinematics alone are used. This implies that the velocity vector can be controlled directly using commands determining the required rotation rates of the velocity vector. Hence, (p, q, r) can be directly controlled using guidance commands. Thus, for most guidance algorithms, we have Figure I.3 replaced by Figure I.11.

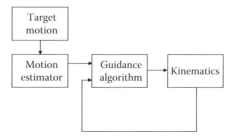

FIGURE I.11
Homing guidance for algorithm development.

If the autopilot is retained, then the analysis becomes more complex but more realistic. One chapter in the book deals with the integrated design of both the autopilot and guidance systems.

Zero-effort miss

In benchmarking guidance algorithms, the zero-effort miss (ZEM) metric is sometimes used to test the effectiveness of the guidance algorithm. The ZEM is obtained by setting the lateral acceleration command to zero at specific times during the engagement and measuring the closest approach distance. This then determines the miss distance for zero maneuver or effort, hence the name zero-effort miss. A graphical interpretation of ZEM can be obtained by redrawing Figure I.5. If the target velocity vector is subtracted from both the target and the missile, then a relative velocity vector centered on the missile is obtained. Projecting a line along this relative velocity vector and constructing an orthogonal line centered on the target to intercept it produces a right-angle triangle with the ZEM as one side, as shown in Figure I.12.

The ZEM can be related to the pointing error angle, which is used for guidance in one of the guidance chapters. Note that the smaller the guidance error, the smaller the ZEM; hence, a guidance algorithm that quickly reduces the guidance error and maintains a small error will have a small ZEM over the engagement. This is especially important for maneuvering targets.

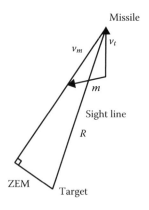

FIGURE I.12
ZEM geometry.

Challenges

From the previous sections, it is clear that several issues arise from the implementation of homing guidance. These are grouped under the following headings.

- *Target tracking.* This is important as the guidance algorithms require accurate information concerning the position, velocity, and acceleration of the target relative to the missile. The reduced capability of the sensor on board the missile, in terms of range and resolution, presents a significant problem. Estimators are mainly used to infer the target motion from the noisy sensor measurements, and these estimators have their own dynamics. This implies that the estimator is not instantaneous and will have an effect on the tracking accuracy. Specifically, the dynamic lag will mean that if a target maneuvers, there will be a delay in producing accurate estimates of the change in position, velocity, and acceleration. Also, because the sensor measures distance (with velocity if the sensor is capable of measuring Doppler), the estimate of velocity and acceleration will involve a form of integration with the associated errors and delays.

- *Guidance algorithms.* As the kinematics and dynamics of the guidance equations are nonlinear, and the trend is toward smaller warheads and more accurate guidance solutions, this is still a challenge. Modern studies are also looking at the optimization of the midcourse guidance phase of the engagement to try to control the geometry of the engagement when target acquisition is attained. Shaping the trajectory in the terminal guidance phase is also important to control both the endgame geometry and aim point. This will enable smaller warheads and a more effective kill mechanism to be used, resulting in a smaller cheaper missile with the potential for less collateral damage as the blast effect is smaller and more contained.

- *Autopilot algorithms.* The dynamics and aerodynamics of modern missiles are very nonlinear. Many modern missiles are required to fly at higher incidences to achieve higher lateral accelerations for more accurate guidance against highly maneuverable targets or have ram and/or scram jet intakes that produce both restrictions on incidence in specific lateral planes of the missile and highly nonlinear aerodynamics. Thus, autopilot design must take into account the inherent nonlinear aerodynamics as well as the high dynamic pressure ranges as the missile velocity changes over the engagement. Traditionally, autopilots have been scheduled over time or speed to take into account the high variability of the aerodynamic effectiveness of the wings and the control surfaces, which can be of the order of 5:1 over the operating envelope of the missile. Modern designs produce autopilots that are explicit functions of speed and incidence to produce a smoothly interpolating autopilot.

- *Implementation issues.* Several problems arise when implementing guidance, autopilot, and estimation algorithms. These involve the choice of tracking sensor, software implementation, fin actuation, motor type, as well as the size and type of warhead. All of these issues are dealt with in the book.

The book is separated into four main contributions. The first deals with the design of autopilots with contributions:

- Robust Autopilot Design for Quasilinear Parameter-Varying Missile Model (A. Tsourdos and B. A. White)
- Polynomial Approach for Design and Robust Analysis of Lateral Missile Control (A. Tsourdos, B. A. White, and L. Bruyere)
- Control Design and Gain Scheduling Using Observer-Based Structures (D. Alazard)
- Adaptive Neural Network–Based Autopilot Design (K. Rajagopal and S. N. Balakrishnan)

These chapters deal with the design of autopilots over the whole flight envelop of Mach number and altitude. They also take into account the effect of changing aerodynamics as functions of incidence as well as Mach number and altitude. This is followed by a set of contributions on guidance. There follows a chapter that looks at the design of homing guidance integrated with autopilot design: Integrated Guidance and Control for Missiles (N. Harl, M. Dancer, S. N. Balakrishnan, E. J. Ohlmeyer, and C. Phillips).

This is followed by six chapters on different aspects and techniques applied to the homing guidance problem. These range from using sliding modes to produce a linear-type behavior for the nonlinear kinematic system equations to differential geometry, and optimal control applied to terminal geometry:

- Higher-Order Sliding Modes for Missile Guidance and Control (Y. B. Shtessel and C. H. Tournes)
- Neoclassical Missile Guidance (P. Gurfil)
- Differential Geometry Applied to Missile Guidance (B. A. White and A. Tsourdos)
- Differential Game-Based Interceptor Missile Guidance (J. Shinar and T. Shima)

- Optimal Guidance Laws with Impact Angle Control (C.-K. Ryoo, M.-J. Tahk, and H. Cho)
- A chapter on the use of differential games, integrating the guidance law with the estimation of the target maneuver, follows: Integrated Design of Estimator and Guidance Law (J. Shinar and T. Shima)

The next contribution deals with the estimation problem using particle filters: Introduction to Particle Filters for Tracking and Guidance (D. Salmond)

The final contribution area deals with practical implementation issues in the design of a missile autopilot and guidance laws: The first chapter deals with the design of digital homing guidance laws and autopilots. The second contribution is a chapter on command to line-of-sight guidance systems, and the third is on the design of an autopilot control system, both of which model sensors, actuators, noise, and uncertainty, as well as deal with other aspects of implementation for a practical system:

- Practical Techniques for Design of Multirate Digital Guidance Laws and Autopilots (C. A. Rabbath, N. Léchevin, and M. Lauzon)
- Design of CLOS Guidance System (G. Hexner and H. Weiss)
- Practical Considerations in Robust Control of Missiles (K. A. Wise)

The practical implementation of the tracking, guidance, and autopilot algorithms into a missile system is an important issue in that the theoretical designs require a lot of systematic reviews when inserting into a practical missile system. All of the components interact in some way and have characteristics that need to be taken into account. These involve such diverse issues as the coupling between the homing sensor and the body motion. Radome aberration can produce apparent changes in the sight line to the target that can destabilize the guidance algorithm. The autopilot sensors (usually rate gyros and lateral accelerometers) also have biases, dynamics, and limits that will affect the implementation. One of the major effects in autopilot design will be the effectiveness of the actuators. For canard systems, the control fins are located at the front of the missile and can cause loss of lift and rolling moments that can severely affect the maneuverability of the missile. Rear control surfaces do not produce such effects, but introduce nonminimum phase effects as the fin movement to produce incidence will initially produce lateral accelerations in the opposite direction to that required, with some destabilization effects. If there are nonsymmetric aerodynamics, sensors, or warheads on the missile, then there is a preferred maneuver direction bank to turn (BTT), as opposed to a symmetric missile, which can maneuver laterally in any direction skid to turn (STT). Both have their implementation problems.

MATLAB®/Simulink® Disclaimer

MATLAB® and Simulink® are registered trademarks of The MathWorks, Inc. For product information, please contact:

The Math Works, Inc.
3 Apple Hill Drive
Natick, MA 01760-2098
Tel: 508-647-7000
Fax: 508-647-7001
E-mail: info@mathworks.com
Web: http://www.mathworks.com

1

Robust Autopilot Design for Quasilinear Parameter-Varying Missile Model

Antonios Tsourdos and Brian A. White

CONTENTS

1.1 Introduction

One of the most popular methods for applying linear time-invariant (LTI) control theory to time-varying and/or nonlinear systems is gain scheduling [1]. This strategy involves obtaining Taylor linearized models for the plant at finitely many equilibria ("set points"), designing an LTI control law ("point design") to satisfy local performance objectives for each point, and then adjusting ("scheduling") the controller gains in real time as the operating conditions vary. This approach has been applied successfully for many years, particularly for aircraft and process control problems. Relatively recent examples (some of which involve modern control design methods) include jet engines [2], active suspensions [3], high-speed drives [4], missile autopilots [5], and VSTOL aircraft [6–8].

Despite past success of gain scheduling in practice, until recently, little has been known about it theoretically as a time-varying and/or nonlinear control technique. Also, determining the actual scheduling routine is more of an art than a science. While *ad hoc* approaches such as linear interpolation and curve fitting may be sufficient for simple static-gain controllers, doing the same for dynamic multivariable controllers can be a rather tedious process.

An early theoretical investigation into the performance of parameter-varying systems can be found in the work of Kamen and Khargonekar [9]. During the 1980s, Rugh and his colleagues developed an analytical framework for gain scheduling using extended linearization [1,10,11]. Also, Shamma and Athans [12–14] introduced linear parameter-varying (LPV) systems as a tool for quantifying such heuristic design rules as "the resulting parameter must vary slowly" and "the scheduling parameter must capture the nonlinearities of the plant." Shahruz and Behtash [15] suggested using LPV systems for synthesizing gain-scheduled controllers, and Shamma and Cloutier [16] have used LPV plant models with μ-synthesis [17–19] for designing missile autopilots.

In this chapter, an autopilot design is described for a realistic model of a tactical missile and robust stability of the closed-loop system investigated. The tail-controlled missile in the cruciform fin configuration [20] is modeled as a second-order quasilinear parameter-varying (QLPV) system. This nonlinear model is obtained from the Taylor linearized model of the horizontal motion by including explicit dependence of the aerodynamic derivatives on a state (sideslip velocity) and external parameters (longitudinal velocity and roll angle). The first contribution is to consider this detailed QLPV (and thus nonlinear) model.

The autopilot design is based on input–output pseudolinearization [21–23]. The design makes Taylor linearization of the closed-loop system independent of the choice of equilibria. Thus, if the operating points are in the vicinity of the equilibria, then one and only one linear model will describe closed-loop dynamics, regardless of the rate of change of the operating points. Simulations for constant lateral acceleration demands good tracking with fast response time. The second contribution is to interpret pseudolinearization as the restriction of feedback linearization [24] to the set of equilibria, and the third is to perform a successful pseudolinearizing design for a QLPV system.

1.2 Preliminaries

The purpose of this section is to present the background information necessary to follow the context, meaning, and methodology of design and stability analysis developed in Section 1.3–1.4. First, a detailed description of the missile under consideration is given in Section 1.2.1. Then several possible models useful for the missile representation are defined in Section 1.2.2. Finally, Section 1.2.3 describes the algorithm of pseudolinearization (and its mathematical and control theoretic contexts), as it is a fundamental tool for Section 1.3.

1.2.1 Missile Model

Missile autopilots are usually designed using linear models of nonlinear equations of motion and aerodynamic forces and moments [25,26]. The objective of this chapter is robust design of a sideslip (yaw) velocity autopilot for a nonlinear missile model. This model describes a reasonably realistic airframe of a tail-controlled tactical missile in the cruciform fin configuration (see Figure 1.1). The aerodynamic parameters in this model are derived from wind-tunnel measurements [20].

The starting point for mathematical description of the missile is the linearized model of the horizontal motion (on the xy-plane in Figure 1.1):

$$\dot{v} = y_v v - Ur + y_\zeta \zeta \quad \text{(translational dynamics)}$$

$$\dot{r} = n_v v + n_r r + n_\zeta \zeta \quad \text{(angular dynamics)}, \tag{1.1}$$

where the variables are defined in Figure 1.1. Here v and r are incremental forms of the sideslip velocity V and body rate R, ζ is the rudder fin deflection, y_v and y_ζ are semi-nondimensional force derivatives due to lateral and fin angle, respectively, and n_v, n_ζ and n_r are semi-nondimensional moment derivatives due to sideslip velocity, fin angle, and body rate, respectively. Finally, U is the longitudinal velocity.

For a trim condition, the derivatives y_v, y_ζ, n_v, n_r, and n_ζ in Equation 1.1 are assumed constant, so that it is a linear model. However, the aerodynamic forces and moments acting on the airframe are nonlinear functions of Mach

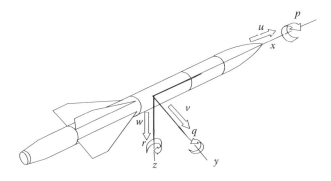

FIGURE 1.1
Airframe axes

number, longitudinal and sideslip velocities, control surface deflection, aerodynamic roll angle, and body rates. Thus, for a wider (nontrim) flight envelope, nonlinear dependence of the derivatives in Equation 1.1 on these variables cannot be ignored. Indeed, Equation 1.1 must be rewritten by making explicit the dependence so that it becomes the following nonlinear model [20,27]:

$$\dot{v} = y_v v - Ur + y_\zeta \zeta$$

$$= \frac{1}{2} m^{-1} \rho V_o S(C_{y_v} v + V_o C_{y_\zeta} \zeta) - Ur$$

$$\dot{r} = n_v v + n_r r + n_\zeta \zeta$$

$$= \frac{1}{2} I_z^{-1} \rho V_o S d \left(\frac{1}{2} dC_{n_r} r + C_{n_v} v + V_o C_{n_\zeta} \zeta \right). \tag{1.2}$$

Here, $m = 125$ kg is the missile mass, $\rho = \rho_o - 0.094h$ is the air density ($\rho_o = 1.23$ kg m^{-3} is the sea level air density and h is the missile altitude), V_o is the total velocity, $S = \pi d^2/4 = 0.0314$ m^2 is the reference area ($d = 0.2$ m is the reference diameter), and $I_z = 67.5$ kgm^2 is the lateral inertia. For the coefficients C_{y_v}, C_{y_ζ}, C_{n_r}, C_{n_v}, and C_{n_ζ}, only discrete data points are available, obtained from wind tunnel experiments. Therefore, an interpolation, involving the Mach number $M \in [0.6, 6.0]$, roll angle $\lambda \in [4.5°, 45°]$, and total incidence $\sigma \in [3°, 30°]$, has been done with the results summarized in Table 1.1.

The total velocity vector $\vec{V}_o$ is the sum of the longitudinal velocity vector $\vec{U}$ and the incremental sideslip velocity vector $\vec{v}$, that is, $\vec{V}_o = \vec{U} + \vec{v}$, with all three vectors lying on the xy-plane (see Figure 1.1). We assume that $U \gg v$,

TABLE 1.1

Coefficients in Nonlinear Model (Equation 1.2)

	Coefficient Name	Interpolated Formula
C_{y_v}	Sideslip normal force	$0.5[(-25 + M - 60\,\|\sigma\|)(1 + \cos 4\lambda) + (-26 + 1.5M - 30\,\|\sigma\|)$ $(1 - \cos 4\lambda)]$
C_{y_ζ}	Fin normal force	$10 + 0.5[(-1.6M + 2\,\|\sigma\|)(1 + \cos 4\lambda) + (-1.4M + 1.5\,\|\sigma\|)$ $(1 - \cos 4\lambda)]$
C_{n_r}	Damping moment	$-500 - 30M + 200\,\|\sigma\|$
C_{n_v}	Sideslip moment	$s_m C_{y_v}$, where $s_m = d^{-1}[1.3 + 0.1M + 0.2(1 + \cos 4\lambda)\,\|\sigma\| +$ $0.3(1 - \cos 4\lambda)\,\|\sigma\| - (1.3 + m/500)]$
C_{n_ζ}	Control moment	$s_f C_{y_\zeta}$, where $s_f = d^{-1}[2.6 - (1.3 + m/500)]$

so that the total incidence σ, or the angle between $\vec{U}$ and $\vec{V}_0$, can be taken as $\sigma = v/V_0$, as $\sin\sigma \approx \sigma$ for small σ. Thus, we have $\sigma = v/V_0 = v/\sqrt{v^2 + U^2}$, so that the total incidence is a nonlinear function of the incremental sideslip velocity and longitudinal velocity, $\sigma = \sigma\,(v,U)$.

The Mach number is obviously defined as $M = V_0/a$, where a is the speed of sound. Since $V_0 = \sqrt{v^2 + U^2}$, the Mach number is a nonlinear function of the incremental sideslip velocity and longitudinal velocity, $M = M\,(v,U)$.

It follows from the above discussion that all coefficients in Table 1.1 can be interpreted as nonlinear functions of three variables: incremental sideslip velocity v, longitudinal velocity U, and roll angle λ.

1.2.2 Taxonomy of Models

Missiles are required to operate over an expanded flight envelope to meet the challenge of highly maneuverable targets. In such a scenario, an autopilot derived from linearization about a single flight condition will be unable to achieve suitable performance for the whole envelope. Thus, there is an inherent tension in autopilot design because of nonlinearity of the missile and linearity of the controller.

Mathematical representation of nonlinear missile dynamics lends itself to several interpretations. The interpretations aim at deriving models that adequately capture missile behavior *and* are practical for systematic control design. A taxonomy of dynamics models is presented in Table 1.2, which should be viewed in conjunction with Equation 1.2 and Table 1.1.

TABLE 1.2

Dynamics of Missile Models and Their Relationships

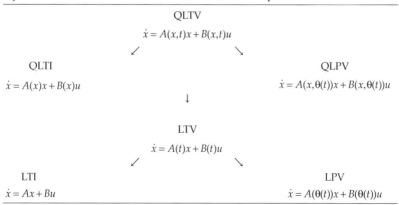

QLTV
$$\dot{x} = A(x,t)x + B(x,t)u$$

QLTI
$$\dot{x} = A(x)x + B(x)u$$

QLPV
$$\dot{x} = A(x,\theta(t))x + B(x,\theta(t))u$$

LTV
$$\dot{x} = A(t)x + B(t)u$$

LTI
$$\dot{x} = Ax + Bu$$

LPV
$$\dot{x} = A(\theta(t))x + B(\theta(t))u$$

Note: $x = x(t)$ is the state, $u = u(t)$ is the input, and $\theta = \theta(t)$ are the external parameters (variables different from x and u).

In the lower left corner of Table 1.2, we have the LTI model:

$$\dot{x} = Ax + Bu$$
$$y = Cx,$$
(1.3)

where $u = u(t) \in \mathbb{R}^m$ is the vector of inputs, $x = x(t) \in \mathbb{R}^n$ is the vector of states, and $y = y(t) \in \mathbb{R}^q$ is the vector of outputs. Finally, A, B, and C are matrices with constant real entries, $A \in \mathbb{R}^{n \times n}$, $B \in \mathbb{R}^{n \times m}$, and $C \in \mathbb{R}^{n \times q}$. This familiar model arises from Taylor linearization about a single flight condition (see Equation 1.1) and is excellent for linear controller design but has rather limited applicability for the whole flight envelope.

Traditionally, satisfactory performance across the flight envelope can be attained by gain scheduling local autopilot controllers to yield a global controller. The global controller is a collection of LTI controllers designed for the corresponding family of LTI models obtained via Taylor linearizations about equilibria. An LTI controller of the collection is switched on when the current operating point of the flight envelope is in the vicinity of the relevant equilibrium. This switching schedule is determined by the scheduling variables, which are "external" in the sense that they are different from state x and input u. A precise mathematical description of the resulting control system has only recently been achieved (see Section 1.1) with LPV models, appearing in the lower right corner of Table 1.2:

$$\dot{x} = A(\theta(t))x + B(\theta(t))u$$
$$y = C(\theta(t))x.$$
(1.4)

The entries of matrices A, B, and C are no longer constant as in the LTI model (Equation 1.3) but are time varying, making LPV models a special case of linear time-varying (LTV) models, as symbolized in Table 1.2. The variation over time is determined by the parameter θ, which is a generalization of scheduling variables. Gain scheduling requires "scheduling on a slow variable," which means that changes in θ should be much slower than changes in x and u. This requirement (violated for a rapidly maneuvering missile) is absent in the LPV model, and hence, it is a generalization of gain scheduling. This motivates recent interest in the LPV approach to autopilot design (see Section 1.2.1), as it promises to preserve the transparency of linear controller design, while reflecting the rapidly changing missile dynamics.

However, the LPV model is still a collection of linear designs, and in each of those, it is impossible to distinguish between real disturbances and normal manifestations of nonlinearity. Hence, any further improvement in performance and robustness can be achieved only by directly acknowledging missile nonlinearity, rather than treating it as nuisance in a linear model. Thus, the nonlinear dynamics must be explicitly incorporated into the mathematical description, but without undue generality. This is done in the upper part of Table 1.2. The topmost model is quasilinear time varying (QLTV), which in the context of Equation 1.2 and Table 1.1 is the most general framework we need:

$$\dot{x} = A(x,t)x + B(x,t)u$$
$$y = C(x,t)x. \tag{1.5}$$

The most important novelty in the QLTV model over the models in the lower part of Table 1.2 is that it is nonlinear, since matrices A, B, and C depend on state x; note also that Equation 1.5 is time varying (A, B, and C depend on time t, as well).

Obviously, the LTV model is a special case of the QLTV model, but more important are the two nonlinear special cases, as illustrated in Table 1.2. The top left one is the quasilinear time-invariant (QLTI) model, obtained from Equation 1.5 by simply dropping the explicit dependence on time t. The top right representation, the QLPV model, is the most relevant one for this paper:

$$\dot{x} = A(x,\theta(t))x + B(x,\theta(t))u$$
$$y = C(x,\theta(t))x. \tag{1.6}$$

Mathematical description of Equation 1.6 is the focus of this paper because close examination of Equation 1.2 and Table 1.1 reveals that with $x \triangleq (v \quad r)'$, $u \triangleq \zeta$, and $\theta \triangleq (U \quad \lambda)'$, Equation 1.2 is of the form Equation 1.6. This is pursued further in Section 1.3.

1.2.3 Approaches to Nonlocal Linearization

This section summarizes three approaches to nonlocal* linearization, focusing on input–output pseudolinearization, as it is used in Section 1.3 for autopilot design. The actual exposition of input–output pseudolinearization is done in Section 1.2.3.3, and Sections 1.2.3.1 and 1.2.3.2 explain the context in which input–output pseudolinearization arose.

1.2.3.1 Linearization via Coordinate Transformation

When faced with a nonlinear ordinary differential equation, for example,

$$\dot{x} = 8x + x^4 \triangleq f(x), \quad x(t_0) = x_0, \tag{1.7}$$

---

* Here "nonlocal linearization" means a linearization of an ordinary differential equation or a control system other than Taylor linearization at a single equilibrium.

where $x: (t_0, \infty) \to \mathbb{R}$, one would like to integrate it or to find explicit solutions in closed form. An obvious approach is to guess a transformation, $z = \Phi(x)$, such that the resulting (transformed) Ordinary Differential Equation for $z = z(t)$ can be integrated by already-known methods. Indeed, in the case of Equation 1.7, $\Phi(x) = x^{-3}$ results in

$$\dot{z} = -24z - 3 \triangleq g(z), \quad z(t_0) = z_0 \triangleq \Phi(x_0), \tag{1.8}$$

which is a linear (affine) equation and can be readily integrated, and then, using Φ^{-1}, the solution of Equation 1.7 is recovered (note that Φ is singular at $x = 0$, though). The geometric meaning of this process is simple and is shown in Figure 1.2: the nonlinear right-hand side (RHS) of Equation 1.7, f, is transformed into the linear (affine) RHS of Equation 1.8, g. That is, what was nonlinear in the $(x, f(x))$-plane becomes linear in the $(z, g(z))$-plane through transformation Φ.

Even for this simple scalar example, it is obvious that guessing Φ, which exists, is differentiable for all x, or *globally*, and has a globally differentiable inverse Φ^{-1}, is highly nontrivial.

A systematic approach to the above problem was proposed in the 1870s by the Norwegian mathematician Sophus Lie [28,29]. One finding of his theory is that only certain types of ordinary differential equations can be transformed globally into linear ones. This is not surprising, since otherwise, all equations would just be linear ones in "wrong" coordinates.

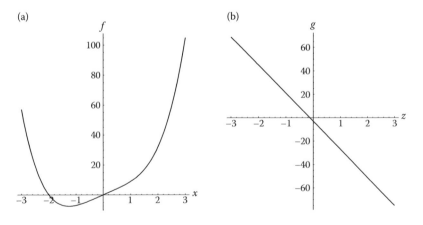

FIGURE 1.2
Linearization of Equation 1.7: (a) RHS f of Equation 1.7 in the $(x, f(x))$-plane; (b) RHS g of Equation 1.8 in the $(z, g(x))$-plane.

1.2.3.2 Input–Output Feedback Linearization

Lie's approach was extended in the 1980s [30,31] to affine nonlinear Single Input Single Output control systems (compare with the QLTI model in Table 1.2):

$$\dot{x} = f(x) + g(x)u, \quad x(t_0) = x_0$$
$$y = h(x),$$

(1.9)

where $f: X \to \mathbb{R}^n$, $g: X \to \mathbb{R}^n$, and $h: X \to \mathbb{R}$ are sufficiently smooth on the open set $X \subset \mathbb{R}^n$. The presence of output y in Equation 1.9 means that the transformation $z = \Phi(x)$ need not linearize the whole of state $x \in X$ but only that part that will be visible from the output; the unobservable dynamics must be stable, though. The presence of control u in Equation 1.9 gives an additional "degree of freedom" for linearization, hence the name *feedback linearization*. In fact, if the system (Equation 1.9) has relative degree $r < n$ in a neighborhood of x^0 (locally at x_0), then it is possible to find $z = \Phi(x)$ such that Equation 1.9 becomes [24]

$$\dot{z}_1 = z_2$$
$$\dot{z}_2 = z_3$$
$$\vdots$$
$$\dot{z}_{r-1} = z_r$$
$$\dot{z}_r = b(z) + a(z)u$$
$$\dot{z}_{r+1} = q_{r+1}(z)$$
$$\vdots$$
$$\dot{z}_n = q_n(z)$$
$$y = z_1,$$

(1.10)

where $a(z) \neq 0$ for all z in the corresponding neighborhood of z^0, and $z^0 = \Phi(x^0)$. If the neighborhood of x_0 coincides with the state space X of Equation 1.9, then the transformation is global (which is usually difficult to obtain). Note that Equation 1.10 is still open loop, that is, u is still present, so the extra "degree of freedom" has not been used yet. A convenient choice, made possible by the affine form of Equation 1.9, is

$$u = \frac{1}{a(z)}(-b(z) + v)$$

(1.11)

with the fictitious input v still to be determined. Given that the resulting rth-order system is linear in state z and input v, it is straightforward to design a stabilizing control law $v = K_1 z_1 + ... + K_r z_r$, where K_i, $i = 1,...,r$, are constants. If the desired output y_d is nonzero, then the tracking error is $e \triangleq y_d - y = y_d - z_1$ and its derivatives $e^{(i)} = y_d^{(i)} - z_{i+1}$, $i = 1,\ldots,r - 1$. Hence, the tracking control law will be $v = K_1 e + \ldots + K_r e^{(r-1)}$. Putting $z = \Phi(x)$ and v in Equation 1.11 gives the nonlinear feedback law in terms of x, so that transformation Φ can be viewed as an auxiliary tool for designing nonlinear feedback control law (Equation 1.11).

Note that the dynamics of order $n - r$, defined by $q_{r+1},\ldots,q_n$ in Equation 1.10, are rendered unobservable, which is theoretically acceptable if the dynamics are stable.

Geometrically, the above algorithm means that the nonlinear surface defined by f and g on the $(n + 1)$-dimensional $(x_1,\ldots,x_n, u)$ space is transformed into another nonlinear surface on the $(n + 1)$-dimensional $(z_1,\ldots,z_n, v)$ space, whose restriction to the $(r + 1)$-dimensional $(z_1,\ldots, z_r, 0,\ldots, 0, v)$ space is a hyperplane. Since the nonlinearity defined by $q_{r+1},\ldots,q_n$ in Equation 1.10 cannot be seen from the output, the hyperplane makes the closed-loop system linear of rth order, when considered from the input–output viewpoint.

1.2.3.3 Input–Output Pseudolinearization

Feedback linearization has several limitations [32], which include applicability to affine systems (Equation 1.9) only and difficulty of extending the methodology to Multiple Input Multiple Output systems. Also, finding a global transformation Φ usually is a daunting problem for real-world applications. Some of these limitations can be overcome if the requirement is relaxed to linearize the system only along its set of equilibria, not the whole state space. Such an approach [21–23] is called *pseudolinearization* and may be viewed as applying the principles of feedback linearization to gain scheduling.

Consider the nonlinear system with m inputs and q outputs:

$$\dot{x} = f(x, u)$$
$$y = h(x), \tag{1.12}$$

where $f: X \times \mathcal{U} \to \mathbb{R}^n$ and $h: X \to \mathbb{R}^q$ are smooth and $X \subset \mathbb{R}^n$ and $\mathcal{U} \subset \mathbb{R}^m$ are open sets. The set of equilibria of Equation 1.12 is assumed to depend on the parameter $p \in P \subset \mathbb{R}^p$, P open, and is denoted as

$$\mathfrak{E}(p) = \{(x_0(p), u_0(p)) \,|\, f(x_0(p), u_0(p)) = 0\}, \tag{1.13}$$

where $x_0: P \to X$ and $u_0: P \to \mathcal{U}$ are at least differentiable. It is important to note that parameter p, unlike θ in Table 1.2, need not be external, that is, p may depend on x and/or u. In particular, p may depend on *both* state x and external parameter θ, a fact that is used in Section 1.3.

Let $\bar{x}(p) \triangleq x - x_0(p)$, $\bar{u}(p) \triangleq u - u_0(p)$, and $\bar{y}(p) \triangleq h(x) - h(x_0(p))$ be the incremental variables arising from Taylor linearization of the open-loop system (Equation 1.12) at an equilibrium from $\mathfrak{E}(p)$. Setting $A(p) \triangleq \partial f / \partial x |_{(x_0(p),u_0(p))}$, $B(p) \triangleq \partial f / \partial u |_{(x_0(p),u_0(p))}$, and $C(p) \triangleq \partial h / \partial x |_{(x_0(p),u_0(p))}$, the corresponding linearized system is

$$\bar{x}(p) = A(p)\bar{x}(p) + B(p)\bar{u}(p)$$
$$\bar{y}(p) = C(p)\bar{x}(p) \tag{1.14}$$

with the additional assumption that Equation 1.14 is completely controllable and observable and has relative degree r for all points from $\mathfrak{E}(p)$ and all $p \in P$.

The problem of input–output pseudolinearization is to find for system (Equation 1.12) the restriction of a transformation $z = \Phi(x)$ to $\mathfrak{E}(p)$ and the restriction of a feedback law $u = k(x,v)^*$ to $\mathfrak{E}(p)$, so that Taylor linearization of the resulting closed-loop system is independent of the choice of equilibrium from $\mathfrak{E}(p)$ and parameter p from P. It should be emphasized that, unlike for feedback linearization in Section 1.2.3.2, we are *not* looking for a *global* trans-formation $z = \Phi(x)$ and a *global* feedback law $u = k(x, v)$. We seek only their restrictions to the parameterized family of curves $\{\mathfrak{E}(p)\}_{p \in P}$, so that we need not find the whole of Φ and the whole of k. This simplifies the design consid-erably, but the resulting control law will be applicable only in the immediate neighborhood of $\{\mathfrak{E}(p)\}_{p \in P}$, not the whole $X \times \mathcal{U}$. However, Taylor lineariza-tion of the resulting closed-loop system will be independent of p and thus of all equilibria of $\{\mathfrak{E}(p)\}$. In this way, if the operating points are in the vicinity of the equilibria, then one and only one linear model will describe closed-loop dynamics, regardless of the rate of change of the operating points. Note that the design cannot guarantee anything beyond the immediate neighborhood of $\{\mathfrak{E}(p)\}_{p \in P}$.

Thus, for a SISO system (Equation 1.14), the restriction of transformation $z = \Phi(x)$ and feedback law $u = k(x,v)$ to $\{\mathfrak{E}(p)\}_{p \in P}$ should give the following Taylor linearization in the (z,v) space:[†]

$$\dot{\bar{z}}_1 = \bar{z}_2$$
$$\dot{\bar{z}}_2 = \bar{z}_3$$
$$\vdots$$
$$\dot{\bar{z}}_{r-1} = \bar{z}_r$$
$$\dot{\bar{z}}_r = \bar{v} \tag{1.15}$$
$$\dot{\bar{z}}_{r+1} = a_{r+1}^T(x_0(p), u_0(p))\bar{z}$$
$$\vdots$$
$$\dot{\bar{z}}_n = a_n^T(x_0(p), u_0(p))\bar{z}$$
$$\bar{y} = \bar{z}_1,$$

where only the n-dimensional vectors $a_{r+1}, \ldots, a_n$ still depend on equilibria from $\mathfrak{E}(p)$ and parameters from P. Since the dynamics defined by $a_{r+1}, \ldots, a_n$ are unobservable (and therefore must be at least stable), the behavior of Equation

---

* With v to be determined as a function of z and reference signal.
† Here, $\bar{z} \triangleq z - \Phi(x_0(p))$ and $\bar{v} \triangleq v - v_0(p)$ with $u_0(p) = k(x_0(p), v_0(p))$.

1.15 from the input–output, $\bar{v} - \bar{y}$, viewpoint is linear of order r and remains the same, no matter what the current values of x_0, u_0, and p are. This should be contrasted with Taylor linearization (Equation 1.14) of the open-loop system (Equation 1.12).

Comparison of Equation 1.15 with Equations 1.10 and 1.11 shows that input–output pseudolinearization may be interpreted as the restriction of feedback linearization to the parameterized family of sets of equilibria $\{\mathfrak{E}(p)\}_{p \in P}$. The focus is on a small portion of the p-parameterized (x,u)-space, that is, on linearization along the parameterized family of curves $\{\mathfrak{E}(p)\}_{p \in P}$. Thus, it suffices to investigate the tangents $\partial \Phi / \partial x \triangleq T$ of Φ and $\partial k / \partial x \triangleq F$ and $\partial k / \partial x \triangleq G$ of k along $\{\mathfrak{E}(p)\}_{p \in P}$, rather than global properties of Φ and k. In particular [22,23], it is required that $T(x_0(p))$ is invertible for all $p \in P$, feedback law $u = k(x,v)$ is smooth and satisfies $u_0(p) = k(x_0(p), v_0(p))$, and $G(x_0(p), v_0(p))$ is invertible for all $p \in P$. The two conditions on k were implicitly used in the third footnote, but explicit knowledge of k—or Φ—is not necessary even there, as $v_0(p)$—and $z_0(p)$—is never explicitly needed. Essentially, this is because the starting point of the design is Equation 1.14, which is then transformed into Equation 1.15).

The formula for the tangent of transformation Φ along $\{\mathfrak{E}(p)\}_{p \in P}$ is [22,23]

$$\bar{z} = T(p)\bar{x} \tag{1.16}$$

with T given by

$$T(p) = \begin{bmatrix} C(p) \\ C(p)A(p) \\ \vdots \\ C(p)A^{r-1}(p) \\ T_{r+1}(p) \\ \vdots \\ T_n(p) \end{bmatrix}, \tag{1.17}$$

where rows T_i, $i = r + 1, \ldots, n$ can be obtained from

$$\begin{bmatrix} T_{r+1}(p) \\ \vdots \\ T_n(p) \end{bmatrix} B(p) = 0, \tag{1.18}$$

which is a system of $n - r$ linear equations in $(n - r)n$ unknowns.

For the tangent of feedback law k along $\{\mathfrak{E}(p)\}_{p \in P}$, the formula is [22,23]

$$\bar{u} = F(p)\bar{x} + G(p)\bar{v}, \tag{1.19}$$

where

$$F(p) = -(C(p)A^{r-1}(p)B(p))^{-1}C(p)A^r(p), \tag{1.20}$$

$$G(p) = (C(p)A^{r-1}(p)B(p))^{-1}. \tag{1.21}$$

Equations 1.16 through 1.21 transform a SISO (Equation 1.14) into Equation 1.15. Design of a stabilizing control law (Equation 1.19) is complete when $\bar{v} = K_1\bar{z}_1 + \ldots + K_r\bar{z}_r$, where K_i, $i = 1,\ldots,r$, are constants. If the desired output $\bar{y}_d$ is nonzero, then the tracking error is $\bar{e} \triangleq \bar{y}_d - \bar{y} = \bar{y}_d - \bar{z}_1$ and its derivatives $\bar{e}^{(i)} = \bar{y}_d^{(i)} - \bar{z}_{i+1}$, $i = 1,\ldots,r-1$. Hence, the tracking control law will be $\bar{v} = K_1\bar{e} + \ldots + K_r\bar{e}^{(r-1)}$. Putting $\bar{z} = T(p)\bar{x}$ and $\bar{v}$ in Equation 1.19 gives the feedback law in terms of $\bar{x}$, so that transformation T can be viewed as an auxiliary tool for designing feedback control law (Equation 1.19). Thus, derived law (Equation 1.19) should be substituted in Equation 1.14.

Geometrically, $\{\mathfrak{E}(p)\}_{p \in P}$ are general curves in the $(n + m)$-dimensional (x, u) space. They are projections of the corresponding $(n + m + n)$-dimensional curves from the $(x, u, f(x, u))$ space, where f is the RHS of Equation 1.12. Taylor linearizations in the $(x, u, f(x, u))$ space at different points of curves $\{\mathfrak{E}(p)\}_{p \in P}$ involve straight lines of different tangents. Mapping Φ and feedback law k transform the general curves of the $(x, u, f(x, u))$ space into a single straight line ℓ,* whose projection on the (z, v) space is also a straight line, l. This line l is the image of curves $\{\mathfrak{E}(p)\}_{p \in P}$ in the (z, v) space under Φ and k. Taylor linearization along l gives one tangent for all points of $\{\mathfrak{E}(p)\}_{p \in P}$, namely, the tangent of ℓ.

A simple illustration is given by the following example:

$$\begin{aligned}
\dot{x}_1 &= x_1 e^{-x_1/4} \cos x_1 + x_2 + u \\
\dot{x}_2 &= x_1 + x_2 \\
y &= x_2.
\end{aligned} \tag{1.22}$$

The set of equilibria $\mathfrak{E} = \{(x_1, x_2, u) \mid x_1 e^{-x_1/4} \cos x_1 + x_2 + u = 0 \text{ and } x_1 + x_2 = 0\}$ is not parameterized for simplicity. The set is a general curve in the (x_1, x_2, u) space, as illustrated in Figure 1.3. Because of the simplicity of example 1.22, global state-space transformation Φ and global feedback control law k can be found, although it is not required by the design algorithm (Equations 1.16 through 1.21). Indeed, Φ defined by $z_1 = x_2$ and $z_2 = x_1 + x_2$ and k given by $u = -x_1 - x_1 e^{-x_1/4} - 2x_2 + v$ linearize $\mathfrak{E}$, as shown in Figure 1.4. In other words, the straight line l in Figure 1.4 is the image of $\mathfrak{E}$ under Φ and k.

---

* A curve in the $(x, u, f(x, u))$ space is fully determined by values of x and u alone. Thus, transformation of (x, u) into (z, v) by $x = \Phi^{-1}(z)$, and thus $u = k(\Phi^{-1}(z), v)$, fully determines the new shape of the curve.

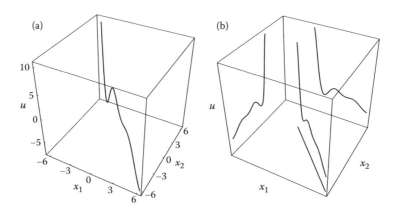

FIGURE 1.3

Set of equilibria $\mathfrak{E}$ for Equation 1.22: (a) resulting curve in the (x_1, x_2, u) space; (b) curve and its projections on the coordinate planes.

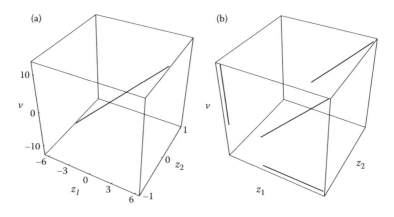

FIGURE 1.4

Linearized set of equilibria for Equation 1.22: (a) resulting straight line in the (z_1, z_2, v) space; (b) straight line and its projections on the coordinate planes.

Example 1.22 is an illustration of feedback linearization restricted to the set of equilibria $\mathfrak{E}$, which effectively is pseudolinearization.

1.3 Augmented Lateral Acceleration Autopilot Design

As explained in Section 1.2.1, the missile model given by Equation 1.2 and Table 1.1 can be represented as

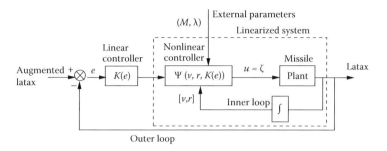

FIGURE 1.5
Block diagram representation of the missile autopilot design.

$$\dot{v} = y_v(v,U,\lambda)v - Ur + y_\zeta(v,U,\lambda)\zeta$$
$$\dot{r} = n_v(v,U,\lambda)v + n_r(v,U,\lambda)r + n_\zeta(v,U,\lambda)\zeta. \tag{1.23}$$

Setting

$$\bar{x} \doteq \begin{bmatrix} v \\ r \end{bmatrix}, \quad p \doteq \begin{bmatrix} v \\ U \\ \lambda \end{bmatrix}, \quad \bar{u} \doteq \zeta \text{ and } \bar{y} \doteq y_v(v,U,\lambda), \tag{1.24}$$

Equation 1.23 leads to the following description:

$$\dot{\bar{x}} = A(p)\bar{x} + B(p)\bar{u}$$
$$\bar{y} = C(p)\bar{x}, \tag{1.25}$$

where

$$A(p) = \begin{bmatrix} y_v(p) & -p_2 \\ n_v(p) & n_r(p) \end{bmatrix}, \quad B(p) = \begin{bmatrix} y_\zeta(p) \\ n_\zeta(p) \end{bmatrix}, \quad C(p) = \begin{bmatrix} y_v(p) & 0 \end{bmatrix}. \tag{1.26}$$

Thus, Equation 1.23 can be seen as the QLPV model,* since p in Equation 1.24 comprises *both* a state (sideslip velocity v) and external parameters (longitudinal velocity U and roll angle λ), so that the matrices (Equation 1.26) depend both on $\bar{x}$ and θ. On the other hand, in a previous paper [33], pseudolinearization has been applied to the missile, and a controller was derived (Figure 1.5). As an example, Figure 1.6 shows the uniform transient behavior of the closed-loop system across the flight envelope of the Horton missile.

---

* QLPV dynamics is $\dot{x} = A(x,0)x + b(x,0)u$.

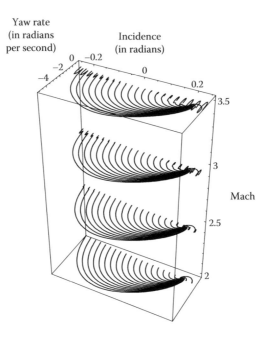

FIGURE 1.6
Constant incidence angle demand illustrates the uniform transient behavior of the Horton missile in the phase portrait.

An accelerometer is used to measure lateral acceleration. If the accelerometer is placed at the missile center of gravity, the resulting system becomes the nonminimum phase, as the accelerometer measures the effect of both the body aerodynamic force and the almost instantaneous fin force. This also has the effect of making the relative degree zero. To overcome both of these effects, an augmented acceleration signal is used. As $a_v = y_v v + y_\zeta \zeta$, augmentation can be obtained by mixing the accelerometer signal with the fin angle to eliminate the dependence of a_v on ζ to give $a_v \approx y_v v$. This approximation will not result in significant error in the control design because the fin force contribution is small in a well-designed airframe. The same effect can be obtained by moving the accelerometer forward from the center of gravity to the center of rotation. This also removes the dependence on ζ and makes the system the minimum phase.

Our design of the augmented lateral acceleration autopilot takes the second-order ($n = 2$) model (Equations 1.25 and 1.26) as the starting point. Noting that the relative degree r is 2, we then use formulae 1.16 through 1.21. Since in our case, $r = n$, we have no need for Equation 1.18, so that

$$T(p) = \begin{bmatrix} C(p) \\ C(p)A(p) \end{bmatrix} = \begin{bmatrix} y_v(p) & 0 \\ y_v^2(p) & -p_2 y_v(p) \end{bmatrix}. \qquad (1.27)$$

The control law will be

$$\bar{u} = F(p)\bar{x} + G(p)\bar{w}, \tag{1.28}$$

where $\bar{w}$ is the fictitious input still to be determined in terms of $\bar{x}$ and the reference signal. Matrix F is

$$F(p) = -(C(p)A(p)B(p))^{-1}C(p)A^2(p)$$

$$= \frac{1}{a(p)}\left[-b_1(p) \quad -b_2(p)\right] \tag{1.29}$$

and scalar G is

$$G(p) = (C(p)A(p)B(p))^{-1} = \frac{1}{a(p)}, \tag{1.30}$$

where

$$b_1(p) = y_v^3(p) - n_v(p)y_v(p)p_2,$$
$$b_2(p) = -y_v^2(p)p_2 - n_r(p)y_v(p)p_2, \tag{1.31}$$
$$a(p) = y_v^2(p)y_\zeta(p) - n_\zeta(p)y_v(p)p_2.$$

Substituting Equations 1.29 and 1.30 to Equation 1.28, the resulting pseudo-linearizing control is

$$\zeta = \frac{1}{a(p)}(-b_1(p)v - b_2(p)r + \bar{w}), \tag{1.32}$$

which still requires defining $\bar{w}$ to ensure tracking.

Let $\bar{e} \doteq a_v^d - a_v$ be the augmented lateral acceleration error, where a_v^d is the augmented lateral acceleration demand and $\dot{\bar{e}} = -\dot{a}_v$ with $\dot{a}_v^d = 0$ so that the demand does not need differentiation, that is, piecewise constant. Then the final form of the control law is

$$\zeta = \frac{1}{a(p) - \dfrac{b_1(p)y_\zeta(p)}{y_v(p)}}\left(-\frac{b_1(p)}{y_v(p)}a_v - b_2(p)r + K_1\bar{e} + K_2\dot{\bar{e}}\right). \tag{1.33}$$

One can get the augmented lateral acceleration directly from the measured output—see the beginning of the section; moreover, using the nonlinear relationship $a_v \approx y_v(p)v$, v can be recovered and hence $\dot{a}$ from $\dot{a} = y_v(p)\dot{v}$ and where $\dot{v}$ is given by Equation 1.23. The constants K_1 and K_2, for instance, of the form $K_1 = \omega_n^2$ and $K_2 = 2\xi\omega_n$, in Equation 1.33 are chosen in such a way that the performance can be achieved for the augmented lateral acceleration, which satisfies that the error equation is $\ddot{e} + 2\xi\omega_n\dot{e} + \omega_n^2\bar{e} = 0$.

Simulation results for 100 m/s² constant demands in lateral acceleration are shown in Figure 1.7. The constants K_1 and K_2 are chosen with $\omega_n = 60$ rad/s

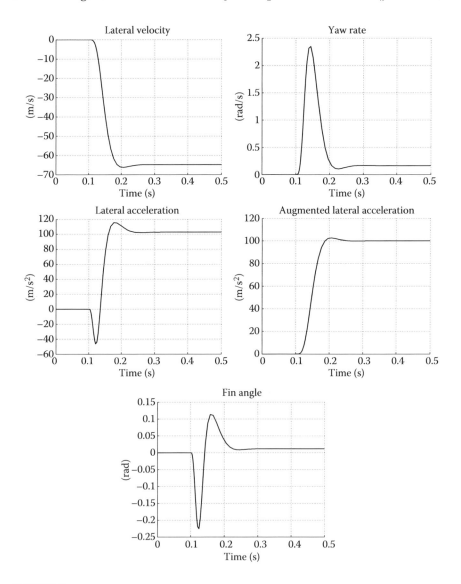

FIGURE 1.7
Simulation of the Horton missile for 100 m/s² of augmented lateral acceleration demand.

and $\xi = 0.7$. This should give a three to four times faster response than in the open loop and achieve a settling time of 0.1 s. The observed steady-state error in lateral acceleration in Figure 1.7 is about 5% of the augmented lateral acceleration.

Recall that the augmented lateral acceleration demand was derived approximately from the normal lateral acceleration demand. The approximation neglected the fin force term $y_\zeta \zeta$. The effect of this term is evident in the nonminimum phase characteristic in normal lateral acceleration a_v, and it is clearly visible in Figure 1.7; note also that the initial fin angle $\zeta < 0$. This is quickly overcome by the sideslip force as incidence builds up. The steady-state error represents the fin force contribution that was neglected along the design process.

1.4 Robust Autopilot Design

1.4.1 Nominal Autopilot Model

The sideslip velocity autopilot closed-loop characteristic equation can be obtained by substituting control law (Equation 1.33) into Equation 1.23. This yields

$$\dot{v} = y_v(p)v - p_2 r + \frac{y_\zeta}{a(p)}\left(-b_1(p)v - b_2(p)r + K_1\bar{e} + K_2\dot{\bar{e}}\right)$$
$$\dot{r} = n_v(p)v + n_r(p)r + \frac{n_\zeta}{a(p)}\left(-b_1(p)v - b_2(p)r + K_1\bar{e} + K_2\dot{\bar{e}}\right). \tag{1.34}$$

Assuming that the derivative $\dot{a}_{v_d}$ of the sideslip velocity demand is zero, we have $\dot{\bar{e}} = -\dot{a}_v$. Then the closed-loop equations (Equation 1.34) can be rewritten as

$$E(p)\dot{\bar{x}} = A_c(p)\bar{x} + B_c(p)v_d$$
$$\bar{y} = C(p)\bar{x} \tag{1.35}$$

with $\bar{x}$, $\bar{y}$, and p as in Equation 1.24, matrix A_c is

$$A_c \doteq \begin{bmatrix} a_{c11} & a_{c12} \\ a_{c21} & a_{c22} \end{bmatrix} \tag{1.36}$$

where

$$a_{c11} = y_v(p)a(p) - y_v(p)y_\zeta(p)(b_1(p) - K_1)$$

$$a_{c12} = -p_2 a(p) - y_\zeta(p)b_2(p)$$

$$a_{c21} = n_v(p)a(p) - y_v(p)n_\zeta(p)(b_1(p) - K_1)$$ (1.37)

$$a_{c22} = n_r(p)a(p) - n_\zeta(p)b_2(p)$$

and matrices E, B_c, and C given as

$$E(p) = \begin{bmatrix} a(p) + y_v(p)y_\zeta(p)K_2 & 0 \\ y_v(p)n_\zeta(p)K_2 & a(p) \end{bmatrix}$$

$$B_c(p) = \begin{bmatrix} y_\zeta(p)K_1 \\ n_\zeta(p)K_1 \end{bmatrix}$$ (1.38)

$$C(p) = \begin{bmatrix} y_v(p) & 0 \end{bmatrix}.$$

In this section, the stability margins associated with parametric uncertainty in aerodynamics parameters y_v, U, n_r, n_v, and n_ζ are obtained along with the performance of the resulting robust autopilot as desired. In previous work [33], the robust analysis of the same missile using Kharitonov-type robustness metrics was performed. While that has given us a good insight of how robust our missile autopilot design is, it was not easily extended to redesign the controller to meet the robustness requirements and guarantee the performance under parametric uncertainty. That is one of the main contributions of this paper.

1.4.2 Robust Performance Design

The pseudolinearization enables us to attain the desired level of performance through the choice of coefficients K_1 and K_2. In this section, the idea is to estimate these coefficients of the pole placement controller according to the desired level of performance by using pole placement criteria. The performance of the system is better stated in the (z, w) space in which the nominal system is a second-order linear time-invariant system. Performance is then achievable; recall that the pseudolinearization has been applied for this purpose. The following work presents a state feedback design through Lyapunov theory for a parametric uncertain system.

The parametric uncertain system that has been investigated represents a system with poor accuracy in its aerodynamic coefficients. Aside from these aerodynamic coefficients, the variation in forward velocity, which is explicit in differential equation 1.23, will also be considered as an uncertainty. Now, the vector of uncertain parameters can be written as $p = [y_v, U, n_v, n_r, n_\zeta]^T$, where p_2, that is, U, represents the forward velocity.

The closed-loop characteristic equation of the augmented acceleration autopilot can be obtained by substituting control law (Equation 1.33) into Equation 1.23. This yields

$$\dot{v} = y_v(p)v - p_2 r$$

$$\dot{r} = n_v(p)v + n_r(p)r + \frac{n_\zeta}{a(p)}(-b_1(p)v - b_2(p)r + K_1\bar{e} + K_2\dot{\bar{e}}) \qquad (1.39)$$

with the same assumptions as in the previous section about $y_\zeta\zeta$ and $\dot{a}_v^d$ neglected. The transformation (Equation 1.16) turns the (x, u) space to the (z, w) space, and then in this new space, the closed-loop equations (Equation 1.39) can be written as

$$
\begin{bmatrix} \dot{z}_1 \\ \dot{z}_2 \end{bmatrix} = \begin{bmatrix} 0 & 1 \\ a_1(p) & a_2(p) \end{bmatrix} \begin{bmatrix} z_1 - a_v^d \\ z_2 \end{bmatrix}
$$

$$
+ \begin{bmatrix} 0 \\ a_c(p) \end{bmatrix} \begin{bmatrix} K_1 & K_2 \end{bmatrix} \begin{bmatrix} z_1 - a_v^d \\ z_2 \end{bmatrix} + \begin{bmatrix} 0 \\ a_i(p) \end{bmatrix} \dot{a}_v^d, \qquad (1.40)
$$

where

$$a_1(p) = -p_2 n_v(p) - n_r(p)y_v(p)$$

$$+ \frac{n_\zeta(p)p_2 b_1(p)}{a(p)} + \frac{n_\zeta(p)y_v(p)b_2(p)}{a(p)}$$

$$a_2(p) = n_r(p) - \frac{n_\zeta(p)b_2(p)}{a(p)} + y_v(p)$$

$$a_c(p) = \frac{p_2 n_\zeta(p)}{a(p)}$$

$$a_i(p) = a_1(p).$$

The nominal system can be written as follows with the assumption that y_ζ is neglected in $a(p)$:

$$\begin{bmatrix} \dot{z}_1 \\ \dot{z}_2 \end{bmatrix} = \begin{bmatrix} 0 & 1 \\ 0 & 0 \end{bmatrix} \begin{bmatrix} z_1 - a_v^d \\ z_2 \end{bmatrix} + \begin{bmatrix} 0 \\ -1 \end{bmatrix} \begin{bmatrix} K_1 & K_2 \end{bmatrix} \begin{bmatrix} z_1 - a_v^d \\ z_2 \end{bmatrix}. \tag{1.41}$$

The performance of the system is then handled by the equivalent to error dynamics equation:

$$\ddot{z}_1 + K_2\dot{z}_1 + K_1(z_1 - a_v^d) = 0,$$

and it is stressed that z_1 is the augmented acceleration.

Before the desired level of performance is described, it has to be noticed that the missile is limited by both its natural behavior and its actuator's performance. The actuators limit the steering of the missile and, consequently, its performance. These actuators are modeled as second-order systems; their characteristic is given by their angle range, ±0.3 rad, and their frequency response, 250 rad/s. Consequently, it has been chosen to keep the response of the pole placement controller within these constraints so that the actuators do not operate above their cutoff frequency. This is done by limiting the frequency response of the pole placement controller to less than 100 rad/s and keeping the damping ratio above the critic, for actuator fatigue and power consumption reasons. Additionally, the desired performance requires that the system perform within 0.1 s of maximum settling time. The resulting D-stability region is shown in Figure 1.8, where the poles have to stay in the cone defined with half-angle $\pi/4$ and with the pole real part in the range between −40 and −100. It is difficult to take into account the physical angle range of the fins, so this will be checked through simulations, *a posteriori*.

Some modified Lyapunov equations are used to state this D-stability region; they turn the usual Lyapunov equation into a family of linear matrix inequalities for which the Linear Matrix Inequality Toolbox [34] for MATLAB® has been used. Assuming a region $\mathcal{D} = \{z \in \mathbb{C} \mid L + Mz + M^T\bar{z} < 0\}$, the matrix A' would have all its eigenvalues in this region $\mathcal{D}$ if there exists a symmetric positive-definite matrix X satisfying the following LMI:

$$(\lambda_{i,j}X + \mu_{i,j}A'X + \mu_{j,i}XA'^T)_{i,j} < 0, \tag{1.42}$$

where $\lambda_{i,j}$ and $\mu_{i,j}$ represent, respectively, the matrix coefficients of L and M. For our system, it has been applied in the case of state feedback control

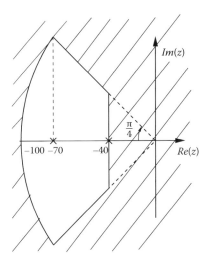

FIGURE 1.8
Robust performance of the pole placement controller for the Horton missile is presented as a
D-stability region. All the poles of the system belong to the "trapezoid" area.

where the above matrix A' can be represented by the closed-loop matrix $A +
BK$. This leads to the LMI

$$(\lambda_{i,j}X + \mu_{i,j}AX + \mu_{j,i}XA^T + \mu_{i,j}BY + \mu_{j,i}Y^TB^T)_{i,j} < 0, \tag{1.43}$$

with $Y = KX$ and X symmetric positive definite.

 The nominal plant considered (Equation 1.41) is an LTI; however, when
it comes to considering parametric uncertainties, the uncertain system is
not anymore so, since it depends on the operating point (Equation 1.40). The
uncertainties are, moreover, involved in a multiaffine form. The uncertain
model has been brought to affine parametric form by introducing some overbounding, and consequently, the uncertain parameter is now extended from
$p = [y_v, U, y_\zeta, n_v, n_r, n_\zeta]^T$ to their two-by-two products as well. After a preliminary study,[*] it appears that some of the aerodynamic coefficients do not have
much influence on the performance, and the design can be simplified using
a limited number of parameters without significant differences. The most
influential parameters have been so identified as $p = [y_v, U, n_\zeta, y_v, \times U, y_v, \times
n_\zeta, U \times n_\zeta]^T$, where

$$y_v = \bar{y}_v \left(1 + \frac{\Delta y_v}{\bar{y}_v}\right)$$

---

[*] The study has been done by establishing the influence of each aerodynamic coefficient using
a similar methodology as now presented for the design.

$$U = \bar{U}\left(1 + \frac{\Delta U}{\bar{U}}\right)$$

$$y_\zeta = \bar{y}_\zeta\left(1 + \frac{\Delta y_\zeta}{\bar{y}_\zeta}\right)$$

$$n_v = \bar{n}_v\left(1 + \frac{\Delta n_v}{\bar{n}_v}\right)$$

$$n_r = \bar{n}_r\left(1 + \frac{\Delta n_r}{\bar{n}_r}\right)$$

$$n_\zeta = \bar{n}_\zeta\left(1 + \frac{\Delta n_\zeta}{\bar{n}_\zeta}\right).$$

The uncertain system from Equation 1.40 has an input matrix. This matrix increases consequently the complexity of the computation, and it is chosen not to take it into account in the following design. Then the following model (Equation 1.44) is used instead where the controller is set for the nominal model:

$$\begin{bmatrix} \dot{z}_1 \\ \dot{z}_2 \end{bmatrix} = \begin{bmatrix} 0 & 1 \\ a_1(p) & a_2(p) \end{bmatrix}\begin{bmatrix} z_1 - a_v^d \\ z_2 \end{bmatrix} + \begin{bmatrix} 0 \\ -1 \end{bmatrix}\begin{bmatrix} K_1 & K_2 \end{bmatrix}\begin{bmatrix} z_1 - a_v^d \\ z_2 \end{bmatrix}. \tag{1.44}$$

where

$$a_1(p) = -\bar{U}n_v\frac{\Delta n_v}{\bar{n}_v} - \bar{y}_v\bar{n}_r\frac{\Delta n_r}{\bar{n}_r} - \bar{y}_v^2\frac{\Delta U}{\bar{U}} + \bar{y}_v^2\frac{\Delta y_v}{\bar{y}_v}$$

$$- \bar{U}n_v\frac{\Delta p_{n_v}^U}{\bar{p}_{n_v}^U} - \bar{n}_r\bar{y}_v\frac{\Delta p_{n_r}^{y_v}}{\bar{p}_{n_r}^{y_v}} + (\bar{n}_v\bar{U} - \bar{y}_v^2)\frac{\Delta p_{n_\zeta}^U}{\bar{p}_{n_\zeta}^U}$$

$$+ (\bar{n}_v\bar{U} + \bar{n}_r\bar{y}_v)\frac{\Delta n_\zeta}{\bar{n}_\zeta} + (\bar{y}_v + \bar{n}_r)\bar{y}_v\frac{\Delta p_{n_\zeta}^{y_v}}{\bar{p}_{n_\zeta}^{y_v}}$$

$$a_2(p) = +\bar{n}_r\frac{\Delta n_r}{\bar{n}_r} + \bar{y}_v\frac{\Delta y_v}{\bar{y}_v} - (\bar{y}_v + \bar{n}_r)\frac{\Delta n_\zeta}{\bar{n}_\zeta}$$

It was chosen to tackle the operating point dependence problem by considering a representative class of uncertain systems, and for each of them, the design of a controller for a parametric uncertain system is processed, so that the nonlinearities of the uncertain system are captured. The coefficients K_1 and K_2 are evaluated at each operating point, (σ, M), by solving LMI equations (Equation 1.43) such that the robust performance of the closed-loop system is achieved at each of these operating points. The set of controllers constitutes then a gain-scheduling controller. The carpet plots of these coefficients are shown in Figure 1.9.

It is a complex problem to design these gains through the whole flight envelope; however, approximating our model of uncertainties with a multiaffine model that captures the nonlinearities enables us to design a pole placement controller at each knot (see carpet plots in Figure 1.9). It can be shown that this linear interpolation between these knots leads to a self-scheduled controller $(K_1(\sigma, M)$ and $K_2(\sigma, M))$, and it would bring more confidence in the robust performance on the full flight envelope. This is checked in the next section through an analysis of the closed-loop system. This process should improve the robust performance of the overall system since the controller is self-scheduled in order to guarantee the robust performance along the flight envelope. It should be emphasized that the methodology developed so far has been of interest when it comes to giving a systematic tool to estimate gains in the pole placement controller, which verifies some performance robustness properties of the transient response. Since the LMI solver gives a solution among solutions, it has been chosen to build a smoother interpolation of these raw controller gains. This leads us to consider the controller gains in Figure 1.10, where robust analysis has been performed in the same lines as for the previous raw controller analysis, which is now presented in the next section.

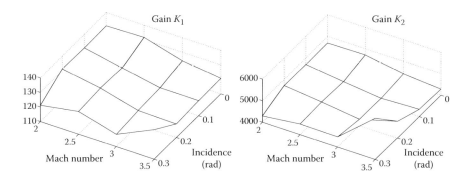

FIGURE 1.9

Gain K_1 (on the left) and K_2 (on the right) of the pole placement controller for the Horton model through the whole flight envelope of the missile ($\lambda = 0°$).

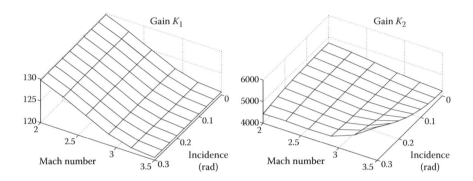

FIGURE 1.10
Smoothened controller gains K_1 (on the left) and K_2 (on the right) of the pole placement controller; see Figure 1.9. Its robustness has been checked for the Horton model through the whole flight envelope of the missile ($\lambda = 0°$).

1.4.3 Robust Performance Analysis

The robust pole placement controller validity is now investigated. Because the self-gain–scheduled controller is built from linear interpolation between chosen gains (K_1 and K_2) and since it has been checked that the representative class of uncertain systems captures the nonlinearities, the self-scheduled controller designed previously is expected to give the whole system some robust performance. However, some of the simplifications in the design (compare Equation 1.40 with Equations 1.41 and 1.44) are then reconsidered in this section. For instance, the uncertain system described by Equation 1.40 becomes, for the closed-loop system,

$$
\begin{bmatrix} \dot{z}_1 \\ \dot{z}_2 \end{bmatrix} = \begin{bmatrix} 0 & 1 \\ a_1^{cl}(p) & a_2^{cl}(p) \end{bmatrix} \begin{bmatrix} z_1 \\ z_2 \end{bmatrix} + \begin{bmatrix} 0 \\ a_i(p) \end{bmatrix} a_v^d \tag{1.45}
$$

where

$$
a_1^{cl}(p) = -K_1 - \bar{U}\bar{n}_v \frac{\Delta n_v}{\bar{n}_v} - \bar{y}_v\bar{n}_r \frac{\Delta n_r}{\bar{n}_r} - (K_1 + \bar{y}_v^2)\frac{\Delta U}{\bar{U}}
$$

$$
+ \bar{y}_v^2 \frac{\Delta y_v}{\bar{y}_v} - \bar{U}\bar{n}_v \frac{\Delta p_{n_v}^U}{\bar{p}_{n_v}^U} - \bar{n}_r\bar{y}_v \frac{\Delta p_{n_r}^{y_v}}{\bar{p}_{n_r}^{y_v}}
$$

$$
+ \left(-K_1 + \bar{n}_v\bar{U} + \bar{n}_r\bar{y}_v \right)\frac{\Delta n_\zeta}{\bar{n}_\zeta}
$$

$$
+ (-K_1 + \bar{n}_v\bar{U} - \bar{y}_v^2)\frac{\Delta p_{n_\zeta}^U}{\bar{p}_{n_\zeta}^U} + (\bar{y}_v + \bar{n}_r)\bar{y}_v \frac{\Delta p_{n_\zeta}^{y_v}}{\bar{p}_{n_\zeta}^{y_v}}
$$

$$a_2^{cl}(p) = -K_2 + \bar{n}_r \frac{\Delta n_r}{\bar{n}_r} + \bar{y}_v \frac{\Delta y_v}{\bar{y}_v}$$

$$- (K_2 + \bar{y}_v + \bar{n}_r) \frac{\Delta n_\zeta}{\bar{n}_\zeta} - K_2 \frac{\Delta U}{\bar{U}} - K_2 \frac{\Delta p_{n_\zeta}^U}{\bar{p}_{n_\zeta}^U}$$

$$a_i(p) = a_1(p).$$

For simplicity in this study, the term $a_i(p)$ is not taken into account in this analysis. This uncertain closed-loop system is then included in a polytope, and the LMI (Equation 1.42) is then solved, the feasibility of which is checked for some 5% uncertainty on each of the parameters of vector p along the whole flight envelope. It is important to note that some of the simplifications along the design have been removed, and the analysis chose a significant difference in the robust properties of the system. However, in this study, the parametric uncertainty boundaries are limited by the most sensible parameter without finding the maximum extent of all parameters; also. further analysis would extend each parameter boundary, for example, via Monte Carlo simulation.

Recall that the augmented acceleration demand is derived from the normal lateral acceleration demand. The effect of the term $y_\zeta \zeta$ is evident in the nonminimum phase characteristic in normal lateral acceleration a_v, and it is clearly visible in Figure 1.11; note also that the initial fin angle $\zeta > 0$. Simulations of the system with a lateral acceleration require 100 m/s². The plain curves represent the nominal system, and the dashed curves represent the system under uncertainties. The transient response behaves as required, respecting the frequency bandwidth, the critic damping, and the settling time. Some prefilter has been added to make sure that the actuators do not saturate. However, for some of the 20% uncertainties (on each parameter), the performance is not achieved, that is, the damping ratio is not achieved, and the actuators are saturated.

1.4.4 Discussions

The performance of the class of uncertain systems is obtained; however, the nonlinearities of the uncertain system make it difficult to capture the whole system with a representative class of parametric uncertain systems. Here, in the context of a polytopic approach (analysis), these nonlinearities can make the polytope very complex and consequently computationally difficult to handle, but in principle, the approximation of the nonlinearities can be as accurate as required. A trade-off between complexity and accuracy is needed. In the robust performance analysis, the chosen class of uncertain systems tries to capture the nonlinearities according to the carpet plot in Figures 1.9 and 1.10, and overbounding has been used in order to get affine

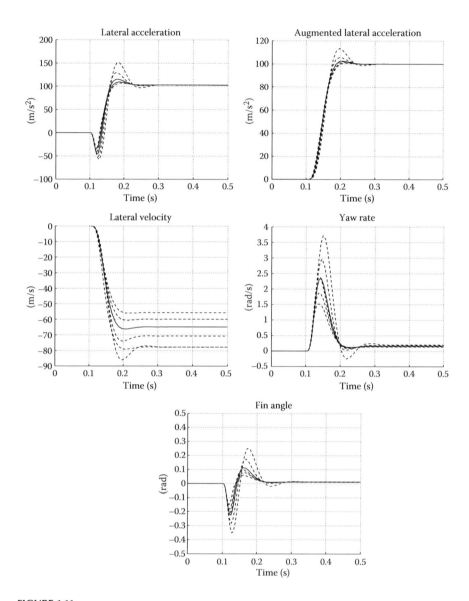

FIGURE 1.11
Simulation of the system for augmented lateral acceleration demand of 100 m/s² with the nominal response (plain curve), 10% and 20% uncertainties on the aerodynamic coefficients.

parametric uncertainties. For big polytopes, current algorithms may fail to prove feasibility of the LMIs and may not bring any answer to our problem via this methodology. Because the Horton model and its uncertain parametric model can be approximated by a multilinear system with respect to incidence and Mach number, the feasibility of the LMI is established to some extent.

The conservativeness of this design/analysis could be reduced by considering the rate of change among the class of uncertain systems (e.g., the rate of change in Mach number). Moreover, in order to keep the presentation of this study simple, some parameters have been omitted both in the design and in the analysis.

1.5 Conclusions

It was shown that a reasonably realistic missile model could be described as a QLPV system. The pseudolinearizing autopilot consists of one controller only, and "scheduling" is done automatically by feedback, giving total independence of the operating point. However, the pseudolinearizing design is (like gain scheduling) valid only in the vicinity of the equilibria. It states performance of the closed-loop system (Figure 1.6), but previous studies have shown not-very-good robust performance. An attempt to achieve the desired transient response for a class of parametric uncertain systems (Figure 1.11) was carried on by estimating the pole placement controller coefficients (Figure 1.10). This is a more systematic tool to tune some coefficients for which the closed-loop system is more likely to be robust. Finally, the robust performance analysis was done to validate this approach.

References

1. W. J. Rugh. Analytical framework for gain scheduling. *IEEE Control Systems Magazine*, 11:799–803, 1993.
2. S. T. Lin and C. M. Lee. Multivariable control of the J-85 turbojet engine for full flight envelope operation. *Journal of Guidance, Control and Dynamics*, 19(4):913–920, 1996.
3. M. N. Tran and D. Hrovat. Application of gain scheduling to design of active suspensions. In *32nd IEEE Conference on Decision and Control*, IEEE, San Antonio, TX, volume 2, pages 1030–1035, 1993.
4. R. W. Beaven, M. T. Wright, S. D. Garvey, and M. I. Friswell. The application of setpoint gain scheduling to high-speed independent drives. *Control Engineering Practice*, 3(11):1581–1585, 1995.
5. R. A. Nichols, R. T. Reichert, and W. J. Rugh. Gain scheduling for H_∞ controllers: A flight control example. *IEEE Transactions on Control Systems Technology*, 1:69–79, 1993.
6. R. A. Hyde and K. Glover. The application of scheduled H_∞ controllers to a VSTOL aircraft. *IEEE Transactions on Automatic Control*, 38(7).1021–1039, 1993.

7. R. A. Hyde and K. Glover. Taking H_∞ control into flight. In *32nd IEEE Conference on Decision and Control*, IEEE, San Antonio, TX, volume 2, pages 1458–1463, 1993.

8. R. A. Hyde and K. Glover. VSTOL first flight on an h_∞ control law. *Computing and Control Engineering Journal*, 6(1):11–16, 1995.

9. E. W. Kamen and P. P. Khargonekar. On the control of linear systems whose coefficients are functions of parameters. *IEEE Transactions on Automatic Control*, 29(1):25–33, 1984.

10. W. T. Baumann and W. J. Rugh. Feedback control of nonlinear systems by extended linearization. *IEEE Transactions on Automatic Control*, 31(1):40–46, 1986.

11. J. Wang and W. J. Rugh. Parameterized linear systems and linearization families for nonlinear systems. *IEEE Transactions on Circuits and Systems*, 34(6):650–657, 1987.

12. J. S. Shamma and M. Athans. Analysis of gain scheduled control for nonlinear plants. *IEEE Transactions on Automatic Control*, 35(8):898–907, 1990.

13. J. S. Shamma and M. Athans. Guaranteed properties of gain scheduled control for linear parameter-varying plants. *Automatica*, 27(3):559–564, 1991.

14. J. S. Shamma and M. Athans. Gain scheduling: Potential hazards and possible remedies. *IEEE Control Systems Magazine*, 13:101–107, 1992.

15. S. M. Shahruz and S. Behtash. Design of controllers for linear parameter varying systems by the gain scheduling technique. *Journal of Mathematical Analysis and Applications*, 168(1):195–217, 1992.

16. J. S. Shamma and J. R. Cloutier. Gain-scheduled missile autopilot design using LPV transformations. *Journal of Guidance, Control and Dynamics*, 16(2):256–263, 1993.

17. G. J. Balas, J. C. Doyle, K. Glover, A. Packard, and R. Smith. *The μ-Analysis and Synthesis Toolbox*. The MathWorks Inc., Cambridge, MA, 1994.

18. A. Packard and J. C. Doyle. The complex structured singular value. *Automatica*, 29(1):71–109, 1993.

19. A. Packard, J. C. Doyle, and G. Balas. Linear, multivariable robust control with a μ perspective. *Journal of Dynamic Systems, Measurement and Control*, 115(2B):426–438, 1993.

20. M. P. Horton. A study of autopilots for the adaptive control of tactical guided missiles. Master's thesis, University of Bath, 1992.

21. C. Reboulet and C. Champetier. A new method for linearizing nonlinear systems: The pseudolinearization. *International Journal of Control*, 40(4):631–638, April 1984.

22. D. A. Lawrence and W. J. Rugh. Input-output pseudolinearization for nonlinear systems. *IEEE Transactions on Automatic Control*, 39(11):2207–2218, November 1994.

23. D. A. Lawrence. A general approach to input-output pseudolinearization for nonlinear systems. *IEEE Transactions on Automatic Control*, 43(10):1497–1501, October 1998.

24. A. Isidori. *Nonlinear Control Systems: An Introduction*. Springer-Verlag, New York, 2nd edition, 1989.

25. M. P. Horton. Autopilots for tactical missiles: An overview. *IMechE: Journal of Systems and Control Engineering*, 209:127–139, 1995.

26. K. A. Wise. Comparison of 6 robustness tests evaluating missile autopilot robustness to uncertain aerodynamics. *Journal of Guidance, Control and Dynamics*, 15(4):861–870, 1992.

27. A. Tsourdos, A. Blumel, and B. A. White. Trajectory control of a nonlinear hom-
    ing missile. In *14th IFAC Symposium on Automatic Control in Aerospace*, 1998.
    Korea.

28. E. L. Ince. *Ordinary Differential Equations*. Longmans, Green and Co. Ltd, London,
    1927.

29. J. G. F. Belinfante and B. Kolman. *A Survey of Lie Groups and Lie Algebras with
    Applications and Computational Methods*. SIAM, Philadelphia, 1972.

30. B. Jakubczyk and W. Respondek. On linearization of control systems. *Bulletin de
    l'Académie Polonaise des Sciences Série des Sciences Mathématiques, Astronomiques et
    Physiques*, 28:517–522, 1980.

31. L. R. Hunt and R. Su. Linear equivalents of nonlinear time-varying systems. In
    N. Levan, editor, *Proceedings of the 4th International Symposium on Mathematical
    Theory of Networks and Systems, August 5–7, 1981, Santa Monica, CA*, pages 119–
    123. Western Period Co., North Hollywood USA, 1981.

32. R. Żbikowski, K. J. Hunt, A. Dzieliński, R. Murray-Smith, and P. J. Gawthrop. A
    review of advances in neural adaptive control systems. Technical Report of the
    ESPRIT NACT Project TP-1, Glasgow University and Daimler-Benz Research,
    1994. Available from FTP server ftp.mech.gla.ac.uk as PostScript file /nact/
    nact_tp1.ps.

33. A. Tsourdos, R. Żbikowski, and B. A. White. Robust Design of Sideslip Velocity
    Autopilot for a Quasi-Linear Parameter-Varying Missile Model. In *Journal of
    Guidance, Control, and Dynamics*, volume 24, pages 287–295. American Institute
    of Aeronautics and Astronautics, Reston, VA, 2001.

34. P. Gahinet, A. Nemirovski, A. J. Laub, and M. Chilali. *LMI Control Toolbox User's
    Guide*. The MathWorks, Natick, MA, 1995.

2

Polynomial Approach for Design and Robust Analysis of Lateral Missile Control

Antonios Tsourdos, Brian A. White, and L. Bruyere

CONTENTS

2.1 Introduction

Polynomial eigenstructure assignment (PEA) [23] is a polynomial approach to eigenstructure assignment (EA) [17,19,21]. For a linear time-invariant (LTI) system, the PEA, similar to the EA, enables the placing of eigenvalues and eigenvectors and thus the shaping of the system response as desired with respect to inputs and outputs. Unlike the classical EA, the PEA enables the development of a solution for the eigenspace based on polynomial matrices. Similar to the EA, this design approach is valid only for LTI systems, which results in designers usually resorting to some interpolation process. However, the design approach developed here makes use of the explicit linear parameter-varying (LPV) parameterization to design directly a suitable LPV controller. Therefore, the approach stays simple, attractive, and comparable with other approaches while constructing an LPV controller [5,6].

Since the initial Kharitonov results for interval polynomials were published, further research work in robust analysis has been motivated. The Kharitonov approach assesses an interval polynomial family's stability by checking the stability of only the four Kharitonov polynomials, thus reducing an infinite problem to a finite one. This result was extended later to more general polynomial families like affine/linear [1] or even to some extent to affine/linear polynomial rational families [2,3]. Although various criteria were formulated in this framework, including H_∞ and D-stability, most results rely on the zero-exclusion criteria, which in practice apply to the value set, while frequencies (or generalized frequencies) are swept across. For more complex families, such as affine/linear polynomial family with polytopic parameter space, one usually needs to resort to the edges theorem, which takes into account the parameter space edges instead of simply its vertices and thus does not lead to a finite polynomial set. For multiaffine/multilinear polynomial families, the approach often preferred is the mapping theorem, which captures the value set of the family of polynomials in its overbounding convex hull. This convex hull can be determined from the parameter box vertices image, and thus, D-stability robustness can be investigated for multilinear parametric uncertain systems using these tools where the value set needs to be evaluated for a sweep across frequencies.

In the framework of the Nyquist stability theorem, however, some extensions have been initiated recently and lead to a finite version, that is, the finite Nyquist theorem (FNT) [7], which requires only a finite number of (generalized) frequency checks to prove the stability (D-stability) of a polynomial. The polynomial family stability can be further stated using an extension of this latter theorem, that is, the finite inclusion theorem (FIT) [7]. Finally, combined with the mapping theorem, the D-stability robustness for multiaffine parametric uncertain systems can be assessed within a finite number of polynomial/frequency checks.

This paper focuses on the design of a missile autopilot and the robust analysis of the resulting flight control system. The tail-controlled missile in the cruciform fin configuration is modeled as a second-order quasilinear parameter-varying (QLPV) system. This nonlinear model is obtained from the Taylor linearized model of the horizontal motion by including explicit dependence of the aerodynamic derivatives on a state (sideslip velocity) and external parameters (longitudinal velocity and roll angle).

The autopilot design is based on the PEA. The design makes the closed-loop system independent of the choice of equilibria. Thus, if the operating points are in the vicinity of the equilibria, then one and only one linear model will describe closed-loop dynamics, regardless of the rate of change of the operating points. Simulations for constant lateral acceleration demands show good tracking with fast response time. The contribution is to interpret the PEA as a new form of dynamic inversion [8,10] and to perform a successful PEA design for a QLPV system.

Parametric stability margins for uncertainty in longitudinal velocity and semi-nondimensional aerodynamic derivatives [2,3,26] are analyzed using the FIT [7]. The analysis shows that the design is fairly robust with respect to parametric uncertainty. The contribution is to perform an effective robustness analysis, despite the involved parametric dependencies.

In the following, the PEA is first described as a design approach for multiple input multiple output (MIMO) LTI systems in the framework of a structured and chosen controller. Robust analysis of single input single output (SISO) LTI systems is presented in the context of the FIT. Finally, the aerospace application of lateral autopilot missile control is taken as an example to illustrate the discussed approaches and is applied in the framework of a missile LPV model. Simulation results show good D-stability while maintaining good robustness with respect to parametric uncertainties.

2.2 Polynomial Eigenstructure Assignment

Assuming the LTI system in the following:

$$\dot{x} = Ax + Bu \tag{2.1a}$$

$$y = Cx + Du \tag{2.1b}$$

the eigenvalue/eigenvector null space equation can be written as a polynomial null space equation as shown by

$$\left[(A - sI) \ B \right] \begin{bmatrix} Z(s) \\ P(s) \end{bmatrix} = 0 \tag{2.2}$$

where s represents both the eigenvalue and the Laplace variable. $Z(s)$ and $P(s)$ represent the eigenvector space and its associated control vector space, respectively. In the sequel, the eigenvectors and/or eigenspace are referred to by their corresponding polynomial matrices.

The open-loop transfer function $G_y(s)$ can be written as

$$G_y(s) = C(sI - A)^{-1} B + D \tag{2.3}$$

and hence becomes a function of the eigenspace and its associated space as described in Equation 2.2. Thus,

$$G_y(s) = CZ(s)P(s)^{-1} + D$$
$$= (CZ(s) + DP(s))P(s)^{-1} \tag{2.4}$$
$$= Z_0(s)P(s)^{-1}$$

where

$$Z_o(s) = (CZ(s) + DP(s)). \tag{2.5}$$

Consider the autonomous system under feedback with matrix gain controller $K(s)$:

$$P(s) = K(s)Z_0(s) \tag{2.6}$$

where the system output is represented by $Z_0(s)$. The closed-loop system becomes

$$G_y^{cl}(s) = G_y(s)K(s)Z_0(s)$$
$$= Z_0(s)P(s)^{-1}P(s) \tag{2.7}$$
$$= Z_0(s)$$

where the open-loop system dynamics is inverted, leaving zero dynamics unaffected.

Using the PEA, specifying a particular controller structure enables the designer to change the dynamics of the open-loop system without having to cancel out the zero dynamics. The next section introduces a particular controller structure presenting sufficient flexibility for the purpose of aerospace application under consideration in this paper.

2.2.1 Controller Structure

The controller structure [23] choice is driven by consideration of improving stability, performance, tracking, sensitivity, and robustness of the system. The controller structure shown in Figure 2.1 gives sufficient flexibility to produce such a design.

The outputs of the system are partitioned into controlled outputs y_c and inner loop outputs y_i. For the missile, controlled outputs are incidence, sideslip velocity, and acceleration, while the inner loop outputs are body rates from rate gyros. Both are required for good performance and robustness. The controller $K_a(s)$ shapes the tracking response of the closed-loop system to the desired demands, the controller $K_u(s)$ shapes the input to the plant, $K_i(s)$ feeds back the extra measurements available, y_i, and $K_c(s)$ feedbacks the controlled output y_c, shaping consequently its transient response. Altogether, gains $K_u(s)$, $K_a(s)$, $K_c(s)$, and $K_i(s)$ compose a dynamic controller. For the case where the order of $K_u(s)$ meets or exceeds the order of the other polynomial matrices $K_a(s)$, $K_c(s)$, and $K_i(s)$, the resulting dynamic controller is proper and thus can be realized.

From the figure, the following system interconnection can be defined, where, for clarity, dependence in s is dropped:

$$u = K_u^{-1}\left(K_a e - K_c y_c - K_i y_i \right) \tag{2.8a}$$

$$y_c = G_c u \tag{2.8b}$$

$$y_i = G_i u \tag{2.8c}$$

$$e = r_1 - y_c \tag{2.8d}$$

$$e_u = r_2 - u \tag{2.8e}$$

where $G_c(s)$ and $G_i(s)$ represent open-loop transfer functions for the controlled and the inner loop outputs, respectively.

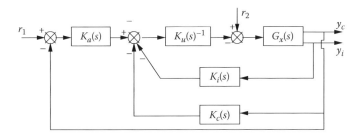

FIGURE 2.1
Controller structure chosen for aerospace applications introducing dynamic gains $K_u(s)$, $K_a(s)$, $K_c(s)$, and $K_i(s)$ with y_c, the controlled outputs, and y_i, the additional, measured outputs.

The transfer functions for the closed-loop system inputs and outputs are given by

$$
\begin{bmatrix} u \\ y_c \end{bmatrix} = \begin{bmatrix} \Delta^{-1}K_u^{-1}K_a & -\Delta^{-1}K_u^{-1}\big((K_c + K_a)G_c + K_iG_i\big) \\ G_c\Delta^{-1}K_u^{-1}K_a & G_c\big(I - \Delta^{-1}K_u^{-1}\big((K_c + K_a)G_c + K_iG_i\big)\big) \end{bmatrix} \begin{bmatrix} r_1 \\ r_2 \end{bmatrix}
$$

(2.9)

where $\Delta = \big(I + K_u^{-1}\big((K_a + K_c)G_c + K_iG_i\big)\big)$ is the transfer function of the open-loop system.

From Equation 2.9, four transfer functions can be defined, as follows:

$$
T_{ur_1}(s) = \Delta^{-1}K_u^{-1}K_a
$$

(2.10a)

$$
T_{ur_2}(s) = -\Delta^{-1}K_u^{-1}\big((K_c + K_a)G_c + K_iG_i\big)
$$

(2.10b)

$$
T_{yr_1}(s) = G_c\Delta^{-1}K_u^{-1}K_a
$$

(2.10c)

$$
T_{yr_2}(s) = G_c\big(I - \Delta^{-1}K_u^{-1}\big((K_c + K_a)G_c + K_iG_i\big)\big)
$$

(2.10d)

where $T_{yr_1}(s)$ is the reference input response of the output, $T_{yr_2}(s)$ is its corresponding disturbance response of the outputs, $T_{ur_1}(s)$ is the actuator response to the reference input, and $T_{ur_2}(s)$ is the actuator response to the disturbance input.

As the outputs are partitioned into controlled and other outputs, the transfer functions $T_{yr_1}(s)$ and $T_{yr_2}(s)$ can be split into transfer functions relating to the controlled outputs and the inner loop outputs. Each of them can be written in the form of Equation 2.4 and thus lead to two eigenvectors composing $Z_0(s)$, namely, $Z_0^c(s)$ and $Z_0^i(s)$, or

$$
Z_0(s) = \begin{bmatrix} Z_0^c(s) \\ Z_0^i(s) \end{bmatrix}.
$$

(2.11)

These relate to their respective open-loop transfer functions, $G_c(s)$ and $G_i(s)$. In fact, for a controllable and observable system, their coprime factorizations are given by

$$
G_c(s) = Z_0^c(s)P(s)^{-1}
$$

(2.12a)

$$G_i(s) = Z_0^i(s)P(s)^{-1}. \tag{2.12b}$$

Using Equations 2.12a and 2.12b, each transfer function developed in Equations 2.10a through 2.10d can be expanded using the eigenvector polynomial forms, $Z_0^c(s)$ and $Z_0^i(s)$, and their common polynomial associate polynomial matrix $P(s)$. The reference input response transfer function, $T_{yr_1}(s)$, is then given by

$$
\begin{aligned}
T_{yr_1}(s) &= Z_0^c P^{-1}\left[I + K_u^{-1}((K_a + K_c)Z_0^c P^{-1} + K_i Z_0^i P^{-1})\right]^{-1} K_u^{-1} K_a \\
&= Z_0^c \left[K_u P + (K_a + K_c)Z_0^c + K_i Z_0^i\right]^{-1} K_a.
\end{aligned}
\tag{2.13}
$$

The remaining transfer functions can also be defined in a similar manner. For example, the actuator response transfer function, $T_{ur_1}(s)$, is given by

$$
\begin{aligned}
T_{ur_1}(s) &= \left(I + K_u^{-1}\left((K_a + K_c)G_c + K_i G_i\right)\right)^{-1} K_u^{-1} K_a \\
&= P\left(K_u P + (K_a + K_c)Z_0^c + K_i Z_0^i\right)^{-1} K_a.
\end{aligned}
\tag{2.14}
$$

The actuator disturbance response transfer function, T_{ur_1}, becomes

$$
\begin{aligned}
T_{ur_1}(s) &= -\left(I + K_u^{-1}\left((K_a + K_c)G_c + K_i G_i\right)\right)^{-1} K_u^{-1}\left((K_c + K_a)G_c + K_i G_i\right) \\
&= -P\left(K_u P + (K_a + K_c)Z_0^c + K_i Z_0^i\right)^{-1}\left((K_c + K_a)Z_0^c + K_i Z_0^i\right)P^{-1}
\end{aligned}
\tag{2.15}
$$

and finally the disturbance response transfer function, T_{yr_2}, becomes

$$
\begin{aligned}
T_{yr_2}(s) &= G_c\left(I - \left(I + K_u^{-1}\left((K_a + K_c)G_c + K_i G_i\right)\right)^{-1} K_u^{-1}\left((K_c + K_a)G_c + K_i G_i\right)\right) \\
&= Z_0^c\left(I - \left(K_u P + (K_a + K_c)Z_0^c + K_i Z_0^i\right)^{-1}\left((K_c + K_a)Z_0^c + K_i Z_0^i\right)\right)P^{-1}.
\end{aligned}
\tag{2.16}
$$

2.2.2 Gain Matrix Structure

The controller structure introduced earlier is often used for aerospace applications and provides enough flexibility to design a closed-loop feedback system. The gains $K_a(s)$, $K_u(s)$, $K_c(s)$, and $K_i(s)$ are complex MIMO transfer functions that define the controller structure and need to be determined. For the PEA, static and dynamic controllers are considered, unlike other EA approaches. Trade-offs between closed-loop properties can thus be performed, and this is made easier by having the closed-loop eigenspace as a design parameter. While the controller structure and order can be adapted to satisfy additional and multiple criteria including stability, performance, decoupling, and robustness, in practice, the designer faces the difficult task of narrowing down the space of suitable controllers. In previous work [22,23], an alternative controller structure is introduced, where each gain function is a coprime factorization [11], which simplifies Equations 2.10a through 2.10d. Singular value decomposition (SVD) of each matrix of the coprime factorization and additional restrictions on the rotation matrices, including static rotation matrices and some identical rotations, led each gain matrix to take the form $K(s) = L\Sigma_L(s)\Sigma_R(s)R$ where $\Sigma_i(s)$ [24], where each component is a rational polynomial diagonal matrix. After an initial placement of the controller poles and zeros, the tuning of gains and rotation matrices is investigated in a similar fashion to root locus for MIMO systems. Finally, additional metrics guide the choices and improve the performance, robustness, sensitivity, decoupling, and actuation dynamics.

In the work presented here, a systematic design tool is to be developed in the framework of LPV systems. The controller structure is chosen such that the controller gains are simply chosen as polynomial matrices, which reduces the complexity of the search space for solutions and is better suited to the available tools for polynomial matrices. For the missile autopilot design, $K_a(s)$ is taken as a pure integrator, $\frac{I}{s}$, to ensure zero steady-state error for a steady demand. This in itself is usual in autopilot design, as the performance requirements, together with possible nonminimum phase zero transfer functions for acceleration control, can lead to slow or unstable designs.

2.2.3 Matching Conditions

When the system satisfies the Kimura condition [16], $n \le r + m - 1$, where r is the number of inputs, m is the number of outputs, and n is the number of states, pole placement for the whole closed-loop system can be performed. For the missile dynamics, this condition is met. Hence, full EA is possible. For the PEA, this condition can be described by defining a desired closed-loop system and matching Equation 2.13 to it. Such a formulation can be written as a coprime factorization:

$$T_y^d(s) = N_d(s)D_d^{-1}(s) \tag{2.17}$$

where $D_d(s)$ is a polynomial matrix with the desired closed-loop eigenstructure, and $N_d(s)$ is a polynomial matrix containing the open-loop zeros. By defining the desired closed-loop eigenstructure in this form, the effect of the system zeros can be managed in that they appear in the closed-loop transfer function. This implies that they are not cancelled out by the controller structure, and the matching conditions (Equation 2.18) must hold. Hence, in the context of full feedback and with the controller structure of Figure 2.1, the feedback loop does not attempt to cancel open-loop zeros but retains them by meeting Equation 2.18a by defining $N_d(s)$. As the controller has a free integrator, it has a steady-state closed-loop gain of 1. Hence, the determinants of $N_d(s)$ and $D_d s$ must be identical; this is reflected in condition 2.18b. Finally, as the system has the same number of inputs as controlled outputs, matching $T_y(s)$ to $T_y^d(s)$ leads to Equation 2.18c:

$$|N_d| = |Z_0^c| \tag{2.18a}$$

$$Z_0^c(0) = D_d(0) \tag{2.18b}$$

$$D_d(s)N_d^+(s) = \left[sK_u(s)P(s) + sK_c(s)Z_0(s) + sK_i(s)Z_0^i(s) + Z_0^c(s) \right] Z_0^{c+}(s) \tag{2.18c}$$

where the superscript + designates the polynomial adjoint matrix. The controller gains thus can be computed from Equation 2.18c by computing the left null space, which takes the form

$$\begin{bmatrix} K_u(s) & K_c(s) & K_i(s) & I \end{bmatrix} \begin{bmatrix} P(s)Z_0^{c+}(s) \\ Z_0^c(s)Z_0^{c+}(s) \\ Z_0^i(s)Z_0^{c+}(s) \\ \dfrac{1}{s}\left(Z_0^c(s)Z_0^{c+}(s) - D_d(s)N_d^+(s) \right) \end{bmatrix} = 0. \tag{2.19}$$

This equation is used to select the controller that matches the desired closed-loop system. Note that if only partial feedback is available, then the closed-loop transfer function is restricted, and not all closed-loop forms are attainable. This issue is dealt with in the paper when actuator dynamics are introduced.

The left polynomial null space is computed by simply reorganizing the polynomial matrices and applying a classic null space algorithm. This formulation enables the designer to specify the desired order for the controller by examining the effect of increasing the controller order on the ability to match the desired closed-loop structure. In fact, each polynomial matrix written in matrix polynomial form:

$$K(s) = K_0 + K_1 s + K_2 s^2 + \cdots + K_n s^n \tag{2.20}$$

can be written in matrix coefficient form K:

$$K = \begin{bmatrix} K_n & \cdots & K_2 & K_1 & K_0 \end{bmatrix}. \tag{2.21}$$

Thus, Equation 2.19 can be rewritten as

$$\begin{bmatrix} K_u(s) & K_c(s) & K_i(s) & I \end{bmatrix} \begin{bmatrix} PZ(s) \\ ZZ^c(s) \\ ZZ^{ic}(s) \\ ZZDN(s) \end{bmatrix} = 0. \tag{2.22}$$

Hence, solving for this null space with a first-order K_u and with constant matrices K_c and K_i gives rise to a fourth-order closed-loop system and hence

$$\begin{bmatrix} (K_u)_1 & (K_u)_0 & (K_c)_0 & (K_i)_0 & I \end{bmatrix} \begin{bmatrix} (PZ)_2 & (PZ)_1 & (PZ)_0 & 0 \\ 0 & (PZ)_2 & (PZ)_1 & (PZ)_0 \\ 0 & (ZZ^c)_2 & (ZZ^c)_1 & (ZZ^c)_0 \\ 0 & (ZZ^{ic})_2 & (ZZ^{ic})_1 & (ZZ^{ic})_0 \\ (ZZDN)_3 & (ZZDN)_2 & (ZZDN)_1 & (ZZDN)_0 \end{bmatrix}$$

$$\tag{2.23}$$

where, for example, $(K_u)_i$ refers to the s^i coefficient matrix. For this system, the left null space of the right-hand matrix in Equation 2.23 is calculated using a standard SVD algorithm and takes the form

$$\begin{bmatrix} X^{u_1} & X^{u_0} & X^{c_0} & X^{i_0} & X^{l_0} \end{bmatrix} \tag{2.24}$$

where $X^{\{u_{1,0},c_0,i_0,l_0\}}$ are constant matrices of appropriate column size, partitioned to match controller matrices' column sizes. The number of rows of the null space in Equation 2.24 determines whether a controller solution exists. If the row dimension of the null space is greater than or equal to the number of system inputs, then a controller having the correct dimensions can be constructed from the null space. There is an extra condition that must be met for existence, that is, X^{l_0} must have full rank. This solution is then row-reduced using Gaussian elimination to take the form

$$\begin{bmatrix} Y_0^{u_1} & Y_0^{u_0} & Y_0^{c_0} & Y_0^{i_0} & I \\ Y_1^{u_1} & Y_1^{u_0} & Y_1^{c_0} & Y_1^{i_0} & 0 \end{bmatrix} \tag{2.25}$$

where $Y_{\{0,1\}}^{\{u_{1,0},c_0,i_0\}}$ are constant matrices of column size equal to their respective controller gain column sizes. The first row ensures the matching to the desired closed-loop system, and if there is a sufficient number of rows in the second row, then a linear combination (in a polynomial sense) added to the first row will also satisfy the matching condition for the desired closed-loop system. If the controller space takes on this form, then additional criteria can be used to include performance, decoupling, and robustness by reference to the remaining transfer functions in Equations 2.10a through 2.10d.

2.2.4 LPV Approach to PEA

EA has many well-established numerical approaches to design, and the main algorithms are surveyed in the work of White [21]. For linear parameter-varying systems, most approaches produce local controller designs that are interpolated over the operating envelope. The resulting controllers all suffer from gain scheduling problems associated with defining the controller, in particular, issues associated with zero and pole interpolation. More recently, work has been done on a multimodel EA that uses embedded models in the controller design and interpolation, which results in high-order controllers. Model order reduction is then used to produce realistic controllers with promising results [18].

In this chapter, the polynomial null space for PEA controllers is computed symbolically using the polynomial framework. The solution produces a generic controller, and thus, the usual gain scheduling problem is replaced by an explicit parameterization matching the parameterization of the LPV system itself. However, for the solution proposed in this chapter, the PEA solution requires exact matching to the desired closed-loop system, and thus, the polynomial matrices $N_d(s, p)$ and $D_d(s, p)$ need to be chosen carefully. This condition will be relaxed in subsequent chapters. The polynomial matrix $D_d(s, p)$ can be selected to be independent of the operating point p, while the polynomial matrix $N_d(s, p)$ is dependent on the open-loop zeros. The software package MATHEMATICA is used to form the polynomial null space and test for matching with the controller structure and order controlled throughout the operating envelope. The approach makes the closed-loop system independent of the current operating point and thus produces a solution similar to dynamic inversion without the need to consider zero dynamics.

2.3 Robust Analysis

In this polynomial framework, parametric robustness is investigated similarly to Kharitonov-type approaches [2,3]. While initial Kharitonov results reduce interval uncertain system stability to four extremal polynomials' stability only, this has been generalized in different aspects to robust stability (D-stability) for polytopic uncertain polynomial families and even multiaffine uncertain systems. The main result is an equivalence between stability of an infinite polynomial family and stability of a finite polynomial family, thus reducing the problem complexity. Two studies by Wise [25,26] are examples of such a successful approach to aerospace applications. However, stability of this reduced (finite) polynomial family usually requires an infinite number of checks in sweeping across frequencies (generalized frequencies).

On the other hand, the recently introduced FNT [7,14] restates the well-known Nyquist stability theorem to a sufficient number of (generalized) frequencies in order to assess polynomial D-stability. To the expense of some conservativeness, this result has been further extended to polynomic polynomial family stability (D-stability) in the FIT [7,12–15]. It then reduces the frequency sweep to a finite number of value sets, each needing to fit in a suitable sector angle.

Depending on the polynomial family type, the value set computation may be intensive. For a multilinear polynomial family, this is usually equivalent to the computation of the parameter space edge image, referring to the edges theorem [3]. However, applying the mapping theorem reduces the complex value set to its convex hull and thus is equivalent to the parameter space vertices' image. Thus, the combination of the mapping theorem with the FIT infers that stability (D-stability) robustness can be achieved within a finite number of polynomial/frequency checks [7] for multilinear polynomial families.

2.3.1 D-Stability for Polynomial Family

This section describes extensions of the well-known Nyquist stability theorem, which simply relates the number of encirclements of the transfer function image for s moving along a closed contour to the relative degree of the transfer function. Similarly, the FNT assesses that all the roots of a polynomial in s are within a specific region of the s-plane with a finite number of checks. This approach should be distinguished from gridding methods since the number of checks can be as low as the order of the polynomial.

Theorem 2.1 (*Finite Nyquist theorem [7]*) Let $p(s) = \sum_{j=0}^{n} \alpha_j s^j$ where $n \geq 0$ and $\alpha_j \in \mathbb{C}$, and let $\Gamma \subset \mathbb{C}$ be a closed Jordan curve such that $int(\Gamma)$ is convex. Then p is of degree n, that is, $\alpha_n \neq 0$, and has all its roots in $int(\Gamma)$ if and only if there exist $m \geq 1$ angles $\theta_k \in \mathbb{R}$ and a counterclockwise sequence of points $s_k \in \Gamma, 1 \leq k \leq m$, such that

$$\forall k \in \mathbb{N}_m^* \quad |\theta_{k+1} - \theta_k| < \pi, \tag{2.26a}$$

$$\forall k \in \mathbb{N}_m^* \quad p(s_k) \neq 0, \tag{2.26b}$$

$$\forall k \in \mathbb{N}_m^* \quad \arg(p(s_k)) \equiv \theta_k \, (mod \, 2\pi), \tag{2.26c}$$

$$| 2\pi n + \theta_1 - \theta_m | < \pi. \tag{2.26d}$$

The notation $\mathbb{N}_m^*$ represents the range of strictly (*) positive integers ($\mathbb{N}$) ranging up to m (m included).

The necessary number of sectors to be considered can be assessed, and in the work of Djaferis [7], for sector angles of $\frac{a}{b}\pi < \pi$, increment angle of $\frac{(b-a)}{b}\pi$, and a polynomial family of order n, the number of necessary sectors to be considered m is given by the relation

$$\frac{2bn}{b-a} \le m \le \frac{2bn}{b-a} + 2 \quad \wedge \quad m \text{ is odd.} \tag{2.27}$$

and for the limit, if the sector angle is π, then the number of sectors is infinite. A simple example of a second-order system with symmetric D-region and using a sector angle of $\frac{3}{4}\pi$ leads to consideration of 9 generalized frequencies, while for an asymmetric D-region, it would be 17 in total. The plots in Figure 2.2 show all sectors for such a case where sector increment is $\frac{\pi}{4}$.

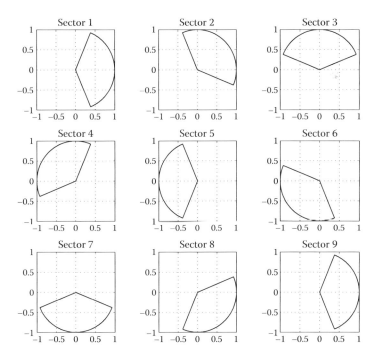

FIGURE 2.2

Selected sectors for application of FIT to a second-order polynomial with $\frac{3}{4}\pi$ sector angle and $\frac{\pi}{4}$ increment and for an investigated symmetric D-stability region.

An example of the FNT is now applied to the following polynomial $p(s)$ (see Equation 2.28 for the plotted contour defined in Figure 2.3):

$$p(s) = (s + 3)(s^2 + 9.5s + 25).\tag{2.28}$$

There are $N_z = 3$ roots inside the chosen contour, and this leads to a net number of counterclockwise encirclements of $N = 3$, which can be checked in Figure 2.3b. To satisfy assumption 2.26a, a sector angle of $\theta_{sec} = \frac{\pi}{4}$ is used, and their corresponding generalized frequencies s_k are plotted on the contours of Figure 2.3. Following the contour, two successive points, for instance, points

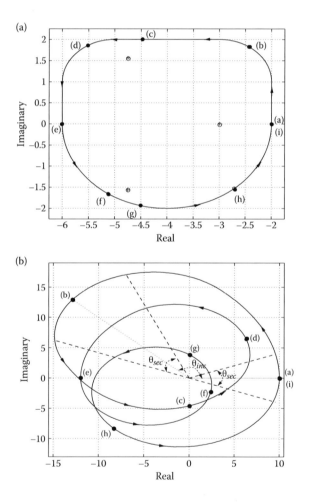

FIGURE 2.3
FNT applied to polynomial defined in Equation (2.8). (a) Roots of $p(s)$; considered contour and selected s_k are plotted. (b) Image of contour by $p(s)$.

labeled (*a*) and the next points (*b*), are separated by a sector angle increment $\theta_{inc} = \pi - \theta_{sec}$. Since there is no symmetry in this example and with the selection chosen, 9 points (*m* = 9) is sufficient to prove the encirclements of all the roots in the contour.

By selecting a sector angle less than π and consequently introducing some conservatism, the previous result can successfully generalize to polynomial families, and the stability (D-stability) of a polynomial family reduces to a finite number of frequency (generalized frequency) checks. These steps are given in the FIT. This theorem is often compared with the finite frequency test theorem, which similarly assesses stability of an affine family of polynomials but, in contrast, requires precise knowledge of the frequencies to be checked.

Theorem 2.2. (Finite inclusion theorem [7]) Consider a polynomic family of polynomials, $\Phi_p(s)$. Further, let $\Gamma \subset \mathbb{C}$ be a closed Jordan curve such that $int(\Gamma)$ is convex. Then for all $q \in \Pi_b$, $p(s, q) \in \Phi(s)$ is of degree n and has all its roots in $int(\Gamma)$ if there exist $m \geq 1$ intervals $(c_k, d_k) \subset \mathbb{R}$ and a counterclockwise sequence of points $s_k \in \Gamma$, $1 \leq k \leq m$, such that

$$\forall k \in \mathbb{N}^*_{m-1} \quad \max(d_{k+1} - c_k, d_k - c_{k+1} \leq \pi) \tag{2.29a}$$

$$\forall k \in \mathbb{N}^*_m \quad p(s_k, \Pi_b) \subset \{re^{j\theta} \mid r > 0, \theta \in [c_k, d_k]\} \tag{2.29b}$$

$$\max(d_m - (c_1 + 2\pi n), (d_1 + 2\pi n) - c_m \leq \pi). \tag{2.29c}$$

This theorem applies to polynomic polynomial families, and for affine/polytopic polynomial families, it becomes an equivalence.

2.3.2 Multilinear Uncertain Systems

The theorem presented earlier still requires evaluation of the value set at some few frequencies, and this may turn out to be complex owing to its polynomial family structure. In fact, for the affine/linear polynomial family, the value set is generated from its parameter space (box case) vertices, and thus, the value set can be obtained from a finite number of polynomials from the polynomial family. However, for a more complex structure or parameter space, there are no such direct results, and this often leads to consideration of the edges theorem or the mapping theorem [2,3]. While the edges theorem extends the previous results to affine/linear polynomial families on polytopic parameter space, the mapping theorem can capture more complex polynomial family structures at the expense of introducing some conservativeness. In fact, the multiaffine polynomial family value set is a subset of its convex hull that can be obtained from the parameter space vertices. Note that the conservatism introduced here can be relaxed to some extent by subdividing the parameter space as many times as required in a recursive manner. Finally, the FIT can be successfully implemented in a finite number of checks for multiaffine polynomial families using the mapping theorem.

Definition 2.1. (Multiaffine/multilinear polynomial family) A multiaffine/multilinear polynomial family, $\Phi_m(s)$, is a family of polynomials in the form of Equation 2.30

$$p(s, \mathbf{q}) = \alpha_0(\mathbf{q}) + \alpha_1(\mathbf{q})s + \alpha_2(\mathbf{q})s^2 + \alpha_3(\mathbf{q})s^3 + \cdots + \alpha_n(\mathbf{q})s^n \qquad (2.30)$$

where $p(s)$ is of order n and $\alpha_i(\mathbf{q})$ are multiaffine/multilinear functions of $\mathbf{q}$, $\forall i \in \mathbb{N}^n$. Commonly, the parameter space $\mathbf{q}$ lies in parameter box Π_b of dimension $m + 1$.

Multiaffine/multilinear denotes here that $\alpha_i(\mathbf{q})$ could be of the form $\alpha_i(\mathbf{q}) = \alpha_i^0 q_0 + \alpha_i^1 q_1 + \alpha_i^2 q_0 q_1 + b_i$ for the parameter box $\Pi_b = [q_0^-, q_0^+] \times [q_1^0, q_1^+]$ and where $\mathbf{q} = [q_0 \; q_1]^T$.

From this definition of polynomial family, some reduction can be obtained with the mapping theorem, and for this purpose, the following definition is necessary to capture the vertex polynomials of the polynomial family.

Definition 2.2. (Vertex polynomials) The vertex polynomials, $\Phi_m^V(s)$, of the multiaffine/multilinear polynomial family, $\Phi_m(s)$, are the family of polynomials described by the vertices of Π_b, Π_b^V. Then

$$\Phi_m^V(s) = \{p(s, \mathbf{q}) \in \Phi_m(s), \quad \mathbf{q} \in \Pi_b^V\} \qquad (2.31)$$

$$\text{where} \quad \Pi_b^V = \{\mathbf{q} \in \Pi_b, \quad \forall j \in \mathbb{N}_m \quad q_j = q_j^- \vee q_j = q_j^+\}.$$

The mapping theorem uses value sets and their convexity to overbound the family uncertain polynomial.

Theorem 2.3. (Mapping theorem) For a multilinear polynomial family, $\Phi_m(s)$, on Π_b, the value set at s^* of the polynomial family, $\Delta(s^*)$, is included in its convex hull, $co(\Delta(s^*))$, and in the convex hull of the value set of its vertex polynomials $\Phi_m^V(s)$, $co(\Delta^V(s^*))$. Therefore

$$\forall s^* \in \mathbb{C} \quad \Delta(s^*) \subset \Delta^V(s^*). \qquad (2.32)$$

This result is important in reducing multilinear parametric problems to linear ones. Similarly, multilinear transfer functions can then be tackled, where the numerator and the denominator are included in their respective convex hulls at the expense of conservativeness. Finally, the division of polytopes gives a set formed of segments and arcs. Nyquist plots, Nichols charts, and Bode plots can be obtained as well.

Although the mapping theorem introduces some conservatism, since the convex could be significantly bigger than the actual value set, a reduction of this conservatism can be performed with the drawback of more computations. The idea is to split the parameter space in subspace and consider each uncertain polynomial family independently. This approach is explained through the following example.

In fact, the multilinear uncertain polynomial $Q(s, q_1, q_2) = p_0(s) + (-6s + 2)$ $q_1 + (-5s -1)q_2 + (10s + 3)q_1q_2$—example from the work of Bhattacharyya et al. [3]—has for a value set $\Delta(s^*)$ at s^* for an interval uncertainty $q_1 \in [0, 1]$ and $q_2 \in [0, 1]$. The value set is given for a specific s^* in Figure 2.4a, with the convex hull of $\Delta(s^*)$, $co(\Delta(s^*))$, for the uncertain box $(q_1, q_2) \in \Pi$. The same uncertain box Π is split into four subboxes denoted Π_{11}, Π_{12}, Π_{21}, and Π_{22}, and Figure 2.4b represents their respective value sets for the same s^*, $\Delta_{11}(s^*)$,

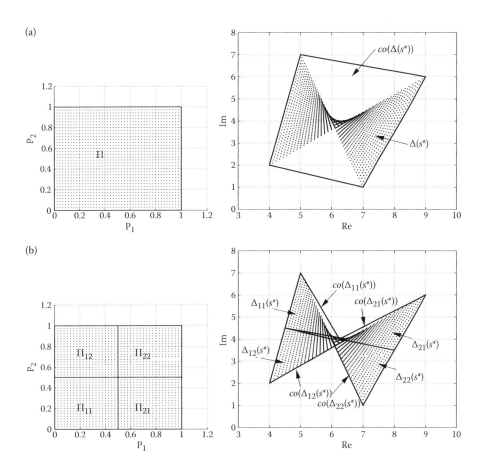

FIGURE 2.4
Although the value set image convexity is lost for multilinear systems, convexity properties can be exploited on separate smaller domains where the conservativeness can be relaxed to some extent. Compare $co(\Delta(s^*))$ with $\cup\, co(\Delta_{ij}(s^*))$. (a) With only one big uncertain domain. (b) Now, uncertainty domain is split in four.

$\Delta_{12}(s^*)$, $\Delta_{21}(s^*)$, and $\Delta_{22}(s^*)$. Finally, their respective convex hulls are generated, $co(\Delta_{11}(s^*))$, $co(\Delta_{12}(s^*))$, $co(\Delta_{21}(s^*))$, and $co(\Delta_{22}(s^*))$, and their union, nonconvex and included in $co(\Delta(s^*))$, is less restrictive than $\Delta(s^*)$ itself.

The major drawback of this approach is the increase in computation involved and the selection of the subspaces. In fact, a poor choice in the splitting of the parameter space may not reduce much the conservativeness of the convex hulls.

2.3.3 Developed Algorithm for D-Stability Robustness

For a sector angle of $\frac{7}{8}\pi < \pi$, a symmetrical D-region (real polynomial), and considering third-order polynomials, 25 sectors should be sufficient. Note that the number of sectors depends on the order of the polynomial but not on the dimension of the parametric uncertainty vector, and although independent, the complexity of the value set needs to be dealt with. Next is a suggested algorithm for such an approach.

- Select a sector angle, θ_{sec} ($\theta_{sec} = \frac{7\pi}{8}$), and a corresponding increment, θ_{inc} ($\theta_{inc} = \frac{\pi}{8}$).
- The combination of θ_{sec} and θ_{inc} gives a condition for the number of sectors, m (this number can be halved for symmetric D-region).
- Generate the sectors, sec_i, $\forall i \in \mathbb{N}_m^*$, $sec_i = [sec_i^-, sec_i^+]$.
- For each sector, $i \in \mathbb{N}_m^*$, find the generalized frequency on the D-contour, s_i, for each middle sector sec_i for which $p(s_i, 0) = \frac{sec_i^- + sec_i^+}{2}$.
- Check that this nominal polynomial $p(s, 0)$ satisfies the FNT at frequencies s_i, so $p(s, 0)$ is D-stable.
- Select an uncertainty box, Π, and an initial weighting α_{ini}, and initialize all $\alpha_i = \alpha_{ini}$.
- For each sector, $i \in \mathbb{N}_m^*$.
- While α_i and s_i still "change."
- Compute the vertices $vtx_i = p(s_i, \alpha_i \Pi)$ from the vertices of Π.
- Compute the convex hull of vtx_i, $co(vtx_i)$.
- Compute the maximum and minimum phases of $co(vtx_i)$, $\phi^-(co(vtx_i))$ and $\phi^+(co(vtx_i))$.
- There are now four cases: If $\phi^-(co(vtx_i)) > sec_i^-$ and $\phi^+(co(vtx_i)) < sec_i^+$, then increase α_i according to absolute phase margin. If $\phi^-(co(vtx_i)) < sec_i^-$ and $\phi^+(co(vtx_i)) > sec_i^+$, then decrease α_i according to absolute phase margin. If $\phi^-(co(vtx_i)) < sec_i^-$ and $\phi^+(co(vtx_i)) < sec_i^+$, then relative phase margin. If $\phi^-(co(vtx_i)) > sec_i^-$ and $\phi^+(co(vtx_i)) > sec_i^+$, then relative phase margin.

- Always recenter s_i according to absolute or relative phase margins.
- End "While" and "For" loops.
- $\alpha = \min_{i \in \mathbb{N}_m^*} (\alpha_i)$.

The algorithm to increase or decrease α_i is based on bisection programming and is not shown on the above algorithm. Upper and lower limits of α_i, α_i^- and α_i^+, respectively, are updated, and the convergence is performed according to bisection approach. So, if α_i can be increased, then the next α_i is taken as $\alpha_i = \dfrac{\alpha_i + \alpha_i^+}{2}$; the same can be applied for decreasing α_i.

To "simplify" the computation of the minimum and maximum phases of the value set convex hull, the zero-exclusion property is first checked, removing consequently the case where minimum and maximum phases reach $\dfrac{sec_i^- + sec_i^+}{2} - \pi$ and $\dfrac{sec_i^- + sec_i^+}{2} + \pi$, respectively. In fact, further simplifications are used since there is no need to determine the exact value set convex hull, and trying to do so usually implies sorting the various vertices in a specific order; instead, one can simply compare vertex phases and keep minimum and maximum vertex phases.

The phase margin is used many times in this algorithm to "recenter" the generalized frequency. First, this generalized frequency without using prior knowledge is taken centered in the sector, and then for each step, the algorithm tries to shift this generalized frequency, taking into account the shape of the value set convex hull. In this attempt, the relative phase margins on the minimum, $sec_i^- - \phi^-(co(vtx_i))$, and on the maximum, $sec_i^+ - \phi^+(co(vtx_i))$, are compared, and the average is used to shift the generalized frequency. While the absolute phase margin performs a similar role, the effects are added instead of averaged. The effect of this approach is to try to center the value set convex hull in the middle of the sector. Note that this process depends fully on the sector selection, which does not adapt in the present algorithm.

Although the shift procedure is efficient, it would require computing each time the underlying polynomial roots and becomes a computational burden. Instead, an approximate mapping between polynomial phase and D-region contour was developed. This integrates a metric-like concept that respects the ratio between midsectors while moving along the D-contour. This approach turned out to be sufficient to reduce the computational burden and lead to a reasonable convergence rate without deadlocks; however, this algorithm relies on some ad hoc parameters at the moment.

Finally, many of the parameters could be adapted as the algorithm runs to improve the results; this encompasses θ_{sec} and θ_{inc}, which could depend on the particular sector of interest and not being uniform across. More practical would be to consider an adaptive algorithm with respect to each individual uncertainty instead of a uniform weighting gain α.

2.4 Missile Lateral Autopilot Design

In this section, the PEA is applied to a quasilinear missile model, named the Horton missile model. The design leads to a linear parameter-varying controller for a sideslip velocity control. Although the underlying system is nonlinear, the closed-loop system is made independent of the operating points and can thus be compared with dynamic inversion approaches.

2.4.1 Missile Model

The Horton missile model describes a realistic airframe of a tail-controlled tactical missile in the cruciform fin configuration, as shown in Figure 2.5. The aerodynamic parameters in this model are derived from wind-tunnel measurements [4,20].

The starting point for mathematical description of the missile is the following nonlinear model [4,9,20] of the horizontal motion (on the xy-plane in Figure 2.5):

$$\dot{v} = y_v(M,\lambda,\sigma)v - Ur + y_\zeta(M,\lambda,\sigma)\zeta$$
$$= \frac{1}{2}m^{-1}\rho V_o S(C_{y_v}v + V_o C_{y_\zeta}\zeta) - Ur$$

$$\dot{r} = n_v(M,\lambda,\sigma)v + n_r(M,\lambda,\sigma)r + n_\zeta(M,\lambda,\sigma)\zeta$$
$$= \frac{1}{2}I_z^{-1}\rho V_o Sd\left(\frac{1}{2}dC_{n_r}r + C_{n_v}v + V_o C_{n_\zeta}\zeta\right) \tag{2.33}$$

where the variables are defined in Figure 2.5. Here, v is the sideslip velocity, r is the body rate, ζ is the rudder fin deflection, y_v and y_ζ are

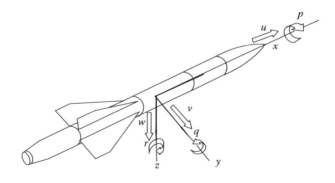

FIGURE 2.5
Missile airframe axes.

the semi-nondimensional force derivatives due to lateral and fin angle, respectively, n_v, n_ζ, and n_r are the semi-nondimensional moment derivatives due to sideslip velocity, fin angle, and body rate, respectively, and finally, U is the longitudinal velocity. Furthermore, $m = 125$ kg is the missile mass, $\rho = \rho_0 - 0.094h$ is the air density ($\rho_0 = 1.23$ kg m^{-3} is the sea level air density and h is the missile altitude in kilometers), V_o is the total velocity (in meters per second), $S = \pi d^2/4 = 0.0314$ m^2 is the reference area ($d = 0.2$ m is the reference diameter), and $I_z = 67.5$ kg m^2 is the lateral inertia. For the coefficients C_{y_v}, C_{y_ζ}, C_{n_r}, C_{n_v}, and C_{n_ζ}, only discrete data points are available, obtained from wind-tunnel experiments. Hence, an interpolation formula, involving the Mach number $M \in [2, 3.5]$, roll angle $\lambda \in [4.5°, 45°]$, and total incidence $\sigma \in [3°, 17°]$, has been calculated, with the results summarized in Table 2.1.

The total velocity vector V_o is the sum of the longitudinal velocity vector $\vec{U}$ and the sideslip velocity vector $\vec{v}$, that is, $V_o = \vec{U} + \vec{v}$, with all three vectors lying on the xy-plane (see Figure 2.5). We assume that $U \gg v$, so that the total incidence σ, or the angle between $\vec{U}$ and V_o, can be taken as $\sigma = v/V_o$, as $\sin \sigma \approx \sigma$ for small σ. Thus, we have $\sigma = v/V_o = v/\sqrt{v^2 + U^2}$, so that the total incidence is a nonlinear function of the sideslip velocity and longitudinal velocity, $\sigma = \sigma(v, U)$. As mentioned earlier, the missile is in cruciform configuration; therefore, symmetry assumptions are invoked. Hence, despite the fact that the nonlinear model coefficients depend only on the absolute value of σ, due to the symmetry assumptions, the coefficients could be used for $v < 0$.

The Mach number is obviously defined as $M = V_o/a$, where a is the speed of sound, $a = 340 - 4h$. Since $V_o = \sqrt{v^2 + U^2}$, the Mach number is also a nonlinear function of the sideslip velocity and longitudinal velocity, $M = M(v, U)$.

It follows from the above discussion that all coefficients in Table 2.1 can be interpreted as nonlinear functions of three variables: sideslip velocity v, Mach number M, and roll angle λ.

TABLE 2.1

Coefficients in Nonlinear Model Equation 2.33

	Interpolated Formula
C_{y_v}	$0.5[(-25 + M - 60\ \lvert\sigma\rvert)(1 + \cos 4\lambda) + (-26 + 1.5M - 30\ \lvert\sigma\rvert)(1 - \cos 4\lambda)]$
C_{y_ζ}	$10 + 0.5[(-1.6M + 2\ \lvert\sigma\rvert)(1 + \cos 4\lambda) + (-1.4M + 1.5\ \lvert\sigma\rvert)(1 - \cos 4\lambda)]$
C_{n_r}	$-500 - 30M + 200\ \lvert\sigma\rvert$
C_{n_v}	$s_m C_{y_v}$, where $s_m = d^{-1}[1.3 + 0.1M + 0.2(1 + \cos 4\lambda)\ \lvert\sigma\rvert + 0.3(1 - \cos 4\lambda)\ \lvert\sigma\rvert - (1.3 + m/500)]$
C_{n_ζ}	$s_f C_{y_\zeta}$, where $s_f = d^{-1}[2.6 - (1.3 + m/500)]$

2.4.2 Missile QLPV Form

Although the system in Equation 2.33 is already in a QLPV form, theoretically, one has to consider the linearization of the nonlinear system around its equilibria. For an equilibrium point (v_0, r_0, ζ_0), it is possible to derive from Equation 2.33 linear models in incremental variables, $\delta v \doteq v - v_0$, $\delta r \doteq r - r_0$, and $\delta \zeta \doteq \zeta - \zeta_0$. In particular, for straight and level flight (with gravity influence neglected), we have $(v_0, r_0, \zeta_0) = (0, 0, 0)$, so that the incremental and absolute variables are numerically identical, although conceptually different. In the rest of this paper, the model defined in Equation 2.34 can refer to the original QLPV form of the missile or its Taylor linearized version, as both forms present a QLPV form. Hence,

$$
\begin{bmatrix} \delta \dot{v} \\ \delta \dot{r} \end{bmatrix} = \begin{bmatrix} y_v(p) & y_r(p) \\ n_v(p) & n_r(p) \end{bmatrix} \begin{bmatrix} \delta v \\ \delta r \end{bmatrix} + \begin{bmatrix} y_\zeta(p) \\ n_\zeta(p) \end{bmatrix} \delta \zeta \tag{2.34}
$$

where p is the scheduling parameter.

Wind-tunnel tests are performed for the modeling of the missile, but these experiments are usually conducted in some quasisteady conditions. This leads to an almost linearized version of the missile where the Mach number, roll angle, and incidence conditions have been included afterward and which in the present case leads to a QLPV representation. Note that for a sideslip velocity controller, the lateral acceleration demand is recast as a sideslip velocity demand via the nonlinear relationship $a = y_v(p)v + y_\zeta(p)\zeta$.

The controller structure presented earlier is now used in the LPV framework, as shown in Figure 2.6, where gain $K_a(s)$ is selected as a pure integrator to ensure zero steady-state error to a step input. This form is shown to be sufficient to obtain a suitable controller with static controller gains. Note that, although the controller is presented in the figure as an LPV controller, it has a quasi-LPV static form.

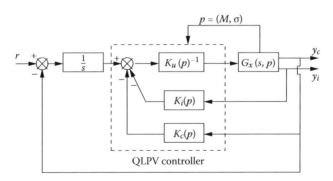

FIGURE 2.6
QLPV controller structure chosen for a lateral velocity controller of the Horton missile includes $K_a(s)$ as a pure integrator, $\dfrac{1}{s}$, and the gains $K_u(s)$, $K_c(s)$, and $K_i(s)$ as scheduled scalar gains.

2.4.3 Performance Objectives

In the context of an LPV system, the PEA presented so far is strictly valid only for the underlying LTI system. However, one may assign the eigenstructure to be (almost) independent of the operating point. This would lead to (almost) identical closed-loop dynamics over the operating envelope. Strict independence is not always possible because the approach does not cancel out open-loop zeros, and these may vary over the operating envelope. The resulting closed-loop dynamics, however, will be very similar to LTI system dynamics, which is often desirable for the designer. Additional zeros may be included in the design to aid in placing the closed-loop system poles.

For static controller gains, the system is of the third-order, that is, second-order plant and first-order pure integrator. The desired characteristic polynomial is of degree 3 and can be written as

$$D_d(s) = c_0 + c_1 s + c_2 s^2 + s^3. \tag{2.35}$$

Under the assumption that the zeros do not influence the response significantly, the closed-loop transient step response can be chosen to produce a peak overshoot less than 5% and settling time less than 0.2 s. The desired characteristic polynomial coefficients are thus chosen as a combination of a second order with natural frequency $w_n = 30$ rad/s and damping ratio $\zeta_n = 0.7$ and an additional pole at -100. The controller does not introduce additional zeros, and there is no attempt to cancel out open-loop zeros; thus $N_d(s)$ is constrained by $Z_0^c(s)$ in condition 2.36a, and it represents the open-loop zeros.

The closed-loop transfer functions in Equations 2.13 and 2.17 match the following LPV conditions:

$$N_d(s,p) = Z_0^c(s,p) \tag{2.36a}$$

$$Z_0^c(0,p) = D_d(0,p) \tag{2.36b}$$

$$D_d(s,p) = \left[sK_u(s,p)P(s,p) + sK_c(s,p)Z_0(s,p) + sK_i(s,p)Z_0^i(s,p) + Z_0^c(s,p) \right] \tag{2.36c}$$

where polynomials $D_d(s, p)$ and $N_d(s, p)$ represent the desired closed-loop transfer function. Note that it is not always possible to achieve exact matching or, in other words, satisfy condition 2.36c, and different controller structure and/or order may be required. However, for the SISO case presented here, this is straightforward, and a static controller achieves matching.

Equation 2.36c can be written as

$$
[K_u(s,p) \quad K_c(s,p) \quad K_i(s,p) \quad I]
\begin{bmatrix}
P(s,p) \\
Z_0^c(s,p) \\
Z_0^i(s,p) \\
\dfrac{1}{s}\left(Z_0^c(s,p) - D_d(s,p)\right)
\end{bmatrix}
= 0
\qquad (2.37)
$$

where the left null space solution is used to construct the controller gains. The null space has to be of sufficient dimension to allow reduction by Gaussian elimination to satisfy the annihilator vector format with last term I. This is in fact equivalent to the condition in Equation 2.36c. Finally, the null space solution may not be limited to a single solution, and thus, any "linear" (in the sense of polynomial) combination of row vectors that produces the annihilator vector format in Equation 2.37 is suitable. From these null space solutions, the controller gains $K_u(s,p)$, $K_c(s,p)$, and $K_i(s,p)$ can be partitioned to produce a suitable controller for the controller structure shown in Figure 2.6.

2.4.4 Sideslip Velocity Controller

Consider a QLPV lateral velocity controller for the Horton missile for the case of state feedback. The output equation for sideslip velocity output is

$$
\begin{bmatrix} y_c \\ y_i \end{bmatrix} =
\begin{bmatrix} 1 & 0 \\ 0 & 1 \end{bmatrix}
\begin{bmatrix} v \\ r \end{bmatrix}.
\qquad (2.38)
$$

This equation sets matrices $Z_0^c(s,p)$ and $Z_0^i(s,p)$ while $Z(s,p)$ and $P(s,p)$ are determined from the state equation. In fact, for this case, in Equations 2.4 and 2.38, $Z_0(s,p)$ is set equal to $Z(s,p)$, and the null space solution of Equation 2.37 is then solved by symbolic computation used to yield the annihilator parameterized polynomial by p. After further row reduction, the null space takes the vector form $[K_u \ K_c \ K_i \ I]^T$.

Equations 2.2 and 2.11 are computed and lead to $Z_0(s, p)$ and $P(s, p)$, which represent the coprime factorization of the system:

$$
Z_0(s,p) =
\begin{bmatrix}
\dfrac{c_0(n_r(p)y_\zeta(p) - y_r(p)n_\zeta(p) - y_\zeta(p)s)}{n_r(p)y_\zeta(p) - y_r(p)n_\zeta(p)} \\[4mm]
\dfrac{c_0(-n_v(p)y_\zeta(p) - y_v(p)n_\zeta(p) - n_\zeta(p)s)}{n_r(p)y_\zeta(p) - y_r(p)n_\zeta(p)}
\end{bmatrix}
$$

$$P(s,p) = \left[\frac{c_0(y_r(p)n_v(p) - y_v(p)n_r(p) + y_v(p)s - n_r(p)s - s^2)}{n_r(p)y_\zeta(p) - y_r(p)n_\zeta(p)} \right]. \tag{2.39}$$

The null space for Equation 2.37 can now be solved as a polynomial equation using the symbolic computation in MATHEMATICA. This produces a multivariable polynomial null space, the polynomial rank of which is 3 since the constituent vectors do not have any null elements and are independent. Any "linear" vector combination in this null space, provided it satisfies the constant last term I as in the vector $[K_u \ K_c \ K_i \ I]$, gives rise to a controller for the system.

The controller gains $K_u(s, p)$, $K_c(s, p)$, and $K_i(s, p)$ are identified by suitable partitioning of the resulting null space. For the present case, static controller gains are sufficient to obtain a matching condition for the desired closed loop, and a unique LPV controller is obtained. The parameterized controller gains are as follows:

$$K_u = (n_\zeta y_r - n_r y_\zeta)(n_\zeta^2 y_r + n_\zeta(y_v - n_r)y_\zeta - n_v y_\zeta^2) \tag{2.40a}$$

$$K_c = c_1 n_\zeta(n_\zeta y_r - n_r y_\zeta) + n_v(n_\zeta y_r - n_r y_\zeta)(n_\zeta y_r - (c_2 + n_r + y_v)y_\zeta)$$
$$+ n_\zeta(n_\zeta y_r y_v^2 - (c_0 + n_r y_v^2)y_\zeta + c_2 y_v(n_\zeta y_r - n_r y_\zeta)) \tag{2.40b}$$

$$K_i = n_\zeta^2 y_r^2 y_v - c_1 n_\zeta y_r y_\zeta - 2n_r^2 n_\zeta y_r y_\zeta - n_v n_\zeta y_r^2 y_\zeta + c_0 y_\zeta^2 + n_r^3 y_\zeta^2$$
$$+ c_2(n_\zeta y_r - n_r y_\zeta)^2 + n_r(n_\zeta^2 y_r^2 - n_\zeta y_r y_v y_\zeta + (c_1 + n_v y_r)y_\zeta^2) \tag{2.40c}$$

where parameterization in p is dropped for ease of presentation, and a common dividing factor given by Equation 2.41 is omitted:

$$c_0(n_\zeta^2 y_r + n_\zeta(y_v - n_r)y_\zeta - n_v y_\zeta^2). \tag{2.41}$$

A direct substitution of the coefficients c_0, c_1, and c_2 of the desired closed-loop denominator of the system as well as of the QLPV semi-nondimensional coefficients $y_v(p)$, $y_r(p)$, $y_\zeta(p)$, $n_v(p)$, $n_r(p)$, and $n_\zeta(p)$ leads to the desired controller Equation 2.42:

$$\zeta = K_u(p)^{-1}\left(\frac{(v - v_d)}{s} - K_c(p)v - K_i(p)r \right). \tag{2.42}$$

2.4.5 Multilinear Parametric Uncertain System

The robust analysis is carried out on the Horton missile using the previous LPV controller. Using the missile model, Equation 2.34, the closed-loop system can be written as in Equation 2.7, where each polynomial vector is dependent on parameter p. Under no uncertainty the closed-loop system would match the desired closed-loop system.

First, the parametric uncertainty vector q is introduced, representing the semi-nondimensional aerodynamic coefficients and longitudinal velocity:

$$q = \begin{bmatrix} y_v & y_r & y_\zeta & n_v & n_r & n_\zeta \end{bmatrix}^T.$$

Each element of q is written as $q_i^0(1 + \delta q_i)$ for each i where q_i^0 corresponds to its nominal value (depending on p in fact) and a relative uncertainty parameter δq_i, which will be the quantity referred to as uncertainty.

From Equation 2.7, the closed-loop system for sideslip velocity output is now written in a simplified form (Equation 2.43), without expanding the nominal controller gains but considering uncertainties in the plant. The uncertainty thus takes a multilinear format:

$$G(s, q) = \frac{a_1(q) + a_2(q)s}{K_u s^3 + a_3(q)s^2 + a_4(q)s + a_5(q)} \tag{2.43}$$

where

$$a_1(q) = q_5 q_3 - q_2 q_6$$

$$a_2(q) = -q_3$$

$$a_3(q) = K_u(q_1 + q_5) - K_c q_3 - K_i q_6$$

$$a_4(q) = K_u(q_2 q_4 - q_1 q_5) + K_c(q_5 q_3 - q_2 q_6) + K_i(q_1 q_6 - q_4 q_3) \tag{2.44}$$

$$a_5(q) = q_5 q_3 - q_2 q_6.$$

One can identify the zero as $z = q_5 - \dfrac{q_2 q_6}{q_3}$. The robust D-stability of the system can be assessed using the denominator of $G(s, q)$ in Equation 2.43. As the nominal system changes over the flight envelope, the coefficients $a_i(q)$ in Equation 2.43 vary accordingly depending on p, $a_i(p, q)$.

The parametric uncertain closed-loop system is therefore checked to be robust D-stable over its full flight envelope. This property ensures basic performance of a missile face to parametric uncertainties. Moreover, the controller design technique generates an LPV controller and by structure ensures zero steady-state error of the controlled output. The closed-loop system in Equation 2.43 is of the third order, and the poles are assessed to belong to the D-region.

2.4.6 Robust D-Stability Analysis Criteria

The D-region has been shaped according to the following requirements:

- Minimum damping ratio of 0.7 (critical damping ratio)
- Maximum rising time of 0.1 s (with rising time taken as there times the time constant)
- Maximum natural frequency of 150 rad/s

The nominal system design places the poles in this D-region (Figure 2.7).

For ease of implementation, the D-region was restricted to polygon contours. Although piecewise linear contour is not smooth and consequently not a Jordan curve, in practice, the FIT is using only a finite number of points describing this curve, so a Jordan curve could be found with these points, and the FIT would apply. In case of symmetry of the D-region across the real axis—this will be the case with real polynomials—computations are halved.

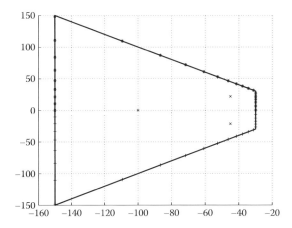

FIGURE 2.7
D-region contour considered for D-stability robustness property.

2.4.7 Applying FIT

The sector angle for this application is $\frac{7}{8}\pi < \pi$, and since the D-region is symmetric and the polynomial for analysis is of the third-order, then only 25 sectors are used. Note that the number of sectors depends on the order of the polynomial but not on the dimension of the parametric uncertainty vector.

Since the uncertainty parameter corresponds to the six independent semi-nondimensional derivatives, the multilinear uncertainty quickly involves a dozen uncertain multilinear coefficients, and it stays difficult to tackle. However, the value sets are successfully reduced by applying the mapping theorem, and thus, the convex hull of the vertex polynomials is used instead (see Figure 2.8). This, however, could imply in some cases some conservatism as discussed in the previous section on the mapping theorem.

The closed-loop system is proved to be robust D-stable (see Figure 2.8a) where for the velocity controller up to ±11.64% of uncertainty on each coefficient is achieved. In fact, these results could be underestimates of the true robust D-stability of the closed-loop system capabilities since conservatism is introduced when sector angles are limited to less than π, the mapping theorem convexifies the value set, and the uncertainty box is taken to be uniform over all components of the parametric uncertainty vector, which mostly depends on the limiting parameter.

Inspection of the form of the convex hull also reveals a few interesting points. In particular, one can notice from Figure 2.8b that Δy_r and Δn_ζ are the main factors in the convex hull size and hence have a strong contribution in "D-destabilizing" the system. Moreover, further works showed that the value set of the D-contour map is almost a polygon where vertices could be computed from the extremal box formed by Δy_r and Δn_ζ, or in other words, the value set

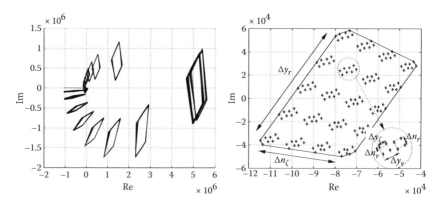

FIGURE 2.8
Application of FIT.

is equivalent to an almost affine uncertainty. This last remark shows that the convex approach used for this analysis should not be so conservative indeed. Finally, Figure 2.8b does not exhibit the influence of Δy_ζ, the uncertainty on fin forces responsible for the nonminimum phase effect in the system.

The D-stability robustness analysis for each individual underlying a closed-loop system—at each operating point of its flight envelope depending on incidence and external parameters—is consequently assessed against real parametric uncertainties present in the missile semi-nondimensional derivatives.

2.4.8 Simulation Results

Simulations have been carried out to check performance, robustness, and when possible, the extent of conservativeness involved. Because the system model does not include the actuator dynamics, the input is prefiltered according to the design performance requirements. Figure 2.9 is the simulation for a missile flying at Mach 2.5 at 6 km altitude with a step lateral acceleration demand from -100 to 100 m s^{-2}. Damping and rising time are achieved according to the design requirements for the nominal system. The dotted curves show the simulations for up to $\pm20\%$ uncertainty in all coefficients. These simulations show good performance and robustness for the sideslip

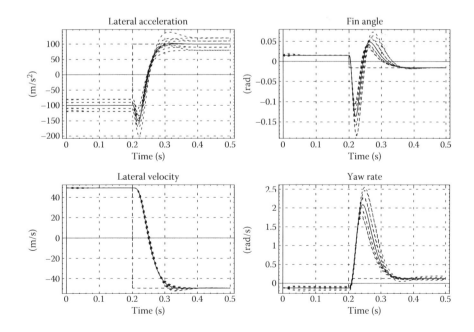

FIGURE 2.9
Simulations with the sideslip velocity controller with up to $\pm20\%$ uncertainty on all six coefficients (only major players were plotted here).

velocity. The lateral acceleration, though, presents an expected steady-state error. This simulation does not guarantee to capture the worst case; however, as discussed in the previous section, the worst cases are highly dependent on uncertainties on y_r and n_ζ, while others play a less important role.

2.5 PEA Including Actuator Dynamics

When actuator dynamics are included, an augmented system must be defined for the case of partial feedback since the actuator states are usually not available for feedback. Hence, for this case, only partial EA is possible for the augmented closed-loop system. Thus, Kimura's condition, $n \leq m + r - 1$, fails, and therefore, not all eigenvalues can be assigned independently.

In this section, the approach investigates the partial feedback case using the LPV framework, where both actuator requirements and/or system capabilities are investigated. The augmented system design should not include any actuator states for feedback leading to a system with partial feedback rather than state feedback. In order to explore the design issues, a full state solution is first computed, and conditions on the feedback structure are imposed to remove the actuator states. Hence, the augmented system with full feedback is calculated initially. Then, the actuator state feedback is removed, which then results in a partial feedback solution. If the system design uses this approach, it enables the designer to assess the effect of the actuator dynamics on the achievable eigenstructure. Hence, trade-offs between the speed of performance of the actuators and the overall system performance can be done. This is of importance as there is limited power and space available in missile structures, and the lowest actuator power solutions are strongly correlated with the actuator bandwidth. The effects of the actuator dynamics on the system's performance are studied by examining two different cases. First, the desired closed-loop system is selected, and the actuator dynamics performance requirements are assessed. Second, specific actuator performance is selected, and the achievable closed-loop system performance is determined. These are successively investigated in the case of lateral velocity, lateral augmented acceleration, and lateral acceleration controller designs for first- and second-order actuator dynamics models.

2.5.1 Matching Conditions with Full Feedback

For this case, the actuator states are fed back using a virtual output, y_l, and the controller structure is redefined including this virtual feedback with its corresponding virtual gain, $K_l(s, p)$ (see Figure 2.10). The controller is designed using the PEA following Equation 2.13, where $K_l(s, p)$ is obtained by suitable partitioning of $K_i(s, p)$. In a similar manner to the $G_i(s, p)$ transfer function,

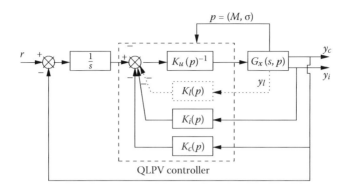

FIGURE 2.10
Specific QLPV controller structure including actuator state feedback, y_l, through gain $K_l(s,p)$.

$G_l(s, p)$ and its associate $Z_0^l(s,p)$ can be introduced, which corresponds to the open-loop system between the inputs and the actuator states. The closed-loop transfer function can be derived from Equation 2.13 as follows:

$$T_y(s,p) = Z_0^c(s,p)\Big[sK_u(s,p)P(s,p) + sK_c(s,p)Z_0^c(s,p) + sK_i(s,p)Z_0^i(s,p)$$
$$+ sK_l(s,p)Z_0^l(s,p) + Z_0^c(s,p)\Big]^{-1}. \tag{2.45}$$

Splitting the overall transfer function into the corresponding controlled output transfer functions, measured output transfer function, and actuator state output transfer function does not affect the flexibility of the PEA approach. MIMO systems can also be dealt with in a similar manner without reducing controller gains to SISO or multiple input single output (MISO) polynomial matrices.

However, conditions 2.36a through 2.36c still need to be satisfied as the desired closed-loop system is required to match the desired performance objectives. This is represented by conditions 2.46a through 2.46c:

$$N_d(s,p) = Z_0^c(s,p) \tag{2.46a}$$

$$Z_0^c(0,p) = D_d(0,p) \tag{2.46b}$$

$$D_d(s,p) = \Big[sK_u(s,p)P(s,p) + sK_c(s,p)Z_0(s,p) + sK_i(s,p)Z_0^i(s,p)$$
$$+ sK_l(s,p)Z_0^l(s,p) + Z_0^c(s,p)\Big] \tag{2.46c}$$

where polynomials $D_d(s, p)$ and $N_d(s, p)$ represent the desired closed-loop transfer function. This can be written as a null space equation.

$$
\begin{bmatrix} K_u(s,p) & K_c(s,p) & K_i(s,p) & K_l(s,p) & I \end{bmatrix}
\begin{bmatrix}
P(s,p) \\
Z_0^c(s,p) \\
Z_0^i(s,p) \\
Z_0^l(s,p) \\
\dfrac{1}{s}\left(Z_0^c(s,p) - D_d(s,p)\right)
\end{bmatrix} = 0
$$

$$(2.47)$$

where the null space of this last vector gives direct access to the controller gains.

2.5.2 Lateral Acceleration Design with Full State Feedback Control

The full feedback design for a lateral acceleration controller performed on the augmented system including actuator dynamics is now considered. First-order actuator dynamics are assumed as follows:

$$\dot{l}(t) + \tau_l l(t) = \tau_l u(t) \tag{2.48}$$

where l represents the fin angle output from the actuator, u is the command input, and τ_l is the time constant of actuator dynamics.

The state equation for this augmented system becomes

$$
\begin{bmatrix} \dot{v} \\ \dot{r} \\ \dot{l} \end{bmatrix} =
\begin{bmatrix}
y_v(p) & y_r(p) & y_\zeta(p) \\
n_v(p) & n_r(p) & n_\zeta(p) \\
0 & 0 & -\tau_l
\end{bmatrix}
\begin{bmatrix} v \\ r \\ l \end{bmatrix} +
\begin{bmatrix} 0 \\ 0 \\ \tau_l \end{bmatrix} u \tag{2.49}
$$

and the output Equation 2.50 includes the lateral acceleration and the yaw rate, together with the actuator state, fin angle l:

$$
y =
\begin{bmatrix}
y_v(p) & 0 & y_\zeta(p) \\
0 & 1 & 0 \\
0 & 0 & 1
\end{bmatrix}
\begin{bmatrix} v \\ r \\ l \end{bmatrix} +
\begin{bmatrix} 0 \\ 0 \\ 0 \end{bmatrix} u. \tag{2.50}
$$

The D matrix normally present for lateral acceleration control vanishes owing to the introduction of actuator dynamics; however, zeros are not affected, and hence, the nonminimum phase zero is still present.

In a similar manner to previous designs for this controllable and observable augmented system, a controller is derived by solving Equation 2.47. The closed-loop system is shown to be of the fourth-order owing to the open-loop system being (second order) with actuator dynamics (first order) and an additional integrator in the control loop (first order) for steady-state accuracy. The full controller then takes the form

$$u = K_u(s,p)^{-1}\left(\frac{(a-a_d)}{s} - K_c(s,p)a - K_i(s,p)r - K_l(s,p)l\right). \qquad (2.51)$$

Simulations are carried out with actuator dynamics having a time constant $\tau_l = 300$. The four closed-loop system poles are chosen to be two poles at -60 and a conjugate pair with natural frequency 30 rad/s and critic damping ratio 0.7. Simulations for the Horton missile are shown in Figure 2.11 and show satisfactory performance.

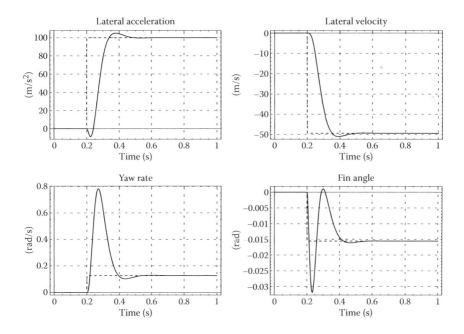

FIGURE 2.11
Simulation of Horton missile controlled in lateral acceleration and including actuator dynamics, in response to a lateral acceleration demand of 100 m s^{-2} with constant Mach number 2.5 and constant altitude 6 km. Pole assignment of the closed-loop system is two poles at -60 and a conjugate pair with natural frequency 30 rad/s and critic damping ratio 0.7.

2.5.3 Lateral Acceleration Design with Partial State Feedback Control

In practice, the actuator states are not available, and thus, only partial feedback is possible. The idea, then, is to remove the virtual actuator feedback by zeroing the gain $K_I(s, p)$, thus constraining the achievable closed-loop system.

Preliminary studies using Nyquist plots using some reference actuator dynamics show that actuator dynamics is usually required to be four to five times faster than the desired closed-loop system. While the studies presented in the paper confirm such results, the approach is more systematic and enables the designer to further analyze the closed-loop system performance as a function of actuator dynamics.

Hence, the full feedback design is amended by additionally requiring the actuator feedback gain, $K_I(s, p)$, to be zero. Doing so reduces the null space solution of Equation 2.47 to a subset of its subspace and hence constrains either actuator dynamics or closed-loop system dynamics. Both cases are investigated.

After row reduction, the null space for Equation 2.47 takes the form

$$\begin{bmatrix} Y_0^{u_0} & Y_0^{c_0} & Y_0^{i_0} & Y_0^{l_0} & I \\ Y_1^{u_0} & Y_1^{c_0} & Y_1^{i_0} & Y_1^{l_0} & 0 \end{bmatrix} \tag{2.52}$$

where any linear combination (in the polynomial sense) of the second row added to the first row defines a possible controller. Specific linear combinations are searched in order to identify the subspace that produces zero gain $K_I(s, p)$. However, this is not always possible, and further conditions on either closed-loop structure or actuator dynamics can be imposed to generate suitable subspaces that contain zero gain solutions for $K_I(s, p)$. For instance, the rank of the polynomial space associated with $K_I(s, p)$ can be controlled by adding constraints on the closed-loop system pole assignment. This task is performed by symbolic computation using MATHEMATICA, and thus, systematic design is achieved for some prespecified closed-loop EA ranges. Such a null space takes the fundamental form

$$[\; Y_0^{u_0} \quad Y_0^{c_0} \quad Y_0^{i_0} \quad 0 \quad I \;] \tag{2.53}$$

with an alternative form given by

$$\begin{bmatrix} Y_0^{u_0} & Y_0^{c_0} & Y_0^{i_0} & 0 & I \\ Y_1^{u_0} & Y_1^{c_0} & Y_1^{i_0} & 0 & 0 \end{bmatrix} \tag{2.54}$$

where any linear combination (in the polynomial sense) of the second row added to the first row spans the space of suitable controllers.

Alternatively, when no suitable controller is found, the designer can choose either to increase the controller order or to change the controller structure. Increasing the controller order may increase the null space and thus the space of suitable controllers. Changing the controller structure affects the design flexibility and eventually the matching capabilities.

2.5.4 Actuator Dynamics of First Order

Assuming actuator dynamics (Equation 2.48) of the first-order augments the system by adding one state variable in a similar manner to Equation 2.49, the output matrix as defined in Equation 2.50 is slightly modified to define lateral augmented acceleration rather than pure lateral acceleration.

Equation 2.47 can then be solved in the form of Equation 2.53 by imposing conditions on the actuator time constant, τ_l, while still achieving the desired closed-loop performance. The minimum actuator dynamics requirement in terms of time constant τ_l is shown in Figure 2.12a for the lateral velocity and lateral augmented acceleration controller and in Figure 2.12b for the lateral acceleration controller, for specific pole assignment of the overall closed-loop system. The pole assignment is chosen to be identical to the previous designs: two poles at −60 and a conjugate pair with natural frequency 30 rad/s and damping ratio 0.7.

The requirements for the lateral velocity design and lateral augmented acceleration design are identical; this is justified by the fact that they are simply related by the aerodynamic derivative $y_v(p)$. However, the lateral acceleration control results shown in Figure 2.12b indicate that this is more demanding on actuator dynamics than the velocity controller. This is partly due to the fact that the nonminimum phase character of the controlled output has a significant part to play in the achievable performance.

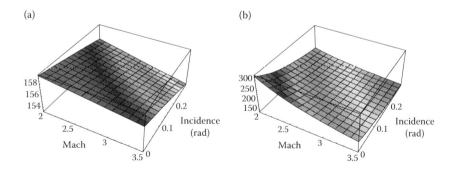

FIGURE 2.12
Actuator dynamics time constant τ_l requirement for first-order actuator imposing fourth-order closed loop with pole assignment: two poles at −60 and a conjugate pair with natural frequency 30 rad/s and damping ratio 0.7. (a) With lateral velocity or lateral augmented acceleration controller. (b) With lateral normal acceleration controller.

Hence, the designer can characterize the actuator dynamics and can choose the maximum τ_l requirement across the flight envelope. For this case, a bandwidth of around 300, which corresponds to a cutoff frequency of around 50 Hz, is a satisfactory solution. This choice ensures that the LPV system performs as required for a controller without actuator state feedback.

Next, the simulation for the lateral acceleration controller for a step input of 250 m s^{-2} is shown in Figure 2.13.

Once the actuator bandwidth has been chosen, it can be fixed. The performance across the flight envelope can be assessed by solving Equation 2.45 while still maintaining $K_l(s, p) = 0$ while varying the closed poles of the system. Hence, the closed-loop system pole locations are not frozen across the flight envelope, and this allows more flexibility in the design than for most dynamic inversion techniques.

In Figure 2.14, pole locations vary slightly along the flight envelope where the first-order actuator dynamics is selected with time constant, $\tau_l = 300$. By design, the dominant conjugate pair is fixed with a natural frequency 30 rad/s and damping ratio 0.7 across the flight envelope, while the two other poles can vary but are always to the left of −60. In fact, selecting a more accurate upper bound for the time constant in the previous analysis would bring the worst case to exactly −60.

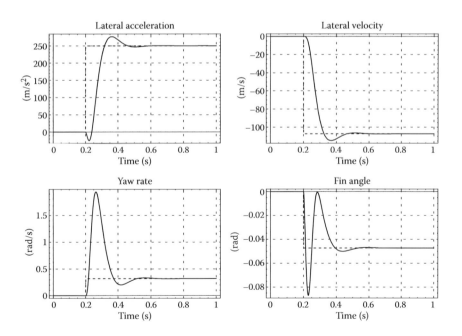

FIGURE 2.13

Lateral acceleration step response of Horton missile including first-order actuator dynamics with time constant, $\tau_l = 300$, for the lateral acceleration controller. Acceleration demand is 250 m s^{-2}, while Mach number is constant at 2.5, and altitude stays constant at 6 km.

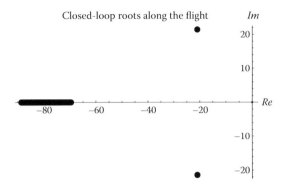

FIGURE 2.14
Scenario of the pole location at Mach 2.5 across the flight envelope for the selected first-order actuator with time constant $\tau_l = 300$.

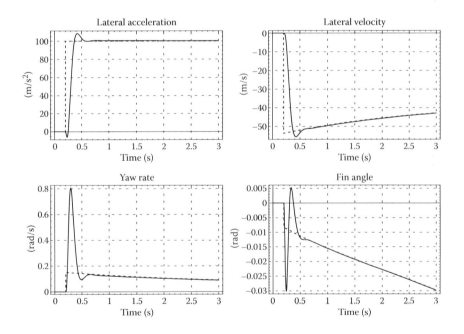

FIGURE 2.15
Lateral acceleration step response of Horton missile including underspecified first-order actuator dynamics of time constant, $\tau_l = 150$, for lateral acceleration controller. Acceleration demand is 100 m s^{-2}, while Mach number varies from Mach 2 to Mach 3.5, and altitude stays constant at 6 km.

Finally, for a slow actuator system with a time constant $\tau_l = 150$, the closed-loop system cannot meet the pole location requirements, and the performance is affected (see Figure 2.15). It can be seen that the actuator has less amplitude in Figure 2.15 than in Figure 2.13; however, there is more acceleration overshoot.

2.5.5 Actuator Dynamics of Second Order

The same approach can be applied to any actuator dynamics model order. For example, actuator dynamics of the second order can be included and partial feedback solutions examined in a similar manner to the first-order actuator case. Actuator dynamics are then modeled as

$$\ddot{l} + 2\zeta_l \omega_l \dot{l} + \omega_l^2 l = \omega_l^2 u \tag{2.55}$$

where l represents the fin angle output, u is the fin angle input, ω_l is the natural frequency, and ζ_l is the damping ratio of the actuator dynamics.

The system is now augmented by two state variables as

$$\begin{bmatrix} \dot{v} \\ \dot{r} \\ \dot{l}_1 \\ \dot{l}_2 \end{bmatrix} = \begin{bmatrix} y_v(p) & y_r(p) & y_\zeta(p) & 0 \\ n_v(p) & n_r(p) & n_\zeta(p) & 0 \\ 0 & 0 & 0 & 1 \\ 0 & 0 & -\omega_l^2 & -2\zeta_l\omega_l \end{bmatrix} \begin{bmatrix} v \\ r \\ l_1 \\ l_2 \end{bmatrix} + \begin{bmatrix} 0 \\ 0 \\ 0 \\ \omega_l^2 \end{bmatrix} u \tag{2.56}$$

and the output equation includes lateral acceleration, yaw rate along with fin angle, and its first derivative to give

$$y = \begin{bmatrix} y_v(p) & 0 & y_\zeta(p) & 0 \\ 0 & 1 & 0 & 0 \\ 0 & 0 & 1 & 0 \\ 0 & 0 & 0 & 1 \end{bmatrix} \begin{bmatrix} v \\ r \\ l_1 \\ l_2 \end{bmatrix} + \begin{bmatrix} 0 \\ 0 \\ 0 \\ 0 \end{bmatrix} u. \tag{2.57}$$

The first row of the output matrix for lateral acceleration as defined in Equation 2.57 is now changed for both lateral velocity or lateral augmented acceleration but omitted here as it has similar structure.

With the same controller structure as previously defined, the null space equation 2.51 remains the same as for the first-order case, although $K_l(s, p)$ is now a vector of dimension 2 rather than 1. Equation 2.47 is solved and results in the form of Equation 2.53 by imposing conditions on the actuator dynamics, ω_l and ζ_l, while requiring the desired closed-loop performance to be achieved. The actuator dynamics requirements in terms of ω_l and ζ_l are presented in Figure 2.16 for the lateral velocity and lateral augmented acceleration controller and in Figure 2.17 for the lateral acceleration controller. The closed-loop system is now a fifth-order system with pole assignment similar to the previous first-order actuator study. Hence, one pole is chosen at −60, with a conjugate pair with natural frequency 30 rad/s and damping ratio 0.7, and the other conjugate pair with natural frequency 110 rad/s and damping ratio 0.6.

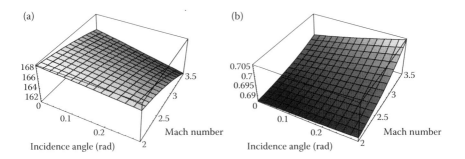

FIGURE 2.16
For lateral velocity or lateral augmented acceleration controller, actuator dynamics requirement on (a) natural frequency ω_l and (b) damping ratio ζ_l imposing fifth-order closed loop
with pole assignment: one pole at -60 and a conjugate pair at natural frequency 30 rad/s and
damping ratio 0.7, and another conjugate pair at natural frequency 110 rad/s and damping
ratio 0.6.

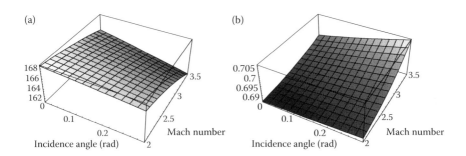

FIGURE 2.17
For lateral acceleration controller, actuator dynamics requirement on (a) natural frequency ω_l
and (b) damping ratio ζ_l imposing fifth-order closed loop with pole assignment: one pole at
-60, a conjugate pair with natural frequency 30 rad/s and damping ratio 0.7, and a conjugate
pair with natural frequency 110 rad/s and damping ratio 0.6.

For the velocity and the lateral augmented acceleration controller case, the
actuator dynamics requirements are around a natural frequency of 160 rad/s
and a damping ratio in the range of 0.5 to 0.7. These requirements seem reasonable since such actuator dynamics are practical. As already mentioned,
the requirements for lateral velocity and lateral augmented acceleration are
identical. As with the first-order actuator, the second-order actuator results
have a more demanding requirement for actuator bandwidth due to the nonminimum phase character of the acceleration output. This is shown in Figure
2.17, and a natural frequency of 240 rad/s for a damping ratio in similar range
of 0.5 to 0.7 is required to meet the closed-loop specifications.

In a similar manner to the first-order actuator dynamics case, once the actuator dynamics is fixed, the designer can assess the closed-loop system performance across the flight envelope by solving Equation 2.45 while still maintaining $K_l(s) = 0$. The performance is estimated from the pole locations, which change slightly along the flight envelope. The pole locations are found to be sensitive to the actuator damping ratio as shown in Figure 2.18a, where the actuator damping ratio sweeps from 0.5 to 0.7 while its natural frequency stays constant at 160 rad/s. By design, the dominant conjugate pair is frozen with a natural frequency of 30 rad/s and damping ratio 0.7 with a pole fixed at −60 across the flight envelope. This results in the other conjugate pair varying significantly from slightly damped to underdamped.

Selecting the same actuator dynamics for lateral acceleration control shows unsatisfactory pole locations in Figure 2.18b. Selecting better actuator dynamics with higher-frequency bandwidth as suggested by previous analysis leads to improved performance. For instance, the pole location for actuator dynamics with natural frequency around 300 rad/s and damping ratio ranging in [0.5, 0.6] and [0.6, 0.7] is shown in Figure 2.19a and b, respectively.

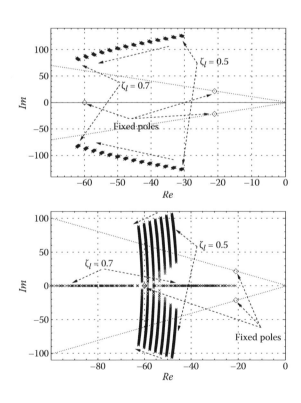

FIGURE 2.18

Scenario of pole location across flight envelope for a panel of second-order actuators with natural frequency 160 rad/s and damping ratio ranging from 0.5 to 0.7.

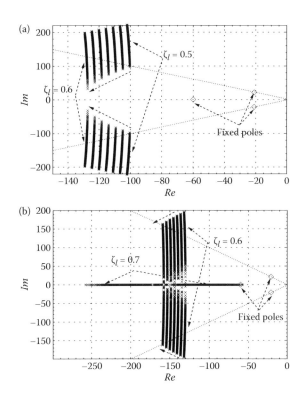

FIGURE 2.19
Scenario of pole location across flight envelope for a panel of second-order actuators with natural frequency 300 rad/s and varying damping ratio for lateral acceleration control. (a) Damping ratio, ζ_l, ranging from 0.5 to 0.6. (b) Damping ratio, ζ_l, ranging from 0.6 to 0.7.

2.6 Discussions and Conclusions

A successful lateral autopilot design was achieved for a realistic quasilinear parameter-varying missile model. This showed that QLPV models are amenable to dynamic inversion design of appropriate controllers using the same parameterization for plants and controllers. The controller was obtained via the PEA. The effects of the actuator bandwidth on the closed-loop system performance were also studied, as were the requirement of dynamics and order of the actuators in order to achieve the desired performance.

The proposed QLPV controller is free of the difficulties associated with gain scheduling, as it consists of one controller only, and "scheduling" is done automatically by feedback, giving total independence of the operating point. Nevertheless, the PEA design is valid only (as gain scheduling is) in the vicinity of the equilibria. The scheduling is directly performed with a QLPV controller and does not need any form of interpolation. The eigenvalues and eigenvectors of the resulting closed-loop system fully characterize its response, and they are suitable to capture the control engineer objectives, stability and performance. In essence, the approach achieves dynamic inversion but in contrast to the involved transformation does not require cancellation of the zero dynamics. It is consequently applicable to nonminimum phase systems as well and thus to a broader class, unlike other dynamic inversion approaches.

The PEA approach using the symbolic solutions enabled trade-off studies to be undertaken that allow the designer to select the appropriate actuator bandwidth to produce satisfactory performance. This is important in missile autopilots as there are usually limits placed on the power and energy available for the actuator.

References

1. Ackermann, J. E. (1980). Parameters space design of robust control systems, *IEEE Transactions on Automatic Control* 25: 1058–1072.
2. Barmish, B. R. (1994). *New Tools for Robustness of Linear Systems*, Macmillan Publishing Company, New York, USA.
3. Bhattacharyya, S. P., Chapellat, H., and Keel, L. H. (1995). *Robust Control, the Parametric Approach*, Prentice Hall, New Jersey, USA.
4. Bruyère, L., Tsourdos, A., Żbikowski, R., and White, B. A. (2002). Robust performance study for lateral autopilot of a quasi-linear parameter-varying missile, *American Control Conference*, Anchorage, Alaska, USA, vol. 1, pp. 226–231.
5. Bruyère, L., White, B. A., and Tsourdos, A. (2003). Dynamic inversion for missile lateral velocity control via polynomial eigenstructure assignment, *AIAA Guidance, Navigation and Control Conference and Exhibit*. AIAA, Reston, VA 2003-5792.
6. Bruyère, L., White, B. A., and Tsourdos, A. (2004). Polynomial approach for design and robust analysis of lateral missile control, International Journal of Systems Science, vol. 37, 8, pp. 585–597.
7. Djaferis, T. E. (1995). *Robust Control Design—a Polynomial Approach*, Kluwer Academic Publishers, Massachusetts, USA.
8. Fossard, A. J. and Normand-Cyrot, D. (eds) (1997). *Nonlinear Systems*, vol. 3, Control, Chapman & Hall, London.
9. Horton, M. P. (1992). *A Study of Autopilots for the Adaptive Control of Tactical Guided Missiles*, Master's thesis, University of Bath.
10. Isidori, A. (1985). *Nonlinear Control Systems*, Springer–Verlag, London, UK.
11. Kailath, T. (1980). *Linear Systems*, Prentice-Hall, London.
12. Kaminsky, R. D. and Djaferis, T. E. (1993). The Finite Inclusions Theorem, *Proceedings of the 32nd IEEE Conference on Decision and Control*, IEEE, New York, USA, vol. 1, pp. 508–518.
13. Kaminsky, R. D. and Djaferis, T. E. (1994). A novel approach to the analysis and synthesis of controllers for parametrically uncertain systems, *IEEE Transactions on Automatic Control* 39(4): 874–876.

14. Kaminsky, R. D. and Djaferis, T. E. (1995). The finite inclusions theorem, *IEEE Transactions on Automatic Control* 40(3): 549–551.
15. Kaminsky, R. D. and Djaferis, T. E. (1996). Application of the finite inclusions theorem to robust analysis and synthesis, *International Journal of Control* 64(3): 511–527.
16. Kimura, H. (1975). Pole assignment by gain output feedback, *IEEE Transaction on Automatic Control* 20(4): 509–516.
17. Liu, G. P. and Patton, R. J. (1998). *Eigenstructure Assignment for Control System Design*, John Wiley & Sons, Chichester.
18. Magni, J.-F. (2002). *Robust Modal Control with a Toolbox for Use with MATLAB*, Kluwer Academic/Plenum, New York, USA.
19. Sobel, K. M., Shapiro, E. Y., and Andry, A. N. (1994). Eigenstructure assignment, *International Journal of Control* 59(1): 13–27.
20. Tsourdos, A., Żbikowski, R., and White, B. A. (2001). Robust design of sideslip velocity autopilot for a quasi-linear parameter-varying missile model, *Journal of Guidance, Control, and Dynamics* 24(2): 287–295.
21. White, B. A. (1995). Eigenstructure assignment: a survey, *IMechE, Systems and Control Engineering* 209(I1): 1–11.
22. White, B. A. (1996). Flight Control of a VSTOL Aircraft Using Polynomial, *IEE UKACC, International Conference on Control 96*, IEE, London, UK, vol. 2, pp. 758–763.
23. White, B. A. (1997). Robust polynomial eigenstructure assignment using dynamic feedback controllers, *IMechE, Systems and Control Engineering* (I1): 35–51.
24. White, B. A. (1998). Robust Flight Control of a VSTOL Aircraft Using Polynomial Matching, *Proceedings of the American Control Conference*, IEEE, New York, USA, pp. 1133–1137.
25. Wise, K. A. (1991). Missile autopilot robustness to uncertain aerodynamics: stability hypersphere radius calculation, *Journal of Guidance, Control and Dynamics* 14(1): 166–175.
26. Wise, K. A. (1992). Comparison of 6 robustness tests evaluating missile autopilot robustness to uncertain aerodynamics, *Journal of Guidance, Control and Dynamics* 15(4): 861–870.

3

Control Design and Gain Scheduling Using Observer-Based Structures

Daniel Alazard

CONTENTS

3.1 Introduction

Observer-based controllers (e.g., linear quadratic Gaussian [LQG] controllers) are quite interesting for different practical reasons and from the implementation point of view. Probably the key advantage of these controller structures lies in the fact that the controller states are meaningful variables as estimates of the physical plant states. It follows that the controller states can be used to monitor (online or offline) the performance of the system. Such a meaningful state allows us also to initialize the state of the controller or to update controller state during control mode switching. Note that this simple property does not hold for general controllers with state-space description:

$$\begin{cases} \dot{x}_K &= A_K x_K + B_K y \\ u &= C_K x_K + D_K y. \end{cases} \tag{3.1}$$

Another well-appreciated advantage comes from the ease of implementation of observer-based controllers. In addition to the plant data, only two static gains (the state-feedback gain and the state-estimator gain) define the entire controller dynamics. In return, this facilitates the construction of gain-scheduled or interpolated controllers. Indeed, assuming the plant model is available in real time, observer-based controllers will only require the storage of these two static gains of lower dimensions instead of the huge set of numerical data in Equation 3.1 to update the controller dynamics at each sample of time. Note that if we are using an interpolating procedure to update the controller dynamics, the general representation in Equation 3.1 is highly questionable from an implementation viewpoint and in many cases will lead to an insuperable computational effort. This was in our opinion a major impediment for a widespread use of modern control techniques such as H_∞ and µ syntheses in realistic applications and particularly for problems necessitating real-time adjustment of the controller gains. These approaches produce high-order controllers expressed under a meaningless state-space realization. Note also that this last point is relevant if a controller reduction has been performed after the design.

To approach this problem, a general procedure is proposed in this chapter to compute an observer-based realization for an arbitrary given controller and a given plant (for both continuous and discrete time cases). Independently of the solver used for the control design, such a procedure allows providing a realization with a meaningful state vector. In the work of Alazard [1] and Cumer et al. [10], it is shown that by observer-based realization, it is also convenient to isolate high level-tuning parameters (potentiometers) in a complex control law. As the observer-based realization exploits the model of the plant, one can also guess that such a realization is very convenient to update the controller to any change in the model or to build a parameter-dependent controller $K(s, \theta)$ from a parameter-dependent model $G(s, \theta)$.

Among other potential advantages of observer-based realization, we would like to point out the possibility of handling actuator saturation constraints by exploiting this information into the prediction equation. Since we do not cover this matter in this document, the reader is referred to [31] and references therein for more details. More theoretical discussions on the implementation of gain-scheduled controllers that exploit information on plant nonlinearities are given in [19] and [20].

The practical solution to handle nonstationary problems (like launch vehicle control design during atmospheric flight) or nonlinear problems consists of designing a family of controllers at various flight instants or various flight conditions and then interpolating (gain scheduling) these various controllers. It is well known that the nonstationary behavior of interpolated control laws depends strongly upon the controller realizations that are interpolated. Observer-based realizations are very attractive from the gain scheduling point of view [25,28]. The main reason is that the controller states are consistent and have physical units if the model on which the observer-based realization is built has physical states. Then, observer-based realizations of given controllers are a good alternative to provide gain-scheduled controllers.

From the control design point of view, the observer-based realization of a controller allows a simple solution to the inverse optimal control problem to be proposed. This solution, called the cross standard form (CSF), is a canonical augmented standard plant whose unique H_∞ or H_2 optimal controller is a given controller [11]. The general idea is to apply the CSF to a given controller in order to set up a standard problem that can be completed to handle frequency-domain H_2 or H_∞ specifications.

In the second section of this chapter, we present the procedure to compute the observer-based realization of a given controller and a given model. The reader will find more details in [3]. The application of this procedure to a very simple missile model is proposed in the third section to illustrate the interest of an observer-based controller for gain scheduling, controller switching, and state monitoring. This application has been chosen for its pedagogic feature: demo files can be downloaded on a Web page for readers to run these illustrations on their own personal computer. In Section 3.4, the CSF is presented and also applied to the same academic example: a low order controller is improved to fulfill a template on its frequency-domain response. The extensions of these results to the discrete-time case are gathered in Section 3.5. In Section 3.6, CSF and gain scheduling using observer-based realizations are applied to the control design for a launch vehicle on the full atmospheric flight envelope [32,33]. Concluding remarks and future works are proposed in the last section.

3.2 Observer-Based Realization of a Given Controller

In this section, we briefly recall central ideas behind the Youla parameterization and show how it can be used to find the state estimator–state feedback structure of an arbitrary compensator associated with a given plant.

Consider the stabilizable and detectable nth-order system $G(s)$ (m inputs and p outputs) with minimal state-space realization:

$$\begin{cases} \dot{x} &= Ax + Bu, \\ y &= Cx + Du. \end{cases} \tag{3.2}$$

The so-called Youla parameterization of all stabilizing compensators built on the general LQG controller structure is depicted in Figure 3.1, where K_c,

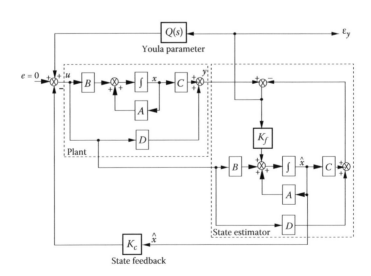

FIGURE 3.1
Observer-based structure and Youla parameterization.

K_f, and $Q(s)$ are, respectively, the state-feedback gain, the state-estimator gain, and the Youla parameter.

The compensator associated with this structure is easily shown to have the following state-space description:

$$
\begin{cases}
\dot{\hat{x}} & = & A\hat{x} + Bu + K_f(y - C\hat{x} - Du) \\
\dot{x}_Q & = & A_Q x_Q + B_Q(y - C\hat{x} - Du) \\
u & = & -K_c\hat{x} + C_Q x_Q + D_Q(y - C\hat{x} - Du)
\end{cases}
\tag{3.3}
$$

where A_Q, B_Q, C_Q, and D_Q are the four matrices of the state-space representation of $Q(s)$ associated with the state variable x_Q. Hereafter, $\hat{x}$ denotes an estimate of the plant state x.

The Youla parameterization principle is based on the fact that the closed-loop transfer function between the input e and the innovation $\varepsilon_y = y - C\hat{x} - Du$ is null (see the work of Luenberger [22] for instance). As a consequence, changing $Q(s)$ leads to various compensators, but the closed-loop transfer function remains unaffected. It is readily shown that this closed-loop transfer function can be represented by the state-space form (Equation 3.4) involving the estimation error $\varepsilon_x = x - \hat{x}$:

$$
\begin{bmatrix} \dot{x} \\ \dot{x}_Q \\ \dot{\varepsilon}_x \end{bmatrix}
=
\begin{bmatrix}
A - BK_c & BC_Q & BK_c + BD_Q C \\
0 & A_Q & B_Q C \\
0 & 0 & A - K_f C
\end{bmatrix}
\begin{bmatrix} x \\ x_Q \\ \varepsilon_x \end{bmatrix}
+
\begin{bmatrix} B \\ 0 \\ 0 \end{bmatrix} e
$$

$$
\varepsilon_y = \begin{bmatrix} 0 & 0 & C \end{bmatrix} \begin{bmatrix} x \\ x_Q \\ \varepsilon_x \end{bmatrix}.
\tag{3.4}
$$

From this representation, the separation principle appears clearly and can be stated in the following terms:

- The closed-loop eigenvalues can be separated into n closed-loop state-feedback poles (spec($A - BK_c$)), n closed-loop state-estimator poles (spec($A - K_f C$)), and the Youla parameter poles (spec(A_Q)).
- The closed-loop state-estimator poles and the Youla parameter poles are uncontrollable by e.
- The closed-loop state-feedback poles and the Youla parameter poles are unobservable from ε_y. The transfer function from e to ε_y always vanishes.

Now let us consider a stabilizing n_Kth-order controller $K_0(s)$ with minimal state-space realization:

$$\begin{cases} \dot{x}_K &=& A_K x_K + B_K y \\ u &=& C_K x_K + D_K y. \end{cases}$$

$$(3.5a)$$
$$(3.5b)$$

In the sequel, the following notations will be used:

$$J_m = (I_m - D_K D)^{-1} \text{ and } J_p = (I_p - DD_K)^{-1}, \tag{3.6}$$

with the following properties:

- $J_m D_K = D_K J_p, J_p D = D J_m$
- $I_m + D_K D J_m = J_m, I_p + J_p D D_K = J_p.$

We are first going to express the compensator state equation (Equation 3.5a) as a Luenberger observer of the variable $z = Tx$. So, we will denote

$$x_K = \hat{z} \tag{3.7}$$

According to Luenberger's formulation [22], this problem can be stated as the search of

$$T \in \mathbb{R}^{n_K \times n}, F \in \mathbb{R}^{n_K \times n_K}, G \in \mathbb{R}^{n_K \times p}$$

such that

$$\dot{\hat{z}} = F\hat{z} + G(y - Du) + TBu \tag{3.8}$$

is an (asymptotic) observer of the variable z, that is, $z - \hat{z}$ vanishes as t goes to infinity. Luenberger has shown that the constraints

$$TA - FT = GC, \text{ and } F \text{ stable}, \tag{3.9}$$

ensure that this holds true. Then, with the output equation (Equation 3.5b), the state-space representation of the compensator reads

$$\begin{cases} \dot{\hat{z}} &=& (F + (TB - GD)C_K)z + (G(I_p - DD_K) + TBD_K)y \\ u &=& C_K \hat{z} + D_K y \end{cases}.$$

$$(3.10)$$

With Equation 3.7, the identification of Equations 3.5 and 3.10 leads to the algebraic relations

$$G = (B_K - TBD_K)J_p \qquad (3.11)$$

$$F = A_K + (B_K D - TB) J_m C_K. \qquad (3.12)$$

These equations with Equation 3.9 guarantee that we are dealing with an observer-based controller. Note that the stability of F (Equation 3.9) is secured whenever the original controller (Equations 3.5a and 3.5b) is stabilizing. Indeed, from Equations 3.2 and 3.10, a closed-loop state-space realization reads

$$\begin{bmatrix} \dot{x} \\ \dot{\hat{z}} \end{bmatrix} = \begin{bmatrix} A + BJ_m D_K C & BJ_m C_K \\ GC + TBJ_m D_K C & F + TBJ_m C_K \end{bmatrix} \begin{bmatrix} x \\ \hat{z} \end{bmatrix}. \qquad (3.13)$$

Let us consider the change of state coordinates involving the estimation error $\varepsilon_z = z - \hat{z}$:

$$\begin{bmatrix} x \\ \hat{z} \end{bmatrix} = \mathcal{M} \begin{bmatrix} x \\ \varepsilon_z \end{bmatrix} \quad \text{with} \quad \mathcal{M} = \begin{bmatrix} I_n & 0 \\ T & -I_{n_K} \end{bmatrix} \quad \text{and} \quad \mathcal{M}^{-1} = \mathcal{M}. \qquad (3.14)$$

The new state-space realization highlights the separation principle:

$$\begin{bmatrix} \dot{x} \\ \dot{\varepsilon}_z \end{bmatrix} = \begin{bmatrix} A + BJ_m (D_K C + C_K T) & -BJ_m C_K \\ 0 & F \end{bmatrix} \begin{bmatrix} x \\ \varepsilon_z \end{bmatrix}. \qquad (3.15)$$

Thus, the set of $n + n_K$ closed-loop eigenvalues includes the n_K eigenvalues of F. Therefore, F is stable if the initial controller is stabilizing.

Substituting Equations 3.11 and 3.12 in the first relation in Equation 3.9, we get

$$(A_K + B_K DJ_m C_K)T - T(A + BJ_m D_K C) - TBJ_m C_K T + B_K J_p C = 0 \qquad (3.16)$$

Thus, the problem is reduced to solving in T the generalized nonsymmetric and rectangular Riccati equation (Equation 3.16) and next to computing F and G using Equations 3.12 and 3.11, respectively.

Equation 3.16 can also be reformulated as

$$[-T \quad I]A_{cl} \begin{bmatrix} I \\ T \end{bmatrix} = 0 \qquad (3.17)$$

where the characteristic matrix A_{cl} associated with the Riccati equation (Equation 3.16) is nothing else but the closed-loop system matrix:

$$A_{cl} := \begin{bmatrix} A + BJ_m D_K C & BJ_m C_K \\ B_K J_p C & A_K + B_K DJ_m C_K \end{bmatrix}. \tag{3.18}$$

The Riccati equation (Equation 3.16) can then be solved by standard invariant subspace techniques that consist of the following:

- Finding an n-dimensional invariant subspace $S := Range(U)$ of the closed-loop system matrix A_{cl}, that is,

$$A_{cl} U = U\Lambda. \tag{3.19}$$

This subspace is associated with a set of n eigenvalues, spec(Λ), among the $n + n_K$ eigenvalues of A_{cl}. Such subspaces are easily computed using Schur factorizations or eigenvalue decompositions of the matrix A_{cl}. See [15] for more details.

- Partitioning the vector U that spans this subspace conformably to the partitioning in Equation 3.18:

$$U = \begin{bmatrix} U_1 \\ U_2 \end{bmatrix}, \quad U_1 \in \mathbb{R}^{n \times n}. \tag{3.20}$$

- Computing the solution

$$T = U_2 U_1^{-1}. \tag{3.21}$$

Narasimhamurthi and Wu [24] have shown that the existence of a solution T satisfying Equation 3.16 is guaranteed whenever the eigenvalues of the Hamiltonian matrix A_{cl} are distinct. In Proposition 3.3, a necessary condition is given for the existence of a solution T. In the general case, however, there are finitely many admissible subspaces S and thus many solutions. Each solution corresponds to a particular choice of n eigenvalues among the set of closed-loop eigenvalues of A_{cl}.

Then, given an nth-order plant and an n_Kth-order compensator, one can compute the linear combination $T_{n_K \times n} x$ of the plant states, which is estimated by the compensator state. An analogous result is also discussed by Bender et al. [8].

The reader will find in http://personnel.supaero.fr/alazard-daniel/demos/demo_obr.html an interactive MATLAB® function cor2tfg to compute the matrices T, F, and G from a given controller K_0 and a given plant G.

3.2.1 Augmented-Order Compensators

In this section, we consider the problem where $n_K \geq n$, and our aim is to find a state-feedback gain K_c, a state-estimator gain K_f, and a dynamic Youla parameter $Q(s)$ with order $n_K - n$, such that the observer-based compensator structure in Figure 3.1 is equivalent to the original controller (Equations 3.5a and 3.5b). We will assume that T has been computed by the previous technique according to an admissible choice of n poles among the $n + n_K$ closed-loop poles. Next, F and G can be computed from Equations 3.11 and 3.12.

Let us consider the Schur decomposition of A_{cl} used to solve in T the Riccati equation (Equation 3.16):

$$A_{cl} = \begin{bmatrix} U_1 & U_3 \\ U_2 & U_4 \end{bmatrix} \begin{bmatrix} \Lambda & * \\ 0 & \Lambda^F \end{bmatrix} \begin{bmatrix} U_1^* & U_2^* \\ U_3^* & U_4^* \end{bmatrix}. \tag{3.22}$$

$\begin{bmatrix} U_1 & U_3 \\ U_2 & U_4 \end{bmatrix}$ is a unitary $(n + n_K) \times (n + n_K)$ matrix with $U_1 \in \mathbb{C}^{n \times n}$, $U_2 \in \mathbb{C}^{n_K \times n}$,

$U_3 \in \mathbb{C}^{n \times n_K}$, and $U_4 \in \mathbb{C}^{n_K \times n_K}$.

From Equations 3.13 and 3.15, we can write

$$\begin{bmatrix} A + BJ_m(D_K C + C_K T) & -BJ_m C_K \\ 0 & F \end{bmatrix} = \begin{bmatrix} I_n & 0 \\ T & -I_{n_K} \end{bmatrix} A_{cl} \begin{bmatrix} I_n & 0 \\ T & -I_{n_K} \end{bmatrix}. \tag{3.23}$$

As $T = U_2 U_1^{-1}$, substituting Equation 3.22 in Equation 3.23, one can derive*

$$F = V\Lambda^F V^{-1} \quad \text{with} \quad V = U_2 U_1^{-1} U_3 - U_4 \tag{3.24}$$

Λ^F is an $n_K \times n_K$ upper triangular matrix, which can be decomposed by blocks with block sizes $n_K - n$ and n. The adequate decomposition of V and V^{-1} allows us to write

$$F = [V_1 \quad V_2] \begin{bmatrix} \Lambda_{11}^F & \Lambda_{12}^F \\ 0 & \Lambda_{22}^F \end{bmatrix} \begin{bmatrix} W_1 \\ W_2 \end{bmatrix} \tag{3.25}$$

---

* Because $\begin{bmatrix} U_1 & U_3 \\ U_2 & U_4 \end{bmatrix}$ is a unitary, it can be shown that $U_4^* = U_4 - U_2 U_1^{-1} U_3$ and $U_3^* + U_4^* U_2 U_1^{-1} = 0$.

with

$$V = [\underbrace{V_1}_{n_K - n} \ \underbrace{V_2}_{n}] \quad \text{and} \quad V^{-1} = \begin{bmatrix} W_1 \\ W_2 \end{bmatrix} \begin{matrix} \}n_K - n \\ \}n. \end{matrix} \tag{3.26}$$

Let us perform the change of variable

$$\hat{z} = \begin{bmatrix} V_1 & V_2 \end{bmatrix} \begin{bmatrix} w_1 \\ w_2 \end{bmatrix} \tag{3.27}$$

in Equations 3.8 and 3.9 and introduce the notations

$$\begin{bmatrix} \widetilde{G}_1 \\ \widetilde{G}_2 \end{bmatrix} = \begin{bmatrix} W_1 \\ W_2 \end{bmatrix} G; \quad \begin{bmatrix} \widetilde{T}_1 \\ \widetilde{T}_2 \end{bmatrix} = \begin{bmatrix} W_1 \\ W_2 \end{bmatrix} T. \tag{3.28}$$

Equations 3.8 and 3.9 then become

$$\begin{cases} \dot{w}_1 &= \widetilde{F}_{11}w_1 + \widetilde{F}_{12}w_2 \quad +\widetilde{G}_1(y - Du) \quad +\widetilde{T}_1 Bu \qquad (3.29a) \\ \dot{w}_2 &= \widetilde{F}_{22}w_2 \quad +\widetilde{G}_2(y - Du) \quad +\widetilde{T}_2 Bu \qquad (3.29b) \end{cases}$$

and

$$\begin{cases} \widetilde{T}_1 A \quad -\widetilde{F}_{11}\widetilde{T}_1 - \widetilde{F}_{12}\widetilde{T}_2 &= \widetilde{G}_1 C \qquad (3.30a) \\ \widetilde{T}_2 A \quad -\widetilde{F}_{22}\widetilde{T}_2 &= \widetilde{G}_2 C. \qquad (3.30b) \end{cases}$$

Now, we will assume that the Schur decomposition has been performed in such a way that $\widetilde{T}_2 = W_2 T$ is nonsingular (in Proposition 3.4, a necessary condition for T to be full column rank is given), and we perform the second change of variable:

$$w_2 = \widetilde{T}_2 \hat{x}. \tag{3.31}$$

From Equations 3.29b and 3.30b, one can derive

$$\dot{\hat{x}} = A\hat{x} + Bu + \widetilde{T}_2^{-1}\widetilde{G}_2(y - C\hat{x} - Du). \tag{3.32}$$

Using now Equations 3.30a and 3.31 to substitute $\widetilde{F}_{12}w_2$ into Equation 3.29a, we get

$$\dot{w}_1 = \widetilde{F}_{11}(w_1 - \widetilde{T}_1\hat{x}) + \widetilde{G}_1(y - C\hat{x} - Du) + \widetilde{T}_1(A\hat{x} + Bu). \tag{3.33}$$

Premultiplying Equation 3.32 by $\widetilde{T}_1$, subtracting it from Equation 3.33, and using the last change of variable

$$w_1 - \widetilde{T}_1\hat{x} = x_Q \tag{3.34}$$

we obtain

$$\dot{x}_Q = \widetilde{F}_{11}x_Q + (\widetilde{G}_1 - \widetilde{T}_1\widetilde{T}_2^{-1}\widetilde{G}_2)(y - C\hat{x} - Du). \tag{3.35}$$

From Equations 3.7, 3.27, 3.31, and 3.34, one can easily derive the global linear transformation between the compensator original state x_K and the new states $\hat{x}$ and x_Q:

$$x_K = \hat{z} = [V_1 \quad T]\begin{bmatrix} x_Q \\ \hat{x} \end{bmatrix}. \tag{3.36}$$

Then, the compensator output equation (Equation 3.5b) can be expressed as

$$u = C_K T\hat{x} + C_K V_1 x_Q + D_K y \tag{3.37}$$

or

$$u = J_m\left[(C_K T + D_K C)\hat{x} + C_K V_1 x_Q + D_K(y - C\hat{x} - Du)\right]. \tag{3.38}$$

The identification of the set of Equations 3.32, 3.35, and 3.38 with Equation 3.3 provides all the parameters for the observer-based controller structure shown in Figure 3.1:

$$K_f = \widetilde{T}_2^{-1}\widetilde{G}_2 = (W_2 T)^{-1} W_2 G \tag{3.39}$$

$$K_c = -J_m(C_K T + D_K C) \tag{3.40}$$

$$A_Q = \widetilde{F}_{11} = W_1 F V_1 \tag{3.41}$$

$$B_Q = \widetilde{G}_1 - \widetilde{T}_1\widetilde{T}_2^{-1}\widetilde{G}_2 = W_1[I_{n_K \times n_K} - T(W_2 T)^{-1} W_2]G \tag{3.42}$$

$$C_Q = J_m C_K V_1 \tag{3.43}$$

$$D_Q = J_m D_K. \tag{3.44}$$

Proposition 3.1

If $n_K = n$, then T is square, and the decomposition (Equation 3.25) of F is such that $V_2 = I_{n \times n}$ and V_1 is empty. Then, Equations 3.39 through 3.44 become

$$K_f = T^{-1}G = (T^{-1}B_K - BD_K)J_p \tag{3.45}$$

$$K_c = -J_m(C_K T + D_K C) \tag{3.46}$$

$$Q(s) = D_Q = J_m D_K. \tag{3.47}$$

This result then specializes to those of Bender and Fowell [7].   ∎

3.2.2 Discussion

There is a combinatoric of solutions according to the choice of the partition of the closed-loop eigenvalues, first, in the computation of matrix T, and second, in the decomposition of matrix F. Hereafter, some rules are proposed to reduce the number of admissible choices.

Proposition 3.2

The n eigenvalues chosen for the computation of the solution T of the Riccati equation (Equation 3.16) using the invariant subspace approach are the n eigenvalues of the closed-loop state feedback associated with the equivalent observer-based controller, that is, spec($A - BK_c$).   ∎

Proof

From Equations 3.18 through 3.20, we have

$$\begin{bmatrix} A + BJ_m D_K C & BJ_m C_K \\ B_K J_p C & A_K + B_K DJ_m C_K \end{bmatrix} \begin{bmatrix} I_{n \times n} \\ T \end{bmatrix} = \begin{bmatrix} I_{n \times n} \\ T \end{bmatrix} U_1 \Lambda U_1^{-1}. \tag{3.48}$$

The first row of this matrix equality reads

$$A + BJ_m(D_K C + C_K T) = U_1 \Lambda U_1^{-1}. \tag{3.49}$$

Using Equation 3.40, we have

$$A - BK_c = U_1 \Lambda U_1^{-1}. \tag{3.50}$$

Thus, the eigenvalues of Λ are the eigenvalues of $A - BK_c$. As a consequence, the n_K remaining eigenvalues are the Luenberger observer poles [i.e., spec(F), see also Equation 3.15], which are shared, in Equation 3.25, between the $n_K - n$ Youla parameter poles [i.e., spec(A_Q)] and the n closed-loop state-estimator poles [i.e., spec($A - K_f C$)].   ∎

Hereafter, we are considering the set of equations (from Equations 3.18, 3.19, and 3.22)

$$\begin{bmatrix} A + BJ_m D_K C & BJ_m C_K \\ B_K J_p C & A_K + B_K DJ_m C_K \end{bmatrix} \begin{bmatrix} U_1 \\ U_2 \end{bmatrix} = \begin{bmatrix} U_1 \\ U_2 \end{bmatrix} \Lambda \tag{3.51}$$

and we shall give a necessary condition, on the choice of the subspace S, for the existence of a solution T (i.e., for U_1 to be invertible).

Proposition 3.3

Consider U_1 and U_2 associated with some n-dimensional invariant subspace S of A_{cl}. Assuming there is some uncontrollable plant eigenvalue that is not in spec($A_{cl}|S$), then U_1 is singular. In other words,

if $\exists \lambda \notin$ spec(Λ) s.t. λ is (A, B) uncontrollable, then U_1 is singular. (3.52)

Proof

Consider the (A, B) pair and let λ denote an uncontrollable eigenvalue with associated left-eigenvector u. That is,

$$u^T [A - \lambda I \mid B] = 0 \tag{3.53}$$

Then, premultiplying Equation 3.51 by $[u^T \ 0]$, we get

$$u^T[(\Lambda \mid BJ_m D_K C)U_1 \mid BJ_m C_K U_2] = u^T U_1 \Lambda. \tag{3.54}$$

From Equations 3.53 and 3.54, it follows that

$$u^T U_1(\Lambda - \lambda I) = 0. \tag{3.55}$$

Thus, if $\lambda \notin \operatorname{spec}(\Lambda)$, then $u^T U_1 = 0$; that is, U_1 is singular.   ∎

We also have a dual property that concerns the column rank of T (i.e., for U_2 to be full column rank). It can be stated as follows.

Proposition 3.4

Consider U_1 and U_2 associated with some n-dimensional invariant subspace S of A_{cl}. Assuming there is some unobservable plant eigenvalue in $\operatorname{spec}(A_{cl}|S)$, then U_2 is column rank deficient. In other words,

if $\exists \lambda \in \operatorname{spec}(\Lambda)$ s.t. λ is (A, C) unobservable, then U_2 is column rank deficient.
$$\tag{3.56}$$
∎

Proof

Omitted for brevity. See Proposition 3.3.   ∎

Propositions 3.3 and 3.4 are quite useful when an observer-based realization for H_∞ or μ controllers must be computed from the standard problem augmented with input and output frequency weights (see the work of Alazard and Apkarian [3] for more details).

Remark 3.1

Among all the admissible choices, the only restriction that can reduce the set of solutions is that complex conjugate pairs of poles cannot be separated if we are seeking state-space representations with real coefficients. Note that such a choice is not always possible. For instance, consider the plant $G(s) = 1/s$ and the compensator $K_0(s) = 2/(s + 2)$. Then, the computation of the state feedback–state estimator form leads to $Q = 0$, $K_c = 1 + i$ (or $1 - i$) and $K_f = 1 - i$ (resp. $1 + i$). Although the gains K_c and K_f are complex, the transfer function of the controller has real coefficients. It can be easily shown that

- If n (model order) is even, then a real solution always exists.
- If n is odd, then a real solution T exists if the number of real eigenvalues in $\operatorname{spec}(A_{cl})$ is at least equal to 1, and a real parameterization (K_c,

K_f, $Q(s)$) exists (in the case $n_K > n$) if the number of real eigenvalues in spec(A_{cl}) is at least equal to 2.

The following selection rules have proved also useful in practical applications of the method:

- Affect the fastest poles to spec(A_Q) in such a way that the Youla parameter acts as a direct feedthrough in the compensator.
- Assign to spec($A - BK_c$) the n closed-loop poles that are the "nearest" from the n plant poles in order to respect the dynamic behavior of the physical plant and reduce the state-feedback gains.
- Assign fast closed-loop poles to spec($A - K_fC$) to have an efficient state estimator.

■

3.2.3 In Brief

The procedure to compute the observer-based form and the dynamic Youla parameter of a given n_Kth-order compensator associated with an nth-order plant ($n_K \geq n$) can be summarized as follows:

- Compute the closed-loop matrix A_{cl} (Equations 3.6 and 3.18) and split up the $n + n_K$ eigenvalues of A_{cl} into three autoconjugate sets:
  - n eigenvalues to be assigned to state-feedback dynamics spec($A - BK_c$)
  - $n_K - n$ eigenvalues to be assigned to the Youla parameter dynamics spec(A_Q)
  - n eigenvalues to be assigned to state-estimator dynamics spec($A - K_fC$)
- Compute a Schur or a diagonal decomposition of A_{cl} (Equation 3.22) such that the eigenvalues are ordered on the diagonal according to the previous choice; that is, spec(Λ) = spec($A - BK_c$) and spec(Λ^F) = spec(A_Q) $\cup$ spec($A - K_fC$).
- Compute T, F, and G with Equations 3.21, 3.12, and 3.11, respectively.
- Compute V, V_1, V_2, W_1, and W_2 with Equations 3.24 and 3.26.
- Compute the sought parameters K_c, K_f, A_Q, B_Q, C_Q, and D_Q using Equations 3.6 and 3.39 through 3.44.

The reader will find a demo file (corresponding to the example proposed in the work of Alazard and Apkarian [3]) and an interactive MATLAB function to compute the observer-based realization for a given controller and a given plant in http://personnel.supaero.fr/alazard daniel/demos/demo_obr.html.

The help of this function is given below (from Alazard and Apkarian [3]; see also OBR2COR and COR2TFG):

```
================================================================
Observer-Based Realization of a given controller
================================================================
[KC,KF,Q,T] = COR2OBR(PLANT,SYS_K) compute a real Observer
Based Realization, that is the Youla parameterization
(defined by Kc, Kf and Q), of a given continuous-time
controller SYS_K for a given continuous-time plant
PLANT in the case:
 NK (SYS_K order) >= N (PLANT order):
 Remarks: * SYS_K, PLANT and Q are defined as SYSTEM matrices,
          * a real solution may not exist,
          * NQ (order of Q) = NK - N.
This function plots the map of closed-loop eigenvalues (red x)
and PLANT open-loop eigenvalues (blue +) in the complex plane.
Then, the user can choose, in an interactive procedure, the
closed-loop eigenvalue distribution between:
  * state feedback dynamics [A-BKc] (blue o),
  * state estimation dynamics [A-KfC] (red o),
  * Youla parameter dynamics (Q) (green o).
Uncontrollable eigenvalues are automatically assigned to [A-BKc].
Unobservable eigenvalues are automatically assigned to [A-KfC].
(the controller SYS_K is assumed to be minimal).
 Auto-conjugate eigenvalues are assigned together.
 T is the transformation matrix between the old and the new state
 space realizations of the controller:
     X_k = T X_hat.
 [KC,KF,Q,T] = COR2OBR(PLANT,SYS_K,TOL) allows a tolerance TOL
 (default: 10^-6) to be taken into account in the unobservable
 and uncontrollable subspaces computation.
```

3.2.4 Reduced-Order Compensator Case

In the case $n_K < n$ [i.e., dim(z) < dim(x)], the observer-based structure shown in Figure 3.1 is no more valid. However, an interesting alternative can be derived using a reduced-order estimator.

It is interesting to point out the case where $[T^T \ C^T]$ is a rank n matrix (i.e., $p + n_K \geq n$). Then, a reduced observer-based realization involving an estimate $\hat{x}$ of the plant state x can be obtained by a linear combination of the compensator state $\hat{z}$, plant input u and output y (see Luenberger [22]):

$$\hat{x} = H_1\hat{z} + H_2(y - Du) \tag{3.57}$$

with the constraint

$$H_1T + H_2C = I_n. \tag{3.58}$$

Then, the separation principle still holds, and a Youla parameterization (with a static parameter D_Q) built on such a reduced-order estimator reads

$$\begin{cases} \dot{\hat{z}} & = & F\hat{z} + G(y - Du) + TBu \\ \hat{x} & = & H_1\hat{z} + H_2(y - Du) \\ u & = & -K_c\hat{x} + D_Q(y - C\hat{x} - Du) \end{cases}$$ (3.59a) (3.59b) (3.59c)

$$\begin{vmatrix} TA - FT & = & GC \\ H_1T + H_2C & = & I_n. \end{vmatrix}$$ (3.60)

As previously, it can be easily shown that the closed-loop poles, with a compensator defined by Equations 3.59 and 3.60, are distributed between the n closed-loop state-feedback poles [spec($A - BK_c$)] and the n_K estimator poles [spec(F)]. Equations 3.16, 3.12, and 3.11, which respectively provide T, F, and G, are still valid. The problem is therefore reduced to computing K_c, H_1, H_2, and D_Q such that (from the identification of Equations 3.59b and 3.59c with Equation 3.5b)

$$\begin{cases} J_m C_K = -(K_c + D_Q C)H_1 \\ J_m D_K = -(K_c + D_Q C)H_2 + D_Q \\ H_1 T + H_2 C = I_n. \end{cases}$$ (3.61a) (3.61b) (3.61c)

It is easily deduced that

$$K_c = -J_m(C_K t + D_K C).$$ (3.62)

This is the same as Equation 3.40, established in the augmented-order compensator case.

To compute H_1, H_2, an D_Q, the following situations can be considered:

- If $\begin{bmatrix} T \\ C \end{bmatrix}^{-1}$ exists (which implies that $n_K + p = n$), then

$$\underset{n_K}{[H_1} \quad \underset{p}{H_2]} = \begin{bmatrix} T \\ C \end{bmatrix}^{-1}$$ (3.63)

and

$$\begin{bmatrix} T \\ C \end{bmatrix} [H_1 \quad H_2] = \begin{bmatrix} TH_1 & TH_2 \\ CH_1 & CH_2 \end{bmatrix} = \begin{bmatrix} I_{n_K} & 0 \\ 0 & I_p \end{bmatrix}.$$ (3.64)

Hence, relationships 3.61a through 3.61c are satisfied for any D_Q, and we can choose $D_U - 0$ without loss of generality.

- If $n_K > n - p$, then there are several solutions (H_1, H_2) satisfying Equation 3.61c. One can choose, for example, the least norm solution (in order to reduce the control gains) using the pseudoinverse of matrix $[T^T C^T]$:

$$\left| \begin{matrix} H_1 \\ H_2 \end{matrix} \right. \quad \begin{matrix} = \\ = \end{matrix} \quad \begin{matrix} [T^T T + C^T C]^{-1} T^T \\ [T^T T + C^T C]^{-1} C^T. \end{matrix} \tag{3.65}$$

Then, from Equation 3.61

$$D_Q = (J_m D_K + K_C H_2)(I_p - C H_2)^{-1}.$$

- If $n_K < n - p$, it can only be stated that, in open loop, the compensator state $\hat{z}$ is an estimate of the linear combination T of the plant state x, that is, the estimation error $\varepsilon_z = Tx - \hat{z}$ tends to 0 with the following dynamics:

$$\varepsilon_z = (A_K + (B_K D - TB) J_m C_K) \varepsilon_z. \tag{3.66}$$

In this case $(n_K < n - p)$, the only way around consists of performing a reduction of the plant until the previous technique is applicable. The compensator is then interpreted as an observer-based compensator associated to the reduced plant.

In the next section, the interest of observer-based realizations of given controllers is highlighted through three examples: plant state monitoring, controller switching, and smooth gain scheduling on an academic second-order missile model.

3.3 Illustrations

The model of a missile between the angle of attack α and the thruster deflection δ can be roughly approximated by the second-order transfer function

$$G(s) = \frac{1}{s^2 - 1}$$

associated with the state-space realization

$$\begin{bmatrix} \dot{\alpha} \\ \ddot{\alpha} \\ \alpha \end{bmatrix} = \begin{bmatrix} 0 & 1 & 0 \\ 1 & 0 & 1 \\ 1 & 0 & 0 \end{bmatrix} \begin{bmatrix} \alpha \\ \dot{\alpha} \\ \delta \end{bmatrix}. \tag{3.67}$$

Let us consider the following stabilizing controller (positive feedback):

$$K_0(s) = -\frac{s^2 + 27s + 26}{s^2 + 7s + 18}.$$

A state-space realization (modal canonical form*) of this controller reads

$$
\begin{bmatrix} \dot{x}_1 \\ \dot{x}_2 \\ \hline \delta \end{bmatrix} =
\left[\begin{array}{cc|c}
-3.5 & 2.398 & 1.027 \\
-2.398 & -3.5 & -1.5 \\
\hline
-17.95 & 1.037 & -1
\end{array} \right]
\begin{bmatrix} x_1 \\ x_2 \\ \hline \alpha \end{bmatrix}.
\tag{3.68}
$$

In this example, the closed-loop dynamics reveals multiple eigenvalues:

$$\mathrm{spec}(A_{cl}) = \{-2, \quad -2, \quad -2, \quad -1\}.$$

Then, there exist two admissible choices to solve in T the Riccati equation (Equation 3.16). The choice $\mathrm{spec}(A - BK_c) = \{-1, \quad -2\}$ and the application of the procedure provide the following parameterization:

$$K_c = [3 \quad 3]; \ K_f = [4 \quad 5]^T; \ Q = -1.$$

Then, the observer-based realization of $K_0(s)$ reads

$$
\begin{bmatrix} \dot{\hat{\alpha}} \\ \dot{\hat{\dot{\alpha}}} \\ \hline \delta \end{bmatrix} =
\left[\begin{array}{cc|c}
-4 & 1 & 4 \\
-6 & -3 & 4 \\
\hline
-2 & -3 & -1
\end{array} \right]
\begin{bmatrix} \hat{\alpha} \\ \hat{\dot{\alpha}} \\ \hline \alpha \end{bmatrix}
\tag{3.69}
$$

associated with the estimated state vector $\hat{x} = [\hat{\alpha}, \hat{\dot{\alpha}}]^T$.

The corresponding MATLAB sequence using functions cor2obr and obr2cor[†] is:

```
G = pck([0 1;1 0],[0;1],[1 0],0);  cor = tf([-1 -27 -26],[1 7 18]);
[a,b,c,d] = ssdata(canon(cor,'modal'));  K = pck(a,b,c,d);
[Kc,Kf,Q] = cor2obr(G,K);  Kob = obr2cor(G,Kc,Kf,Q)
```

A demo file for the following illustrations is also available at http://personnel.supaero.fr/alazard-daniel/demos/demo_obr.html.

---

* Such a canonical form can be easily obtained using the MATLAB macrofunction canon (version 6.5.1). For later versions, canon(SYS,'modal') provides a different state-space realization.
† See http://personnel.supaero.fr/alazard-daniel/demos/demo_obr.html.

3.3.1 Illustration 1: Plant State Monitoring

Figures 3.2 and 3.3 plot the closed-loop state responses (missile and controller states) to initial conditions on missile states [$\alpha(t = 0) = 1$ rad and $\dot{\alpha}(t = 0) = -1$ rad/s]. Figure 3.2 is obtained when the first controller realization (Equation 3.68) is used, while Figure 3.3 is obtained with the observer-based realization (Equation 3.69). For both simulations, the missile state

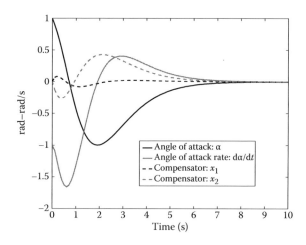

FIGURE 3.2
Responses to initial conditions on missile states—modal canonical realization of $K_0(s)$ (Equation 3.68).

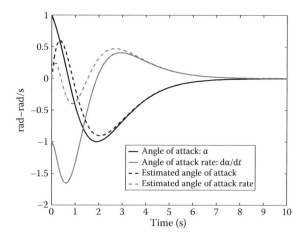

FIGURE 3.3
Responses to initial conditions on missile states—observer-based realization of $K_0(s)$ (Equation 3.69).

responses are the same because the initial conditions are the same and the input–output behavior of the controller is independent of its realization. However, one can see in Figure 3.2 that there is no straightforward relation between controller states and missile states (α and $\dot{\alpha}$), while Figure 3.3 highlights that (after the transient response of the state estimator) the controller states of the observer-based realization are good estimates of missile states and can be used to monitor missile states for offline or inline analysis (for failure diagnosis purposes, for instance). As the plant states are meaningful variables, [α (in radians) and $\dot{\alpha}$ (in radians per second)], one can also conclude that the state-feedback gain K_c has a physical dimension $K_c = [3 \text{ rad/rad } 3 \text{ s}]$, while the dimension of the various components of realization (Equation 3.68) is not defined.

3.3.2 Illustration 2: Controller Switching

Let us consider a second stabilizing controller

$$K_1(s) = -\frac{1667s + 2753}{s^2 + 27s + 353}$$

and let us assume that the control law must switch from controller K_0 to controller K_1 at time $t = 5$ s. This new controller increases closed-loop dynamic performances required, for instance, during the final flight phase (just before the impact). Indeed, the closed-loop dynamics is now

$$\text{spec}(A_{cl}) = \{-3, \quad -4, \quad -10 + 10i, -10-10i\}.$$

Note that the structure of this new controller K_1 is quite different from the previous one (the direct feedthrough term is null in K_1). An observer-based parameterization for $K_1(s)$ reads*

$$K_c = [13 \ 7]; \ K_f = [20 \ 201]^T; \ Q = 0.$$

The state vector initialization of the second controller K_1 with the value of the state vector of the first controller at the switch time (5 s) can create an undesirable transient response (see Figure 3.4 when modal canonical realizations are used for K_0 and K_1). The meaningful state of the observer-based realizations of both controllers allows us to initialize correctly the second controller and so allows the transient response on the attitude $\alpha(t)$ to be reduced in a significant way (see Figure 3.5).

---

* This observer-based parameterization corresponds to the choice affecting the two real closed-loop eigenvalues (i.e., −3 and −4) to the state feedback dynamics.

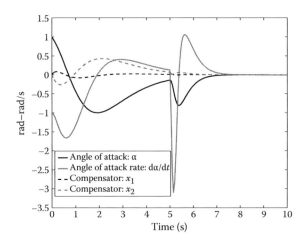

FIGURE 3.4
Responses to initial conditions and switch from $K_0(s)$ to $K_1(s)$ at time $t = 5$ s—modal canonical realizations of $K_i(s)$.

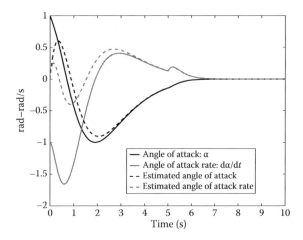

FIGURE 3.5
Responses to initial conditions and switch from $K_0(s)$ to $K_1(s)$ at time $t = 5$ s—observer-based realizations of $K_i(s)$.

3.3.3 Illustration 3: Smooth Gain Scheduling

Now, let us assume that one wishes to interpolate the controller from K_0 to K_1 over 5 s. The linear interpolation of the four state-space matrices of modal canonical realizations provides a nonstationary controller $K(s, t)$ whose frequency response with respect to (w.r.t.) time t is depicted in Figure 3.6. One can notice that this response is nonmonotonous at low frequency, and one

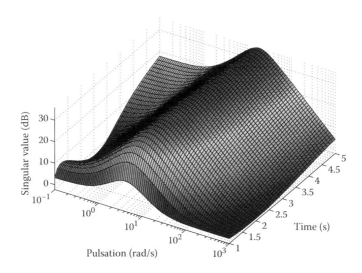

FIGURE 3.6
$K(s,t)$: singular value w.r.t. time.

can also easily check that, at time 2 s, the controller $K(s, 2)$ does not stabilize the plant $G(s)$.

The interpolation of the four state-space matrices of observer-based realizations of K_0 and K_1 provides a smoother interpolation (see Figure 3.7). One can also check that this new interpolated controller stabilizes $G(s)$ for all time $t \in [0, 5 \text{ s}]$.

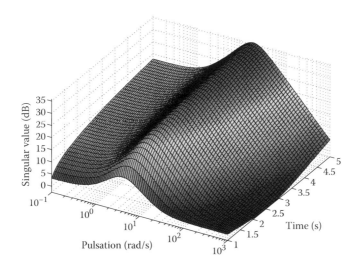

FIGURE 3.7
$K_{observer\text{-}based}(s,t)$: singular value w.r.t. time.

3.4 Cross Standard Form

In most practical applications, the control design problem can be expressed in the following terms: Is it possible to improve a given controller (often, a simple low-order controller designed upon a particular know-how or good-sense rules) to meet additional H_2 or H_∞ specifications? Or in other terms, is it possible to take into account a given controller (which meets some closed-loop specifications) in a standard H_2 and H_∞ control problem? To address this problem, the notion of CSF is introduced in this section for a given nth-order plant and an arbitrary given stabilizing n_Kth-order controller. The CSF can be seen as a solution for both inverse H_∞ and H_2 optimal control problems, that is, the CSF is a standard augmented problem whose *unique* H_∞ and H_2 optimal controller is an arbitrary given controller. The CSF is directly defined by the four state-space matrices of the plant, the four state-space matrices of the given controller, and the solution T to the general nonsymmetric Riccati equation (Equation 3.16) introduced in Section 3.2 to compute the observer-based realization of a given controller for a given plant. The CSF can be applied to full-order, low-order, or augmented-order controllers.

The interest for inverse optimal control problems motivates many works [13,14,17,18,23,26]. The practical interest of such solutions lies in the possibility of mixing various approaches or taking into account different kinds of specifications [27,29,30]. In the particular case of the H_∞ optimal control problem, the various contributions address restrictive cases: state-feedback controller in [14] and single-input single-output controller and specific sensitivity problem in [17]. But a solution for the general case (multi-input multi-output, dynamic output feedback of arbitrary order) has never been stated. This general case is addressed in [26]: for a given weight system $W(s)$ and a given controller $K(s)$, the problem is to find all the plants $G(s)$ such that $\| F_l(F_l(W,G),K) \|_\infty < \gamma$ (see Figure 3.8). Note that the problem considered in this section is different since the plant $G(s)$ [that is, the lower right-hand transfer matrix of the standard augmented plant $P = F_l(W, G)$]

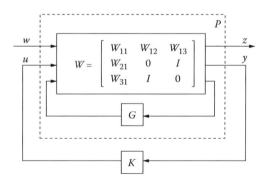

FIGURE 3.8
Block diagram of standard plant P, weight function W, model G, and controller K.

is given and corresponds to the model of the plant between the control input and the measured output.

The convex closed-loop technique [9] seems also an attractive approach to take into account a given controller and additional H_2 or H_∞ constraints. However, such an approach needs a Youla parameterization of the controller and so is limited to full-order (observer-based) controllers. Furthermore, this approach leads to a very high-order controller.

In Section 3.4.1, the CSF is defined as a solution to H_2 and H_∞ inverse optimal control problems, for an nth-order linear time invariant (LTI) system and an n_Kth-order stabilizing LTI controller. In Section 3.4.2, an analytical expression of a CSF is proposed for low-order controllers ($n_K \leq n$), and the existence of such a CSF is discussed. In Section 3.4.3, this new CSF is extended for augmented-order controllers and so encompasses previous results presented in [5]. Finally, the missile second-order example is used in Section 3.4.4 to highlight the way to use CSF to take into account an initial low-order compensator and a frequency-domain specification in an augmented standard problem.

3.4.1 Definitions

The general standard plant between exogenous input w, control input u, controlled output z, and measurement output y is denoted by

$$P(s) = \begin{bmatrix} P_{zw}(s) & P_{zu}(s) \\ P_{yw}(s) & P_{yu}(s) \end{bmatrix},$$

with corresponding state-space realization

$$P(s) := \left[\begin{array}{c|cc} A_p & B_1 & B_2 \\ \hline C_1 & D_{11} & D_{12} \\ C_2 & D_{21} & D_{22} \end{array} \right]. \tag{3.70}$$

Let us consider again the plant $G(s)$ defined in Equation 3.2 and the stabilizing initial controller $K_0(s)$ defined by Equations 3.5a and 3.5b.

Definition 3.1: Inverse H_2 Optimal Problem

Find a standard plant $P(s)$ such that

- $P_{yu}(s) = G(s)$
- K_0 stabilizes $P(s)$
- $K_0(s) = \mathrm{argmin}_{K(s)} \| F_l(P(s), K(s)) \|_2$

(namely, $K_0(s)$ minimizes $\| F_l(P(s), K(s)) \|_2$). ∎

Definition 3.2: Inverse H_∞ Optimal Problem

Find a standard plant $P(s)$ such that

- $P_{yu}(s) = G(s)$
- K_0 stabilizes $P(s)$
- $K_0(s) = \mathrm{argmin}_{K(s)} \| F_l(P(s), K(s)) \|_\infty$ ∎

Definition 3.3: Cross Standard Form

If the standard plant $P(s)$ is such that the four conditions,

- C1: $P_{yu}(s) = G(s)$
- C2: K_0 stabilizes $P(s)$
- C3: $F_l(P(s), K_0(s)) = 0$
- C4: K_0 is the unique solution of the optimal H_2 or H_∞ problem $P(s)$

are met, then $P(s)$ is called the CSF associated with the system $G(s)$ and the controller $K_0(s)$ and will be denoted by $P_{CSF}(s)$ in the sequel. ∎

By construction, the CSF solves the inverse H_2 optimal problem and the inverse H_∞ optimal problem. Note that the uniqueness condition C4 is relevant in our context since we are looking for an H_2 or H_∞ design to recopy a given controller.

3.4.2 Low-Order Controller Case ($n_K \leq n$)

The following proposition provides a general analytical characterization of the CSF.

Proposition 3.5

For a given stabilizable and detectable nth-order system $G(s)$ (Equation 3.2) and a given stabilizing n_Kth-order controller $K_0(s)$ with $n_K < n$ (Equation 3.5), a CSF reads

$$P_{CSF}(s) := \left[\begin{array}{c|ccc} A & T^\# B_K - BD_K & B \\ \hline -C_K T - D_K C & D_K DD_K - D_K & I_m - D_K D \\ C & I_p - DD_K & D \end{array} \right] \quad (3.71)$$

where T is the solution of the generalized Riccati equation (Equation 3.16) and where $T^{\#}$ is a right inverse* of T (such that $TT^{\#} = I_{n_K}$).   ■

Proof

From Equation 3.71, it is obvious that conditions C1 and C2 are met. A state-space realization of $F_l(P_{CSF}, K_0)$ associated with state vector $[x^T, \ x_K^T]^T$ reads

$$
\left[
\begin{array}{cc|c}
A + BJ_m D_K C & BJ_m C_K & T^{\#}B_K \\
B_K J_p C & A_K + B_K DJ_m C_K & B_K \\
\hline
-C_K T & C_K & 0
\end{array}
\right]
$$

where J_m and J_p are defined in Equation 3.6. Let us consider the change of state coordinates (already defined in Equation 3.14):

$$
\mathcal{M} = \mathcal{M}^{-1} =
\begin{bmatrix}
I_n & 0 \\
T & -I_{n_K}
\end{bmatrix}
\tag{3.72}
$$

where T is a solution of Equation 3.16 and $TT^{\#} = I_{n_K}$. The new state-space realization of $F_l(P_{CSF}, K_0)$ reads

$$
\left[
\begin{array}{cc|c}
A + BJ_m(D_K C + C_K T) & -BJ_m C_K & T^{\#}B_K \\
0 & A_K + (B_K D - TB)J_m C_K & 0 \\
\hline
0 & -C_K & 0
\end{array}
\right].
\tag{3.73}
$$

Thus, the $n + n_K$ stable closed-loop eigenvalues are composed of

- n eigenvalues of $A + BJ_m(D_K C + C_K T)$, which are unobservable by the controlled output z of P_{CSF}
- n_K eigenvalues of $A_K + (B_K D - TB)J_m C_K$, which are uncontrollable by the exogenous input w of P_{CSF}

Thus, condition C3 is met:

$$
F_l(P_{CSF}(s), K_0(s)) = 0.
$$

In the next section, it is shown that it is always possible to find a right inverse $T^{\#}$ of T such that the uniqueness condition C4 is met, and that ends the proof.   ■

The general block diagram associated with P_{CSF} is depicted in Figure 3.9.

---

[a] See also Proposition 3.6.

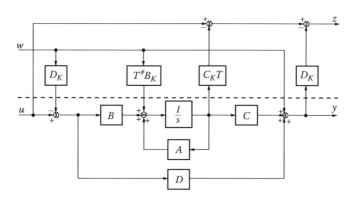

FIGURE 3.9
Block diagram of CSF $P_{CSF}(s)$ (case $n_K \leq n$).

One can notice that the CSF is a one-block problem and can be seen as a combination of well-known output estimation problem and disturbance feedforward (DF) problem [35]. Thus, if both cross transfers $[P_{zu}(s)$ and $P_{yw}(s)]$ are minimum phase (no zero in the closed right half plane), then both H_2 and H_∞ syntheses converge toward the same H_∞ performance index (γ) [34]. However, for the standard problem P_{CSF}, one can state that $\gamma = 0$ and that both syntheses are exactly equal.

3.4.2.1 Uniqueness Condition

The uniqueness condition (C4) can be proven considering the H_2-optimal controller of P_{CSF}: first of all, to vanish the direct feedthrough between exogenous inputs and controlled outputs in P_{CSF}, a simple change of variable ($u \leftarrow u - D_K y$) is performed to transform P_{CSF} into the problem $\overline{P_{CSF}}(s)$:

$$\left[\begin{array}{c|cc} A + BJ_m D_K C & T^\# B_K & BJ_m \\ \hline -C_K T & 0 & I_m \\ J_p C & I_p & DJ_m \end{array} \right] \tag{3.74}$$

and thus

$$F_l(P_{CSF}, K) = F_l(\overline{P_{CSF}}, K - D_K),$$

$$\operatorname*{argmin}_{K} \| F_l(P_{CSF}, K) \| = \operatorname*{argmin}_{K} \| F_l(\overline{P_{CSF}}, K) \| + D_K.$$

In the work of Doyle et al. [12] and Zhou et al. [35], it is demonstrated that a standard problem P has a unique H_2-optimal controller if and only if P is a regular problem, that is, in our case, if cross transfers

$$P_{zu}(s) := \left[\begin{array}{c|c} A + BJ_mD_KC & BJ_m \\ \hline -C_KT & I_m \end{array} \right] \text{ and } P_{yw}(s) := \left[\begin{array}{c|c} A + BJ_mD_KC & T^\#B_K \\ \hline J_pC & I_p \end{array} \right] \text{ have no}$$

invariant zeros on the $j\omega$-axis. It is clear that the n zeros of $P_{zu}(s)$ are the n eigenvalues of $\phi_{zu} = A + BJ_m(D_KC + C_KT)$ [ϕ_{zu} is the dynamic matrix of $P_{zu}^{-1}(s)$] and, considering Equation 3.73, belong to the set of $n + n_K$ closed-loop eigenvalues and thus are stable by assumption. Thus, $P_{zu}(s)$ has no zeros on the $j\omega$-axis.

The problem of the zeros of $P_{yw}(s)$ is more complex: the n zeros of $P_{yw}(s)$ are the n eigenvalues of $\phi_{yw} = A + BJ_m D_K C - T^\#B_KJ_p C$ [ϕ_{yw} is the dynamic matrix of $P_{yw}^{-1}(s)$]. Then, premultiplying ϕ_{yw} by $N = [T^\# \ T^\perp]$, postmultiplying by $N^{-1} = [T^T \ T^\perp]^T$, and using Equation 3.16, it becomes

$$N^{-1}\phi_{yw}N = \left[\begin{array}{cc} A_K + (B_KD - TB)J_mC_K & 0 \\ \star & T^{\perp^T}(A + BJ_mD_KC - T^\#B_KJ_pC)T^\perp \end{array} \right].$$

The n zeros of $P_{yw}(s)$ are therefore composed of

- n_K eigenvalues of $A_K + (B_KD - TB)J_mC_K$. Considering Equation 3.73, these eigenvalues belong to the set of $n + n_K$ closed-loop eigenvalues and thus are stable by assumption.
- $n - n_K$ eigenvalues of $\varphi(T^\#) = T^{\perp^T}(A + BJ_mD_KC - T^\#B_KJ_pC)T^\perp$ whose location in the complex plane is discussed in the following proposition.

Proposition 3.6

It is always possible to find a right inverse $T^\#$ of T such that all the $n - n_K$ eigenvalues of $\varphi(T^\#)$ (and thus all the n zeros of the cross transfer P_{yw}) are not on the $j\omega$-axis. ∎

Proof

The set of right-inverse matrices of T can be parameterized in the following way:

$$T^\# - T^+ + T^\perp X$$

where X is an $(n - n_K) \times n_K$ matrix of free parameters. Then,

$$\varphi(T^\#) = \varphi(X) = T^{\perp^T}(A + BJ_mD_KC)T^\perp - XB_KJ_pCT^\perp. \tag{3.75}$$

Thus, X allows the $n - n_K$ eigenvalues of φ to be assigned in the s-plane. The computation of X is in fact an eigenvalue assignment problem by a state feedback X^T on the pair $(T^{\perp^T}(A + BJ_mD_KC)^TT^\perp, (B_KJ_pCT^\perp)^T)$. ∎

Thus, Proposition 3.6 allows us to state that $P_{zu}(s)$ has no zeros on the $j\omega$-axis. Thus, $P_{CSF}(s)$ is regular, and $K_0(s)$ is the unique solution of the H_2-optimal problem P_{CSF}.

As $F_l(P_{CSF}, K_0) = 0$, all controller solutions of the H_∞-optimal problem are also solutions of the H_2-optimal problem. Thus, $K_0(s)$ is also the unique solution of the H_∞-optimal problem P_{CSF}.

3.4.2.2 Existence of CSF

Proposition 3.7

The nonexistence of a full-row rank matrix T solution of the generalized nonsymmetric Riccati equation (Equation 3.16) implies the nonexistence of a CSF for $G(s)$ and $K_0(s)$. ∎

Contrariwise Proof

Let us assume that a regular CSF exists for the strictly proper stabilizing controller $K_0(s) - D_K$ and for the stabilizable and detectable modified system $\overline{G}(s)$ (such a change of variable is not restrictive):

$$\overline{G}(s) := \left[\begin{array}{c|c} A + BJ_mD_KC & BJ_m \\ \hline J_pC & DJ_m \end{array} \right].$$

Then, it is shown in the work of Doyle et al. [12] that the unique solution K_{H_2} of the corresponding H_2-optimal problem involves a state-feedback gain K_c and a state-estimator gain K_f [according to the structure depicted in Figure 3.1 with $Q(s) = 0$]. The n-th-order state-space realization of such a controller associated with the state vector $\hat{x}$ reads

$$\overline{K}_{H_2} := \left[\begin{array}{c|c} A + BJ_mD_KC - BJ_mK_c - K_fJ_pC + K_fDJ_mK_c & K_f \\ \hline -K_c & 0 \end{array} \right]. \tag{3.76}$$

As the solution is unique, $\overline{K_{H_2}}(s) = K_0(s) - D_K$. Thus, the state-space realization (Equation 3.76) is nonminimal if $n_K < n$. Therefore, a projection matrix $S_{n_K \times n}$ (full-row rank) exists such that $x_K = S\hat{x}$ and

$$S(A + BJ_m D_K C - BJ_m K_c - K_f J_p C + K_f DJ_m K_c) = A_K S$$

$$SK_f = B_K$$

$$-K_c = C_K S.$$

Thus, S solves the following equation:

$$S(A + BJ_m D_K C) + SBJ_m C_K S - B_K J_p C - (A_K + B_K DJ_m C_K)S = 0. \qquad (3.77)$$

This equation is exactly the same as the Riccati equation (Equation 3.16) in T. Thus, if T (or S) does not exist, then the CSF for given $\bar{G}(s)$ and $K_0(s) - D_K$ [or $G(s)$ and $K_0(s)$] does not exist. ∎

Remark 3.2

This last proposition highlights that the controller $\hat{K}(s)$ provided by H_2 or H_∞ design on P_{CSF} is nonminimal. It can be shown that the $n - n_K$ nonminimal dynamics in $\hat{K}(s)$ are assigned to the eigenvalues of $\varphi(X)$ (Equation 3.75) and thus can be assigned by a suitable choice of X (see example in Section 3.4.4). ∎

3.4.3 Augmented-Order Controller Case ($n_K > n$)

In the case $n_K > n$, the CSF is directly defined from the three parameters K_c, K_f, and $Q(s)$ of the observer-based realization of $K_0(s)$ (see Figure 3.10 and the work of Alazard et al. [5] for the proof). These parameters can be computed using the procedure presented in Section 3.2.3.

3.4.4 Illustration

The results of this section are illustrated on the missile example $G(s)$ presented in Section 3.3. Let us consider the system described in Equation 3.67 and the initial controller

$$K_0(s) = \frac{-23s - 32}{s + 12} = -23\frac{s + 1.391}{s + 12} := \left[\begin{array}{c|c} -12 & 4 \\ \hline 61 & -23 \end{array}\right].$$

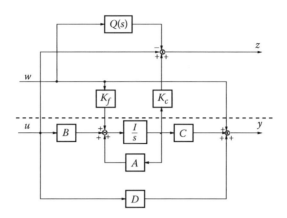

FIGURE 3.10
Block diagram of CSF P_{CSF} (case $n_K > n$).

The only real solution T of Equation 3.16 reads

$$T = [0.32787 - 0.032787].$$

Let us choose $T^\# = T^+$. Then, the CSF (Equation 3.71) reads

$$P_{CSF} := \left[\begin{array}{cc|cc} 0 & 1 & 12.079 & 0 \\ 1 & 0 & 21.792 & 1 \\ 3 & 26 & 23 & 1 \\ 1 & 0 & 1 & 0 \end{array} \right].$$

It is easy to check that the optimal H_∞ controller reads

$$K_\infty(s) = -23 \frac{(s + 1.391)(s + 2.079)}{(s + 12)(s + 2.079)}$$

The corresponding MATLAB sequence using function cor2tfg* is:

```
a=[0 1;1 0]; b=[0;1]; c=[1 0]; d=0; AK=-12; BK=4; CK=61; DK=-23;
T=cor2tfg(pck(a,b,c,d),pck(AK,BK,CK,DK)) Tm1=pinv(T);
plant=pck(a, [Tm1*BK-b*DK b], [-CK*T-DK*c;c],...
    [-DK+DK*d*DK,eye(size(d,2))-DK*d; eye(size(d,1))-d*DK d]);
K=hinfsyn(plant,1,1,0,1000,0.01); [ak,bk,ck,dk]=unpck(K);
zpk(ss(ak,bk,ck,dk))
```

Furthermore, Equation 3.75 reads

$$\varphi(X) = -2.0792 - 0.39801X \text{ and } \varphi(246.02294) = -100.$$

---

* See http://personnel.supaero.fr/alazard-daniel/demos/demo_obr.html.

Then the choice

$$T^{\#} = T^{+} + 246.0229T^{\perp} = [27.5 \quad 244.5]^{T}$$

leads to a new P_{CSF} and a new optimal H_{∞} controller:

$$K_{\infty}(s) = -23 \frac{(s+1.391)(s+100)}{(s+12)(s+100)} .$$

In both designs, K_{∞} is not minimal, and $K_{\infty} = K_{0}$.

3.4.4.1 Improving K_0 with Frequency-Domain Specification

In fact, K_0 has been designed to assign the dominant closed-loop eigenvalues to $-1 \pm i$. Indeed,

$$\text{poles of } \frac{1}{1 - K_0(s)G(s)} = \{-1+i, -1-i, -10\}.$$

The magnitude of the frequency-domain response of $K_0(s)$ is plotted in Figure 3.11 (solid line). Now, let us assume that we want the controller to have a roll-off behavior beyond 10 rad/s and must fulfill the low-pass template also depicted in Figure 3.11 (gray patch). Such a specification can be formulated to attenuate missile flexible modes, which are not taken into account in the design model $G(s)$.

This specification can be handled, in H_{∞} framework, in weighting the closed-loop transfer from a disturbance on the plant output (measurement

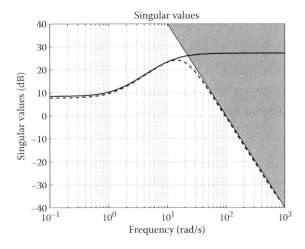

FIGURE 3.11
Frequency-domain responses (magnitude) of $K_0(s)$ (solid line), $K(s)$ (dashed line), and template (gray patch).

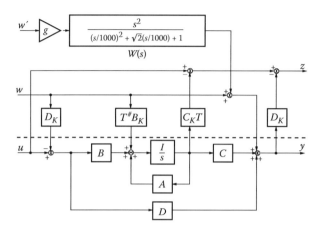

FIGURE 3.12
Augmented CSF to take into account roll-off specification (with $T^# = [27.5 \quad 244.5]^T$).

noise) to the plant input u.* It is obvious that, in the standard problem associated to the CSF (see Figure 3.9), the plant input u is directly linked to the controlled output z. Then, in order to take into this frequency-domain specification, one can augment this standard problem with a noise w' acting on the measurement y and weighted by a second-order high-pass filter (in order to get a −40 dB/dec roll-off behavior). The augmented CSF is then depicted in Figure 3.12. The high-pass filter $W(s)$ is in fact a second-order derivative filter whose poles $\left(\dfrac{-1000}{\sqrt{2}}(-1 \pm i)\right)$ are introduced for properness reasons.

The gain g is tuned by a trial-and-error procedure. The tuning $g = 0.02$ provides a fourth-order H_∞ optimal controller $K(s)$ whose frequency response is depicted in Figure 3.11 (dashed line). The template is now fulfilled, and one can check that the closed-loop dominant dynamics is assigned to the nominal values $-1 \pm i$. Indeed,

$$\text{poles of } \frac{1}{1 - K(s)G(s)} = \{-1 \pm i, -9.7947, -14.272 \pm 12.985\,i, -1424.5\}.$$

3.5 Discrete-Time Case

Techniques presented in Sections 3.2 and 3.4 in the continuous-time case are now extended to the discrete-time case (proofs are omitted for brevity).

---

* Such a transfer reads $K(I_m - KG)^{-1}$ (with positive feedback).

The discrete-time plant $G(z)$ (order n) is defined as

$$
\begin{cases}
x(k+1) & = & Ax(k) + Bu(k) \\
y(k) & = & Cx(k) + Du(k).
\end{cases}
\tag{3.78}
$$

The discrete-time controller $K_0(z)$ (order n_k) is defined as

$$
\begin{cases}
x_K(k+1) & = & A_K x_K(k) + B_K y(k) \\
u(k) & = & C_K x_K(k) + D_K u(k).
\end{cases}
\tag{3.79}
$$

Two classical implementation structures of discrete-time observer-based controllers can be used: the predictor and the estimator structures.

3.5.1 Discrete-Time Predictor Form

The predictor form is described by

$$
\begin{cases}
\hat{x}(k/k) & = & A\hat{x}(k/k-1) + Bu(k) & \text{prediction} \\
\hat{x}(k+1/k) & = & \hat{x}(k/k) + K_f(y(k) - C\hat{x}(k/k-1) - Du(k)) & \text{correction} \\
u(k+1) & = & -K_c \hat{x}(k+1/k) & \text{control.}
\end{cases}
\tag{3.80}
$$

This case is analogous to the continuous-time one. The construction procedure is therefore the same. It provides the parameters K_c^p, K_f^p, A_Q^p, B_Q^p, C_Q^p, and D_Q^p of the Youla parameterization associated with the predictor form whose state-space representation reads

$$
\begin{cases}
\hat{x}(k+1/k) & = & A\hat{x}(k/k-1) + Bu(k) + K_f^p(y(k) - C\hat{x}(k/k-1) - Du(k)) \\
x_Q(k+1) & = & A_Q^p x_Q(k) + B_Q^p(y(k) - C\hat{x}(k/k-1) - Du(k)) \\
u(k) & = & -K_c^p \hat{x}(k/k-1) + C_Q^p x_Q(k) + D_Q^p(y(k) - C\hat{x}(k/k-1) - Du(k)).
\end{cases}
\tag{3.81}
$$

3.5.2 Discrete-Time Estimator Form

The estimator structure of an observer-based controller is now described as

$$
\begin{cases}
\hat{x}(k+1/k) & = & A\hat{x}(k/k) + Bu(k) & \text{prediction} \\
\hat{x}(k+1/k+1) & = & \hat{x}(k+1/k) + K_f\left(y(k+1) - C\hat{x}(k+1/k) - Du(k+1)\right) & \text{correction} \\
u(k+1) & = & -K_c\hat{x}(k+1/k+1) & \text{control.}
\end{cases}
$$

$$(3.82)$$

In contrast to the previous case, this discrete-time estimator controller exhibits a direct feedthrough between $y(k)$ and $u(k)$, but the separation principle still holds: the closed-loop transfer function between the input reference and the innovation $y(k) - C\hat{x}(k/k-1) - Du(k)$ is zero, and the closed-loop poles can be split into the closed-loop state-feedback poles $[\mathrm{spec}(A - BK_c)]$, which are unobservable from the innovation, and the closed-loop state-estimator poles $[\mathrm{spec}(A(I - K_fC))]$, which are uncontrollable by the reference input. The Youla parameterization associated with this structure is depicted in Figure 3.13 and reads

$$
\begin{cases}
\hat{x}(k+1/k) & = & A\hat{x}(k/k-1) + Bu(k) + AK_f\left(y(k) - C\hat{x}(k/k-1) - Du(k)\right) \\
x_Q(k+1) & = & A_Q x_Q(k) + B_Q\left(y(k) - C\hat{x}(k/k-1) - Du(k)\right) \\
u(k) & = & -K_c\hat{x}(k/k-1) + C_Q x_Q(k) \\
& & +(D_Q - K_c K_f)\left(y(k) - C\hat{x}(k/k-1) - Du(k)\right).
\end{cases}
$$

$$(3.83)$$

We know from Sections 3.2 and 3.5.1 how to compute all the parameters $(K_c^p, K_f^p, A_Q^p, B_Q^p, C_Q^p,$ and $D_Q^p)$ of the predictor form and the corresponding Youla parameterization, from a given compensator (A_K, B_K, C_K, D_K) and a given plant (A, B, C, D). As a consequence, the parameters $(K_c, K_f, A_Q, B_Q, C_Q,$ and $D_Q)$ of the equivalent estimator form can be obtained by direct identification of the representations in Equations 3.81 and 3.83. This yields

$$
K_c = K_c^p, \quad K_f = A^{-1}K_f^p,
$$
$$
A_Q' = A_Q^p, \quad B_Q = B_Q^p, \quad C_Q = C_Q^p, \quad D_Q = D_Q^p + K_c^p K_f^p. \tag{3.84}
$$

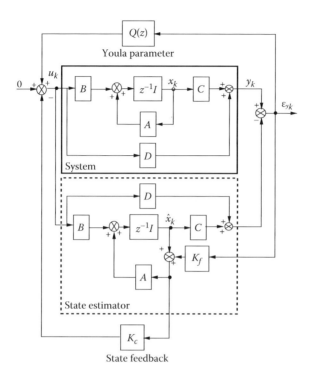

FIGURE 3.13
Discrete-time Youla parameterization using state-estimator structure [where $\hat{x}_k = \hat{x}(k/k-1)$].

3.5.3 Discrete-Time CSF

In the case of low-order controller ($n_K \leq n$), the general expression for the CSF (Equation 3.71) is valid for the discrete-time case.

In the case of the augmented-order controller ($n_K \geq n$), it is possible to define the CSF associated with an estimator form of the controller (Equation 3.83). This CSF reads

$$P_{CSF}(z) := \left[\begin{array}{cc|cc} A & 0 & AK_f & B \\ 0 & A_Q & B_Q & 0 \\ \hline K_c & -C_Q & -D_Q+K_cK_f & I_m \\ C & 0 & I_p & D \end{array}\right]. \tag{3.85}$$

The block diagram associated with this CSF is depicted in Figure 3.14.

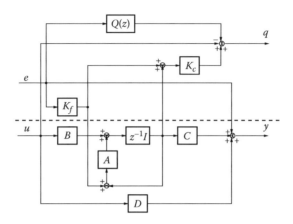

FIGURE 3.14
Discrete-time CSF.

3.6 Launch Vehicle Control Problem

CSF and gain scheduling using observer-based realization are illustrated in this section on a control design problem for a launch vehicle (representative of a strategic missile).

3.6.1 Description

This application considers the launch vehicle inner control loop.

According to Figure 3.15, the following notation is used:

- i: the launch vehicle angle of attack
- ψ: the deviation angle around axis w.r.t. the guidance attitude reference
- V_a and V_r: the absolute and relative velocity, respectively
- w: the wind velocity
- β: the thruster angle of deflection
- $\dot{z}$: the lateral drift rate

The rigid behavior is modeled by a third-order system with state vector $:x^r = [\psi \quad \dot{\psi} \quad \dot{z}]^T$. This rigid model strongly depends on two uncertain dynamic parameters, A_6 (aerodynamic efficiency) and K_1 (thruster efficiency).

From Figure 3.15 and under small angle assumption, one can derive the angle-of-attack equation:

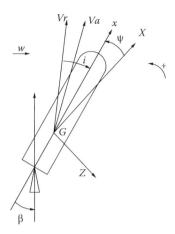

FIGURE 3.15
Launch vehicle simplified representation.

$$i = \psi + \frac{\dot{z} - w}{V}.$$
(3.86)

The discrete-time validation model considered in this section [i.e., the full-order model $G_f(z)$] is characterized by the rigid dynamics, the dynamics of thrusters (order 2), sensors (order 2), and the first five bending modes (order 10). The launch vehicle is aerodynamically unstable. Finally, the characteristics of bending modes are uncertain (four uncertain parameters per mode).

3.6.2 Objectives

The available measurements are the attitude angle (ψ) and rate ($\dot{\psi}$). The control signal is the thruster deflection angle β. Launch vehicle control objectives for the whole atmospheric flight phase are as follows:

- Performance with respect to disturbances (wind). The angle-of-attack peak, in response to the typical wind profile $w(t)$, must stay within a narrow band ($\pm i_{max}$). This wind profile is plotted in Figure 3.16 (dashed plot) and corresponds to a worst-case wind encountered during launches with a strong gust when aerodynamic pressure is maximal.

- Closed-loop stability with sufficient stability margins. This involves constraints on the rigid mode and also on the flexible modes. In fact, the first flexible mode is "naturally" phase controlled (collocation between sensors and the actuator), while other flexible modes must be gain controlled (roll-off). Thus, the peaks associated with the

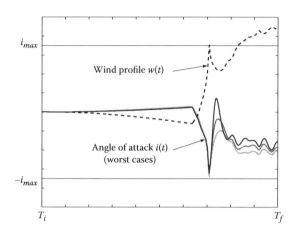

FIGURE 3.16
Angle of attack $i(t)$ (solid) obtained with $K_1(z)$ and wind profile $w(t)$ (dashed, normalized unit).

flexible modes (except for the first) on the frequency response of the loop gain $[L(s) = K(s)G(s)]$ must stay below a specified level X_{dB} for all parametric configurations (see Figure 3.21 as an example). From the synthesis point of view, the flexible modes are not taken into account in the synthesis model. However, a roll-off behavior with a cutoff frequency between the first and the second flexible modes must be specified in the synthesis.

• Delay margin must be greater than one sampling period.

All these objectives must be achieved for all configurations in the uncertain parameter space (22 uncertain parameters including aerodynamics coefficient, propulsion efficiency, and bending mode characteristics), particularly in some identified worst cases where the combination of parameter extremal values is particularly critical. In this paper, the robustness analysis is limited to these worst cases as experience has shown that they are quite representative of the robustness problem. A more complete μ-analysis is presented in the work of Imbert [16].

3.6.3 Launch Vehicle Control Design

The approach proposed to satisfy all these stationary objectives proceeds in two steps: the first one aims to satisfy time-domain specification (angle-of-attack constraint), and the second one is an H_∞ synthesis based on the CSF allowing the frequency-domain specifications (roll-off, stability margins) to be met.

The models used for the synthesis are discrete-time models including a zero-order hold.

3.6.3.1 First Synthesis: Nonconventional LQG/LTR Synthesis

3.6.3.1.1 State Feedback on Rigid Model

The standard control problem is characterized by two controlled outputs i and $\dot{z}$, two measurements ψ and $\dot{\psi}$, one control signal β, and one exogenous input w (disturbance). This standard problem reads

$$
\begin{bmatrix} \dot{x}^r \\ i \\ \dot{z} \\ \psi \\ \dot{\psi} \end{bmatrix} = \left[\begin{array}{c|cc} A & B_1 & B_2 \\ \hline C_1 & D_{11} & D_{12} \\ C_2 & D_{21} & D_{22} \end{array} \right] \begin{bmatrix} x^r \\ w \\ \beta \end{bmatrix}.
\tag{3.87}
$$

Then, the gain K_d is computed such that the discrete-time control law $\beta_k = -K_d\, x_k^r$ minimizes the following continuous-time LQ criterion:

$$
J = \int_0^\infty \left(\alpha \dot{z}^2 + i^2 + r\beta^2 \right) dt = \int_0^\infty \left(x^{r^T} Q x^r + \beta^T R \beta + 2 x^{r^T} N \beta \right) dt
\tag{3.88}
$$

with

$$
Q = C_1^T \begin{bmatrix} 1 & 0 \\ 0 & \alpha \end{bmatrix} C_1, \quad R = r, \quad N = 0_{3\times 1}.
$$

The model and the performance index are discretized by taking into account the zero-order hold at the input β_k:

$$
J_d = \sum_{k=1}^\infty (x_k^{r^T} Q_d x_k^r + \beta_k^T R_d \beta_k + 2 x_k^{r^T} N_d \beta_k)
\tag{3.89}
$$

for the discrete-time model $x_{k+1}^r = A_d x_k^r + B_{2_d} \beta_k$. The matrices $(A_d, B_{2_d}, Q_d, N_d,$ and $R_d)$ involving the matrix exponential are computed using Van Loan's formula [21].
Adopting the notation

$$
K_d = [K_\psi, K_{\dot\psi}, K_{\dot z}],
\tag{3.90}
$$

the gain K_d can be used to build a servo-loop of the measured variable ψ, that is,

$$
\beta_k = K_\psi(\psi_{ref_k} - \psi_k) - K_{\dot\psi} \dot\psi_k - K_{\dot z} \dot z_k
\tag{3.91}
$$

where ψ_{ref_k} is the input reference.

3.6.3.1.2 Augmented State with Wind Dynamics

The wind dynamics is modeled by a stable first-order filter and is then discretized with the zero-order hold method:

$$w_{k+1} = A_w w_k + \tilde{w}_k.$$

This DF model introduces a new tuning parameter A_w. The discrete-time augmented problem corresponding to the state vector $x^a = [x^y, w]^T$ then reads

$$
\begin{bmatrix} x_{k+1}^a \\ i_k \\ \dot{z}_k \\ \psi_k \\ \dot{\psi}_k \end{bmatrix}
=
\left[\begin{array}{cc|c|c}
A_d & B_{1_d} & 0 & B_{2_d} \\
0 & A_w & I & 0 \\
\hline
C_1 & D_{11} & 0 & D_{12} \\
C_2 & D_{21} & 0 & D_{22}
\end{array}\right]
\begin{bmatrix} x_k^a \\ \tilde{w}_k \\ \beta_k \end{bmatrix}
=
\left[\begin{array}{c|c|c}
A_d^a & B_{1_d}^a & B_{2_d}^a \\
\hline
C_1^a & 0 & D_{12} \\
C_2^a & 0 & D_{22}
\end{array}\right]
\begin{bmatrix} x_k^a \\ \tilde{w}_k \\ \beta_k \end{bmatrix}
$$

(3.92)

with $B_{1_d} = \int_0^{T_s} e^{A\eta} B_1 \, d\eta$.

In order to compute the new state-feedback gain K_d^a associated with the augmented state x^a, Equation 3.91 is used with ψ_{ref_k} such that the angle of attack due to disturbance w is cancelled (see Equation 3.86), that is,

$$\psi_{ref_k} = \frac{w_k - \dot{z}_k}{V}.$$

Then, the term $\dfrac{\dot{z}_k}{V}$ is ignored because it can introduce nonstabilizing couplings in the lateral motion. Finally, the gain K_d^a is obtained as

$$K_d^a = \begin{bmatrix} K_d & -\dfrac{K_\psi}{V} \end{bmatrix}.$$

(3.93)

Following this procedure, the LQ state-feedback closed-loop dynamics is stable and satisfies

$$\text{spec}(A_d^a - B_{2_d}^a K_d^a) = \text{spec}(A_d - B_{2_d} K_d) \cup \text{spec}(A_w).$$

3.6.3.1.3 *Kalman's Filter with LTR Tuning*

To compute the gain G_d^a of Kalman's filter on the augmented model (A_d^a, $B_{2_d}^a$, C_2^a, D_{22}), an LTR tuning is proposed. It is well known that stability margins of the LQ state feedback are degraded when Kalman's filter is introduced in the control loop. The LTR procedure allows these stability margins to be recovered [6]. Thus, the state noise is composed of two disturbing signals: one on the wind model input ($\tilde{w}$) and one on the control input β through a gain $\sqrt{\rho}$ (LTR effect):

$$
W = \begin{bmatrix} \rho B_2 B_2^T & 0 \\ 0 & I \end{bmatrix} \quad \text{and} \quad V = v \begin{bmatrix} 1 & 0 \\ 0 & \omega_f^2 \end{bmatrix}.
$$

W and V are the covariance matrices of continuous-time noises on the state vector (x^a) and the measurement vector ($[\psi, \dot{\psi}]^T$), respectively. Therefore, the Kalman filter tuning depends on three parameters: ρ (LTR weighting), v (measurement-to-state-noise ratio), and ω_f (in radians per second; rate-to-position-measurement-noise ratio). ω_f represents the frequency beyond which it is better to integrate the rate measurement $\dot{\psi}$ to estimate the position $\hat{\psi}$ rather than to use the measurement position ψ directly.

The covariance matrices, W_d and V_d, of discrete-time noises on the state vector and the measurement vector are also discretized using Van Loan's formulae.

This nonconventional LQG/LTR design yields a fourth-order compensator $K_1(z)$ involving the gains K_d^a and G_d^a and the augmented model (A_d^a, $B_{2_d}^a$, C_2^a, D_{22}) and is defined by Equation 3.83 without Youla parameter $Q(z)$. The results obtained so far are presented in Figures 3.16 and 3.17.

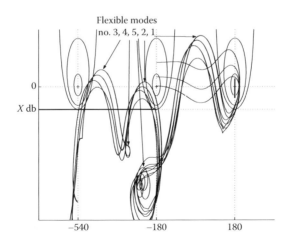

FIGURE 3.17

$K_1(z)G_j(z)$: Nichols's plots for worst cases.

In Figure 3.16, it can be observed that the performance requirements (angle of attack) are quite satisfied for all worst cases. In Figure 3.17, one can also note that the template for low-frequency stability margins is satisfied (this template is depicted in Figure 3.17 with the vertical line on the first critical point on the right-hand side), and the first flexible mode remains between two critical points for all worst cases (phase control). However, the roll-off effect is not strong enough: the template for gain margins on flexible mode numbers 2 and 3 (depicted in Figure 3.17 with the horizontal line at X dB) is not satisfied in any case. Note that Nichols plots are obtained with discrete-time transfers: it appears that flexible modes 4 and 5 are aliasing between flexible modes 1 and 3. These modes are not significant for the control design.

3.6.3.2 Second Synthesis: H_∞ Synthesis Using CSF for Frequency-Domain Specifications

In order to satisfy this last frequency-domain requirement, an H_∞ synthesis is performed on the standard problem depicted in Figure 3.18.

This standard problem can be described as follows:

- Between inputs $[e \quad u]^T$ and outputs $[q_2 \quad y]^T$, one can recognize the CSF presented in Section 3.4, which will inflect the solution toward the previous pure performance compensator (LQG/LTR design).

- The output q_1 is introduced to specify the roll-off behavior with a second-order filter $F(z)$ in order to fulfill the gain margin template on flexible mode numbers 2 and 3.

The output q_1 in fact weighs the second-order derivative of the control signal u. The frequency-domain response of $F(z)$ is depicted in Figure 3.19. This

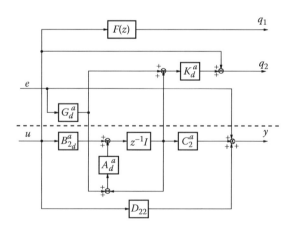

FIGURE 3.18
$P_f(z)$: setup for the final H_∞ synthesis.

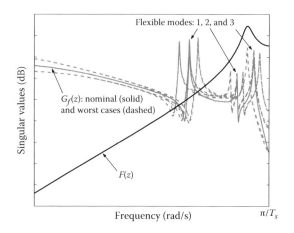

FIGURE 3.19
Singular values: $F(z)$ (black) and $G_f(z)$ (gray).

response exhibits a wide hump centered on the flexible modes 2 and 3. This hump frames peak variations of flexible modes 2 and 3 for all worst cases.

Then, the H_∞ synthesis provides a sixth-order compensator $K_2(z)$. Analysis results are displayed in Figures 3.20 and 3.21. The time-domain performance specification is still met (Figure 3.20). Figure 3.21 shows that stability margins are good enough for all worst cases, and the roll-off behavior is now quite satisfactory.

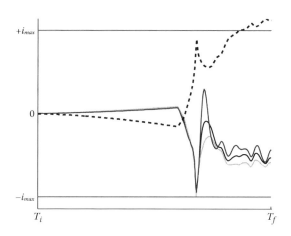

FIGURE 3.20
Angle of attack $i(t)$ (solid) obtained with $K_2(z)$ and wind profile $w(t)$ (dashed).

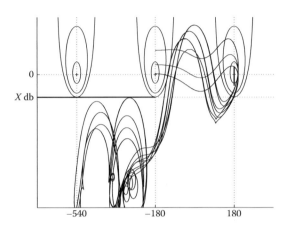

FIGURE 3.21
$K_2(z)G_f(z)$: Nichols's plots for worst cases.

3.6.4 Gain Scheduling

The previous stationary design has been applied for various instants t^i along the flight envelope. The H_∞ solver that has been used is the MATLAB macro-function dhinfric because it provides the best index γ among the various algorithms proposed in the various MATLAB toolboxes. The drawback of this algorithm lies in the fact that the solution $K_2^i(z)$ is not the central DGKF solution. Because of multiple variable changes performed to increase numerical conditioning in Riccati equations, the realization of the solution has no physical meaning. The linear interpolation of the four matrices (A_K^i, B_K^i, C_K^i, and D_K^i) provides a nonstationary compensator noted $K_2(z,t)$ with an awkward behavior as can be seen from the evolution of the singular value of $K_2(z,t)$ as a function of time t during the atmospheric flight (Figure 3.22).

This problem can be easily mastered using observer-based realizations. Thus, an observer-based realization of each compensator $K_2^i(z)$ is computed using the approach presented in Section 3.2. The model used in this realization is the transfer between u and y of the standard problem $P_f(z)$ (see Figure 3.18). The main difficulty with this approach is that the observer-based realization is not unique and depends on the way the closed-loop dynamics $F_l(P_f^i(z), K_2^i(z))$ is split between the state-feedback dynamics and the state-estimation dynamics. Considering the particular structure of the standard problem $P_f(z)$, this difficulty is easily overcome.

Let $\begin{bmatrix} A_F & B_F \\ C_F & D_F \end{bmatrix}$ be a realization of the weighting filter $F(z)$. Then, the realization of the augmented plant $P_f(z)$ depicted in Figure 3.18 is given as

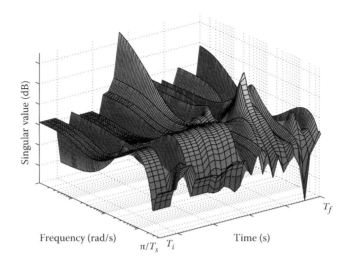

FIGURE 3.22
$K_2(z,t)$: singular value w.r.t. time.

$$P_f(z) := \left[\begin{array}{ccc|c} A_d^a & 0 & A_d^a G_d^a & B_{2_d}^a \\ 0 & A_F & 0 & B_F \\ 0 & C_F & 0 & D_F \\ K_d^a & 0 & K_d^a G_d^a & 1 \\ C_2^a & 0 & I_{2\times2} & D_{22} \end{array}\right] = \left[\begin{array}{c|cc} \mathcal{A} & \mathcal{B}_1 & \mathcal{B}_2 \\ \mathcal{C}_1 & \mathcal{D}_{11} & \mathcal{D}_{12} \\ \mathcal{C}_2 & I_{2\times2} & D_{22} \end{array}\right].$$

One can also derive

$$\mathrm{spec}(\mathcal{A} - \mathcal{B}_1\mathcal{C}_2) = \mathrm{spec}(A_d^a(I - G_d^a C_2^a)) \cup \mathrm{spec}(A_F).$$

The first term $[\mathrm{spec}(A_d^a(I - G_d^a C_2^a))]$ represents the stable dynamics of the Kalman filter previously designed. The second term $[\mathrm{spec}(A_F)]$ stands for the roll-off filter dynamics, which must be chosen to be stable. It can be shown that our standard problem $P_f(z)$ is a pure DF problem (see the work of Zhou et al. [35] and the appendix in the work of Voinot et al. [33]) and that half of the closed-loop dynamics of $F_l(P_f(z), K_2(z))$ will be assigned to $\mathrm{spec}(\mathcal{A} - \mathcal{B}_1\mathcal{C}_2)$ for any value of the final index γ. This dynamics must be assigned to the state-estimation dynamics when one wants to find the equivalent observer-based realization of the compensator $K_2(z)$ using the procedure proposed in Section 3.2. Then, the observer-based realization becomes unique.

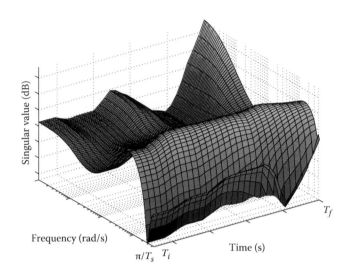

FIGURE 3.23
$K_{LQG}(z,t)$: singular value w.r.t. time.

$$\text{Let us note} \quad \left[\begin{array}{c|c} A_{LQG}^i & B_{LQG}^i \\ \hline C_{LQG}^i & D_{LQG}^i \end{array}\right] = \left[\begin{array}{c|c} \mathcal{A}^i - \mathcal{B}_2^i \mathcal{K}_c^i - \mathcal{K}_f^i \mathcal{C}_2^i + \mathcal{K}_f^i \mathcal{D}_{22}^i \mathcal{K}_c^i & \mathcal{K}_f^i \\ \hline \mathcal{K}_c^i & \mathcal{D}_Q \end{array}\right] \quad \text{the}$$

observer-based realization of each compensator $K_2^i(z)$. The linear interpolation of the four new matrices (A_{LQG}^i, B_{LQG}^i, C_{LQG}^i, and D_{LQG}^i) provides a new nonstationary compensator noted $K_{LQG}(z,t)$. The evolution of the singular value of $K_{LQG}(z,t)$ w.r.t. time t is presented in Figure 3.23. This response is significantly smoother than the one in Figure 3.22.

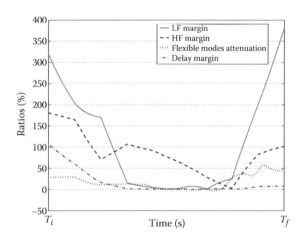

FIGURE 3.24
Obtained-margin-to-desired-margin ratios w.r.t. time.

Figure 3.24 depicts the evolution of the stability margins during the whole atmospheric flight for all worst cases. Obtained-margin-to-desired-margin ratios (in percent) are plotted w.r.t. time for the low-frequency gain margin (LF margin: above the right-hand critical point in the Nichols chart), the high-frequency gain margin (HF margin: under the right-hand critical point in the Nichols chart), the attenuation of the flexible modes below X_{dB} (corresponding to the horizontal line in the Nichols chart), and the delay margin. One can notice that the specifications are met at each instant of the flight (ratios must be positive to fulfill specifications).

3.7 Conclusions

In this chapter, a procedure to compute observer-based structures for arbitrary controllers was proposed. This technique was based upon the resolution of a generalized nonsymmetric Riccati equation. Necessary conditions were given for the solvability of this equation in terms of observability and controllability properties of the plant. The interest of observer-based realization for gain scheduling, controller switching, and state monitoring was highlighted on a very simple example. Demo files are available for readers who wish to practice.

Further work is still needed to exploit the multiplicity of choices in the distribution of the closed-loop poles between the closed-loop state-feedback poles, the closed-loop state-estimator poles, and the Youla parameter poles. This problem is particularly important to smoothly interpolate or schedule a family of state-feedback gains and state-estimator gains for practical problems requiring some gain-scheduling strategy. The usefulness of these controller structures to handle input saturation constraints also deserves investigation.

The CSF was presented here as a particular solution of the inverse optimal control problem. The CSF can be used to mix various synthesis techniques in order to satisfy a multiobjective problem. Indeed, the general idea is to design a first controller to meet some specifications, mainly performance specification. Then, the CSF is applied on this first solution to initialize a standard problem, which will be completed to handle frequency-domain or parametric robustness specifications. This heuristic approach is very interesting when the control law designer wants to

- Take into account a first controller based on *a priori* know-how and physical considerations
- Access modern optimal control framework to manage frequency-domain robustness specifications and the trade-offs between these various specifications

A multiobjective control design procedure based on the CSF was proposed in the work of Alazard et al. [5] and illustrated on an academic mixed-sensibility (two channels) control problem. Realistic applications of this approach in the field of aeronautics (flight control law design) were described by Alazard [2] and Alazard et al. [4] and, in this chapter, in the launch vehicle control design.

Nomenclature

The following notations will be used all throughout this chapter.

A^T	A transposed
A^+	Moore–Penrose pseudoinverse of matrix A
$A^\perp$	Orthonormal basis for the null space of A
spec(A)	Set of eigenvalues for a square matrix A
I_n	$n \times n$ identity matrix
$\mathbb{R}$	Set of real numbers
$\mathbb{C}$	Set of complex numbers
i	$\sqrt{-1}$
$\dot{x}$	Time derivation ($\dot{x} = \mathrm{d}x/\mathrm{d}t$)
s	Laplace variable
LQG	Linear quadratic Gaussian
$F_l(P, K)$	Lower linear fractional transformation of P and K
$\|G(s)\|_2$	H_2 norm of the stable system $G(s)$
$\|G(s)\|_\infty$	H_∞ norm of the stable system $G(s)$
$G(s) := \begin{bmatrix} A & B \\ \hline C & D \end{bmatrix}$	Shorthand for $G(s) = C(sI - A)^{-1} B + D$

References

1. D. Alazard. Extracting physical tuning potentiometers from a complex control law: Application to flexible aircraft flight control. In *AIAA, Guidance, Navigation and Control Conference*, Montreal, Canada, August 2001.
2. D. Alazard. Robust H_2 design for lateral flight control of a highly flexible aircraft. *Journal of Guidance, Control and Dynamics*, 25(3):502–509, 2002.
3. D. Alazard and P. Apkarian. Exact observer-based structures for arbitrary compensators. *International Journal of Robust and Non-Linear Control*, 9:101–118, 1999.
4. D. Alazard, C. Cumer, and F. Delmond. Improving flight control laws for load alleviation. In *Proceedings of the ASCC2006, 6th Asian Control Conference*, Bali, Indonesia, July 2006.

5. D. Alazard, O. Voinot, and P. Apkarian. A new approach to multiobjective control design from the viewpoint of the inverse optimal control problem. In *Proceedings of the SSSC'04, IFAC Symposium on Systems Structure and Control,* Oaxaca, Mexico, December 2004. Elsevier.

6. M. Athans. A tutorial on the lqg/ltr method. In *Proc. of American Control Conference,* pp. 1289–1296, Seattle, WA, 1986. IEEE. http://dspace.mit.edu/handle/1721.1/2923.

7. D. J. Bender and R. A. Fowell. Computing the estimator-controller form of a compensator. *International Journal of Control,* 41:1565–1575, 1985.

8. D. J. Bender, R. A. Fowell, and A. F. Assal. Estimating the plant state from the compensator state. *IEEE Transactions on Automatic Control,* 31:964–967, 1986.

9. S. Boyd and L. Vanderberghe. *Convex Optimization.* Cambridge University Press, Cambridge, United Kingdom, 2003.

10. C. Cumer, F. Delmond, D. Alazard, and C. Chiappa. Tuning of observer-based controllers. *Journal of Guidance, Control and Dynamics,* 27(4):607–615, 2004.

11. F. Delmond, D. Alazard, and C. Cumer. Cross standard form: A solution to improve a given controller with h_2 and h_∞ specifications. *International Journal of Control,* 79(4):279–287, 2006.

12. J. C. Doyle, K. Glover, P. D. Khargonekar, and B. A. Francis. State space solutions to standard H_2 and H_∞ control problem. *IEEE Transactions on Automatic Control,* 34(8):831–847, 1989.

13. T. Fujii. A new approach to the LQ design from the viewpoint of the inverse regulator problem. *IEEE Transactions on Automatic Control,* AC32(11):995–1004, 1987.

14. T. Fujii and P. P. Khargonekar. Inverse problems in H_∞ control theory and linear-quadratic differential games. In *Proceedings of the 27th Conference on Decision and Control,* pp. 26–31, Austin, TX, December 1988. IEEE.

15. G. H. Golub and C. F. Van Loan. *Matrix—Computations. Studies in the Mathematical Sciences,* Second edition, J. Hopkins University Press, 1989.

16. N. Imbert. Robustness analysis of a launcher attitude controller via μ-analysis. In *15th IFAC Symposium on Automatic Control in Aerospace,* Bologna, 2–7 September 2001. Elsevier.

17. P. P. Khargonekar, K. E. Lenz, and J. C. Doyle. When is a controller H_∞ optimal? *Mathematics of Control, Signals and Systems,* 1:107–122, 1988.

18. R. E. Kalman. When is a linear control optimal? *Transactions of the ASME, Journal of Basic Engineering,* 86:51–60, 1964.

19. I. Kaminer, A. M. Pascoal, P. P. Khargonekar, and E. E. Coleman. A velocity algorithm for the implementation of gain-scheduled controllers. *Automatica,* 31(8):1185–1191, 1995.

20. D. A. Lawrence and W. J. Rugh. Gain scheduling dynamic linear controllers for a nonlinear plant. *Automatica,* 31(3):381–290, 1995.

21. C. F. Van Loan. Computing integrals involving the matrix exponential. *IEEE Transactions on Automatic Control,* AC-23(3):395–404, 1978.

22. D. G. Luenberger. An introduction to observers. *IEEE Transactions on Automatic Control,* AC-16:596–602, 1871.

23. B. P. Molinari. The stable regulator problem and its inverse. *IEEE Transactions on Automatic Control,* AC-18(5):454–459, 1973.

24. N. Narasimhamurthi and F. F. Wu. On the Riccati equation arising from the study of singularly perturbed systems. *IEEE Joint Automatic Control Conference,* San Francisco, 1977.

25. P. C. Pellanda, P. Apkarian, and D. Alazard. Gain-scheduling through continuation of observer-based realizations - applications to H_∞ and μ controllers. In *Proc. 39th IEEE Conf. on Decision and Control*, pp. 2787–2792, Sydney, 12–15 December 2000. IEEE.

26. N. Sebe. A characterization of solutions to the inverse H_∞ optimal control problem. In *Proceedings of the 40th Conference on Decision and Control*, pp. 273–278, Orlando, FL, December 2001. IEEE.

27. T. Shimomura and T. Fujii. Strictly positive real H_2 controller synthesis from the viewpoint of the inverse problem. In *Proceedings of the 36th Conference on Decision and Control*, pp. 1014–1019, San Diego, CA, December 1997. IEEE.

28. D. J. Stilwell and W. Rugh. Interpolation of observer state feedback controllers for gain scheduling. *IEEE Transactions on Automatic Control*, 44(6):1225–1229, 1999.

29. K. Sugimoto. Partial pole-placement by LQ regulator: An inverse problem approach. *IEEE Transactions on Automatic Control*, 43(5):706–708, 1998.

30. K. Sugimoto and Y. Yamamoto. New solution to the inverse regulator problem by the polynomial matrix method. *International Journal of Control*, 45-5:1627–1640, 1987.

31. S. Tarbouriech and G. Garcia. Control of uncertain systems with bounded inputs. *Lectures Notes in Control and Information Sciences*, Vol. 227, Springer Verlag, 1997.

32. O. Voinot, D. Alazard, A. Piquereau, and A. Biard. A robust multi-objective synthesis applied to launcher attitude control. In *15th IFAC Symposium on Automatic Control in Aerospace*, Bologna, 2–7 September 2001. Elsevier.

33. O. Voinot, D. Alazard, P. Apkarian, S. Mauffrey, and B. Clément. Launcher attitude control: Discrete-time robust design and gain-scheduling. *IFAC Control Engineering Practice*, 11:1243–1252, 2003.

34. K. Zhou. Comparison between H_2 and H_∞ controllers. *IEEE Transaction on Automatic Control*, 37(8):1261–1265, 1992.

35. K. Zhou, J. C. Doyle, and K. Glover. *Robust and Optimal Control*. Prentice Hall, Upper Saddle River, NJ, 1996.

4

Adaptive Neural Network–Based Autopilot Design

Karthikeyan Rajagopal and S. N. Balakrishnan

CONTENTS

4.1 Introduction

Traditionally, a missile autopilot is designed using linear control approaches. The plant model is linearized around a given trim point and then used for the resultant controller design [4, 15]. The initial design is then carried out by assuming that no coupling exists between the roll, pitch, and yaw axes. Thus, the controllers are designed individually for each axis. The pitch, yaw, and roll channel control system parameters are selected based on relative stability margins (Bode methods) and missile response time requirements. The guidance loop provides the required pitch and roll axis acceleration commands, whereas the yaw axis control loop operates in a regulator mode. In order to obtain consistent performance throughout the operational envelope of the missile, gain scheduling is used. Generally, control system parameters are scheduled with respect to slowly varying parameters like Mach number, dynamic pressure, altitude, and weight. Interpolation techniques are used to obtain the controller gains at intermediate points. The final control system parameters are chosen based on a six-degrees-of-freedom nonlinear simulation analysis.

System performance obtained from the classical design can further be enhanced by retuning the controller parameters based on a multivariable analysis [19]. An alternative approach is to use modern control theory based on state-space formulations that can explicitly take into account the coupling effects in designing the controller and can thus handle the multivariable nature of the problem. One such methodology can be found in the work of Williams [18], where the controller design for the pitch/yaw channels is carried out using linear quadratic Gaussian (LQG) theory by considering the roll rate as an exogenous input. The controller gains for the pitch/yaw channel are scheduled with respect to both dynamic pressure and roll rate. Since all the states may not be available for control computation, LQG theory with loop transfer recovery is used for estimating the unmeasured states. The controller for the roll channel is a typical Single Input Single Output controller designed using pole placement techniques. Further, to improve the tracking performance methodologies based on integral control, techniques like robust servo linear quadratic regulator can be considered [20]. All the methodologies discussed above use a linear model for the controller design and rely on gain scheduling to make the controller work globally. Another type of gain scheduling approach that does not require linearization of the model at the trim point is called the linear parameter varying (LPV) approach [16]. In this approach, the missile dynamics are converted to a quasi-LPV form via a state transformation. The varying parameter is generally an endogenous variable like angle of attack; hence, the transformation is quasilinear. Once the quasi-LPV model is available, nonconservative solutions can be obtained using μ synthesis. Although a large amount of literature is available on the application of the LPV methodology to missile systems, its practical application is restricted because of the difficulty in using complex missile models.

Nonlinear control methods provide an effective alternative for linear design approaches since they allow us to analytically design full envelope controllers without resorting to gain scheduling. Moreover, they can explicitly take care of aerodynamic and kinematic nonlinearities. One effective technique that can provide a solution for nonlinear regulator problems is the state-dependent Riccati equation (SDRE) method [5]. In this approach, the equations of motion of the missile are converted into a linear-like structure, and linear optimal control methodologies are used for the synthesis of nonlinear control. This technique can be applied to a broad class of systems where the cost function is quadratic. Cloutier and Stansbery [6] used SDRE method for nonlinear Missile autopilot design. But the SDRE method requires solving an algebraic Riccati equation at each sample time, which makes it computation intensive. A technique that is similar to the SDRE method but does not require online computation is the $\theta - D$ technique [21]. In this approach, an intermediate variable θ is used, which aids in finding an approximate solution for the Hamilton–Jacobi–Bellman equation. The resulting nonlinear controller is a closed-form solution and so eliminates the need for solving online any algebraic equation.

One of the simplest nonlinear design methodologies is based on dynamic inversion (DI). In this approach, the nonlinearities are directly cancelled using a negative feedback, and they are replaced with the desired dynamics. The desired dynamics is usually generated using a linear reference model, and typically only one such model is required for the full operation envelope of the missile. The nonlinear DI technique is easy to implement and thus offers many practical advantages over other nonlinear design techniques, but successful implementation involves addressing some critical issues. One of the most critical issues is the fact that the performance of the controller heavily depends on the precise prior knowledge of the plant. The controller has to deal with uncertainties that arise because of modeling errors, as well as unmodeled dynamics because of flexible structural modes, sensors, actuators, and any in flight hardware failure. In this regard, a significant amount of research is being carried out toward evolving robust DI techniques. The most common approach for providing robustness to the DI technique is to account for uncertainties in the outer loop control design. Adams [1] and Adams and Banda [2] first proposed using the DI technique for the inner loop control and using μ synthesis for the outer loop control. McFarland and D'Souza [10] also proposed using the combination of DI and μ synthesis in missile autopilot design. But it is always problematic to quantitatively state how robust a controller design is in a nonlinear setting. Another practical problem with the DI technique is the assumption that full-state feedback is always available; this assumption is very restrictive in the case of agile missiles because of weight considerations. Although there are many theoretical extensions to the DI technique that assume only output feedback is available, their practical usage must still be evaluated. The DI technique cannot be directly used for nonminimum phase systems, and all tail-controlled missiles are nonminimum phase when the controlled variables of the missile are its pitch and yaw accelerations. By controlling the attitude of the missile directly instead of the body accelerations, the above problem can be overcome. But the guidance law has to be modified to derive relationships between the attitude of the missile and the commanded accelerations.

Real-time compensation of the control signal for modeling inaccuracies or even unexpected in-flight hardware failures is possible with various adaptive control algorithms. Adaptive control techniques can generally be classified into direct and indirect approaches. In the indirect approach, unknown plant parameters are explicitly estimated, and the control is designed assuming the estimates as the true plant parameters. In the direct approach, the control parameters are directly tuned to cancel the unknown plant nonlinearity. In recent years, research on the model reference direct adaptive control techniques (MRAC) has gained significant attention. The objective of the MRAC is to make the plant mimic the transient response of a reference model even if unknown nonlinearities are present. The difference between measured states and the reference state trajectory is used in estimating the uncertainty. Recently, neural networks have emerged as a major tool for explicitly estimating the uncertainties in real time. Narendra and Parthasarathy [11] successfully used neural networks for identification and control of nonlinear dynamical systems. For robot control, Lewis [8] used online neural networks for approximating the uncertainties in real time and provided Lyapunov stability analysis that showed the boundedness of the tracking error and weights. Calise [3] and Kim and Calise [7] applied neural network–based direct adaptive control architecture for flight control. Sharma [17] and Wise [20] used neural network–based adaptive controllers to handle modeling inaccuracies in guided munitions and missile dynamics, respectively.

McFarland and Calise [9] combined the DI technique and neural network–based adaptive controller for the control of an agile anti-air missile. The authors used an inner/outer loop architecture with the outer loop controlling the attitude of the missile aided by an inner loop that controls the missile body rates. The study uses single hidden layer neural networks in the inner loop for taking care of the uncertainties, and approximate DI is used for generating the fin deflection commands. A Lyapunov theory–based simple weight update rule is used for updating the weights in real time.

One of the major concerns with adaptive control techniques is their robustness to unmodeled dynamics. When dealing with agile missiles, fast adaptation in estimating the uncertainties is needed to ensure the stability of the system and good performance. Achievement of such an objective calls for large adaptation rates. Large adaptive gains, however, may induce high-frequency oscillations in the adaptive control signal, which, in turn, can excite the unmodeled dynamics of the missile leading to instability. Various adaptive control techniques have been proposed for solving the above problem. Nguyen et al. [12,13] proposed an optimal adaptive control law modification that minimizes the tracking error and also improves robustness allowing for large adaptation rate. Rajagopal et al. [14] proposed the use of a general observer structure instead of a reference model for uncertainty estimation. This methodology separates the design of the nominal closed-loop dynamics from the estimation error dynamics. As a result, the estimation error dynamics can be made much faster than the nominal system dynamics, which allows for large adaptive gains. In this chapter, we will see the application of direct model reference adaptive control technique proposed in [14] for the control of a generic air–air missile. The theory behind the derivation of the weight update rule is presented. Simulation results and the performance of the methodology at different altitude conditions are presented to demonstrate the effectiveness of the adaptive control methodology.

4.2 Nonlinear Air–Air Missile Model

The nonlinear equations of motion of a generic air–air missile model, used for simulations in this chapter, is given by

$$
\dot{V} = -\frac{\bar{q}S}{m} C_A \cos\alpha \cos\beta + \frac{\bar{q}S}{m} C_{Y_o} \sin\beta - \frac{\bar{q}S}{m} C_{N_o} \sin\alpha \cos\beta - g\sin\gamma + \frac{\cos\alpha \cos\beta}{m} T
$$
$$
+ \frac{\bar{q}S}{m}[(C_{Y_{\delta p}} \sin\beta - C_{N_{\delta p}} \sin\alpha \cos\beta)\delta p - (C_{Y_{\delta p}} \sin\beta - C_{N_{\delta p}} \sin\alpha \cos\beta)\delta q
$$
$$
- (C_{Y_{\delta r}} \sin\beta - C_{N_{\delta r}} \sin\alpha \cos\beta)\delta r]
$$

$$(4.1)$$

$$
\dot{\alpha} = q - p\cos\alpha \tan\beta - r\tan\beta \sin\alpha + \frac{g}{V\cos\beta} \cos\gamma \cos\mu - \frac{\bar{q}S}{MV\cos\beta} C_{N_o} \cos\alpha
$$
$$
+ \frac{\bar{q}S}{MV\cos\beta} C_A \sin\alpha - \frac{\sin\alpha}{MV\cos\beta} T - \frac{\bar{q}S\cos\alpha}{MV\cos\beta} (C_{N_{\delta p}} \delta p - C_{N_{\delta q}} \delta q - C_{N_{\delta r}} \delta r)
$$

$$(4.2)$$

$$
\dot{\beta} = p\sin\alpha - r\cos\alpha + \frac{g}{V} \cos\gamma \sin\mu + \frac{\bar{q}S}{MV} C_{N_o} \sin\alpha \sin\beta + \frac{\bar{q}S}{MV} C_A \cos\alpha \sin\beta
$$
$$
+ \frac{\bar{q}S}{MV} C_{Y_o} \cos\beta - \frac{\cos\alpha \sin\beta}{MV} T + \frac{\bar{q}S}{MV}[(C_{Y_{\delta p}} \cos\beta + C_{N_{\delta p}} \sin\alpha \cos\beta)\delta p
$$
$$
- (C_{Y_{\delta q}} \cos\beta - C_{N_{\delta q}} \sin\alpha \sin\beta)\delta q - (C_{Y_{\delta r}} \cos\beta - C_{N_{\delta r}} \sin\alpha \sin\beta)\delta r]
$$

$$(4.3)$$

$$
\dot{\mu} = p\frac{\cos\alpha}{\cos\beta} + r\frac{\sin\alpha}{\cos\beta} + \frac{T}{mV}[\sin\alpha \tan\gamma \sin\mu - \cos\alpha \sin\beta \tan\gamma \cos\mu + \sin\alpha \tan\beta]
$$
$$
+ \frac{\bar{q}S}{mV}[\ C_A \sin\alpha(\tan\gamma \sin\mu + \tan\beta) + C_A \cos\alpha \cos\mu \sin\beta \tan\gamma
$$
$$
+ C_{Y_o} \cos\beta \tan\gamma \cos\mu + C_{N_o} \sin\alpha \sin\beta \tan\gamma \cos\mu]
$$
$$
+ \frac{\bar{q}S}{MV}(C_{N_{\delta p}} \delta p - C_{N_{\delta q}} \delta q - C_{N_{\delta r}} \delta r)(\cos\alpha(\tan\gamma \sin\mu + \tan\beta)
$$
$$
+ \sin\alpha \sin\beta \tan\gamma \cos\mu) + \frac{\bar{q}S}{MV}(C_{N_{\delta p}} \delta p - C_{N_{\delta q}} \delta q - C_{N_{\delta r}} \delta r)\cos\beta \tan\gamma \cos\mu
$$
$$
- \frac{g}{V} \cos\gamma \cos\mu \tan\beta
$$

$$(4.4)$$

$$\dot{p} = \frac{\bar{q}Sd^2}{2I_xV}(C_{l_p}p\cos\alpha - C_{n_r}r\sin\alpha) + \frac{\bar{q}Sd}{I_x}(C_{l_o}\cos\alpha - C_{n_o}\sin\alpha)$$

$$+ \frac{\bar{q}Sd}{I_x}[(C_{l_{\delta p}}\cos\alpha - C_{n_{\delta p}}\sin\alpha)\delta p - (C_{l_{\delta q}}\cos\alpha - C_{n_{\delta q}}\sin\alpha)\delta q \qquad (4.5)$$

$$- (C_{l_{\delta r}}\cos\alpha - C_{n_{\delta r}}\sin\alpha)\delta r] + \frac{I_y - I_z}{I_x}qr$$

$$\dot{q} = \frac{I_z - I_x}{I_y}pr + \frac{\bar{q}Sd^2}{2I_yV}C_{m_q}q + \frac{\bar{q}Sd}{I_y}(C_{m_o} + C_{m_{\delta p}}\delta p - C_{m_{\delta q}}\delta q - C_{m_{\delta r}}\delta r) \qquad (4.6)$$

$$\dot{r} = \frac{I_x - I_y}{I_z}pq + \frac{\bar{q}Sd^2}{2I_zV}(C_{l_p}p\sin\alpha - C_{n_r}r\cos\alpha) + \frac{\bar{q}Sd}{I_z}(C_{l_o}\sin\alpha + C_{n_o}\cos\alpha)$$

$$+ \frac{\bar{q}Sd}{I_z}[(C_{l_{\delta p}}\sin\alpha + C_{n_{\delta p}}\cos\alpha)\delta p - (C_{l_{\delta q}}\sin\alpha + C_{n_{\delta q}}\cos\alpha)\delta q \qquad (4.7)$$

$$- (C_{l_{\delta r}}\sin\alpha + C_{n_{\delta r}}\cos\alpha)\delta r]$$

Definitions of parameters used in Equations 4.1 through 4.7 are tabulated below:

Parameter	Definition	Parameter	Definition
α	Angle of attack	h	Missile flight altitude
β	Sideslip angle	I_x, I_y, I_z	Moments of inertia about body frame
μ	Bank angle about the velocity vector	$\delta p, \delta q, \delta r$	Aileron, elevator, and rudder fin deflections
γ	Flight path angle	C_A	Axial force coefficient
p, q, r	Body-frame roll, pitch, and yaw	C_Y	Side force coefficient
V	Missile speed	$C_z(-C_N)$	Normal force coefficient
M	Mach number	C_{N_o}, C_{Y_o}	Normal and side force coefficients with zero fin
m	Missile mass	$C_{N_{\delta p}}, C_{N_{\delta q}}, C_{N_{\delta r}}$	Side force coefficients with respect to aileron, elevator, and rudder fin, respectively
$\bar{q}$	Dynamic pressure	C_l	Roll moment coefficient
S	Missile cross-sectional area	C_m	Pitch moment coefficient
d	Missile diameter	C_n	Yaw moment coefficient
g	Gravity	C_{l_o}	Roll moment coefficient with zero fin deflections

C_{l_p}	Roll moment coefficient with respect to roll rate	C_{n_o}	Yaw moment coefficient with zero fin deflections
$C_{l_{\delta p}}, C_{l_{\delta q}}, C_{l_{\delta r}}$	Roll moment coefficient with respect to aileron, elevator, and rudder fin, respectively	C_{n_q}	Roll moment coefficient with respect to roll rate
C_{m_o}	Pitch moment coefficient with zero fin deflections	$C_{n_{\delta p}}, C_{n_{\delta q}}, C_{n_{\delta r}}$	Roll moment coefficient with respect to aileron, elevator, and rudder fin, respectively
C_{m_q}	Pitch moment coefficient with respect to pitch rate	a_y^I, a_z^I	Acceleration along the inertial y and z axes
$C_{m_{\delta p}}, C_{m_{\delta q}}, C_{m_{\delta r}}$	Pitch moment coefficient with respect to aileron, elevator, and rudder fin, respectively	T	Thrust

4.3 Missile Autopilot Control Loop Design

The acceleration commands from the missile guidance law are converted into corresponding attitude commands, that is, angle of attack, sideslip, and bank angle commands, using the bank-to-turn/skid-to-turn (BTT/STT) logic described in this section. The control objective is to make the missile follow a reference trajectory given by a second-order reference model of the following form in frequency domain:

$$x_r(s) = \frac{\omega_n^2}{s^2 + 2\zeta\omega_n s + \omega_n^2} x_c(s) \qquad (4.8)$$

where $x_r = [\alpha_r \ \beta_r \ \mu_r]^T$ represents the reference trajectory for the attitude angles, $x_c = [\alpha_c \ \beta_c \ \mu_c]^T$ represents the commanded trajectory for the attitude angles obtained using BTT/STT logic, and ω_n, ζ are second-order model parameters that can be chosen appropriately to get the desired response time and damping characteristics.

Now to compute the required fin deflections so that the plant can closely follow the reference angles, the missile autopilot design has a two-loop structure.

1) The outer loop converts the attitude commands into corresponding body rate commands for the inner loop.
2) The inner loop converts the body rate commands into fin deflection commands for the actuator control loop.

4.3.1 Outer Loop Control Design

For the nominal controller design of the outer loop, the nonlinear equations of motion (Equations 4.2 through 4.4) are linearized about the states of interest, that is, angle of attack, sideslip angle, and bank angle. Thus, the linearized equations of motion for the outer loop are given by

$$\dot{x}_{ol} = A_{ol}x_{ol} + B_{ol}u_{ol} + f_{ol}(x, u) \tag{4.9}$$

where $A_{ol} \in R^{3\times3}$, $B_{ol} \in R^{3\times3}$, $x_{ol} = [\alpha \ \beta \ \mu]^T$, $u_{ol} = [p \ q \ r]^T$, and $f_{ol}(x, u) \in R^{3\times1}$ represent the effects of higher-order terms neglected during the linearization process and any other nonlinearity that is unknown during the process of modeling. The parameters x and u represent the typical missile states and control inputs. The linearized state and control matrices for the outer loop evaluated at x_o with $x_{ol} = [0 \ 0 \ 0]^T$ are given by

$$A_{ol} = \begin{bmatrix} \dfrac{\bar{q}S}{mV}C_A - \dfrac{T}{mV} & 0 & 0 \\ 0 & \dfrac{\bar{q}S}{mV}C_A & \dfrac{g}{V} \\ 0 & -\dfrac{g}{V} & 0 \end{bmatrix}_{at \ x_0}, B_{ol} = \begin{bmatrix} 0 & 1 & 0 \\ 0 & 0 & -1 \\ 1 & 0 & 0 \end{bmatrix}_{at \ x_0}. \tag{4.10}$$

The control signal u_{ol} is computed by using the DI technique. Let $e_{ol} = x_r - x_{ol}$ so $\dot{e}_{ol} = \dot{x}_r - \dot{x}_{ol}$, and the nominal controller for the case without any nonlinearity is given by

$$u_{ol_{n\,om}} = B_{ol}^{-1}(\dot{e}_{ol} - A_{ol}x_{ol}). \tag{4.11}$$

The unknown nonlinearities are estimated using online neural networks through a "modified state observer" (MSO). Let us assume that it is possible to approximate the nonlinearity within a given approximation error $\varepsilon \in R^{3\times1}$. In an ideal case

$$f_{ol}(x, u) = W^T\phi(x_{ol}) + \varepsilon \tag{4.12}$$

where $W \in R^{8\times3}$ is the ideal neural network weight, and $\phi(x_{ol}) \in R^{8\times1}$ is the basis function of the neural network. For our simulation, the basis function is taken as the Kronecker product of states of interest x_{ol}. So

$$\phi(x_{ol}) = [1 \ \alpha \ \beta \ \mu \ \alpha\beta \ \beta\mu \ \alpha\mu \ \alpha\beta\mu]^T. \tag{4.13}$$

Since we will not know the ideal neural network weights, we have to esti-mate the weights online. So the estimated uncertainty is given by

$$\hat{f}_{ol}(x, u) = \hat{W}^T \phi(x_{ol}) \tag{4.14}$$

where $\hat{W}$ is the estimated neural network weight. To estimate the neural net-work weights, the MSO of the following form is constructed:

$$\dot{\hat{x}}_{ol} = A_{ol} x_{ol} + B_{ol} u_{ol} + \hat{W}^T \phi(x_{ol}) + K_2(x_{ol} - \hat{x}_{ol}). \tag{4.15}$$

The estimation error $e = x_{ol} - \hat{x}_{ol}$ is used in updating the estimated neural network weights. The time derivative of estimation error is given by

$$\dot{e} = -K_2 e + \tilde{W}^T \phi(x_{ol}) + \varepsilon \tag{4.16}$$

where $\tilde{W} = W - \hat{W}$.

To ensure that the estimation error and neural network weights remain bounded, the weight update law is derived by considering a Lyapunov-like function:

$$V = e^T P e + tr(\tilde{W}^T \Gamma^{-1} \tilde{W}) \tag{4.17}$$

where tr is the trace operator, P is a positive-definite matrix, and Γ is the adaptive gain matrix. The weight update rule considered for updating the estimated neural network weights is given by

$$\dot{\hat{W}} = \Gamma(\phi(x_{ol}) e^T P - \sigma \hat{W}). \tag{4.18}$$

In Equation 4.18, σ is the robustness factor. Now the time derivative of the Lyapunov function is given by

$$\dot{V} = \dot{e}^T P e + e^T P \dot{e} + tr(\dot{\tilde{W}}^T \Gamma^{-1} \tilde{W}) + tr(\tilde{W}^T \Gamma^{-1} \dot{\tilde{W}}). \tag{4.19}$$

By using the trace identity $tr(x^T y) = tr(yx^T)$ and substituting estimation error dynamics Equation 4.16 in Equation 4.19, we get

$$\dot{V} = e^T(-K_2^T P - PK_2)e + 2e^T P(\tilde{W}^T \phi(x_{ol}) + \varepsilon) + 2tr(\tilde{W}^T \Gamma^{-1} \dot{\tilde{W}}). \tag{4.20}$$

Since $\tilde{W} = W - \hat{W}$ and $\dot{\tilde{W}} = -\dot{\hat{W}}$, we substitute the weight update rule (Equation 4.18) in the above equation to yield

$$\dot{V} = e^T(-K_2^T P - PK_2)e + 2e^T P(\tilde{W}^T \phi(x_{ol}) + \varepsilon) - 2tr(\tilde{W}^T(\phi(x_{ol})e^T P - \sigma\hat{W})) \tag{4.21}$$

$$\dot{V} = -e^T Qe + 2e^T P(\tilde{W}^T \phi(x_{ol}) + \varepsilon) - 2tr(\tilde{W}^T \phi(x_{ol})e^T P) + 2tr(\sigma\tilde{W}^T \hat{W}) \tag{4.22}$$

$$\dot{V} = -e^T Qe + 2e^T P\varepsilon + 2\sigma tr(\tilde{W}^T \hat{W}) \tag{4.23}$$

where $-Q = -(K_2^T P + PK_2)$ is a solution of a Lyapunov equation. Now to simplify Equation 4.23 using the properties of the trace operator

$$tr(\tilde{W}^T \hat{W}) \le \|W\|_F \|\hat{W}\|_F - \|\hat{W}\|_F^2 \tag{4.24}$$

$$\le W_{max} \|\hat{W}\|_F - \|\hat{W}\|_F^2.$$

In Equation 4.24, $\|W\|_F$ is the Frobenius norm, and W_{max} is the upper bound on the ideal weight, that is, $\|W\|_F \le W_{max}$. By using the norms of the terms on the right-hand side, an inequality can be obtained as

$$\dot{V} \le -\lambda_{min}(Q)\|e\|^2 + 2\|e\|\lambda_{max}(P)\varepsilon_n + 2\sigma(W_{max}\|\hat{W}\|_F - \|\hat{W}\|_F^2). \tag{4.25}$$

Here $\lambda_{min}(Q)$ is the minimum eigenvalue of the Q matrix, $\lambda_{max}(P)$ is the maximum eigenvalue of P, and ε_n is the upper bound on the approximation error, $\|\varepsilon\| \le \varepsilon_n$.

Now to show that the estimated neural network weights and the estimation errors remain bounded, it is necessary to show that when the trajectory of e and $\hat{W}$ starts outside a compact set, the time derivative of the Lyapunov function is negative so that the trajectory reaches a compact set and remains within this compact set for all time. This is shown by deriving upper bounds on the estimation error and the estimated neural network weights.

4.3.1.1 Upper Bound on Estimation Error

To estimate an upper bound on the estimation error, square on $\|\hat{W}\|$ is completed as

$$\dot{V} \le -\lambda_{min}(Q)\|e\|^2 + 2\|e\|\lambda_{max}(P)\varepsilon_n - 2\sigma\left(\|\hat{W}\| - \frac{W_{max}}{2}\right)^2 + 2\sigma\left(\frac{W_{max}}{2}\right)^2. \tag{4.26}$$

Hence, for $\dot{V}$ to be less than or equal to zero

$$\lambda_{\min}(Q)\|e\|^2 - 2\|e\|\lambda_{\max}(P)\varepsilon_n - 2\sigma\left(\frac{W_{\max}}{2}\right)^2 \geq 0 \tag{4.27}$$

which gives a bound on the estimation error as shown below:

$$\|e\| \geq \beta \tag{4.28}$$

where

$$\beta = \frac{\lambda_{\max}(P)\varepsilon_n + \sqrt{(\lambda_{\max}(P)\varepsilon_n)^2 + 2\sigma\lambda_{\min}(Q)\left(\frac{W_{\max}}{2}\right)^2}}{\lambda_{\min}(Q)}. \tag{4.29}$$

4.3.1.2 Upper Bound of Neural Network Weights

In order to estimate an upper bound on $\|\hat{W}\|$, the square on $\|e\|$ needs to be completed in

$$\dot{V} \leq -\lambda_{\min}(Q)\|e\|^2 + 2\|e\|\lambda_{\max}(P)\varepsilon_n + 2\sigma\left(W_{\max}\|\hat{W}\|_F - \|\hat{W}\|_F^2\right) \tag{4.30}$$

which results in the following inequality:

$$\dot{V} \leq -\lambda_{\min}(Q)\left(\|e\| - \frac{\lambda_{\max}(P)}{\lambda_{\min}(Q)}\varepsilon_n\right)^2 + \frac{(\lambda_{\max}(P))^2}{\lambda_{\min}(Q)}\varepsilon_n^2 + 2\sigma\left(W_{\max}\|\hat{W}\|_F - \|\hat{W}\|_F^2\right). \tag{4.31}$$

Hence, for $\dot{V}$ to be less than or equal to zero

$$2\sigma\left(W_{\max}\|\hat{W}\|_F - \|\hat{W}\|_F^2\right) + \frac{(\lambda_{\max}(P))^2}{\lambda_{\min}(Q)}\varepsilon_n^2 \leq 0 \tag{4.32}$$

$$\text{or } 2\sigma\|\hat{W}\|_F^2 - 2\sigma W_{\max}\|\hat{W}\|_F - \frac{(\lambda_{\max}(P))^2}{\lambda_{\min}(O)}\varepsilon_n^2 \geq 0. \tag{4.33}$$

This gives a bound on the neural network weights above which $\dot{V}$ will be negative:

$$\|\hat{W}\|_F \geq \frac{2\sigma W_{max} + \sqrt{4\sigma^2 W_{max}^2 + 8\sigma \frac{(\lambda_{max}(P))^2}{\lambda_{min}(Q)}\varepsilon_n^2}}{4\sigma} \qquad (4.34)$$

$$\|\hat{W}\|_F \geq \zeta/2 \qquad (4.35)$$

where

$$\zeta = W_{max} + \sqrt{W_{max}^2 + 2\frac{(\lambda_{max}(P))^2}{\sigma\lambda_{min}(Q)}\varepsilon_n^2}. \qquad (4.36)$$

Hence, from Equations 4.28 and 4.35, it can be concluded that the estimation error and neural network weights are uniformly bounded.

4.3.1.3 Outer Loop Controller Design

The outer loop controller is designed by taking into consideration the estimated uncertainties:

$$u_{ol} = B_{ol}^{-1}(\dot{e}_{ol} - A_{ol}x_{ol} - W^T\phi(x_{ol})). \qquad (4.37)$$

Note that the controller (Equation 4.37) designed using DI techniques allows us to totally eliminate gain scheduling and still provides consistent handling qualities all through the flight envelope.

4.3.2 Inner Loop Control Design

The demanded pitch, yaw, and roll rates from the outer loop are realized by using only a standard state feedback-based controller in the inner loop. The reason is the reference error e_{ol}-based outer loop DI adaptive controller will automatically compensate for uncertainties in the inner loop. The states of interest in the inner loop are $x_{il} = [p \; q \; r]^T$, and the control inputs are $u_{il} = [\delta p \; \delta q \; \delta r]^T$. For the controller design, first the nonlinear equations of motion (Equations 4.5 through 4.7) are linearized about the states of interest x_{il}. Then, the nominal controller is designed using the LQR theory. The linearized equations of motion for the inner loop are given by

$$\dot{x}_{il} = A_{il}x_{il} + B_{il}u_{il} + f_{il}(x,u) \qquad (4.38)$$

where $A_{il} \in R^{3\times3}$, $B_{il} \in R^{3\times3}$, and $f_{il}(x,u) \in R^{3\times1}$ represent the effect of higher-order terms neglected during the linearization process and any other nonlinearity that is unknown during the process of modeling. The linearized state and control matrices for the inner loop evaluated at x_o with $x_{il} = [0\ 0\ 0]^T$ and $x_{ol} = [0\ 0\ 0]^T$ are given by

$$
A_{il} = \begin{bmatrix} \dfrac{\bar{q}Sd^2}{2I_xV}C_{l_p} & 0 & 0 \\[2ex] 0 & \dfrac{\bar{q}Sd^2}{2I_yV}C_{m_q} & 0 \\[2ex] 0 & 0 & \dfrac{\bar{q}Sd^2}{2I_zV}C_{n_r} \end{bmatrix}_{at\ x_0},
$$

(4.39)

$$
B_{il} = \begin{bmatrix} \dfrac{\bar{q}Sd}{I_x}C_{l_{\delta p}} & -\dfrac{\bar{q}Sd}{I_x}C_{l_{\delta q}} & -\dfrac{\bar{q}Sd}{I_x}C_{l_{\delta r}} \\[2ex] \dfrac{\bar{q}Sd}{I_y}C_{m_{\delta p}} & -\dfrac{\bar{q}Sd}{I_y}C_{m_{\delta q}} & -\dfrac{\bar{q}Sd}{I_y}C_{m_{\delta r}} \\[2ex] \dfrac{\bar{q}Sd}{I_z}C_{n_{\delta p}} & -\dfrac{\bar{q}Sd}{I_z}C_{n_{\delta q}} & -\dfrac{\bar{q}Sd}{I_z}C_{n_{\delta r}} \end{bmatrix}_{at\ x_0}.
$$

To improve the transient performance of the inner loop, roll and pitch rate states are augmented with integral states. Let

$$
x_a = [p\ p_I\ q\ q_I\ r]^T
$$

(4.40)

where

$$
p_I = \int_0^t (p - p_c)d\tau
$$

(4.41)

$$
q_I = \int_0^t (q - q_c)d\tau.
$$

(4.42)

The LQR controller is designed to minimize the following objective function with the appropriate state and control weighting matrices Q and R, respectively:

$$
J = \int_0^\infty (x_a^T Q x_a + u_{il}^T R u_{il})dt.
$$

(4.43)

The linear controller obtained for the inner loop is of the form

$$u = -K_{il}[p - p_c \; p_I \; q - q_c \; q_I \; r - r_c]^T \tag{4.44}$$

where $K_{il} \in R^{3\times5}$ is the optimal gain matrix.

4.3.3 BTT/STT Command Logic

The logic for converting the commanded accelerations to commanded attitude angles of missiles is a hybrid of BTT and STT commands. Depending upon the phase of missile flight, one or both sets of logic are used. The BTT command is used in the midcourse and terminal phases of flight to prevent the air breathing engine from flaming out. STT is used for small acceleration commands, which when executed in BTT mode may ask for large roll rates. As the missile approaches the endgame phase and passes a preset time-to-go threshold, both BTT and STT logic are executed simultaneously over a preselected time interval to improve transient responses.

Depending on the magnitude of commanded accelerations, the BTT mode is divided into three regions in the inertial frame (I):

Condition	Mode	Attitude Command
$a_c^I \leq 0.1g$	STT control	$\alpha_c = \dfrac{a_{z_c} m / \bar{q}S - (C_z - C_{Z_\alpha}\alpha)}{C_{Z_\alpha}}$
		$\beta_c = \dfrac{a_{y_c} m / \bar{q}S - (C_Y - C_{Y_\beta}\beta)}{C_{Y_\beta}}$
		$\mu_c = \mu$
$0.1g \leq a_c^I \leq g$	Reduced BTT control	$\alpha_c = \dfrac{\|a_c\| m / \bar{q}S - |(C_z - C_{Z_\alpha}\alpha)|}{|C_{Z_\alpha}|}\cos(\mu_{full} - \mu_c)$
		$\beta_c = \dfrac{\|a_c\| m / \bar{q}S - |(C_Y - C_{Y_\beta}\beta)|}{C_{Y_\beta}}\sin(\mu_{full} - \mu_c)$
		$\mu_c = \mu + \left[\tan^{-1}\dfrac{-a_{y_c}^I}{a_{z_c}^I} - \mu\right]\dfrac{a_c^I}{g}$
$a_c^I \geq g$	BTT control	$\alpha_c = \dfrac{\|a_c\| m / \bar{q}S - |(C_z - C_{Z_\alpha}\alpha)|}{|C_{Z_\alpha}|}$
		$\beta_c = 0$
		$\mu_c = \tan^{-1}\left[\dfrac{-a_{y_c}^I}{a_{z_c}^I}\right]$

Here $a_c^I = [a_{y_c}^I \quad a_{z_c}^I]^T$ is the commanded acceleration in the inertial frame, and $a_c = [a_{y_c} \quad a_{z_c}]^T$ is the commanded acceleration in the body frame. As mentioned previously, during the endgame phase, for a certain preselected time interval, both BTT and STT modes are executed simultaneously, and attitude commands are generated using the following formula:

$$
\begin{bmatrix} \alpha_c \\ \beta_c \\ \mu_c \end{bmatrix} = \rho \begin{bmatrix} \alpha_c \\ \beta_c \\ \mu_c \end{bmatrix}_{STT} + (1-\rho) \begin{bmatrix} \alpha_c \\ \beta_c \\ \mu_c \end{bmatrix}_{BTT}
\tag{4.45}
$$

where the parameter ρ varies linearly from 0 to 1 over the time interval.

4.4 Simulation Studies

Simulation studies were carried out with a six-degrees-of-freedom nonlinear missile model to investigate the potential of the MSO-based adaptive control methodology. As explained earlier, the linear nominal controllers for the inner and outer loops are designed *only* for the initial conditions. To compensate for the uncertainties that arise due to neglected nonlinearities and change in environmental conditions, the MSO with the Lyapunov theory–based weight update rule is used in the outer loop. Simulations were carried out at the following three altitude conditions:

Mach No.	Altitude (ft.)
2.7	100
2.7	20,000
2.7	30,000

Figures 4.1 through 4.10 show the simulation results obtained for the case of $M = 2.7$ and $h = 20,000$ ft. It can be observed from Figures 4.1 through 4.3 that even though the nominal controller was designed only for the initial conditions, with the aid of an adaptive controller in the outer loop, the actual system was able to closely track the commanded attitude angles. The maximum transient in sideslip angle (Figure 4.2) is seen to be of the order of 1°, which is acceptable. The demanded body rates (Figures 4.4 through 4.6) from the outer loop do not exceed the imposed hard saturation limit (200°/s) at any time. Examining the inner control loop performance, the actual body

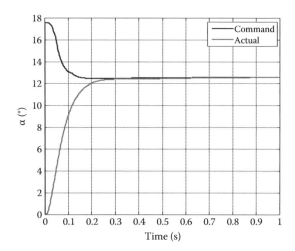

FIGURE 4.1
Angle of attack ($M = 2.7$, $h = 20,000$ ft.).

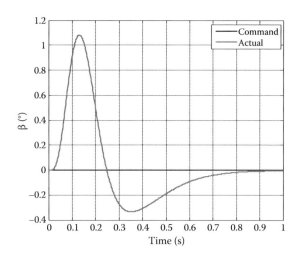

FIGURE 4.2
Sideslip angle ($M = 2.7$, $h = 20,000$ ft.).

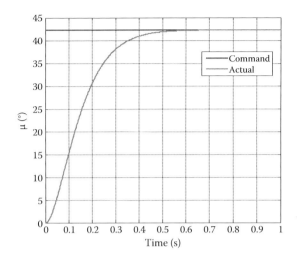

FIGURE 4.3
Bank angle ($M = 2.7$, $h = 20,000$ ft.).

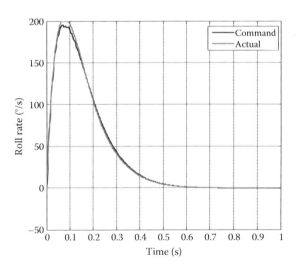

FIGURE 4.4
Roll rate ($M = 2.7$, $h = 20,000$ ft.).

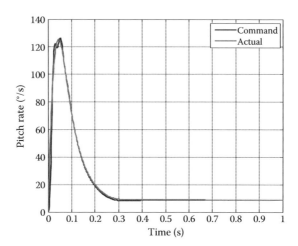

FIGURE 4.5
Pitch rate ($M = 2.7$, $h = 20,000$ ft.).

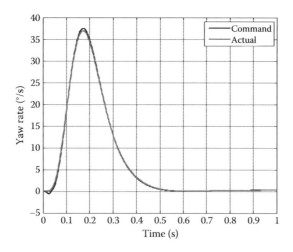

FIGURE 4.6
Yaw rate ($M = 2.7$, $h = 20,000$ ft.).

rates are seen to closely follow the commanded rates (Figures 4.4 through 4.6). It is important to note that the inner control loop has only a single linear controller designed at one nominal condition, but it is able to provide satisfactory performance during the entire flight. Figure 4.7 shows the history of fin deflections over the entire simulation.

Figures 4.8 through 4.10 compare the estimated uncertainty output by the neural networks with the actual uncertainty. It can be observed that the estimated nonlinearity closely follows the actual nonlinearity, and the estimates do not have any oscillatory behavior. Absence of oscillations is also reflected in the smooth commanded body rates generated by the outer loop.

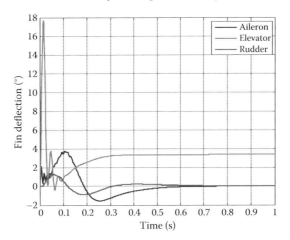

FIGURE 4.7
Fin deflection ($M = 2.7$, $h = 20,000$ ft.).

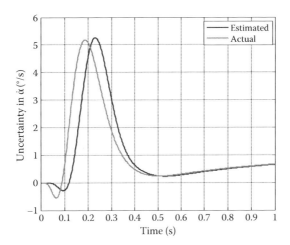

FIGURE 4.8
Uncertainty in $\dot{\alpha}$ ($M = 2.7$, $h = 20,000$ ft.).

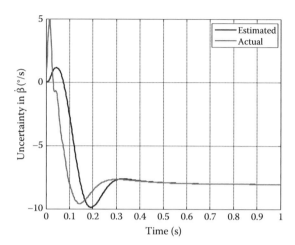

FIGURE 4.9
Uncertainty in $\dot{\beta}$ ($M = 2.7$, $h = 20{,}000$ ft.).

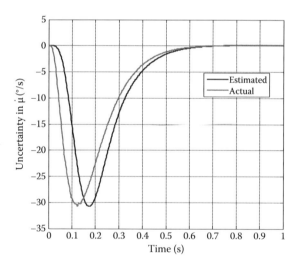

FIGURE 4.10
Uncertainty in $\dot{\mu}$ ($M = 2.7$, $h = 20{,}000$ ft.).

4.4.1 Performance at Various Altitude Conditions

In order to analyze the performance of the autopilot with off-nominal flight conditions, two different altitude conditions were considered: one with a minimum altitude of 100 ft. and another with a higher altitude of 30,000 ft. Figures 4.11 through 4.13 compare the performance of the outer loop in tracking the commanded attitudes at various altitude conditions. It can be

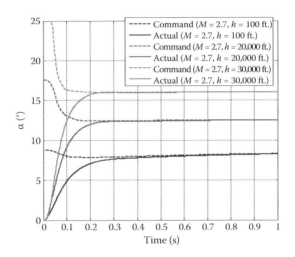

FIGURE 4.11
Angle of attack at various altitudes.

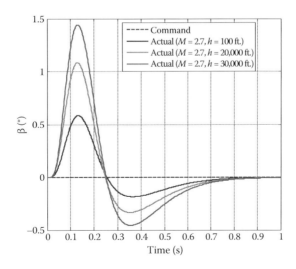

FIGURE 4.12
Sideslip angle at various altitudes.

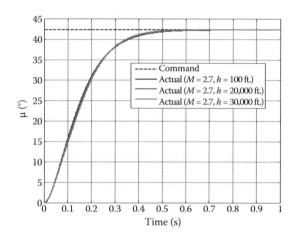

FIGURE 4.13
Bank angle at various altitudes.

observed that the actual trajectories closely follow the commanded trajectories. The inner loop is able to provide stable performance even with varying flight conditions as seen from the body rate plots (Figures 4.14 through 4.16). Figures 4.17 and 4.18 show the fin deflection histories at $h = 100$ ft. and $h = 30,000$ ft., respectively. The uncertainty estimation is quite robust again with no oscillations observed in the uncertainty estimate in any of the considered flight conditions (Figures 4.19 through 4.21).

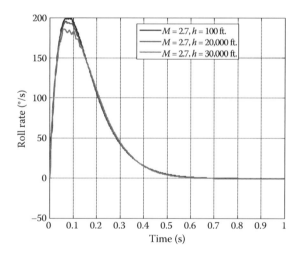

FIGURE 4.14
Roll rate at various altitudes.

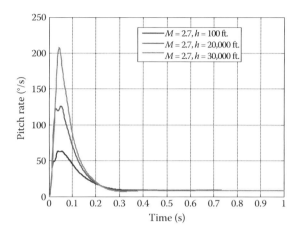

FIGURE 4.15
Pitch rate at various altitudes.

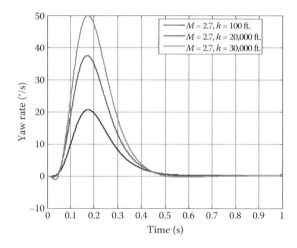

FIGURE 4.16
Yaw rate at various altitudes.

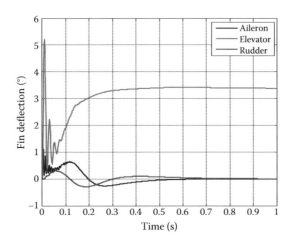

FIGURE 4.17
Fin deflection ($M = 2.7$, $h = 100$ ft.).

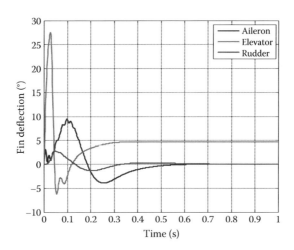

FIGURE 4.18
Fin deflection ($M = 2.7$, $h = 30{,}000$ ft.).

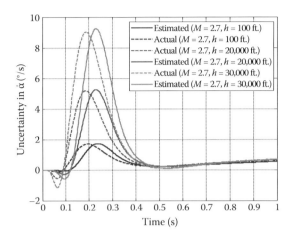

FIGURE 4.19
Uncertainty in $\dot{\alpha}$ at various altitudes.

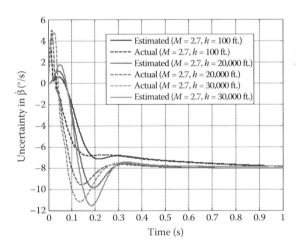

FIGURE 4.20
Uncertainty in $\dot{\beta}$ at various altitudes.

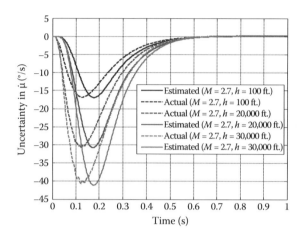

FIGURE 4.21
Uncertainty in $\dot{\mu}$ at various altitudes.

4.5 Conclusions

This chapter presented the application of adaptive neurocontrollers for missile autopilot design. The proposed methodology used the DI technique for nonlinear controller design and MSO for uncertainty estimation. The combination of DI controller along with MSO uncertainty estimator eliminates the need for gain scheduling to obtain consistent performance. Also, the observer structure of MSO allowed for separate design of estimation error dynamics from nominal dynamics design, thus allowing for fast adaptation. Simulation studies were carried out with a six-degrees-of-freedom nonlinear missile model varying the altitude conditions. Steady performance obtained at both nominal and off-nominal conditions demonstrated the potential of the proposed methodology for missile autopilot designs.

Acknowledgment

Development of the MSO adaptive controller was partially supported by a grant from NASA.

References

1. Adams, R. J., Buffington, J. M., Sparks, A. G., and Banda, S. S. (1994). *Robust Multivariable Flight Control*. London: Springer-Verlag.
2. Adams, R. J. and Banda, S. S. (1993). Robust flight control design using dynamic inversion and structured singular value synthesis. *IEEE Transcations on Control Systems Technology*, 1(2), 80–92.
3. Calise, A., Lee, S., and Sharma, M. (1998). Direct Adaptive Reconfigurable Control of a Tailless Fighter Aircraft. *Proceedings Guidance, Navigation, and Control Conference*. Boston, AIAA.
4. Cloutier, J., Evers, J. H., and Feely, J. J. (1988). An assessment of air to air missile guidance and control technology. *Proceedings of the American Control Conference*, pp. 133–142. IEEE, Atlanta, GA.
5. Cloutier, J., D'Souza, C. N., and Mracek, C. P. (1996). Nonlinear regulation and nonlinear h-infinity control via the state dependent Riccati equation technique. *Proceedings 1st International Conference On Nonlinear Problems in Aviation and Aerospace*, pp. 117–130. Embry-Riddle Aeronautical University Press, Daytona Beach, FL.
6. Cloutier, J. and Stansbery, D. T. (2001). Nonlinear Hybrid Bank to Turn/Skid to Turn Missile Autopilot Design. *AIAA Guidance, Navigation and Control Conference and Exhibit*. Montreal, Canada.
7. Kim, B. and Calise, A. J. (1997). Nonlinear flight control using neural networks. *Journal of Guidance, Control and Dynamics*, 20(1), 26–33.
8. Lewis, F., Yesildirek, A., and Liu, K. (1996). Multilayer neural net robot controller with guaranteed tracking performance. *IEEE Transactions on Neural Networks*, 7(2), 388–399.
9. McFarland, B. and Calise, A. J. (2000). Neural networks and adaptive nonlinear control of agile antiair missiles. *Journal of Guidance, Control and Dynamics*, 23(3) 547–553.
10. McFarland, M. and D'Souza, C. N. (1994). Missile flight control with dynamic inversion and structured singular value synthesis. *Proceedings of the AIAA Guidance, Navigation and Control Conference*, pp. 544–550, Scottsdale, AZ.
11. Narendra, K. and Parthasarathy, K. (1990). Identification and control of dynamical systems using neural networks. *IEEE Transactions on Neural Networks*, 1(1), 4–27.
12. Nguyen, N., Krishnakumar, K., and Boskovic, J. (2008). An optimal control modification to model-reference adaptive control for fast adaptation. *AIAA Guidance, Navigation and Control Conference and Exhibit*. Honolulu, HI.
13. Nguyen, N. (2009). Robust optimal adaptive control method with large adaptive gain. *AIAA Infotech Aerospace Conference*. Moffett Field, CA.
14. Rajagopal, K., Mannava, A., and Balakrishnan, S. N. (2009). Neuroadaptive model following controller for non-affine, nonsquare aircraft system. *AIAA Guidance, Navigation Control conference and exhibit*. Chicago, IL.
15. Ridgely, B. and McFarland, M. B. (1999). Tailoring theory to practice in tactical missile control. IEEE Control Systems Magazine, 19(6), 49–55.
16. Shamma, J. and Cloutier, J. (1993). Gain-scheduled missile autopilot design using linear parameter varying transformation. *Journal of Guidance, Control and Dynamics*, 16(2), 256–263.

17. Sharma, M. (2006). Application and flight testing of an adaptive autopilot on precision guided munitions. *AIAA Guidance Navigation and Control Conference and Exhibit*. Keystone, CO.
18. Williams, D. E., Friendland, B., and Madiwale, A. N. (1987). Modern control theory for design of autopilots for bank to turn missiles. *Journal of Guidance, Control and Dynamics*, 10(4), 378–386.
19. Wise, K. (1988). Maximizing performace and stability robustness in a conventional bank to turn missile autopilot design. *AIAA Missile Systems Science Conference*. Monterey, CA.
20. Wise, K. (2008). Robust stability analysis of adaptive missile autopilots. *AIAA Guidance, Navigation and Control Conference and Exhibit*. Honolulu, HI.
21. Xin, M., Balakrishnan, S. N., Stansbery, D. T., and Ohlmeyer, E. J. (2004). Nonlinear missile autopilot design with theta-D technique. *Journal of Guidance, Control and Dynamics*, 27(3), 406–417.

5

Integrated Guidance and Control for Missiles

Nathan Harl, Michael Dancer, S. N. Balakrishnan,
Ernest J. Ohlmeyer, and C. Phillips

CONTENTS

5.1 Introduction

Interceptor systems for missile defense engagements, by virtue of their mission objectives, demand extremely accurate performance from all their components. This exacting requirement in turn needs guidance, control, and estimation systems that will guarantee intercept against highly nonlinear and complex target maneuvers or in seek-and-destroy missions. Therefore, careful attention should be paid to the development of guidance laws and control strategies that go into the interceptors. There has been a lot of work in the published literature on missile guidance, control, and estimation problems [1,4,6,8,14].

Integrated guidance and control (IGC) design is an emerging trend in missile technology. This is a response to meet the need for improving the accuracy of interceptors and extend their kill envelope. Current and past practices in industry have been to design guidance and control systems separately and then integrate them to form the complete missile. The whole system usually consists of a range of technologies from classical control theory to optimal estimation and control. These subsystems typically had different bandwidths. Despite the fact that this paradigm has been applied successfully on many systems, it can be argued that the overall system performance can be improved if the design exploits the synergy between the various missile subsystems. Furthermore, an IGC design will eliminate the number of iterations needed in the separate guidance-control design.

In 1984, Lin and Yueh [11] first addressed the application of an IGC scheme to a homing missile. An optimal controller was designed to combine the conventionally separated guidance law design and autopilot design into one framework by minimizing a quadratic cost function subject to intercept dynamics. The IGC design resulted in better RMS miss, the terminal angle of attack, the pitch rate, and the control surface "flapping" rate in the presence of unmodeled errors. In 1992, Evers et al. [7] extended the concepts presented in the work of Lin and Yueh [11] to include a first-order Markov model for target acceleration. The resulting IGC law was expected to be less sensitive to the errors in estimating the current target acceleration.

Menon and Ohlmeyer [13] employed the feedback linearization method in conjunction with the linear quadratic regulator technique to design a set of nonlinear IGC laws for homing missiles. Their IGC design was presented in three formulations, which were based upon three different guidance objectives. A 6-degrees-of-freedom (DOF) nonlinear dynamic model of an air-to-air homing missile was simulated, and each of three IGC schemes achieved a specific favorable performance over the other. A disadvantage of the feedback linearization technique is that it could cancel beneficial nonlinearities. The IGC concept was further used by Menon and Ohlmeyer [12] by employing the state-dependent Riccati equation (SDRE) [5] technique. A command generator was used to prevent actuator saturation, and the goal included meeting a terminal aspect angle constraint. The design was evaluated based on a 6-DOF nonlinear missile model with nonmaneuvering and weaving targets. Numerical results demonstrated the feasibility of the IGC design for the next-generation high-performance missile systems. At the Johns Hopkins Applied Physics Laboratory (publication restricted), Palumbo and Jackson [15] have been working on IGC schemes using finite-time SDRE-based schemes with assumptions on the state evolution for easier solution to SDRE. However, solving the SDRE online *is very time-consuming*. In 2004, Xin et al. [23] applied the θ-D method, a method that yields near optimal closed-form solutions to nonlinear optimal control problems to the IGC design based on the same nonlinear missile model as in the work of Cloutier et al. [5] and achieved some good preliminary results. Compared to the SDRE approach, the θ-D controller gives a closed-form solution and, therefore, is easy to implement.

There have also been several approaches to solving the IGC problem by using the sliding mode control (SMC) technique. Shkolnikov et al. [16] used first-order SMC to develop an IGC system for a homing interceptor. They investigated the line-of-sight (LOS) rate and the transversal relative velocity component as possible sliding surfaces and developed guidance and control laws based on methods previously developed by Brown et al. [2] and Shtessel and Buffington [17]. An inner-loop/outer-loop approach was used, with the outer loop creating a commanded pitch rate that was tracked in the inner loop via the pitch fin deflection. Thus, the authors' approach involved the creation of two separate control systems; a direct relationship between the control input (fin deflection) and the control objective (LOS rate), however, is not obtained. Shtessel and Shkolnikov [18] later modified their approach using second-order sliding mode rather than the first-order one as was used in the study of Shkolnikov et al. [16].

In Shima et al.'s paper [19], the zero-effort miss distance is used as the sliding surface. Unlike the previous sliding mode approaches to IGC, the authors formulated the guidance and control systems together in one loop and established a direct relationship between the control input and guidance objective. Performance of the integrated approach was compared to that of two different inner-loop/outer-loop designs, and it was shown that the integrated controller led to superior results. The authors later applied their IGC approach to the particular cases of dual control missiles [9] and missiles with on–off actuators [10].

In this chapter, a general analysis of the IGC problem is first presented using the θ-D controller from [23]. First, a *body-based* IGC formulation is introduced, in which the fin deflections control both the rotational and translational motions of a missile. While the body-based formulation can produce acceptable miss distances, it is unfortunately very sensitive to how the various weights are tuned. To bypass the tuning difficulty, a more physically intuitive formulation of the dynamics, called the *velocity-based* IGC (VIGC), is developed.

After describing the body-based and velocity-based IGC formulations, a *new* solution technique to IGC named the sliding mode IGC (SMIGC) is presented. This approach exploits the finite-time reaching phase of the sliding mode technique to ensure that a desired constraint will be achieved in a finite time. Sliding surfaces are chosen as functions of the predicted impact point (PIP) heading error. In comparison to the previous sliding mode approaches to IGC by Shtessel and Idan et al., SMIGC makes several different contributions. First, by finding an exact expression for the impact of target acceleration on the heading error rate, SMIGC is able to account for the target acceleration in a novel fashion and yield small miss distances in the presence of agile targets. Next, implementing the SMIGC approach only requires that the *bound* of the target acceleration be known. In this work, results from three-dimensional engagements are presented. Control laws are developed for the vertical and horizontal planes that can be used to intercept a target that is maneuvering in each of the planes with a simple roll autopilot. Finally, the SMIGC approach is implemented in a high-fidelity 6-DOF model.

5.2 Body-Based IGC Formulation

This section describes the development of the IGC framework using equations of motion (EOMs) developed in the missile body frame. Results of the application of the θ-D nonlinear control technique [23] to these EOMs are shown.

5.2.1 Equations of Motion

In the IGC development to follow, the following EOMs are used for the missile:

$$\dot{p} = M_x / I_x \tag{5.1}$$

$$\dot{q} = [M_y - (I_x - I_z)pr]/I_y \tag{5.2}$$

$$\dot{r} = [M_z - (I_y - I_x)qp]/I_z \tag{5.3}$$

$$\dot{u}^m = F_x/m - qw^m + rv^m \tag{5.4}$$

$$\dot{v}^m = F_y/m - ru^m + pw^m \tag{5.5}$$

$$\dot{w}^m = F_z/m - pv^m + qu^m \tag{5.6}$$

where I_x, I_y, and I_z are the principle moments of inertia, m is the missile mass, F_x, F_y, and F_z are the aerodynamic forces acting on the missile, and M_x, M_y, and M_z are the aerodynamic moments. Note that for an axisymmetric missile, $I_y = I_z$. The superscripts m in Equations 5.4 through 5.6 indicate quantities relating to the missile.

Similarly for the target, the following EOMs are used:

$$\dot{u}^t = a_x^t - qw^t + rv^t \tag{5.7}$$

$$\dot{v}^t = a_y^t - ru^t + pw^t \tag{5.8}$$

$$\dot{w}^t = a_z^t - pv^t + qu^t \tag{5.9}$$

Note that the superscript t indicates quantities that are related to the target, and the quantities u, v, and w are the target velocities represented in the missile body x, y, and z axes, respectively. The controller states defined in the next section will use the relative position and velocity of the missile relative to the target (i.e., $v = v^m - v^t$, $y_b = y_b^m - y_b^t$, etc.).

The aerodynamic forces and moments for the missile are modeled as

$$F_x/m = \eta_F \left(c_{A_0} + c_{A_\alpha} |\alpha| + c_{A_\beta} |\beta| + c_{A_{\alpha\beta}} |\alpha\beta| \right) \tag{5.10}$$

$$F_y/m = \eta_F c_{Y_\beta} \beta \tag{5.11}$$

$$F_z/m = \eta_F c_{N_\alpha} \alpha \tag{5.12}$$

$$M_x = \eta_\tau c_{l_\delta} \delta_p \tag{5.13}$$

$$M_y = \eta_\tau \left(c_{m_\alpha} \alpha + c_{m_\delta} \delta_q \right) \tag{5.14}$$

$$M_z = \eta_\tau \left(c_{n_\beta} \beta + c_{n_\delta} \delta_r \right) \tag{5.15}$$

where $\eta_F = \bar{q}S/m$ and $\eta_\tau = \bar{q}Sl$.

Before the θ-D method can be applied, an extra state is needed in order to turn the system EOMs into the needed linear-like structure. This extra state is defined by

$$\dot{\lambda} = -a_\lambda \lambda \tag{5.16}$$

The extra state is simply a mathematical artifice that is used to factor the system EOMs without introducing singularities. Each time the control is calculated, λ is set to 1.

The EOMs can be cast in the linear-like form

$$\dot{x} = A(x)x + B(x)\delta \tag{5.17}$$

where the system state is selected to be $x = [\, p\; q\; r\; y_b\; z_b\; v\; w\; \lambda\,]^T$ and the input is $\delta = [\, \delta_p\; \delta_q\; \delta_r\,]^T$. The system state–dependent matrices are given by

$$A(x) = \begin{bmatrix} 0 & I_1 r & I_1 q & 0 & 0 & 0 & 0 & 0 \\ I_2 r & 0 & I_2 p & 0 & 0 & 0 & 0 & \eta_\tau c_{m_\alpha} \alpha / I_y \lambda \\ I_3 q & I_3 p & 0 & 0 & 0 & 0 & 0 & \eta_\tau c_{n_\beta} \beta / I_z \lambda \\ z_b & 0 & -x_b & 0 & 0 & 1 & 0 & 0 \\ -y_b & x_b & 0 & 0 & 0 & 0 & 1 & 0 \\ w & 0 & -u & 0 & 0 & 0 & 0 & \left(\eta_F c_{Y_\beta} \beta - a_y^t \right) / \lambda \\ -v & u & 0 & 0 & 0 & 0 & 0 & \left(\eta_F c_{N_\alpha} \alpha - a_z^t \right) / \lambda \\ 0 & 0 & 0 & 0 & 0 & 0 & 0 & -a_\lambda \end{bmatrix} \tag{5.18}$$

$$B(x) = \begin{bmatrix} \eta_\tau c_{l_\delta} / I_x & 0 & 0 \\ 0 & \eta_\tau c_{m_\delta} / I_y & 0 \\ 0 & 0 & \eta_\tau c_{n_\delta} / I_z \\ 0 & 0 & 0 \\ 0 & 0 & 0 \\ 0 & 0 & 0 \\ 0 & 0 & 0 \\ 0 & 0 & 0 \end{bmatrix} \tag{5.19}$$

5.2.2 Cost Function

To implement the guidance law into the IGC formulation, an appropriate cost function is used. The guidance law will attempt to cease motion in the LOS. This can be accomplished by driving the cross product between the relative position and velocity to zero. The cost function will place weights on the following values: $y = [p \ q \ r \ \Delta_x \ \Delta_y \ \Delta_z]^T$, where Δ_x, Δ_y, and Δ_z are the vector components of the cross product. These values are related to the state vector by

$$y = H(x)x \tag{5.20}$$

where

$$H(x) = \begin{bmatrix} 1 & 0 & 0 & 0 & 0 & 0 & 0 & 0 \\ 0 & 1 & 0 & 0 & 0 & 0 & 0 & 0 \\ 0 & 0 & 1 & 0 & 0 & 0 & 0 & 0 \\ 0 & 0 & 0 & 0 & 0 & -z_b & y_b & 0 \\ 0 & 0 & 0 & 0 & u & 0 & x_b & 0 \\ 0 & 0 & 0 & u & 0 & x_b & 0 & 0 \end{bmatrix} \tag{5.21}$$

The final cost function is then

$$J = \int_0^\infty (y^T W y + \delta^T R \delta)\, dt = \int_0^\infty (x^T H^T W H x + \delta^T R \delta)\, dt = \int_0^\infty (x^T Q x + \delta^T R \delta)\, dt \quad (5.22)$$

where the cost on the states, $Q = H^T W H$, results nicely from the "output" regulation.

5.2.3 Body-Based IGC Results

This section contains simulation results for the missile body-based IGC formulation using the θ-D nonlinear control technique. A description of the θ-D technique can be found in [23]. Through simulation testing for different launch scenarios, it was determined that the body-based IGC formulation was very sensitive to weight tuning. The reasoning for this is discussed in the following section. When the weights are properly tuned, the body-based IGC scheme is capable of intercepting a weaving target as can be seen in Figure 5.1. In this scenario, the initial range is approximately 10,000 ft, and the target is weaving with a frequency of 1 Hz and max acceleration of 10 g's. The fin deflections for this scenario are shown in Figure 5.2.

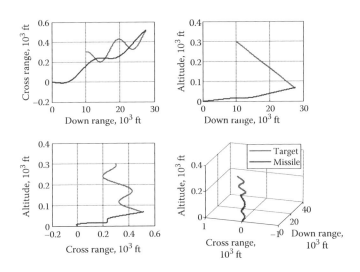

FIGURE 5.1
Interception of weaving target using body-based IGC formulation.

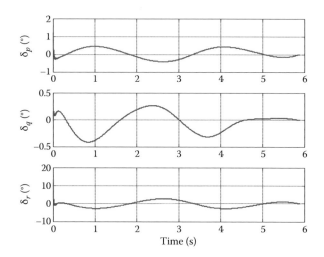

FIGURE 5.2
Fin deflections using body-based IGC.

5.2.3.1 Weight Sensitivity

It was found through simulation testing that the body-based IGC formulation is overly sensitive to the chosen weights in the cost function. To understand the source of this sensitivity, a simulation was run with ill-tuned weights and the motion restricted to the vertical plane. The results of this simulation are shown in Figure 5.3.

From the results in Figure 5.3, it is evident that the missile is actually turning in the opposite direction to that needed for intercepting the target. To

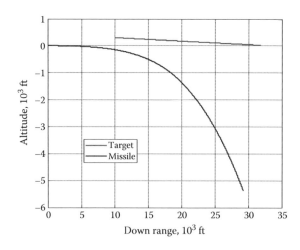

FIGURE 5.3
Sensitivity demonstration simulation.

see what is happening in this case, the effects of position and velocity errors were individually eliminated by zeroing their respective error components in the control law. The results of these two simulations are shown in Figures 5.4 and 5.5. From these figures, it appears that the effects of z_b and w errors are competing with each other to steer the missile toward the target. Figure 5.6 shows the same scenario with the effect of z_b error restricted to 80% of its normal effect. These simulations indicate that the competing effect of z_b and w errors is the underlying source of the extreme weight sensitivity of the body-based IGC formulation.

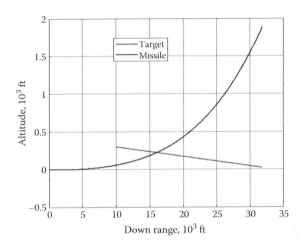

FIGURE 5.4
Sensitivity simulation without z_b error.

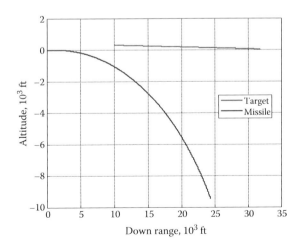

FIGURE 5.5
Sensitivity simulation without w error.

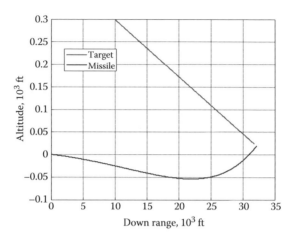

FIGURE 5.6
Sensitivity simulation with 80% z_b error.

5.2.3.2 Discussion of Missile Physics

With the IGC system developed in the missile body frame, the dynamics
are derived so that the fin deflections control both rotational and translational motions of the missile. However, in actual missile physics, the fins primarily adjust the orientation of the missile, and then the resulting incident
flow produces forces that provide translational motion for the missile. In the
body-based IGC formulation, this dominant effect of fin deflection results in
the rotational errors, y_b and z_b, to be regulated more effectively than velocity errors, v and w. This phenomenon results in the missile actually *sliding
around* the target instead of correctly steering toward the target as is demonstrated in Figure 5.7.

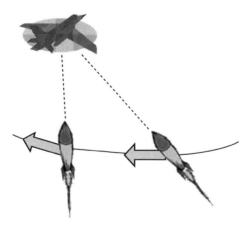

FIGURE 5.7
Missile engagement physics.

5.3 Velocity-Based IGC Formulation

In the previous section, a body-based IGC formulation was developed. It became evident that such a formulation was unable to accurately capture the physics behind missile motion. In this section, an alternate IGC formulation based on the missile velocity, as opposed to the missile body frame, is described. The VIGC is first derived in a two-dimensional scenario and later extended to full three-dimensional missile engagements.

5.3.1 Missile IGC Design in Vertical Plane

The goal of any missile guidance and control scheme is to steer the missile so as to impact the target in a finite period of time. If the target is traveling with a constant direction and the missile is traveling in the desired direction for impact, then the PIP will be fixed in space. Given these circumstances, the goal of the missile guidance and control scheme can be construed as steering the missile velocity toward a fixed point in space: the PIP.

Figure 5.8 shows the geometry of the missile/target intercept problem. The missile is controlled via pitch fin deflections, which creates a torque on the missile. When the missile rotates, the incoming air impacts the missile with an angle of attack, α, which in turn creates a restoring moment and a normal force, a_N, which turns the missile. The combined angular acceleration produced from the pitch moment acting on the missile is denoted α_y. The location of the PIP with respect to the missile is denoted by the coordinates x_v and z_v, which lie along and normal, respectively, to the missile velocity vector. The pitch angle, θ, defines the orientation of the missile in the pitch plane.

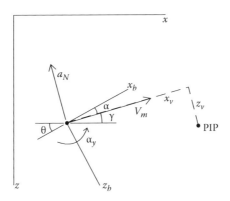

FIGURE 5.0
Missile intercept geometry.

Since the goal of the missile guidance and control scheme is to turn the missile so that the velocity vector points at the PIP, the missile intercept problem can be formulated as driving the coordinate, z_v, to zero. The derivatives of x_v and z_v are

$$\dot{x}_v = V_m - \dot{\gamma} z_v \tag{5.23}$$

$$\dot{z}_v = \dot{\gamma} x_v \tag{5.24}$$

where $\gamma = \theta - \alpha$ is the missile flight path angle. Equation 5.24 must now be differentiated until the pitch fin deflection, δ_q, appears as a result of moments on the missile and not just from translational forces. The missile normal acceleration, a_N, assuming constant missile velocity, V_m, is given by

$$a_N = \dot{\gamma} V_m \tag{5.25}$$

In this missile model, it is assumed that the normal acceleration is only a function of α and is normal to the velocity vector. While this in general is not true, it is a reasonable approximation for preliminary analysis. Equation 5.25 then reduces to

$$\dot{\gamma} = (\bar{q} S c_{N_\alpha} / m V_m) \alpha = k_{N_\alpha} \alpha \tag{5.26}$$

where k_{N_α} is constant given the current assumptions, $\bar{q} = \rho V_m^2 / 2$ is the dynamic pressure, S is the reference area, and m is the missile mass.

Differentiating Equations 5.24 and 5.26 a second time results in

$$\ddot{z}_v = \ddot{\gamma} x_v + \dot{\gamma} \dot{x}_v = \ddot{\gamma} x_v - \dot{\gamma}^2 z_v + \dot{\gamma} V_m \tag{5.27}$$

$$\ddot{\gamma} = k_{N_\alpha} \dot{\alpha} = k_{N_\alpha} (q - \dot{\gamma}) \tag{5.28}$$

where Equation 5.23 and the definition of γ have been used in the right equality of Equations 5.27 and 5.28, respectively. Also note that $\dot{\theta} = q$, the missile pitch rate. Since the pitch fin deflection, δ_q, has still not appeared, differentiate Equations 5.27 and 5.28 to obtain

$$\dddot{z}_v = (\dddot{\gamma} - \dot{\gamma}^3) x_v - 3\dot{\gamma}\ddot{\gamma} z_v + 2\ddot{\gamma} V_m \tag{5.29}$$

$$\dddot{\gamma} = k_{N_\alpha} (\alpha_y - \ddot{\gamma}) \tag{5.30}$$

In the missile model, it is assumed that the aerodynamic pitch moment consists of only two terms: a moment produced by the pitch fin deflection, δ_q, and a restoring moment produced by α. Thus, the pitch acceleration is given by

$$\alpha_y = (\bar{q}Slc_{m_\alpha}/I_y)\alpha + (\bar{q}Slc_{m_\delta}/I_y)\delta_q = k_{m_\alpha}\alpha + k_{m_\delta}\delta_q \qquad (5.31)$$

with the assumption that k_{m_α} and k_{m_δ} are constant, and l is the reference length. Using Equations 5.30 and 5.31 in Equation 5.29 results in

$$\ddot{z}_v = (k_{N_\alpha}k_{m_\alpha}\alpha + k_{N_\alpha}k_{m_\delta}\delta_q - k_{N_\alpha}\ddot{\gamma} - \dot{\gamma}^3)x_v - 3\ddot{\gamma}\dot{\gamma}z_v + 2\ddot{\gamma}V_m \qquad (5.32)$$

which now contains the pitch fin deflection. It is desired that z_v be asymptotically stable, so δ_q should be chosen so that the closed-loop dynamics for z_v is

$$\dddot{z}_v + k_{\ddot{z}}\ddot{z}_v + k_{\dot{z}}\dot{z}_v + k_z z_v = 0 \qquad (5.33)$$

where $k_{\ddot{z}} > 0$, $k_{\dot{z}} > 0$, and $k_z > 0$. By comparing Equations 5.32 and 5.33, the desired pitch fin deflection is found to be

$$\delta_q = \left[3\ddot{\gamma}\dot{\gamma}z_v - (k_{m_\alpha}\dot{\gamma} - k_{N_\alpha}\ddot{\gamma} - \dot{\gamma}^3)x_v - 2\ddot{\gamma}V_m - k_{\ddot{z}}\ddot{z}_v - k_{\dot{z}}\dot{z}_v - k_z z_v\right]/k_{N_\alpha}k_{m_\delta}x_v \quad (5.34)$$

5.3.1.1 Results

The control given by Equation 5.34 was simulated to validate its performance. The missile was initially traveling downrange with a velocity of 4850 ft/s, and the PIP was selected to be 10,000 ft downrange at an altitude of 300 ft. Figure 5.9 shows the resulting time history for z_v when the closed-loop eigenvalues are selected to be –8, –12, and –16. It is clear that the fin deflection control has successfully driven z_v to zero after approximately 1 s. The resulting missile trajectory is shown in Figure 5.10. During the first second of the simulation, the missile turns toward the PIP and then travels in a straight line until it reaches the PIP.

Figures 5.11 and 5.12 show the missile pitch angle and angle of attack histories, respectively, with time. Notice that the missile actually pitches up to an angle of a little over 12° before pitching down to the steady value of approximately 2°. This initial pitch up maneuver is required so that the induced angle of attack will be sufficient to turn the missile in the desired direction

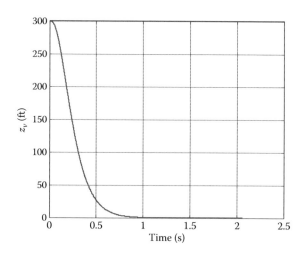

FIGURE 5.9
Time history of z_v.

Once the missile velocity vector approaches the desired flight direction, the missile is able to pitch down since less of an angle of attack is required to finish the turn. The pitch fin deflection used to perform the maneuver is shown in Figure 5.13. A positive initial fin deflection is required to pitch the missile to 12°. Notice that after the initial fin deflection, the control level

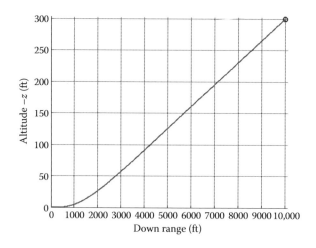

FIGURE 5.10
Missile trajectory.

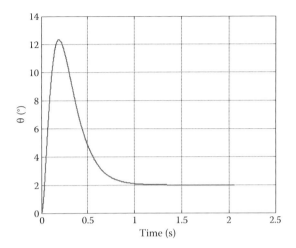

FIGURE 5.11
Pitch angle time history.

drops, but the control never becomes negative. Since the aerodynamic pitch moment included a restoring moment due to angle of attack, the final pitch down maneuver of the missile is primarily accomplished with this restoring moment, and the pitch fin deflection simply serves to regulate the rate of the pitch down maneuver.

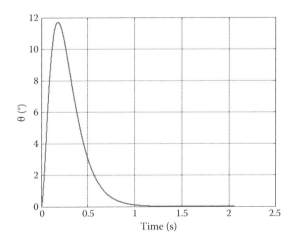

FIGURE 5.12
Angle-of-attack time history.

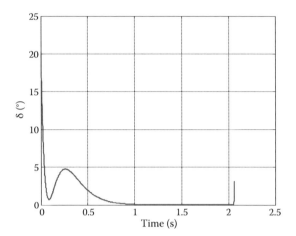

FIGURE 5.13
Pitch fin deflection.

5.3.2 Location of PIP

The new IGC formulation steers the missile toward a fixed point in space, the PIP, which will minimize the final miss distance. In order for the IGC formulation to work effectively, the PIP needs to be calculated with a high degree of accuracy.

5.3.2.1 Intercept Geometry

When the missile is correctly traveling toward the PIP, there exists a collision triangle, which is depicted in Figure 5.14. In this configuration, the distance

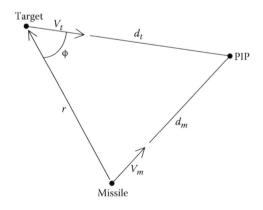

FIGURE 5.14
Collision triangle geometry.

traveled by the missile, d_m, in a certain period of time, t_{go}, will cause the missile to reach the same point as the target, which has traveled a distance d_t in the same period of time. The current range between the missile and target is r.

The first step is to determine the distance traveled by both the missile and target in the yet unknown time t_{go}. Consider a general aerodynamic body traveling in a straight line with a drag that is proportional to the square of the vehicle velocity:

$$a = dv/dt = -kv^2 \tag{5.35}$$

Using separation of variables, the velocity with respect to time is

$$v = dx/dt = v_0/(1 + kv_0 t) \tag{5.36}$$

which, after using separation of variables a second time, results in the position versus time:

$$x = x_0 + \frac{1}{k}\ln(1 + kv_0 t) \tag{5.37}$$

Equation 5.37 can be used to determine the distance traveled by the missile and target in the time, t_{go}, with the initial velocities of the missile and target being V_m and V_t, respectively. By using k_m and k_t as the proportionality constants for the missile and target, respectively, the distances are

$$d_m = \frac{1}{k_m}\ln(1 + k_m V_m t_{go}) \tag{5.38}$$

$$d_t = \frac{1}{k_t}\ln(1 + k_t V_t t_{go}) \tag{5.39}$$

Now by using the law of cosines, the time-to-go, t_{go}, can be calculated. Noting the geometry in Figure 5.14, the law of cosines results in

$$d_m^2 = r^2 + d_t^2 - 2rd_t \cos\phi \tag{5.40}$$

Substituting Equations 5.38 and 5.39 gives an expression that can be solved to determine t_{go}:

$$r^2 - 2\frac{r\cos\phi}{k_t}\ln(1 + k_t V_t t_{go}) + \left[\frac{1}{k_t}\ln(1 + k_t V_t t_{go})\right]^2 - \left[\frac{1}{k_m}\ln(1 + k_m V_m t_{go})\right]^2 = 0 \tag{5.41}$$

Once the value for t_{go} has been calculated, the location of the PIP can easily be constructed by determining the distance that will be traveled by the target, d_t.

5.3.2.2 Approximate Solutions

It is difficult to find a closed-form solution to Equation 5.41; however, some interesting approximations can be made to the expression. Consider the Taylor series expansion for $\ln(1 + x)$:

$$\ln(1+x) = -\sum_{n=1}^{\infty} \frac{(-1)^n}{n} x^n \tag{5.42}$$

which converges if $x \in (-1,1]$. Using Equation 5.42 to approximate the distance traveled by the missile and target to first-degree accuracy results in

$$d_m = V_m t_{go} \tag{5.43}$$

$$d_t = V_t t_{go} \tag{5.44}$$

which is simply the result that would be obtained if it were assumed that no drag acted upon either the missile or target. Using Equations 5.43 and 5.44 in Equation 5.40 results in a quadratic expression for t_{go}:

$$r^2 + (-2rV_t \cos\phi)t_{go} + (V_t^2 - V_m^2)t_{go}^2 = 0 \tag{5.45}$$

It is important to determine when the first-order approximation is valid. Since the Taylor series approximation given by Equation 5.42 is an alternating series, the truncation error is less than the magnitude of the first truncated term. With this observation, consider the percent error in the approximated travel distance

$$p_e = t_e/(d + t_e) \le M/(d + M) \tag{5.46}$$

where d is the approximate distance and t_e is the truncation error, which is bounded by $t_e \le M$. For the first-order approximation, $d = vt$ and $M = kv^2t^2/2 = a_D t^2/2$, which gives $p_e \le a_D t/(2v + a_D t)$. To ensure that the distance is calculated to within 10% of its actual value, the acceleration due to drag must satisfy the limiting condition

$$a_D \le \frac{2\,v}{9\,t} \tag{5.47}$$

Next consider if the distances are approximated with second-order accuracy. In this case

$$d_m = V_m t_{go} - \frac{1}{2} a_D^m t_{go}^2 \tag{5.48}$$

$$d_t = V_t t_{go} - \frac{1}{2} a_D^t t_{go}^2 \tag{5.49}$$

which is the result that would be obtained if a constant drag was assumed to act on both the missile and target. In this case, a quartic expression for t_{go} results:

$$r^2 + (-2rV_t \cos\phi)t_{go} + (ra_D^t \cos\phi + V_t^2 - V_m^2)t_{go}^2 + (V_m a_D^m - V_t a_D^t)t_{go}^3$$
$$+ \frac{1}{4}\left[(a_D^t)^2 - (a_D^m)^2\right]t_{go}^4 = 0 \tag{5.50}$$

Again it is important to consider the error in the second-order approximation. Following along the same lines as was done for the first-order case, to ensure 10% accuracy in the distance calculation, the drag acceleration is restricted by

$$a_D \leq \frac{1}{2}\frac{v}{t} \tag{5.51}$$

5.3.3 Extension to Three Dimensions

The IGC scheme developed in Section 3.1 makes use of a general inertial frame for the purpose of developing the EOMs for x_v and z_v. By making use of the generality of the inertial frame, one can select the instantaneous missile body frame for calculating the fin deflection. This is the key observation to extending the two-dimensional results to three-dimensional scenarios. Each time the control is calculated, the instantaneous missile body frame is used as a virtual inertial frame. Then the x–z and x–y planes serve as the two-dimensional planes for calculating δ_q and δ_r, respectively.

5.3.3.1 Results

Figure 5.15 shows the resulting trajectories when using the VIGC formulation. In this scenario, the initial range is approximately 30 km with the missile initially traveling at 2 km/s and the target at 3 km/s. The initial target aspect angle is 10°, and the missile's initial heading error is 8°. The scenario is

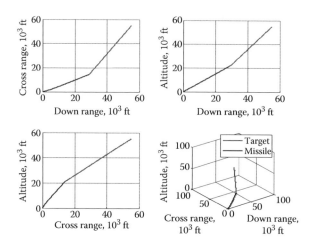

FIGURE 5.15
Target interception using VIGC.

three-dimensional with the missile having a velocity component of approximately 174 ft/s perpendicular to the initial collision plane.

The final miss distance for the scenario depicted in Figure 5.15 was less than 3×10^{-5} ft. Such a low miss distance is obtainable since the missile was modeled as traveling in a straight line with no weave. Figure 5.16 shows the resulting fin deflections. Note that a 15° hard limit was placed on the pitch and yaw fin deflections. Finally, Figures 5.17 and 5.18 show the resulting total angle of attack/aerodynamic roll angle and the missile angular velocity components, respectively, in the missile body frame.

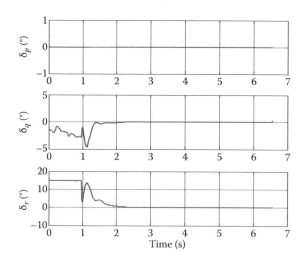

FIGURE 5.16
Fin deflections using VIGC.

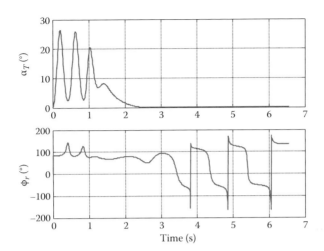

FIGURE 5.17
Missile total angle of attack using VIGC.

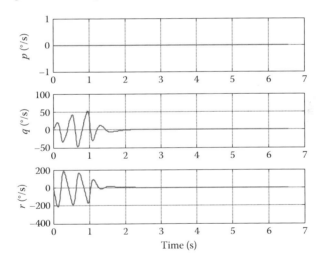

FIGURE 5.18
Missile angular velocities using VIGC.

5.4 Sliding Mode Integrated Guidance and Control

In this section, a sliding mode approach to IGC, called SMIGC, is described. The goal of the SMIGC method is to obtain expressions for the pitch and yaw fin deflections that will send the heading errors, δ_α and δ_β, to zero in a *finite* time. Once the heading errors are zero, the missile will be on a collision course toward the PIP (and hence the target), and a hit will be achieved. Also, in the presence of a maneuvering target, the SMIGC approach only requires the *bound* of the target acceleration's perturbation on the missile heading error. In order to achieve an accurate approximation for the bound of the target acceleration perturbation, an analysis is performed to determine the effects of target acceleration on the PIP heading error. The end result of this section is a set of control laws for the horizontal and vertical planes that can be used to engage maneuvering targets.

5.4.1 Effect of Target Acceleration on PIP Heading Error

It is assumed that the target can only accelerate normal to its direction of motion. Consequently, the target velocity direction can change but not its magnitude. Let the components of the target's normal acceleration in the α and β planes be taken as A_{t_α} and A_{t_β}, respectively. In terms of these normal acceleration components, the target acceleration vector can be expressed in the inertial $\hat{\mathbf{x}} - \hat{\mathbf{y}} - \hat{\mathbf{z}}$ frame as

$$\mathbf{a}_{tgt} = [A_{t_\alpha} \sin \gamma_t + A_{t_\beta} \sin \phi_t]\hat{\mathbf{x}} - A_{t_\beta} \cos \phi_t \hat{\mathbf{y}} - A_{t_\alpha} \cos \gamma_t \hat{\mathbf{z}} \qquad (5.52)$$

where γ_t and ϕ_t are the target's flight path angle and heading angle, respectively.

Now the expression for the rate of change of the PIP location can be found as

$$\dot{\mathbf{r}}_{PIP} = \mathbf{a}_{tgt} t_{go} - \mathbf{v}_{tgt} \frac{(\mathbf{r}_{PIP} - \mathbf{r}_{msl}) \cdot \mathbf{a}_{tgt} t_{go}}{(\mathbf{r}_{PIP} - \mathbf{r}_{msl}) \cdot \mathbf{v}_{rel}} \qquad (5.53)$$

where

$$\mathbf{r}_{PIP} = \mathbf{r}_{tgt} + \mathbf{v}_{tgt} t_{go} \qquad (5.54)$$

$$\mathbf{v}_{rel} = \mathbf{v}_{tgt} - \mathbf{v}_{msl} \qquad (5.55)$$

and t_{go} is the *time-to-go*. Substituting Equations 5.52, 5.54, and 5.55 into Equation 5.53 and carrying out the required vector operations, Equation 5.53 becomes

$$\dot{\mathbf{r}}_{PIP} = \left[\left(t_{go} \sin \gamma_t - \frac{A}{C} \dot{x}_t \right) A_{t_\alpha} + \left(t_{go} \sin \phi_t - \frac{B}{C} \dot{x}_t \right) A_{t_\beta} \right] \hat{\mathbf{x}}$$

$$+ \left[-\frac{A}{C} \dot{y}_t A_{t_\alpha} - \left(t_{go} \cos \phi_t + \frac{B}{C} \dot{y}_t \right) A_{t_\beta} \right] \hat{\mathbf{y}}$$

$$+ \left[\left(-t_{go} \cos \gamma_t - \frac{A}{C} \dot{z}_t \right) A_{t_\alpha} + \left(-\frac{B}{C} \dot{z}_t \right) A_{t_\beta} \right] \hat{\mathbf{z}}$$

$$(5.56)$$

where

$$A = (x_t - x_m + \dot{x}_t t_{go})t_{go} \sin \gamma_t - (z_t - z_m + \dot{z}_t t_{go})t_{go} \cos \gamma_t \tag{5.57}$$

$$B = (x_t - x_m + \dot{x}_t t_{go})t_{go} \sin \phi_t - (y_t - y_m + \dot{y}_t t_{go})t_{go} \cos \phi_t \tag{5.58}$$

$$C = (x_t - x_m + \dot{x}_t t_{go})(\dot{x}_t - \dot{x}_m) + (y_t - y_m + \dot{y}_t t_{go})(\dot{y}_t - \dot{y}_m) + (z_t - z_m + \dot{z}_t t_{go})(\dot{z}_t - \dot{z}_m). \tag{5.59}$$

As expected, Equation 5.56 shows that in the case of a nonmaneuvering target, the PIP location will remain constant. Next, it is desired to see the effect of target acceleration on the derivatives of the two heading errors, δ_α and δ_β. These derivatives can be found in a straightforward fashion to be

$$\dot{\delta}_\alpha = \frac{V_m}{R_\alpha} \sin \delta_\alpha + k_{N_\alpha} \alpha + \frac{1}{R_\alpha} (\dot{\mathbf{r}}_{PIP} \cdot \hat{\mathbf{z}}_r) \tag{5.60}$$

$$\dot{\delta}_\beta = \frac{V_m}{R_\beta} \sin \delta_\beta + k_{Y_\beta} \beta + \frac{1}{R_\beta} (\dot{\mathbf{r}}_{PIP} \cdot \hat{\mathbf{y}}_r) \tag{5.61}$$

where the $\hat{\mathbf{z}}_r$ and $\hat{\mathbf{y}}_r$ axes are as defined in Figures 5.1 and 5.2. The third terms in Equations 5.60 and 5.61 account for the effect of the target's normal acceleration on the rate of change of heading errors through Equation 5.56. In order to simplify Equations 5.60 and 5.61, note that

$$\hat{\mathbf{z}}_r = -\sin(\gamma_m - \delta_\alpha)\hat{\mathbf{x}} + \cos(\gamma_m - \delta_\alpha)\hat{\mathbf{z}} \tag{5.62}$$

$$\hat{\mathbf{y}}_r = -\sin(\phi_m - \delta_\beta)\hat{\mathbf{x}} + \cos(\phi_m - \delta_\beta)\hat{\mathbf{y}} \tag{5.63}$$

Substitution of Equations 5.56, 5.62, and 5.63 into Equations 5.60 and 5.61 leads to

$$\dot{\delta}_\alpha = \frac{V_m}{R_\alpha} \sin \delta_\alpha + k_{N_\alpha} \alpha + \frac{D}{R_\alpha} A_{t_\alpha} + \frac{E}{R_\alpha} A_{t_\beta} \tag{5.64}$$

$$\dot{\delta}_\beta = \frac{V_m}{R_\beta} \sin \delta_\beta + k_{Y_\beta} \beta + \frac{F}{R_\beta} A_{t_\alpha} + \frac{G}{R_\beta} A_{t_\beta} \tag{5.65}$$

where

$$D = \left(-t_{go}\sin\gamma_t + \frac{A}{C}\dot{x}_t\right)\sin(\gamma_m - \delta_\alpha) - \left(t_{go}\cos\gamma_t + \frac{A}{C}\dot{z}_t\right)\cos(\gamma_m - \delta_\alpha) \quad (5.66)$$

$$E = \left(-t_{go}\sin\phi_t + \frac{B}{C}\dot{x}_t\right)\sin(\gamma_m - \delta_\alpha) - \frac{B}{C}\dot{z}_t\cos(\gamma_m - \delta_\alpha) \quad (5.67)$$

$$F = \left(-t_{go}\sin\gamma_t + \frac{A}{C}\dot{x}_t\right)\sin(\phi_m - \delta_\beta) - \frac{A}{C}\dot{y}_t\cos(\phi_m - \delta_\beta) \quad (5.68)$$

$$G = \left(-t_{go}\sin\phi_t + \frac{B}{C}\dot{x}_t\right)\sin(\phi_m - \delta_\beta) - \left(t_{go}\cos\phi_t + \frac{B}{C}\dot{y}_t\right)\cos(\phi_m - \delta_\beta) \quad (5.69)$$

Equations 5.64 and 5.65 are exact expressions for the rate of change of the heading errors if the target is maneuvering. The variables D, E, F, and G have somewhat complex expressions in terms of the target and missile positions and velocities; however, for the control law implementation, only the *bounds* of those terms are used.

5.4.2 α-Plane SMIGC Control Law

This section describes the SMIGC derivation in the α-plane, with the end result being a control law for the pitch fin deflection. In this analysis, the target acceleration is considered to be unknown but bounded.

Recall from Equation 5.64 that the derivative of the heading error in the α-plane is

$$\dot{\delta}_\alpha = \frac{V_m}{R_\alpha}\sin\delta_\alpha + k_{N_\alpha}\alpha + \frac{D}{R_\alpha}A_{t_\alpha} + \frac{E}{R_\alpha}A_{t_\beta}$$

Now noting that the derivatives of the angle of attack α and pitch rate q are given by

$$\dot{\alpha} = -k_{N_\alpha}\alpha + q \quad (5.70)$$

$$\dot{q} = k_{m_\alpha}\alpha + k_{m_\delta}\delta_q \quad (5.71)$$

and making the assumption that the target acceleration does *not* affect higher-order derivatives of the heading error, Equation 5.13 can be differentiated until the pitch fin deflection δ_q appears, leading to

$$\dddot{\delta}_\alpha = \left(\frac{V_m}{R_\alpha}\right)^2 \sin(2\delta_\alpha) + \frac{V_m}{R_\alpha}\cos\delta_\alpha k_{N_\alpha}\alpha + k_{N_\alpha}(q - k_{N_\alpha}\alpha) \qquad (5.72)$$

$$\dddot{\delta}_\alpha = 2\left(\frac{V_m}{R_\alpha}\right)^2 \sin(3\delta_\alpha) + 3\left(\frac{V_m}{R_\alpha}\right)^2 \cos(2\delta_\alpha)k_{N_\alpha}\alpha - \frac{V_m}{R_\alpha}\sin\delta_\alpha k_{N_\alpha}^2\alpha^2$$
$$+ \frac{V_m}{R_\alpha}\cos\delta_\alpha k_{N_\alpha}(q - k_{N_\alpha}\alpha) + k_{N_\alpha}\left[k_{M_\alpha}\alpha + k_{M_\delta}\delta_q - k_{N_\alpha}(q - k_{N_\alpha}\alpha)\right] \qquad (5.73)$$

It is now desired to derive an expression for the pitch fin deflection δ_q that will send the heading error δ_α to zero in a finite time. An appropriate second-order system as a sliding surface s_α to achieve this objective is given by

$$s_\alpha = \ddot{\delta}_\alpha + 2\zeta\omega\dot{\delta}_\alpha + \omega^2\delta_\alpha = 0 \qquad (5.74)$$

where ζ and ω are tuning parameters. In this work, the time-to-go is calculated as

$$t_{go} = \left|\frac{r}{V_r}\right| = -\frac{r}{V_r} \qquad (5.75)$$

Also, it should be noted that when calculating Equation 5.74, the target acceleration terms in the $\ddot{\delta}_\alpha$ term are *ignored*. Ignoring the acceleration terms (for control calculations and not in the 6-DOF simulation test bed) was found to have a negligible effect on the final results.

The sliding surface can be differentiated to yield

$$\dot{s}_\alpha = \dddot{\delta}_\alpha + 2\zeta\omega\ddot{\delta}_\alpha + \omega^2\dot{\delta}_\alpha = H - I\delta_q + \omega^2\left(\frac{D}{R_\alpha}A_{t_\lambda} + \frac{E}{R_\alpha}A_{t_\phi}\right) \qquad (5.76)$$

where

$$
H = 2\left(\frac{V_m}{R_\alpha}\right)^3 \sin(3\delta_\alpha) + 3\left(\frac{V_m}{R_\alpha}\right)^2 \cos(2\delta_\alpha)k_{N_\alpha}\alpha - \frac{V_m}{R_\alpha}\sin\delta_\alpha k_{N_\alpha}^2\alpha^2
$$

$$
+ \frac{V_m}{R_\alpha}\cos\delta_\alpha k_{N_\alpha}(-k_{N_\alpha}\alpha + q) + k_{N_\alpha}\left[-k_{N_\alpha}(-k_{N_\alpha}\alpha + q) + k_{m_\alpha}\alpha\right]
$$

$$
+ 2\zeta\omega\left[\left(\frac{V_m}{R_\alpha}\right)^2 \sin(2\delta_\alpha) + \frac{V_m}{R_\alpha}\cos\delta_\alpha k_{N_\alpha}\alpha + k_{N_\alpha}(-k_{N_\alpha}\alpha + q)\right]
$$

$$
+ \omega^2\left(\frac{V_m}{R_\alpha}\sin\delta_\alpha + k_{N_\alpha}\alpha\right)
$$

(5.77)

$$
I = -k_{N_\alpha}k_{m_\delta}
$$

(5.78)

Equation 5.76 shows that the sliding surface has a relative degree of 1, so a traditional first-order sliding mode (1-sliding mode) approach can be used to solve the control problem. Also, it is seen that in contrast to the sliding surface (Equation 5.23), the target acceleration terms are allowed to enter the sliding surface's derivative. Any uncertainties in the derivative are assumed to be dealt with by the control δ_q.

In order to derive an expression for the control δ_q, a candidate Lyapunov function, V, is chosen as

$$
V = \frac{1}{2}s_\alpha^2
$$

(5.79)

The derivative of the Lyapunov function of Equation 5.79 can then be found as

$$
\dot{V} = s_\alpha \dot{s}_\alpha = s_\alpha\left[H - I\delta_q + \omega^2\left(\frac{D}{R_\alpha}A_{t_\lambda} + \frac{E}{R_\alpha}A_{t_\phi}\right)\right]
$$

(5.80)

It is now desired to select a value for the control input δ_q such that the Lyapunov function (and hence the sliding surface s_α) will reach zero in a finite time. In this regard, δ_q is chosen as

$$
\delta_q = \frac{1}{I}(\delta_{q_{eq}} + v)
$$

(5.81)

where $\delta_{q_{eq}}$ is the "equivalent control"

$$\delta_{q_{eq}} = H \tag{5.82}$$

and v is an "extra" control term that is required in order to send the Lyapunov function to zero and deal with the uncertainty terms involving the target acceleration.

Substituting Equations 5.81 and 5.82 into Equation 5.80 leads to

$$\dot{V} = s_\alpha \left[\omega^2 \left(\frac{D}{R_\alpha} A_{t_\lambda} + \frac{E}{R_\alpha} A_{t_\phi} \right) - v \right] \tag{5.83}$$

In order to make Equation 5.83 negative, the variable v is chosen as

$$v = \eta \, \text{sat}(s_\alpha) + \left(\left| \frac{D(t_0)}{R_\alpha(t_0)} A_{t_\lambda \max} \right| + \left| \frac{E(t_0)}{R_\alpha(t_0)} A_{t_\phi \max} \right| \right) \text{sat}(s_\alpha) \tag{5.84}$$

In Equation 5.84, the initial values of D, E, and R_α are used due to the fact that the initial value of each corresponds to its maximum value. Also, the maximum values of the target's normal accelerations are used. Substituting Equation 5.84 into Equation 5.83 and simplifying leads to

$$\dot{V} \le -\eta |s_\alpha| < 0 \tag{5.85}$$

Equation 5.85 shows that the Lyapunov function's derivative is negative for all $s_\alpha \ne 0$, thus guaranteeing that the sliding surface of Equation 5.74 will be reached in a finite time given by

$$t_r \le \frac{|s_\alpha(t_0)|}{\eta} \tag{5.86}$$

Note that the sliding surface reaching time will be less than or equal to some maximum value of $|s_\alpha(t_0)|/\eta$. In the absence of the unknown target acceleration, the less than or equal to sign in Equation 5.86 becomes equality, which means that there will be a specific reaching time for the value of η that is selected.

It is interesting to observe that Equation 5.86 can be rearranged to solve for the value of η required to achieve a *desired* reaching time t_r. The surface of Equation 5.74 will go to zero at a time less than or equal to t_r, but the heading error δ_α will only go to zero *after* the sliding surface is reached, so some amount of time must be spent on the surface $s_\alpha = 0$. It is therefore important that the reaching time be less than the initial time-to-go to allow time for the heading error to go to zero. In order to achieve this condition, η is chosen as

$$\eta = \frac{\left|s_\alpha(t_0)\right|}{t_{go}(t_0) - 3} \qquad (5.87)$$

The value of η in Equation 5.87 ensures that the sliding surface will be reached in a time that is less than or equal to 3 s before the initial t_{go}. It was found that this choice for the reaching time provides ample time for the heading error to become extremely small.

Substituting Equations 5.82, 5.84, and 5.87 into Equation 5.81, the expression for the pitch fin deflection is obtained as

$$\delta_q = \frac{1}{I}\left[H + \frac{\left|s_\alpha(t_0)\right|}{t_{go}(t_0) - 3}\,\text{sat}(s_a) + \left(\left|\frac{D(t_0)}{R_\alpha(t_0)}\,A_{t_\lambda \max}\right| + \left|\frac{E(t_0)}{R_\alpha(t_0)}\,A_{t_\phi \max}\right|\right)\text{sat}(s_a)\right] \qquad (5.88)$$

The control law of Equation 5.88 will ensure that the α-plane heading error δ_α will go to zero in a time that is less than the initial time-to-go in the presence of a maneuvering or nonmaneuvering target. For the calculation of the control law, the variables needed are the instantaneous values of the heading error, the missile velocity, the angle of attack, the pitch rate, and the initial time-to-go. Finally, the saturation function $\text{sat}(s_a)$ in Equation 5.88 is used to prevent chattering.

5.4.3 β-Plane SMIGC Control Law

This section outlines the derivation of an expression for the yaw fin deflection δ_r for control in the β-plane. Steps in this development are fairly similar to the α-plane SMIGC derivation, so the derivation will not be as elaborate as in the previous section.

Recall from Equation 5.65 that the heading error derivative in the β-plane is

$$\dot{\delta}_\beta = \frac{V_m}{R_\beta}\sin\delta_\beta + k_{Y_\beta}\beta + \frac{F}{R_\beta}A_{t_\alpha} + \frac{G}{R_\beta}A_{t_\beta}$$

Noting that the derivatives of the sideslip angle β and the yaw rate r are

$$\dot{\beta} = -k_{Y_\beta}\beta - r \qquad (5.89)$$

$$\dot{r} = k_{n_\beta}\beta + k_{n_\delta}\delta_r \qquad (5.90)$$

and again making the assumption that the target acceleration does not affect the higher-order derivatives of the heading error, Equation 5.65 is differentiated until the yaw fin deflection δ_r appears as

$$\dddot{\delta}_\beta = \left(\frac{V_m}{R_\beta}\right)^2 \sin(2\delta_\beta) + \frac{V_m}{R_\beta}\cos\delta_\beta k_{Y_\beta}\beta + k_{Y_\beta}(-r - k_{Y_\beta}\beta) \tag{5.91}$$

$$\dddot{\delta}_\beta = 2\left(\frac{V_m}{R_\beta}\right)^2 \sin(3\delta_\beta) + 3\left(\frac{V_m}{R_\beta}\right)^2 \cos(2\delta_\beta)k_{Y_\beta}\beta - \frac{V_m}{R_\beta}\sin\delta_\beta k_{Y_\beta}^2\beta^2$$

$$+ \frac{V_m}{R_\beta}\cos\delta_\beta k_{Y_\beta}(-r - k_{Y_\beta}\beta) + k_{Y_\beta}\left[-k_{N_\beta}\beta - k_{N_\delta}\delta_r - k_{Y_\beta}(-r - k_{Y_\beta}\beta)\right] \tag{5.92}$$

As was seen with the heading error in the α-plane, the control input appears in the third derivative of δ_β.

A sliding surface is now chosen as

$$s_\beta = \ddot{\delta}_\beta + 2\zeta\omega\dot{\delta}_\beta + \omega^2\delta_\beta = 0 \tag{5.93}$$

Next, the sliding surface is differentiated to yield

$$\dot{s}_\beta = \dddot{\delta}_\beta + 2\zeta\omega\ddot{\delta}_\beta + \omega^2\dot{\delta}_\beta = J - K\delta_r + \omega^2\left(\frac{F}{R_\beta}A_{t_\lambda} + \frac{G}{R_\beta}A_{t_\phi}\right) \tag{5.94}$$

where

$$J = 2\left(\frac{V_m}{R_\beta}\right)^3 \sin(3\delta_\beta) + 3\left(\frac{V_m}{R_\beta}\right)^2 \cos(2\delta_\beta)k_{Y_\beta}\beta + \frac{V_m}{R_\beta}\cos\delta_\beta k_{Y_\beta}^2\beta^2$$

$$+ \frac{V_m}{R_\beta}\cos\delta_\beta k_{Y_\beta}(-k_{Y_\beta}\beta - r) + k_{Y_\beta}\left[k_{Y_\beta}(k_{Y_\beta}\beta + r) - k_{n_\beta}\beta\right]$$

$$+ 2\zeta\omega\left[\left(\frac{V_m}{R_\beta}\right)^2 \sin(2\delta_\beta) + \frac{V_m}{R_\beta}\cos\delta_\beta k_{Y_\beta}\beta + k_{Y_\beta}(-k_{Y_\beta}\beta - r)\right]$$

$$+ \omega^2\left(\frac{V_m}{R_\beta}\sin\delta_\beta + k_{Y_\beta}\beta\right) \tag{5.95}$$

$$K = k_{Y_\beta}k_{n_\delta} \tag{5.96}$$

As was seen in the α-plane analysis, Equation 5.94 indicates that the sliding surface has a relative degree of 1.

In order to derive an expression for δ_r, a candidate Lyapunov function is chosen as

$$V = \frac{1}{2} s_\beta^2 \tag{5.97}$$

The Lyapunov function's derivative is then found as

$$\dot{V} = s_\beta \dot{s}_\beta = s_\beta \left[J - K\delta_r + \omega^2 \left(\frac{F}{R_\beta} A_{t_\lambda} + \frac{G}{R_\beta} A_{t_\phi} \right) \right] \tag{5.98}$$

Now it is simply necessary to select δ_r such that the Lyapunov function, and hence s_β, will be guaranteed to go to zero in a finite time in the presence of unknown target acceleration. This finite time must be less than the initial t_{go} so that the heading error δ_β will have enough time to become negligibly small according to the second-order system in Equation 5.93. Similar to the development in the previous section, an expression for δ_r that can satisfy these conditions is found as

$$\delta_r = \frac{1}{K} \left[J + \frac{|s_\beta(t_0)|}{t_{go}(t_0) - 3s} \operatorname{sat}(s_\beta) + \left(\left| \frac{F(t_0)}{R_\beta(t_0)} A_{t_\lambda \max} \right| + \left| \frac{G(t_0)}{R_\beta(t_0)} A_{t_\phi \max} \right| \right) \operatorname{sat}(s_\beta) \right] \tag{5.99}$$

The yaw fin deflection control law of Equation 5.99 guarantees that the sliding surface of Equation 5.93 will go to zero in a finite time that is less than or equal to 3 s before the initial t_{go} in the presence of a maneuvering or nonmaneuvering target.

5.5 SMIGC Results and Analysis

In this section, numerical results from the application of the SMIGC approach to a missile-intercept problem are presented for a set of stressed three-dimensional engagement scenarios. In each scenario, the missile must correct a large heading error in a small amount of time. Also, it is assumed that the engagement scenarios are post burn-out scenarios. In the inertial x–y–z frame, for these scenarios, the initial position vectors of the target and missile are $\mathbf{r}_{tgt} = [55{,}000 \text{ ft } 55{,}000 \text{ ft } 105{,}000 \text{ ft}]^T$ and $\mathbf{r}_{msl} = [0\ 0\ 50{,}000 \text{ ft}]^T$, and their initial velocity magnitudes are taken as V_{tgt} = 9842 ft/s and V_{msl} = 6561 ft/s. Also, it is assumed that the target is weaving at 5 g's in both planes at a frequency of 1 Hz. For this set of initial conditions, two different values for the initial heading error were chosen: 15° and 30°. In all simulations, the values of ζ and ω in Equations 5.74 and 5.93 are set to 0.707 and 5, respectively, and the boundary layer ε used in the saturation function is chosen as 0.001. The results from these simulations are shown in Figures 5.3 through 5.10.

Figures 5.19 and 5.20 show the missile and target trajectories for the 15° and 30° initial heading error cases. It can be seen that the SMIGC approach performs well in both cases, yielding miss distances of 0.1331 and 0.471 ft,

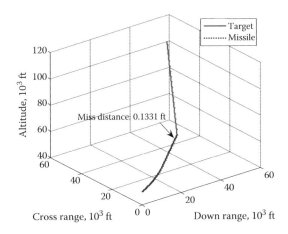

FIGURE 5.19
Missile and target trajectories with 15° initial heading error.

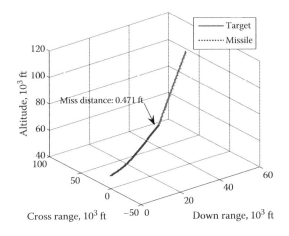

FIGURE 5.20
Missile and target trajectories with 30° initial heading error.

respectively. From Figures 5.21 and 5.22, it can be seen that the fin deflections and normal accelerations tend to increase as the initial heading error is
increased, although all the values stay within safe bounds. Note that the yaw
fin deflections are consistently higher than the pitch fin deflections. It makes
sense that the yaw fin deflections are higher since the engagement is primarily in the horizontal plane. These figures also indicate that, for both cases,
the deflections and normal accelerations all exhibit the same pattern of an
initially large value that tends toward zero as the heading error goes to zero.
For the 15° and 30° heading error cases, respectively, the maximum fin deflection magnitudes are 10° and 23°, and the maximum normal accelerations are

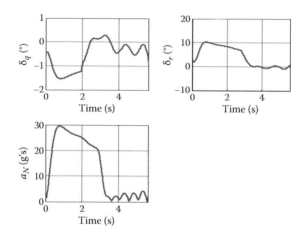

FIGURE 5.21
Fin deflections and normal acceleration with 15° initial heading error.

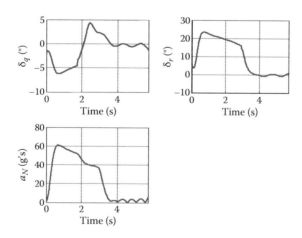

FIGURE 5.22
Fin deflections and normal acceleration with 30° initial heading error.

30 and 60 g's. The angle of attack and sideslip angle histories can be found in Figures 5.23 and 5.24, and it can be seen that the values exhibit the same trends as the fin deflections and normal accelerations. In particular, for each case, the missile initially rises to a large angle of attack and sideslip angle, and then the angles approach zero as the heading error goes to zero. For the 15° and 40° heading error cases, respectively, the maximum angle-of-attack magnitudes are 2.8° and 10°, and the maximum sideslip angles are 13.5° and 23°. Finally, Figures 5.25 and 5.26 show that the sliding surfaces and heading errors are successfully sent to zero before the end of the engagement scenarios as desired. It should be pointed out that the heading error histories

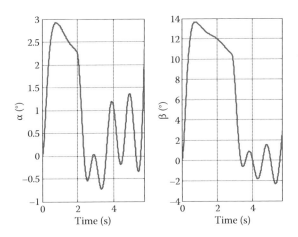

FIGURE 5.23
Angle of attack and sideslip angle with 15° initial heading error.

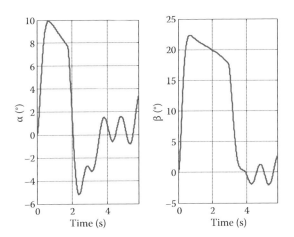

FIGURE 5.24
Angle of attack and sideslip angle with 30° initial heading error.

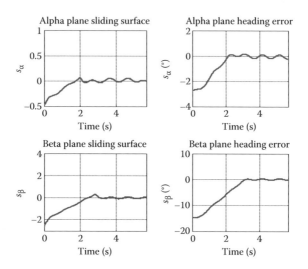

FIGURE 5.25
Sliding surfaces and heading errors with 15° initial heading error.

show oscillations near zero. Corresponding oscillations can be observed in fin deflections, angle of attack, and pitch rate histories toward the end. A low pass filter could be used in control calculations to smooth these oscillations. An alternate method is to use higher-order sliding mode techniques. It was, however, felt that the current simpler structure does an adequate job and leads to very low miss distances, and therefore, there was no need to go for more sophisticated options. It should also be observed that a saturation function is

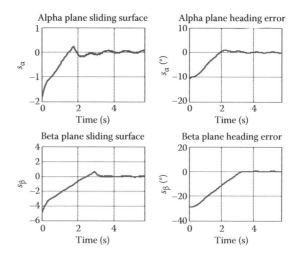

FIGURE 5.26
Sliding surfaces and heading errors with 30° initial heading error.

used, and only the bound of the target acceleration is known. Hence, while it cannot be guaranteed that the heading error will be sent exactly to zero, it *can* be guaranteed that the heading error can be made arbitrarily small.

5.6 Summary and Conclusions

In this chapter, the missile guidance and control designs were wrapped into a single IGC subsystem. The importance of selecting a proper and meaningful formulation was brought out. It was shown how the body-based IGC formulation was unable to fully capture the complex physics behind missile motion. Since the velocity-based IGC method incorporates the missile physics, it is much more effective in a wide range of engagement scenarios.

A new IGC method, called the sliding mode IGC (SMIGC), was presented. A promising aspect of this approach is that it does not require exact information about the target acceleration to be implemented but only its *bounds*. Results of the SMIGC approach with a 6-DOF nonlinear missile model were presented for a few taxing engagement scenarios. From the results, it appears that SMIGC can yield a favorable hit-to-kill accuracy against agile targets. Furthermore, the control laws were shown to result in moderate but reasonable fin deflection histories.

References

1. Anderson, G. M., "Comparison of optimal control and differential game intercept missile guidance laws," *Journal of Guidance and Control*, 4(2), 109–115, March–April, 1981.
2. Brown, M., Shtessel, Y., and Buffington, J., "Finite reaching time continuous sliding mode control with enhanced robustness," *Proceedings of the AIAA Guidance, Navigation, and Control Conference*, Denver, CO, August 14–18, 2000.
3. Bryson, A. E. and Ho, Y. C., *Applied Optimal Control*, Hemisphere Publishing Co., New York 1975.
4. Cloutier, J. R., "Time-to-go-less guidance with cross-channel couplings," AIAA Missile System Conference, Monterey, CA, 1996.
5. Cloutier, J. R., D'Souza, C. N., and Mracek, C. P., "Nonlinear regulation and nonlinear H_∞ control via the state-dependent Riccati equation technique," *Proceedings of the First International Conference on Nonlinear Problems in Aviation and Aerospace*, Embry-Riddle Aeronautical University Press, Daytona Beach, FL, May 1996.
6. Cloutier, J. R. and Stansbery, D. T., "Nonlinear, hybrid bank-to-turn/skid-to-turn missile autopilot design," AIAA Guidance, Navigation, and Control Conference and Exhibit, August 6–9, Montreal, Canada, 2001.

7. Evers, J. H., Cloutier, J. R., Lin, C. F., Yueh, W. R., and Wang, Q., "Application of integrated guidance and control schemes to a precision guided missile," *Proceedings of American Control Conference*, pp. 3225–3230, IEEE, Chicago, IL, 1992.

8. Hall, K. R., "Development and comparison of estimation algorithms for airborne missiles," AD-A141 535/5, Defense Technical Information Centre, 1983.

9. Idan, M., Shima, T., and Golan, O., "Integrated sliding mode autopilot—guidance for dual control missiles," Presented as paper AIAA-2005-6455 at the AIAA Guidance, Navigation, and Control Conference and Exhibit, San Francisco, CA, August 15–18, 2005.

10. Koren, A., Idan, M., and Golan, O., "Integrated sliding mode guidance and control for missile with on–off actuators," *Journal of Guidance, Control, and Dynamics*, 31(1), 204–214, 2008.

11. Lin, C. F. and Yueh, W. R., "Optimal controller for homing missiles," *Proceedings of American Control Conference*, IEEE, San Diego, CA, 1984.

12. Menon, P. K. and Ohlmeyer, E. J., "Integrated design of agile missile guidance and control systems," *Proceedings of the 7th Mediterranean Conference on Control and Automation*, Haifa, Israel, IEEE, June 28–30, 1999.

13. Menon, P. K. and Ohlmeyer, E. J., "Nonlinear integrated guidance and control laws for homing missiles," *AIAA Guidance, Navigation and Control Conference and Exhibit*, Montreal, Canada, August 6–9, 2001.

14. Ohlmeyer, E. J., "Standard missile guidance laws," NSWCDD Internal Report, Code G23, Naval Surface Warfare Center, Dahlgren Division, Dahlgren, VA, April 1994.

15. Palumbo, N. F. and Jackson, T. D., "Development of a fully integrated missile guidance and control system: a state dependent Riccati differential equation approach," *Proceedings of the IEEE Conference on Control Applications*, IEEE, Hawaii, HI, August 1999.

16. Shkolnikov, I., Shtessel, Y., and Lianos, D., "Integrated guidance-control system of a homing interceptor—sliding mode approach," Presented as paper AIAA-2001-4218 at the AIAA Guidance, Navigation, and Control Conference and Exhibit, Montreal, Canada, August 6–9, 2001.

17. Shtessel, Y. and Buffington, J., "Continuous sliding mode control," *Proceedings of the American Control Conference*, IEEE, Philadelphia, PA, 1998.

18. Shtessel, Y. and Shkolnikov, I., "Integrated guidance and control of advanced interceptors using second order sliding modes," *Proceedings of the 42nd IEEE Conference on Decision and Control*, IEEE, Maui, Hawaii, December 2003.

19. Shima, T., Idan, M., and Golan, O., "Sliding-mode control for integrated missile autopilot guidance," *Journal of Guidance, Control and Dynamics*, 29(2), 250–260, March–April 2006.

20. Song, T. L. and Speyer, J. L., "A stochastic analysis of a modified gain extended Kalman filter with application to bearings only measurements," *IEEE Transactions on Automatic Control*, AC-30(10), 940–949, October 1985.

21. Xin, M., "A new method for suboptimal control of a class of nonlinear systems," Ph.D. Dissertation, University of Missouri-Rolla, December 2002.

22. Xin, M. and Balakrishnan, S. N., "Missile longitudinal autopilot design using a new suboptimal nonlinear control method," *IEEE Proceedings on Control Theory Applications*, 150(6), 577–584, 2003.

23. Xin, M. and Balakrishnan, S. N., Stansbery, D. T., and Ohlmeyer, E. J., "Nonlinear missile autopilot design with Theta-D technique," *Journal of Guidance, Control and Dynamics*, Accepted and in print, 27(3), 406–417, 2004.
24. Xin, M., Balakrishnan, S. N., and Ohlmeyer, E., "Integrated guidance and control of missiles with Theta-D method," accepted for presentation at the 2004 IFAC Symposium on Aerospace, St. Petersburg, Russia. 16th IFAC symposium on Automatic Control in Aerospace, June 2004.

6

Higher-Order Sliding Modes for Missile Guidance and Control

Yuri B. Shtessel and Christian H. Tournes

CONTENTS

6.1 Introduction

6.1.1 Technical Challenges

The flight domain of aerospace vehicles has enlarged considerably over the last decades. As a consequence, the development of accurate plant models required for designing control laws becomes more and more expensive and difficult. Since the performance of Guidance Navigation and Control Algorithms developed using conventional control techniques is only as good as an underlying plant model used in their design, the robustness of the design presents an increasingly difficult technical challenge aggravated by the increasingly larger flight domain.

The challenge exists in the robust steering of the vehicles using a variety and possibly concurrent use of divert mechanisms such as aerodynamic lift, booster/sustainer orientation, and divert actuators, and use of either continuous actuators or discontinuous actuators in the presence of uncertain shock waves and actuator malfunctions. The concurrent use of several different divert mechanisms poses a major technical challenge for the design of the autopilot. This problem is aggravated by the fact that the different divert mechanisms exert disturbing effects on each other. This is the case when thrusters are fired; the fuel mass ejected creates a thickening of the boundary layer, which in turn modifies the shock wave system around the vehicle thereby altering the characteristics of the lift. Likewise, the pressure system governed by the shock wave alters the static pressure at the exit of the nozzle and thereby modifies corresponding specific impulses. Research results have shown that the relative degree of multiplicative disturbances may reach up to 30% [1].

6.1.2 Why Sliding Mode Control?

Control in the presence of uncertainty is one of the main topics of modern control theory. In the formulation of any control problem, there is always a discrepancy between the actual plant dynamics and its mathematical model used for the controller design. These discrepancies (or mismatches) mostly come from external disturbances, unknown plant parameters, and unmodeled dynamics. Designing control laws that provide the desired closed-loop system performance in the presence of these disturbances/uncertainties is a very challenging task for a control engineer. This has led to intense interest in the development of so-called nonlinear robust control methods [26], which are supposed to solve this problem. In spite of the extensive and successful development of robust adaptive control, H_∞ control, and back-stepping techniques [26], sliding mode control (SMC) [2, 23] remains, probably, the most successful approach in handling bounded uncertainties/disturbances and unmodeled dynamics.

Historically, sliding modes were discovered as a special mode in variable structure systems (VSSs). VSSs comprise a variety of structures. Certain rules are developed to switch between the structures in current time to achieve a suitable system performance, whereas using only a single fixed structure from the set of controllers could even be unstable. The result is VSS, which may be regarded as a combination of subsystems where each subsystem has a fixed control structure and is valid for specified regions of system behavior. It appeared to be that the closed-loop system may be designed to possess new properties not present in any of the constituent substructures alone. Furthermore, in a special mode, named a sliding mode, these properties include robustness to certain (so-called matched) external disturbances and model uncertainties, as well as to unmodeled dynamics. Achieving reduced order dynamics of the compensated system in a sliding mode (termed partial dynamical collapse) is also a very important useful property of sliding modes. The development of these novel ideas began in the Soviet Union in the late 1950s. The idea of SMC is based on the introduction of a "custom-designed" function, named the sliding variable. As soon as the properly designed sliding variable becomes equal to zero, it defines the sliding manifold (or the sliding surface in the linear case). The proper design of the sliding variable yields suitable closed-loop system performance while the system's trajectories belong to the sliding manifold. The idea of SMC is to steer the system trajectory to the properly chosen sliding manifold and then maintain motion on the manifold thereafter by means of control, thus exploiting the main features of the sliding mode: its insensitivity to external and internal disturbances matched by the control, ultimate accuracy, and finite-time reaching of the transient.

The SMC design approach consists of two components [2]. The first involves the design of a switching function so that the system motion on the sliding manifold (termed the sliding motion) satisfies the design specifications. The second is concerned with the selection of a control law, which will make the sliding manifold attractive to the system state in the presence of external and internal disturbances/uncertainties. Note that this control law is not necessarily discontinuous.

SMC-based observers allow estimation of the system's states in the presence of unknown external disturbances, which can also be explicitly reconstructed online by the observer [2].

The already matured classical SMC theory received a significant boost in the beginning of the 1990s: when a new "higher-order" paradigm was introduced [10, 11, 14]. The introduction of this new paradigm was dictated by the following reasons:

1. The classical sliding mode design approach requires the system relative degree to be equal to 1 with respect to the sliding variable. This can seriously constrain the choice of the sliding variable.
2. Also, very often, a sliding mode controller yields high-frequency switching control action that leads to the so-called "chattering effect," which is difficult to avoid or attenuate.

These intrinsic difficulties of classical SMC are mitigated by the higher-order sliding mode (HOSM) controllers [10–12, 14] that are able to drive to zero not only the sliding variable but also its $k - 1$ successive derivatives (kth-order sliding mode). The novel approach is effective for arbitrary relative degrees, and the well-known chattering effect is significantly reduced since the high-frequency control switching is "hidden" in the higher derivative of the sliding variable.

When implemented in discrete time, HOSM provides sliding accuracy proportional to the kth power of the time increment, which makes HOSM an enhanced-accuracy robust control technique. Since only the kth derivative of the sliding manifold is proportional to the high-frequency switching control signal, the switching amplitude is well attenuated at the sliding manifold level, which significantly reduces chattering.

The unique power of the approach is revealed by the development of practical arbitrary-order real-time robust exact differentiators, whose performance is proved to be asymptotically optimal in the presence of small Lebesgue-measurable input noises. The HOSM differentiators are used in advanced HOSM-based observers for estimation of the system's phase state in the presence of unknown external disturbances, which are also reconstructed online by the observers. In addition, HOSM-based parameter observers have been developed as well.

The combination of a HOSM controller with the above-mentioned HOSM-based differentiator produces a robust and exact output-feedback controller [10, 11, 14]. No detailed mathematical models of the plant are needed. SMC of arbitrary smoothness can be achieved by artificially increasing the relative degree of the system, significantly attenuating the chattering effect.

The practicality of the classical SMC and HOSM control and observation techniques is demonstrated by a large variety of applications that include direct current-to-direct current (DC/DC) and alternate current-to-direct current (AC/DC) power converters, control of AC and DC motors and generators, robotic control, and aircraft and *missile guidance and control* [3–9, 11, 13, 15–18, 25]. Integrated guidance and control is considered in the work of Sweriduk et al. [27] and Xin et al. [28] using optimal control techniques that usually lack robustness. Robust integrated guidance and automatic pilot using traditional SMC [2] was studied in the work of Idan and Shima [3]. The integrated controller uses a canard control to steer a first sliding variable representing the zero effort miss to zero, while a second, model-based sliding variable, representing pitch acceleration contributions of the angle of attack and tail control, is also driven to zero. Traditional SMC is used for developing robust to target maneuvers guidance laws for missile interceptors [3–9]. The approximation of traditional SMC by saturation functions is employed [3] to smooth out the guidance command by a price of losing robustness.

Development of the *smooth* robust guidance law to target maneuvers [11, 13] is essential for effectively following this law by autopilot and also for integrating guidance and autopilot [3, 5, 13].

The structure of this work is as follows. The fundamentals of traditional SMC are presented in Section 6.2. Section 6.3 presents the fundamentals of higher (second) order sliding mode (HOSM/SOSM) control and presents multiple HOSM/SOSM algorithms. Section 6.4 discusses the fundamental properties of HOSM/SOSM control. Section 6.5 presents the test case used to illustrate various algorithms, with discussion on the interception strategy and the notional ballistic missile interceptor, steered by continuous actuators during boost steered and by divert and attitude actuators during terminal flight. Section 6.6 presents the overall control architecture and the design of a smooth guidance law. Section 6.7 presents the inversion process that interrelates the inner loop inputs with the outer loop acceleration commands and several autopilot designs operating on continuous and discontinuous actuators. Finally, Section 6.8 presents the conclusions.

6.2 Fundamentals of Traditional SMC

Assume that a dynamic system is described by a system of differential equations:

$$\dot{x} = f(x) + B(x)u + \Delta f(x,t), \quad x \in \mathbb{R}^n, u \in \mathbb{R}^m \tag{6.1}$$

where x is a state; u is control; $B(x) \in \mathbb{R}^{n \times m}$ and $f(x) \in \mathbb{R}^n$ are known vector field and a matrix, correspondingly; $\Delta f(x, t)$ is matched by control [i.e., there exists $\lambda \in \mathbb{R}^m$ so that $\Delta f(x, t) = B(x)\lambda$] unknown norm-bounded vector field $\|\Delta f(x,t)\| \leq \tilde{L}$. The problem is to design the control function u that drives $x \to 0$ in the presence of unknown bounded disturbance/uncertainty $\Delta f(x, t)$.

In order to address the problem using *traditional SMC*, a *sliding variable* $\sigma = \sigma(x) \in \mathbb{R}^m$ is introduced so that stirring $x(t)$ to the *sliding surface*

$$\sigma(x) = 0 \tag{6.2}$$

yields $x \to 0$ as time increases.

Definition 6.1

The control $u = u(x)$ in Equation 6.1 that drives the state variable x to the sliding surface (Equation 6.2) in finite time t_r and keeps them there thereafter in the presence of the bounded disturbance/uncertainty $\Delta f(x, t)$, is called traditional *SMC* and *an ideal sliding mode* is said to be taking place in the system (Equation 6.1) for all $t > t_r$ [2]. ■

The system dynamics in the sliding mode is described by Equation 6.1 after substituting *equivalent control* that is defined as follows:

$$\dot{\sigma}(x) = G(x)(f(x) + B(x)u_{eq} + B(x)\lambda) = 0 \to$$

$$u_{eq} = -\left[G(x)B(x)\right]^{-1} G(x)f(x) - \lambda, \quad G(x) = \frac{d\sigma}{dx}. \tag{6.3}$$

Remark 6.1

It is clear from Equation 6.3 that for the existence of the sliding mode, it is necessary that $\det[G(x)B(x)] \neq 0$. This condition means that the sliding variable dynamics is supposed to have a vector-relative degree $r = \underbrace{\left[1,1,\ldots,1\right]}_{m}$. ■

Finally, the system's dynamics in the sliding modes becomes [2, 23]

$$\begin{cases} \dot{x} = f(x) - B(x)\left[G(x)B(x)\right]^{-1} G(x)f(x) \\ \sigma(x) = 0. \end{cases} \tag{6.4}$$

Apparently, that dynamics (Equation 6.4) is of a reduced $n - m$ order that is named *a partial dynamical collapse* of the system (Equation 6.1) in the sliding mode. Also, the sliding mode dynamics (Equation 6.4) is insensitive to the matched bounded disturbance/uncertainty $\Delta f(x, t)$.

The trajectory $x(t)$ of the system (Equation 6.1) can be stirred to the sliding surface (Equation 6.2) in finite time t_r and kept there thereafter by means of the SMC:

$$u = -\left[G(x)B(x)\right]^{-1}\left[G(x)f(x) + (\rho + L)\frac{\sigma}{\|\sigma\|}\right], \quad \|G(x)\Delta f(x)\| \le L, \rho > 0. \quad (6.5)$$

The SMC (Equation 6.5) is discontinuous on the sliding surface (Equation 6.2). Furthermore, since $\frac{\sigma}{\|\sigma\|} \to \text{sign}(\sigma)$ as $\sigma \to 0$ in finite time $t_r \le \frac{\|\sigma(0)\|}{\rho}$, the SMC (Equation 6.5) changes its sign in the vicinity of $\sigma = 0$. This fact could yield a control chattering (zigzag motion) if the control sign does not change its sign exactly on $\sigma = 0$. Traditional SMC achieves insensitivity to matched disturbances by high-frequency switching of the control. While high-frequency switching of the control is perfectly acceptable when it applies to electrical or electronic circuits, it is not a viable solution for aerospace vehicles.

A possible remedy is introducing a boundary layer with high gain control [26], that is, replacing $\frac{\sigma}{\|\sigma\|} \to \frac{\sigma}{\|\sigma\| + \varepsilon}, \varepsilon > 0$, but this approximation is unfortunately detrimental to robustness.

A phase portrait of a sliding mode with a zigzag motion (chattering) in a generic second-order system is illustrated in Figure 6.1.

In order to attenuate chattering, it is desirable to hide a discontinuous high-frequency switching portion of the SMC $\frac{\sigma}{\|\sigma\|}$ under the second- or higher-order derivative of the sliding variable. The second- and higher-order SMC techniques are invented to achieve this goal [10, 11, 14].

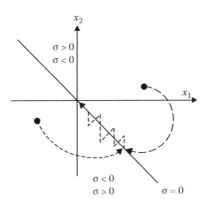

FIGURE 6.1
Traditional SMC.

6.3 Fundamentals of HOSM/SOSM Control

Assume that a dynamic system is described by a system of differential equations (Equation 6.1) with $m = 1$, that is, with a scalar control function u. The sliding variable dynamics is derived:

$$\dot{\sigma}(x) = G(x)f(x) + G(x)B(x)u + G(x)\Delta f(x, t). \tag{6.6}$$

Suppose that $G(x)B(x) = 0$; then differentiating Equation 6.6, we obtain

$$\ddot{\sigma}(x) = \underbrace{\varphi_0(x) + \Delta\varphi_0(x, t)}_{\varphi_0(x,t)} + b(x)u \tag{6.7}$$

where

$$\varphi_0(x) = \dot{G}(x)f(x) + G(x)\frac{df(x)}{dx}f(x)$$

$$\Delta\varphi_0(x, t) = G(x)\frac{df(x)}{dx}\Delta f(x, t) + \dot{G}(x)\Delta f(x, t) + G(x)\Delta\dot{f}(x, t) \tag{6.8}$$

$$b(x) = G(x)\frac{df(x)}{dx}B(x)$$

where the control u drives simultaneously $\sigma \to 0$ and $\dot{\sigma} \to 0$ in finite time.

Definition 6.2

Considering the nonlinear sliding manifold (Equation 6.2), the control $u = u(x)$ in Equations 6.1 and 6.6 that drives $\sigma \to 0$ and $\dot{\sigma} \to 0$ in finite time $\bar{t}_r$ and keeps them there thereafter in the presence of a bounded disturbance $\Delta f(x, t)$ is called the *second-order sliding mode (SOSM) control*, and *an ideal SOSM is said to be taking place* in system 6.1 for all $t > \bar{t}_r$ [10]. ∎

Remark 6.2

SOSM control handles a finite-time control problem in Equation 6.7 of relative degree 2 unlike traditional SMC that handles a finite-time single input control problem in Equation 6.3 of a relative degree 1. ∎

The SOSM control is illustrated in Figure 6.2.

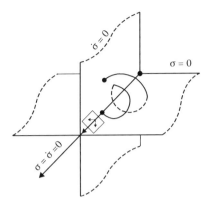

FIGURE 6.2
Second-order SMC.

Assume that

$$0 < K_m \le \frac{\partial}{\partial u} \ddot{\sigma} \le K_M \quad |\ddot{\sigma}|_{u=0} \le C \tag{6.9}$$

holds globally for some K_m, K_M, $C > 0$. Then Equations 6.7 and 6.9 generate the differential inclusion

$$\ddot{\sigma} \in [-C, C] + [K_m, K_M]u. \tag{6.10}$$

Most two-sliding controllers may be considered as controllers for Equation 6.9 steering $\sigma, \dot{\sigma} \to 0$ in (preferably) finite time. Such controllers are obviously robust with respect to any perturbations preserving Equation 6.9.

Hence, the problem is to find a feedback control law:

$$u = u(\sigma, \dot{\sigma})$$

such that all the trajectories of Equations 6.9 and 6.10 converge in finite time to the origin $\sigma = \dot{\sigma} = 0$ of the phase plane $\sigma, \dot{\sigma}$.

Several of the simplest and most popular controllers solving this problem are presented below.

6.3.1 Twisting Controller

The twisting controller [10, 14] was historically the first two-sliding controller. It is defined by the formula

$$u = -r_1 \text{sign}(\sigma) - r_2 \text{sign}(\dot{\sigma}), \quad r_1 > r_2 > 0 \tag{6.11}$$

where r_1 and r_2 satisfy the conditions

$$(r_1 + r_2)K_m - C > (r_1 - r_2)K_M + C$$

$$(r_1 - r_2)K_m > C$$

(6.12)

It is proven [7, 14] that the controller (Equations 6.11 and 6.12) guarantees the appearance of SOSM $\sigma = \dot{\sigma} = 0$ in Equation 6.7 attracting the trajectories in finite time:

$$T \le \frac{|\dot{\sigma}(0)|}{(1-q)[K_m(r_1 - r_2) - C]}, q = \left[\frac{K_M(r_1 - r_2) + C}{K_M(r_1 + r_2) - C}\right]^{1/2} < 1.$$

(6.13)

6.3.2 Control Algorithm with Prescribed Convergence Law

It is known [10, 14] that a solution of a differential equation

$$\dot{\sigma} + \lambda|\sigma|^{1/2} \operatorname{sign}(\sigma) = 0$$

(6.14)

and its derivative converge to zero ($\sigma, \dot{\sigma} \to 0$) in finite time. The idea is to stabilize Equation 6.14 using σ-dynamics of relative degree 2 given by Equation 6.7 by means of traditional SMC. It yields to the SOSM controller with *prescribed convergence law* [7, 14]:

$$u = -\rho \cdot \operatorname{sign}\left[\dot{\sigma} + \lambda|\sigma|^{1/2} \operatorname{sign}(\sigma)\right], \quad \rho, \lambda > 0, \rho K_m - C > \lambda^2/2$$

(6.15)

that drives $\sigma, \dot{\sigma} \to 0$ in Equation 6.7 in finite time.

6.3.3 SOSM Control Based on Nonlinear Dynamic Sliding Manifold

In order to avoid differentiation of σ in Equation 6.15, SOSM control u can be also designed based on nonlinear dynamic sliding manifold (NDSM) as follows [24, 25]:

$$u = -\bar{\rho}\operatorname{sign}(J), \quad J = \sigma + \chi$$

$$\dot{\chi} = \lambda|\sigma|^{1/2} \operatorname{sign}(\sigma) - \beta|J|^{1/2} \operatorname{sign}(J)$$

(6.16)

where J is the nonlinear dynamic sliding variable; the coefficients $\beta > \lambda > 0$ and $\bar{\rho} > 0$ are sufficiently large. We will call the surface $J = 0$ described by the system 6.15 the NDSM.

It is proven [24, 25] that the control law (Equation 6.15) provides for the finite-time convergent second-order sliding mode $J = \dot{J} = 0$.

As soon as the second-order sliding mode, $J = \dot{J} = 0$, is established, the sliding variable second-order dynamics are described by Equation 6.14 and converges to zero in finite time, which is equal to $2\sqrt{e(t_J)}/\beta_2$, where t_J is the moment when NDSM dynamics reach zero, that is, $J = \dot{J} = 0$.

It is worth noting that all algorithms, the twisting, the prescribed convergence law, and the SOSM/NDSM algorithm, are high-frequency switching control laws. However, only the second derivative of the sliding variable σ is discontinued while continuity of $\sigma, \dot{\sigma}$ is retained. It can be used to generate commands to the "on–off" actuators in the systems of relative degree 2. In order to achieve a given frequency of control switching, the sliding variable σ or NDSM J can be mixed with a dither signal of a given frequency. In this case, the control functions 6.11, 6.15, and 6.16 will be pulse width modulated.

When divert acceleration is continuous, which is the case with aerodynamic divert, or when the components of booster/sustainer thrust perpendicular to the velocity commanded acceleration are instrumental to the divert acceleration, acceleration commands need to be smooth.

Assume that the inner loop autopilot commands are pitch rate commands. In such a case, the autopilot needs to calculate prescribed attitude rates through a process called inversion discussed in Section 6.6. Commanded rotation angles from velocity axes to body axes are calculated from commanded acceleration, and the calculation of corresponding body rates requires a differentiation. Consequently, for prescribed attitude rates to be continuous, corresponding prescribed acceleration commands used in their calculation need to be smooth.

The next three subsections are dedicated to continuous/smooth SOSM control that can be used for designing guidance commands.

6.3.4 Quasi-Continuous Control Algorithm

An important class of HOSM controllers comprises the so-called *quasi-continuous* controllers, featuring *continuous* control everywhere except at the two-sliding mode $\sigma = \dot{\sigma} = 0$ itself. Since the two-sliding condition requires the simultaneous fulfillment of two exact equalities, generally, the trajectories never hit the two-sliding set. Hence, in practice, the condition $\sigma = \dot{\sigma} = 0$ is never fulfilled, and the control remains a continuous function at all times. The larger the noises and switching imperfections become, the worse the accuracy is and the slower the changing rate of u. As a result, chattering is significantly reduced. The following is the two-sliding controller with such features [10]:

$$u = -\rho \cdot \frac{\dot{\sigma} + \beta |\sigma|^{1/2} \operatorname{sign}(\sigma)}{|\sigma| + \beta |\sigma|^{1/2}}. \tag{6.17}$$

This control is continuous everywhere except at the origin. It vanishes on the parabola $\dot{\sigma} + \beta |\sigma|^{1/2} \operatorname{sign}(\sigma) = 0$. For sufficiently large ρ, there exist numbers $\rho_1, \rho_2 : 0 < \rho_1 < \beta < \rho_2$ such that all trajectories enter the region between the curves $\dot{\sigma} + \rho_i |\sigma|^{1/2} \operatorname{sign}(\sigma) = 0, i = 1, 2$ and cannot leave.

6.3.5 Supertwisting Controller

Consider once more the σ-dynamics described by a system of relative degree 1 and suppose that

$$\dot{\sigma}(x) = \varphi(x, t) + b(x)u, \quad \sigma, u \in \mathbb{R}. \tag{6.18}$$

Furthermore, suppose that for some positive constants C, K_M, K_m, U_m, q

$$|\dot{\varphi}| \leq C, \quad 0 \leq K_m \leq b(x) \leq K_M, \quad |\varphi/b| < q U_M, \quad 0 < q < 1. \tag{6.19}$$

The following continuous controller, named *supertwisting controller* [14], does not need measurements of $\dot{\sigma}$. Specifically, define [10, 14]

$$u = -\varpi_1 |\sigma|^{1/2} \operatorname{sign}(\sigma) + u_1, \quad \dot{u}_1 = \begin{cases} -u, & |u| > U_M, \\ -\varpi_2 \operatorname{sign}(\sigma), & |u| \leq U_M. \end{cases} \tag{6.20}$$

With $K_m \varpi_2 > C$ and ϖ_1 sufficiently large, the *supertwisting* controller (Equation 6.20) provides a two-sliding mode $\sigma = \dot{\sigma} = 0$ attracting trajectories in finite time. The control u enters within finite time the segment $[-U_M, U_M]$ and stays there.

In particular, the terms ϖ_1, ϖ_2 can be calculated based on the upper bound $|\dot{\varphi}| \leq C$ of the first time derivative of the unknown right-hand side term as

$$\varpi_1 = \frac{1.5}{K_m} \sqrt{C}, \varpi_2 = \frac{1.1}{K_m} C. \tag{6.21}$$

It is the only known continuous controller that drives the output of the relative degree 1 system 6.18 to 0 ($\sigma, \dot{\sigma} \to 0$) in finite time in presence of uncertainties and disturbances. In other words, it can be used instead of a traditional sliding mode controller in order to avoid chattering.

The supertwisting control (Equation 6.20) is also used in SMC differentiators that can be used, in particular, for differentiating σ in the SOSM control laws (Equations 6.11, 6.15, and 6.17).

Let the generic signal $f(t)$ consist of a bounded Lebesgue-measurable noise with unknown features and an unknown base signal $f_0(t)$ with $\left|\ddot{f}_0\right| \leq L$ having a known Lipschitz constant $L > 0$. The problem is to find real-time robust estimations of $\dot{f}_0(t)$, with $\dot{f}_0(t)$ being exact in the absence of a measurement noise.

The proposed supertwist–based differentiator has a form

$$\dot{z}_0 = -\lambda_1 \left|z_0 - f(t)\right|^{1/2} \text{sign}(z_0 - f(t)) + z_1$$
$$\dot{z}_1 = -\lambda_0 \text{sign}(z_0 - f(t))$$

(6.22)

where z_1 can be taken as the differentiator output [10, 14].

It is known [14, 15] that in the absence of noise, for any $\lambda_0 > L$ and for λ_1 satisfying the condition $\dfrac{2(\lambda_0 + L)^2}{\lambda_1^2(\lambda_0 - L)} < 1$, $z_1 \to \dot{f}_0, z_0 \to f_0$ in finite time.

In a case of the measurement noise that satisfies the inequality $|f(t) - f_0(t)| \leq \varepsilon$, the following inequalities are established in finite time for some positive constants μ_1, μ_2, μ_3 depending exclusively on the parameters of the differentiator [14, 15]:

$$\left|z_0 - f_0(t)\right| \leq \mu_1 \varepsilon, \quad \left|z_1 - \dot{f}_0(t)\right| \leq \mu_2 \varepsilon^{1/2}, \quad \left|v - \dot{f}_0(t)\right| \leq \mu_3 \varepsilon^{1/2}.$$

(6.23)

6.3.5.1 Comparison of Supertwisting and Traditional SMC

In order to illustrate a difference between traditional SMC and SOSM performances, supertwisting $u = -2|\sigma|^{1/2} \text{sign}(\sigma) - 3\int \text{sign}(\sigma) d\tau$ and traditional SMC $u = -3\text{sign}(\sigma)$ were used to drive a sliding variable σ in Equation 6.18 to zero with $\varphi(x, t) = \sin(2t)$, $b(x) = 1$. The results of the simulations are shown in Figures 6.3 through 6.5.

Both sliding variables reach zero in finite time in *presence of unknown bounded disturbance*. However, SOSM supertwisting control achieves this goal being continuous, whereas traditional SMC is a discontinuous high-frequency switching function.

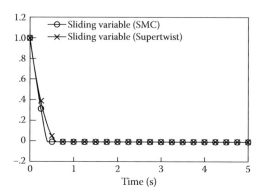

FIGURE 6.3
Sliding variables.

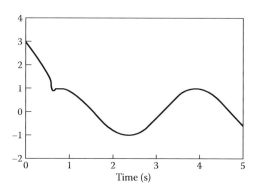

FIGURE 6.4
SOSM supertwisting control function.

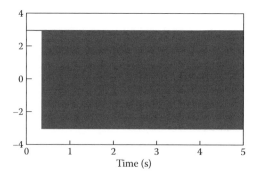

FIGURE 6.5
Traditional sliding mode control function.

6.3.6 Smooth SOSM Control

Consider once more the σ-dynamics described by a system of relative degree 1 in Equation 6.18. The problem that is addressed in this section is to design *smooth* control u that drives $\sigma, \dot{\sigma} \to 0$ [smooth SOSM (SSOSM)] in finite time.

Remark 6.3

This control is supposed to be a very good candidate for designing a guidance law that is robust to target maneuvers. ∎

The drift term $\varphi(x, t)$ is to be cancelled by means of a special observer to be developed further. The prescribed compensated σ-dynamics in Equation 6.18 is chosen as [11]

$$\begin{cases} \dot{x}_1 = -\alpha_1 |x_1|^{(p-1)/p} \operatorname{sign}(x_1) + x_2, \\ \dot{x}_2 = -\alpha_2 |x_1|^{(p-2)/p} \operatorname{sign}(x_1), \quad \sigma = x_1. \end{cases} \tag{6.24}$$

It is known [11] that if $p \geq 2$, $\alpha_1, \alpha_2 > 0$ in system 6.22, then $x_1, x_2 \to 0$ or $\sigma, \dot{\sigma} \to 0$ in finite time [22].

6.3.6.1 Nonlinear Disturbance Observer/Differentiator

The sliding variable dynamics (Equation 6.18) are sensitive to the unknown bounded drift term $\varphi(x, t)$ that can be estimated using the HOSM observer [10, 11].

Let the variables $\sigma(t)$ and $u(t)$ be available in real time, and $|\varphi^{(m-1)}| \leq L$ has a known Lipschitz constant $L > 0$. The control function $u(t)$ is Lebesgue measurable.

Consider the following HOSM observer [10]:

$$\begin{cases} \dot{z}_0 = v_0 + b(x)u, \\ v_0 = -\lambda_0 L^{1/(m+1)} |z_0 - \sigma|^{m/(m+1)} \operatorname{sign}(z_0 - \sigma) + z_1, \\ \dot{z}_1 = v_1, \\ v_1 = -\lambda_1 L^{1/m} |z_1 - v_0|^{(m-1)/m} \operatorname{sign}(z_1 - v_0) + z_2, \\ \dots\dots\dots\dots\dots\dots\dots\dots\dots\dots\dots\dots\dots\dots\dots\dots\dots\dots\dots\dots\dots \\ \dot{z}_{m-1} = v_{m-1}, \\ v_{m-1} = -\lambda_{m-1} L^{1/2} |z_{m-1} - v_{m-2}|^{1/2} \operatorname{sign}(z_{m-1} - v_{m-2}) + z_m, \\ \dot{z}_m = -\lambda_m L \operatorname{sign}(z_m - v_{m-1}). \end{cases} \tag{6.25}$$

It is proven in [10] that if $\sigma(t)$ and $u(t)$ are measured with some Lebesgue-measurable noises bounded respectively by $\varepsilon > 0$ and $k\varepsilon^{(m-1)/m}$, and $k > 0$ is any fixed constant, then the following inequalities are established in finite time for some positive constants μ_i, η_i depending exclusively on k and the choice of parameters:

$$\begin{cases} \left| z_0 - \sigma(t) \right| \leq \mu_0 \varepsilon \\ \dotfill \\ \left| z_i - \varphi^{(i-1)}(t) \right| \leq \mu_i \varepsilon^{(m-i+1)/(m+1)}, i = 1, \ldots, m \\ \left| v_j - \varphi^{(j)}(t) \right| \leq \eta_j \varepsilon^{(m-j)/(m+1)}, j = 0, \ldots m-1 \end{cases} \qquad (6.26)$$

where the parameters λ_i are being chosen sufficiently large in the reverse order.

In particular, in the absence of input noises, the exact equalities are established in finite time:

$$z_0 = \sigma(t), z_1 = \varphi, \ldots, z_i = v_{i-1} = \varphi^{(i-1)} \ i = 1, \ldots m. \qquad (6.27)$$

It is worth noting that parameters λ_i can be chosen recursively so that parameters $\lambda_0, \ldots, \lambda_k$, which are valid for $m = k$, can serve (after changing the notation) as $\lambda_1, \ldots, \lambda_{k+1}$ with $m = k + 1$, which means that only λ_0 is to be assigned. The simulation-checked set 8, 5, 3, 2, 1.5, 1.1 is sufficient for the observer design with $m \leq 5$ [10].

6.3.6.2 Disturbance Cancellation

The prescribed compensated σ-dynamics (Equation 6.24) with $p = m + 1$, $m \geq 1$, α_1, $\alpha_2 > 0$ is easily provided using the HOSM disturbance observer (Equation 6.26) via control u [11]:

$$\begin{cases} u = -z_1 - \alpha_1 \left| \sigma \right|^{m/(m+1)} \text{sign}(\sigma) + w \\ \dot{w} = -\alpha_2 \left| \sigma \right|^{(m-1)/(m+1)} \text{sign}(\sigma). \end{cases} \qquad (6.28)$$

When exact measurements are available, z_1 becomes equal to $\varphi(x, t)$ in a finite time, and the σ-dynamics are described by the finite-time stable system 6.24 thereafter.

In particular, for $m = 2$, SSOSM control law (Equations 6.25, 6.27, and 6.28) becomes [11]

$$\begin{cases} u = -z_1 - \alpha_1 \left| \sigma \right|^{2/3} \text{sign}(\sigma) + w \\ \dot{w} = -\alpha_2 \left| \sigma \right|^{1/3} \text{sign}(\sigma) \end{cases} \qquad (6.29)$$

with

$$
\begin{cases}
\dot{z}_0 = v_0 + b(x)u \\
v_0 = -\lambda_0 L^{1/2} \left| z_0 - \sigma \right|^{2/3} \text{sign}(z_0 - \sigma) + z_1 \\
\dot{z}_1 = v_1 \\
v_1 = -\lambda_1 L^{1/2} \left| z_1 - v_0 \right|^{1/2} \text{sign}(z_1 - v_0) + z_2 \\
\dot{z}_2 = -\lambda_2 L \text{sign}(z_2 - v_1)
\end{cases}
\tag{6.30}
$$

where $z_1 \to \varphi(x, t)$ in finite time.

The controls Equations 6.28 and 6.30 can be interpreted as SSOSM controls since, being smooth, they provide finite-time convergence $\sigma, \dot{\sigma} \to 0$.

It is worth noting that the supertwisting control (Equation 6.20) solves the same problem; however, it is only continuous but not smooth as the controls Equation 6.28 or Equation 6.30.

Remark 6.4

The generalized kth-order HOSM control algorithms u that can handle the kth-order sliding variable dynamics of the form [10]

$$
\sigma^{(k)} = \varphi(x, t) + b(x)u
\tag{6.31}
$$

driving the sliding variable and its $(k - 1)$ consecutive derivatives to zero $(\sigma, \dot{\sigma}, \ldots, \sigma^{(k-1)} \to 0)$ in finite time are available [10]. Study of HOSM algorithms for $k \geq 3$ is beyond the scope of this work. ∎

6.4 Discussion on Properties of Traditional and Higher-Order SMC

Concluding our study of traditional and higher-order SMC, we could summarize their properties as follows.

(1) *Insensitivity to matched disturbances.* This property of HOSM control is common for both traditional SMC and SOSM. Different elements of the interceptor/kill vehicle (KV; such as the observers, the guidance, and autopilot) can operate in an integrated fashion where a controller automatically compensates for the disturbances created by other elements without having to estimate their corresponding disturbing effects.

(2) *Dynamical collapse.* Unlike traditional SMC that guarantees only partial dynamical collapse (reduction of system's order in the sliding mode by 1 that is achieved in finite time), HOSM can achieve full dynamical collapse (reduction of system's dynamics to algebraic equations in finite time) since it can handle the system's dynamics of an arbitrary relative degree. For instance, SOSM control achieves a reduction of system's order by 2. This is much more than an academic distinction; it means that at least in the absence of noise, when the sliding mode is reached, the transfer function of inner loops with relative degree greater than 1 could become an identity.

(3) *Continuous/smooth guidance laws.* SOSM/HOSM controllers can yield continuous and even smooth controls that are applicable in multiple-loop integrated guidance/autopilot control laws. This is especially important in our application where the guidance generates acceleration commands, the time derivatives thereof that are used in the autopilot to calculate commanded attitude rates. Should the commands generated by the guidance not be smooth, that would cause their time derivatives used to calculate the autopilot's attitude rate reference profiles to be discontinuous, which is something the designer needs to avoid.

(4) *Continuous/discontinuous actuators.* SOSM/HOSM techniques are nonlinear robust control techniques. Unlike designs based on linear control laws, which, when discontinuous actuators such as on–off actuators must be used, require the redesign of the initial control law into a discontinuous control law that approximates the effects of the initial control law, HOSM design produces directly, when need arises, an on–off pulse-width-modulated control law that achieves the same level of accuracy as a linear control law.

6.5 Mathematical Model and Problem Formulation

6.5.1 Interception Geometry

The missile-target engagement geometry is shown in Figure 6.6.

Relative kinematics along and perpendicular to the line of sight (LOS) are represented hereafter as [5, 11, 13]

$$\begin{cases} \dot{r} = V_{\|}, \\ \dot{V}_{\|} = V_{\perp}^2/r + \Gamma_{\|} + \sin(\gamma - \lambda)\Gamma, \\ \dot{\lambda} = V_{\perp}/r, \\ \dot{V}_{\perp} = -V_{\|}V_{\perp}/r + \Gamma_{\perp} - \cos(\gamma - \lambda)\Gamma, \end{cases} \tag{6.32}$$

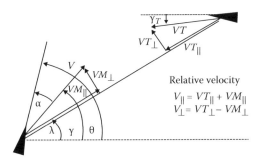

FIGURE 6.6
Interception geometry.

where r is the range along the LOS, λ is the LOS angle, γ is a missile flight path angle, $\dot{\lambda} = \omega_\lambda$ (in radians per square meter) is the LOS rate, $V_\perp = r\omega_\lambda$ (in meters per second) is a transversal component of relative velocity in the reference frame rotating with the LOS, Γ is the missile normal acceleration, and $\Gamma_\parallel$ and $\Gamma_\perp$ (disturbances; in meters per square second) are projections of bounded target acceleration along and orthogonal to the LOS. The target model used normal sinusoidal maneuvers as well as square wave maneuvers.

6.5.2 Missile Model

We are going to illustrate applications of HOSM/SOSM control to the missile guidance, navigation and control (GN&C) in the case of a missile steered by aerodynamic lift, divert, quasi-center of gravity (CG) on–off divert thrusters, and an orientable booster/sustainer as shown in Figure 6.7.

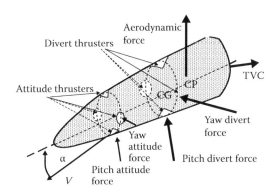

FIGURE 6.7
Proposed missile.

The dynamics of such missile steered by combined effects of divert actuators and pitch maneuver are given by [13, 16–19]:

$$\dot{\alpha} = q - Z_\alpha(1+\bar{d}_\alpha)\alpha + \frac{g}{V}\cos(\gamma) - Z_\delta(1+\bar{d}_\delta)\cos(\alpha)\delta - Z_\Lambda(1+\bar{d}_\Lambda)\cos(\alpha)\Delta - \cos(\alpha)Z_\varsigma\varsigma$$

(6.33)

$$\dot{q} = M_\alpha(1+\bar{d}_\alpha)\alpha + M_q q + M_\Lambda(1+\bar{d}_\Lambda)\Delta + M_\delta(1+\bar{d}_\delta)\delta + M_\varsigma\varsigma$$ (6.34)

$$\dot{\gamma} = Z_\alpha(1+\bar{d}_\alpha)\alpha - \frac{g}{V}\cos(\gamma) + Z_\delta(1+\bar{d}_\delta)\cos(\alpha)\delta + Z_\Lambda(1+\bar{d}_\Lambda)\cos(\alpha)\Delta + \cos(\alpha)Z_\varsigma\varsigma$$

(6.35)

where $Z_\alpha = \dfrac{\rho S V C_{L_\alpha}}{2m}, Z_\delta = \dfrac{T_{\max\delta}}{mV}, Z_\Lambda = \dfrac{T_{\max\Lambda}}{mV}, Z_\varsigma = \dfrac{T_s}{mV}, M_\delta = \dfrac{a_\delta T_{\max\delta}}{I_{yy}}, M_\Lambda =$

$\dfrac{a_\Lambda T_{\max\Lambda}}{I_{yy}},$ and $M_\varsigma = \dfrac{a_s T_s}{I_{yy}}$, and α, γ, and q represent the angle of attack, flight path angles (in radians), and pitch rate (in radians per second), respectively; V is the longitudinal velocity of a missile (meters per second). The cumulative disturbances $\bar{d}_\alpha$, $\bar{d}_\delta$, and $\bar{d}_\Lambda$ represent the unknown interactions between attitude thruster jets, divert thruster jets, and shockwaves, as well as bounded slow-varying perturbations/uncertainties in the stability derivatives. Here it is assumed that $1+\bar{d}_i > 0, i = \alpha, \delta, \Lambda$. Random disturbance samples $\bar{d}_\alpha$, $\bar{d}_\delta$, and $\bar{d}_\Lambda$ are uniformly distributed in intervals $\pm D_\alpha$, $\pm D_\delta$, and $\pm D_\Lambda$, respectively.

Actuator dynamics of divert and attitude actuators and continuous attitude actuator are given by

$$\dot{\Delta} = \frac{1}{\tau_\Lambda}(-\Delta + u_\Lambda); \quad \dot{\delta} = \frac{1}{\tau_\delta}(-\delta + u_\delta); \quad \dot{\varsigma} = \frac{1}{\tau_\varsigma}(-\varsigma + u_\varsigma)$$ (6.36)

where δ and Δ are the normalized attitude and divert actuator forces, respectively; ς represents the continuous attitude actuator deflection, and u_Λ, u_δ, and u_ς are the control inputs to the actuators.

Missile acceleration normal to the velocity vector is related to the flight path angle rate, without account for gravity, as follows:

$$\Gamma = \dot{\gamma} \cdot V.$$ (6.37)

The problem is in designing the feedback control law in terms of $\mathbf{u} = \{u_\Lambda, u_\delta, u_\varsigma\}^T$ that provides intercepting a maneuvering target by an impact via driving $r \to 0$ as time increases or via providing for $|r(t^*)| \leq r^0$ (the r^0 value is to be defined based on the size of the target) that implies the zero intercept at $t_{int} \leq t^*$ in presence of bounded model uncertainties/disturbances $\bar{d}_\alpha$, $\bar{d}_\delta$, and $\bar{d}_\Lambda$ and bounded unknown target accelerations $\Gamma_\|$ and $\Gamma_\perp$. The zero-intercept option can be interpreted as achieving uniform ultimate boundedness (UUB).

6.5.3 Interception Strategy

The following intercept strategy [5, 11, 13] that yields a direct hit (zero intercept) is proposed:

$$V_\perp = c_0\sqrt{r} \tag{6.38}$$

where $c_0 > 0$ is some constant.

The viability of the intercept strategy (Equation 6.38) is studied in the following theorem.

Theorem 6.1

Assume that

(a) The intercept strategy in Equation 6.38 is enforced by means of the control law $\mathbf{u} = \{u_\Delta, u_\delta, u_\varsigma\}^T$.

(b) $|\Gamma_\|| \leq \Gamma_\|^{LIM}$ and $|\Gamma| \leq \Gamma_{max}$; then there exist the parameter values $V_\|(0) < 0$ and $c_0 > 0$ that make the condition $|r(t^*)| \leq r^0$ valid at given time $t_{int} \leq t^*$. ∎

Proof

Assumption (a) yields the following compensated engagement kinematics (Equation 6.32):

$$\begin{cases} \dot{r} = V_\| \\ \dot{V}_\| = c_0^2 + \Gamma_\| + \sin(\gamma - \lambda)\Gamma \\ \dot{\lambda} = V_\perp/r. \end{cases} \tag{6.39}$$

Integrating Equation 6.39 taking into account assumption (b) and $|\sin(\gamma - \lambda)| < c_1 < 1$, the following inequality is obtained:

$$r(t) \leq r(0) + V_{\parallel}(0)t + \frac{c_0^2 + \Gamma_{\parallel}^{LIM} + c_1\Gamma_{\max}}{2}t^2. \tag{6.40}$$

The minimal value of $r(t)$ is identified as

$$r(t^*) \leq r(0) - \frac{V_{\parallel}^2(0)}{2\left(c_0^2 + \Gamma_{\parallel}^{LIM} + c_1\Gamma_{\max}\right)} \tag{6.41}$$

and zero intercept is achieved at

$$t^* = -\frac{V_{\parallel}(0)}{\left(c_0^2 + \Gamma_{\parallel}^{LIM} + c_1\Gamma_{\max}\right)}. \tag{6.42}$$

The parameters $V_{\parallel}(0) < 0$ and $c_0 > 0$ that yield zero intercept, that is, $|r(t^*)| \leq r^0$, can be selected to meet the condition

$$\left|V_{\parallel}(0)\right| \geq \left[2\left(c_0^2 + \Gamma_{\parallel}^{LIM} + c_1\Gamma_{\max}\right)\left(r(t^*) - r^0\right)\right]^{1/2}. \tag{6.43}$$

∎

The theorem is proven.

The problem of intercepting a maneuvering target via enforcing the intercept strategy (Equation 6.38) is reformulated in terms of SMC. The goal is to design the control law

$$\mathbf{u} = \{u_A, u_\delta, u_c\}^T \tag{6.44}$$

that drives the sliding variable

$$\sigma = V_\perp - c_0\sqrt{r} \tag{6.45}$$

to zero on the trajectory of the system (Equations 6.38 through 6.42) in finite time.

The multiple-loop integrated guidance-autopilot design is accomplished using SSOSM control laws for control and observation as per Equations 6.29 and 6.30.

The problem is in designing the feedback control law in terms of $\mathbf{u} = \{u_\Delta, u_\delta, u_\varsigma\}^T$ that provides intercepting a maneuvering target by an impact via driving $r \to 0$ as time increases or via providing for $|r(t^*)| \leq r^0$ (the r^0 value is to be defined based on the size of the target) that implies the zero intercept at $t_{int} \leq t$ in presence of bounded model uncertainties/disturbances $\bar{d}_\alpha$, $\bar{d}_\delta$, and $\bar{d}_\Lambda$ and bounded unknown target accelerations $\Gamma_\parallel$ and $\Gamma_\perp$. The zero-intercept option can be interpreted as achieving UUB.

6.6 Control Architecture

The integrated architecture shown in Figure 6.8 consists of a seeker, a guidance subsystem (outer control loop), and autopilot subsystems (inner control loop). We present thereafter, without loss of generality, only the subsystems governing the motions in the pitch plane.

The seeker subsystem performs bore sight pointing in the LOS direction and estimates the rate of change of the LOS and the target acceleration normal to the LOS.

The guidance (outer loop) subsystem consists of a guidance and inversion that calculate the flight path angle, angle-of-attack pitch rate commands, and eventually, axial thrust commands. Unlike traditional guidance techniques such as proportional navigation (PN) [20] that only achieves collision condition at the end of the interception, the proposed HOSM technique enforces collision condition or any other prescribed guidance throughout the entire interception, which reduces the peak interceptor to target acceleration advantage.

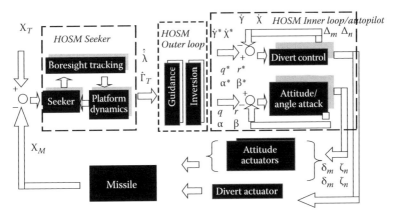

FIGURE 6.8
Control architecture.

The autopilot (inner loop) subsystem includes flight-path-angle, pitch-rate, and angle-of-attack autopilots. The angle-of-attack autopilot is used initially; the angle of attack is continuously steered via continuous actuators. During the end game, the flight-path-angle autopilot and the pitch-attitude autopilot are used concurrently. Hit-to-kill accuracy and short time responses are achieved by controlling the flight path angle via divert actuators. The concurrent tracking by the pitch-rate autopilot of the commanded pitch maneuver steers the angle of attack; the corresponding divert lift creates a "cooperative disturbance" to the flight-path-angle autopilot [18]. In summary, the flight-path autopilot provides the "time constant," and the angle-of-attack maneuver provides a substantial increase in lateral acceleration.

6.6.1 Outer (Guidance) Loop SSOSM Controller Design

In this subsection, the chosen sliding variable (Equation 6.45) is driven to zero in finite time via SSOSM acceleration control Γ in Equations 6.29 and 6.30 format. The σ-dynamics are identified as

$$\dot{\sigma} = g\left(V_{\parallel}(t), V_{\perp}(t), r(t), \Gamma_{\perp}(t)\right) - \cos(\gamma - \lambda)\Gamma \tag{6.46}$$

where

$$g(V_{\parallel}(t), V_{\perp}(t), r(t), \Gamma_{\perp}(t)) = -V_{\parallel}V_{\perp}/r + \Gamma_{\perp} - c_0 V_{\parallel}/(2\sqrt{r})$$

$$\dot{g}(V_{\parallel}(t), V_{\perp}(t), r(t), \Gamma_{\perp}(t)) = -\frac{(\dot{V}_{\parallel}V_{\perp} + V_{\parallel}\dot{V}_{\perp})r - V_{\parallel}^2 V_{\perp}}{r^2} + \dot{\Gamma}_{\perp} - \frac{c_0(\dot{V}_{\parallel}r - V_{\parallel}^2)}{2r\sqrt{r}}. \tag{6.47}$$

Next, assuming the variables $V_{\parallel}(t)$, $V_{\perp}(t)$, and $r(t)$ are measured, the function $g(V_{\parallel}(t), V_{\perp}(t), r(t), \Gamma_{\perp}(t))$ is differentiable with a known Lipchitz constant $L > 0$, which is estimated in the Appendix; the target acceleration transversal to LOS $\Gamma_{\perp}(t)$ can be estimated by the HOSM observer (Equation 6.29).

Apparently, in the absence of input noises, we obtain $\hat{\Gamma}_{\perp} = \Gamma_{\perp}$ after a finite time of a transient process in the HOSM observer (Equation 6.29). If σ and $\cos(\gamma - \lambda)\Gamma + V_{\parallel}V_{\perp} + c_0 V_{\parallel}/(2\sqrt{r})$ are measured with some Lebesgue-measurable noises bounded by $\varepsilon > 0$ and $\varepsilon^{2/3}$, respectively, then [10, 11]

$$|z_1(t) - \Gamma_{\perp}| \leq \mu\varepsilon^{2/3}, \mu > 0. \tag{6.48}$$

The guidance law that drives $\sigma \to 0$ in finite time is designed in terms of a control input $\Gamma^*(t)$ using SSOSM controls 6.29 and 6.30 with $p = 3$ and $m = 2$:

$$\Gamma^* = \frac{1}{\cos(\lambda - \gamma)}(\alpha_1 |\sigma|^{1/2} \text{sign}(\sigma) + \alpha_2 \int |\sigma|^{1/3} \text{sign}(\sigma) d\tau - N'\frac{V_{\perp}V_{\parallel}}{r} - \frac{c_0 V_{\parallel}}{2\sqrt{r}} + \hat{\Gamma}_{\perp}),$$

$$N' = 1. \tag{6.49}$$

Remark 6.5

The term $-N' \dfrac{V_{\|}V_{\perp}}{r} = -N'V_{\|}\omega_{\lambda}$ is known as PN guidance [20]. Rewriting Equation 6.49 in the form

$$\begin{cases} \Gamma^* = \dfrac{1}{\cos(\lambda - \gamma_M)}(-N'\dfrac{V_r V_{\lambda}}{r} + U_d + \hat{\Gamma}_{\perp}), \quad N' = 4 \\ U_d = \alpha_1 |\sigma|^{1/2} \operatorname{sign}(\sigma) + \alpha_2 \int |\sigma|^{1/3} \operatorname{sign}(\sigma)\,d\tau - \dfrac{c_0 V_{\|}}{2\sqrt{r}} \end{cases} \tag{6.50}$$

one can interpret the SSOSM guidance law as pseudo-PN guidance with the term U_d providing the finite-time convergence in absence of the measurement noise. The guidance coefficient $N' = 1$ in Equation 6.49 is taken to be equal to $N' = 4$ in Equation 6.50 in order to have an adequate comparison of the SSOSM guidance law given by Equation 6.50 and the traditional APN guidance law. ■

6.6.2 SSOSM Guidance Simulation Results

The engagement kinematics in Equation 6.32 was simulated using the proposed SSOSM guidance in Equation 6.50. Figure 6.6 represents interceptor and target maneuvers, where once the guidance sliding variable reaches a close vicinity of zero as shown in Figure 6.9, the interceptor literally mimics target maneuvers as shown in Figures 6.10 and 6.11.

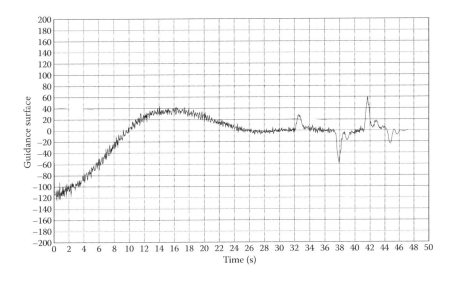

FIGURE 6.9
Guidance sliding variable.

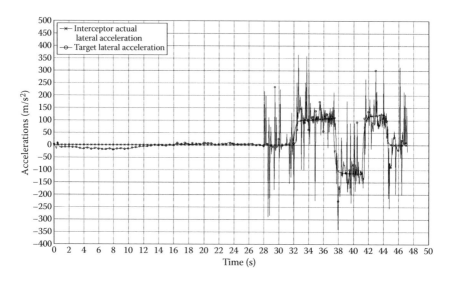

FIGURE 6.10
Target and interceptor accelerations normal to LOS (step maneuvers).

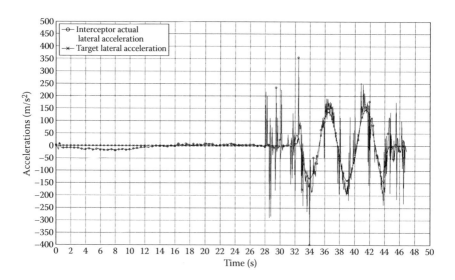

FIGURE 6.11
Target and interceptor accelerations normal to LOS (sine-wave maneuvers).

One can note that during the first part of the trajectory, where continuous divert is applied, interceptor acceleration is continuous. The spikes of interceptor acceleration in the latter part of the scenario are not caused by some unwanted chattering of the control but simply reflect the discontinuous operation of divert actuators.

6.7 Autopilot

Proposed control architecture is a multinested architecture represented in Figure 6.5 where the HOSM guidance is the outer loop and the automatic pilot represents the inner loop. The process used to calculate inner-loop commands from the output of the guidance loop is called the inversion process. It is presented in Section 6.7.1. We will present in Section 6.7.2 the design of a flight-path-angle autopilot and the designs of a pitch-rate autopilot and angle-of-attack autopilot in Sections 6.7.3 and 6.7.4, respectively.

We use the angle-of-attack autopilot during boost flight when divert is continuous, and the attitude is controlled by continuous actuator and the pitch-rate autopilot during terminal flight when the attitude is controlled by on–off actuators. The nested autopilot architecture is shown in Figure 6.12. The block "Command Profiles" represents inversion process that generates commanded flight path angle and flight path rate profiles, pitch rate, and angle-of-attack profiles. The block "Divert Control" represents the flight path autopilot. The angle-of-attack autopilot is used during boost phase, and the pitch-rate autopilot is used during terminal flight.

6.7.1 Inversion

Command profiles γ^*, $\dot{\gamma}^*$, and $\ddot{\gamma}^*$ are computed in real time in the inversion process as follows:

$$\dot{\gamma}^*(t) = \frac{\Gamma^*(t)}{V}, \ddot{\gamma}^*(t) = \frac{\dot{\Gamma}^*(t)}{V}, \gamma^*(t) = \frac{1}{V}\int_0^t \Gamma^*(\tau)\,d\tau, \gamma(0) = \gamma_0 \qquad (6.51)$$

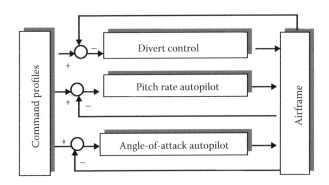

FIGURE 6.12
Architecture of nested autopilot.

where $\Gamma^*(t)$ is assumed to be differentiable. Second, the angle-of-attack command profile α^* and its derivative are computed in real time assuming full knowledge of stability derivative Z_α and $\dot{\gamma}^*(t)$, while nullifying direct effect of attitude and divert actuators δ and Δ in Equations 6.33 through 6.36:

$$\alpha^* = \begin{cases} \dfrac{1}{Z_\alpha}\left[\dot{\gamma}^* + \dfrac{g}{V}\cos(\gamma^*)\right], & if \left|\dfrac{1}{Z_\alpha}\left[\dot{\gamma}^* + \dfrac{g}{V}\cos(\gamma^*)\right]\right| \le \alpha_{max} \\[4mm] \alpha_{max}, & if \left|\dfrac{1}{Z_\alpha}\left[\dot{\gamma}^* + \dfrac{g}{V}\cos(\gamma^*)\right]\right| > \alpha_{max} \\[4mm] -\alpha_{max}, & if \left|\dfrac{1}{Z_\alpha}\left[\dot{\gamma}^* + \dfrac{g}{V}\cos(\gamma^*)\right]\right| < -\alpha_{max} \end{cases}$$ (6.52)

$$\dot{\alpha}^* = \begin{cases} \dfrac{1}{Z_\alpha}\left[\ddot{\gamma}^* - \dfrac{g}{V}\dot{\gamma}^*\sin(\gamma^*) - \dfrac{g\dot{V}}{V^2}\cos(\gamma^*)\right], & if \left|\alpha^*\right| < \alpha_{max} \\[4mm] 0, & \text{otherwise.} \end{cases}$$ (6.53)

The corresponding pitch rate command q^* is calculated in real time as the sum of commanded flight path angle rate profile $\dot{\gamma}^*(t)$:

$$q^* = \dot{\alpha}^* + \dot{\gamma}^*.$$ (6.54)

The pitch rate command profile q^* is supposed to be followed by aerodynamic actuator control u_δ. Clearly $q \to q^*$ implies approximately following $\alpha \to \alpha^*$ while creating a cooperative disturbance term $Z_\alpha\alpha$ in Equations 6.33 through 6.35, and thus, owing to the robustness of SOSM, accurate tracking of commanded angle of attack $\alpha \to \alpha^*$ is not required.

Remark 6.6

It is worth noting that tracking α^* does not imply an accurate tracking of $\dot{\gamma}^*$ since the purpose of the attitude on–off control is only to generate an aerodynamic maneuver, which in effect is a "cooperative disturbance" that alleviates the divert control effort.

Finally, the difference between γ^* and γ is steered to zero by divert control u_Δ in the presence of this cooperative disturbance thereby increasing significantly (up to 100%) the missile overall divert maneuver capability.

6.7.2 Second-Order NDSM-Based Flight Path Angle Autopilot

Equations 6.33 through 6.36 have vector relative degree [2, 2, 2] with respect to the vector output $y = \{\gamma, q, \alpha\}^T$. This calls for SOSM algorithms that are able to drive corresponding sliding variables and their derivatives to zero in finite time.

Remark 6.7

It is assumed that the missile mathematical model (Equations 6.33 through 6.35) is of a minimum or slightly nonminimum phase. ∎

The flight path angle sliding variable is introduced as

$$\sigma_\gamma = \varepsilon_\gamma + \varpi \int_0^t \varepsilon_\gamma(\tau)\,d\tau, \quad \varepsilon_\gamma = \gamma^* - \gamma, \quad \varpi = 50 \text{ rad/s.} \tag{6.55}$$

Equation 6.55 shows that, once the sliding surface $\sigma_\gamma = 0$ is achieved in finite time, $\varepsilon_\gamma \to 0$ asymptotically. The following σ_γ input–output dynamics are derived:

$$\ddot{\sigma}_\gamma = f_\gamma(t) - b_\Delta u_\Delta \tag{6.56}$$

where

$$\left\{ \begin{aligned} & f_\gamma(t) = \ddot{\gamma}^* + \varpi\dot{\gamma}^* + [Z_\alpha(1+\bar{d}_\alpha) - Z_\delta(1+\bar{d}_\delta)\sin(\alpha)\delta - \\ & Z_\Delta(1+\bar{d}_\Delta)\sin(\alpha)\Delta]\dot{\alpha} + \left(\frac{g}{V}\sin(\gamma) - \varpi\right)\dot{\gamma} + Z_\delta(1+\bar{d}_\delta)\cos(\alpha)\dot{\delta} - \\ & \frac{Z_\Delta(1+d_\Delta)\cos(\alpha)}{\tau_\Delta}\Delta, \quad b_\Delta = \frac{Z_\Delta(1+\bar{d}_\Delta)\cos(\alpha)}{\tau_\Delta}. \end{aligned} \right.$$

It is assumed that the disturbance $f_\gamma(t)$ is bounded in an operational domain Ω_γ: $|f_\gamma(t)| \le L_\gamma$, as well as $|d_\Delta| < 1$, $Z_\Delta > 0$, and $|\alpha| \le 0.5$, and then $b_\Delta > 0$, $b'_\Delta < b_\Delta < b''_\Delta$, and the SOSM/NDSM-based divert thrust controller is designed in a form (Equation 6.16) as

$$\left\{ \begin{aligned} & \dot{\chi}_\gamma = \xi_\gamma |\sigma_\gamma|^{0.5}\,\text{sign}(\sigma_\gamma) - \eta_\gamma |J_\gamma|^{0.5}\,\text{sign}(J_\gamma) \\ & J_\gamma = \chi_\gamma + \sigma_\gamma, \quad u_\Delta = \rho_\gamma \cdot \text{sign}(J_\gamma). \end{aligned} \right. \tag{6.57}$$

6.7.3 Second-Order NDSM-Based Pitch Rate Autopilot

Following the SOSM/NDSM control design technique, the pitch rate sliding variable is introduced:

$$\sigma_q = \varepsilon_q + \varpi' \int_0^t \varepsilon_q(\tau) d\tau, \quad \varepsilon_q = q^* - q, \quad \varpi' = 20 \text{ rad/s}. \tag{6.58}$$

Equation 6.58 shows that, once the sliding surface $\sigma_q = 0$ is achieved at the finite time, the pitch rate tracking error ε_q converges to zero asymptotically according to the eigenvalue of $\sigma_q = 0$. Differentiating twice Equation 6.58 gives

$$\ddot{\sigma}_q = f_q(t) - b_\delta u_\delta \tag{6.59}$$

where

$$\begin{cases} f_q(t) = \ddot{q}^* + \varpi' \dot{q}^* - M_\alpha(1 + \bar{d}_\alpha)\dot{\alpha} - (M_q + \varpi)\dot{q} - \\ M_\Delta(1 + \bar{d}_\Delta)\dot{\Delta} + \dfrac{M_\delta d_\delta}{\tau_\delta}(1 + \bar{d}_\delta)\delta; \quad b_\delta = \dfrac{M_\delta(1 + \bar{d}_\delta)}{\tau_\delta}. \end{cases}$$

It is worth noting that the thrust vectoring terms containing ς are removed since thrust vectoring is not used altogether with attitude actuators. One can easily show that the disturbance $f_q(t)$ is bounded in an operational domain $\Omega_q : |f_q(t)| \le L_q$. Since it is assumed that $|\bar{d}_\delta| < 1$ and $M_\delta > 0$, then $b_\delta > 0$, $b_\delta' < b_\delta < b_\delta''$, and the SOSM/NDSM control in Equation 6.16 can be employed for stabilizing σ_q and its derivative $\dot{\sigma}_q$ at zero in finite time. The corresponding SOSM/NDSM-based attitude controller is given by

$$\begin{cases} \dot{\chi}_q = \xi_q |\sigma_q|^{0.5} \text{sign}(\sigma_q) - \eta_q |J_q|^{0.5} \text{sign}(J_q) \\ J_q = \chi_q + \sigma_q, \quad u_\delta = \bar{\rho}_q \cdot \text{sign}(J_q). \end{cases} \tag{6.60}$$

6.7.4 SOSM Supertwist-Based Angle-of-Attack Autopilot

It is worth noting that, in reality, max $|Z_\delta\delta + Z_\Delta\Delta| \ll$ max $|q|$ and max $|M_\delta| \gg$ max $|M_\Delta|$. It means that the missile angle of attack α is mostly governed by the pitch rate q, which itself is controlled by the attitude on–off actuator δ. The sliding surface σ_α is defined as

$$\sigma_\alpha = \varepsilon_\alpha + \varpi'' \int_0^t \varepsilon_\alpha d\tau, \quad \varepsilon_\alpha = \alpha^* - \alpha, \quad \varpi'' = 20 \text{ rad/s}. \tag{6.61}$$

Differentiating Equation 6.60 twice with respect to time yields

$$\ddot{\sigma}_\alpha = f_\alpha(t) - b_\varsigma u_\varsigma \tag{6.62}$$

where one can show that similar to Equation 6.46, $f_\alpha(t)$ is a bounded variable while the term b_ς is defined by $b_\varsigma = M_\varsigma/\tau_\varsigma$. The angle-of-attack continuous control is supposed to be continuous while being robust to the disturbances. The supertwisting continuous control is designed in accordance with Equation 6.20:

$$\begin{cases} u_\varsigma = \dfrac{1}{b_\varsigma}\left[-\xi_\alpha \left| S_\alpha \right|^{1/2} \text{sign}(S_\alpha) + w_\varsigma \right] \\ \dot{w}_\varsigma = -\eta_\alpha \, \text{sign}(S_\alpha) \end{cases} \tag{6.63}$$

where the auxiliary sliding variable is introduced as

$$S_\alpha = \dot{\sigma}_\alpha + c_\alpha \sigma_\alpha, \quad c_\alpha > 0. \tag{6.64}$$

6.7.5 Integrated SOSM Guidance-Autopilot Simulation Results

The designed HOSM guidance autopilot was simulated against the maneuvering target. The plots in Figure 6.13 reveal that the commanded flight path angle is followed very accurately. The result shows that the actual flight path rate is always superimposed to commanded value. One can see the separation between the initial phase of the flight, where the interceptor is steered

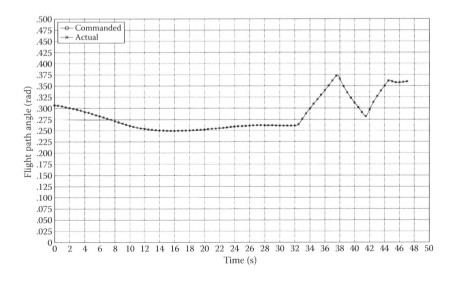

FIGURE 6.13
Tracking the flight path angle.

by continuous angle-of-attack maneuver, with the last phase of the flight, where on–off divert actuators are used. The ripples in Figure 6.14 are the results of the on–off operation of the actuators.

Figure 6.15 shows the tracking of the pitch rate. During the first part of the interception, the attitude steering is achieved by continuous control,

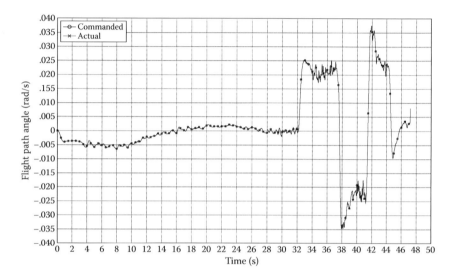

FIGURE 6.14
Tracking the flight path rate.

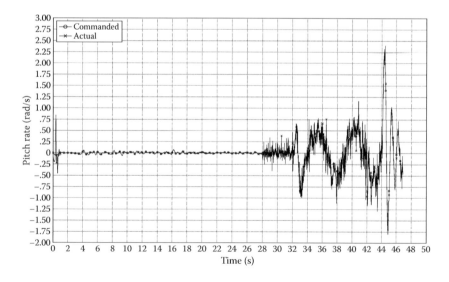

FIGURE 6.15
Pitch rate tracking performance.

whereas in the last part, it is achieved by on–off operation of attitude actuators. Figure 6.16 represents the tracking of the angle of attack. During the first part of the interception, the autopilot tracks angle-of-attack commands, and not surprisingly, perfect accuracy is achieved. During the second part of the interception, the autopilot tracks pitch rate commands. Interceptor pitch rate spikes observed in Figure 6.15 are not the result of some chattering of the control but are caused by on–off pulse-width-modulation (PWM) operation of attitude actuators. Figure 6.17 represents the composite plot of attitude commands. During the initial part of the interception, it represents continuous commands that are very small in magnitude, while in the last portion of the flight, it represents attitude on–off commands. For the interceptor to accelerate very rapidly as required, the sustainer thrust is very large. Figure 6.18 represents the corresponding actuator responses. As a consequence, continuous attitude actuator deflections represented in the early portion of Figure 6.18 are very small and almost unnoticeable. One can note that the actuators are often commanded to shut down well before reaching the maximum amplitude.

One important advantage of the HOSM design when PWM actuators are used is that the control can be designed directly to operate with on–off actuators. Traditional control design produces an initial linear control design, which then needs to be redesigned to work with PWM actuators. In the redesigning process, it is assumed that the pulses are rectangular. As this is not, by far, the case in our missile application as shown in Figure 6.18, this would yield the results that traditional control techniques would achieve to not be very good.

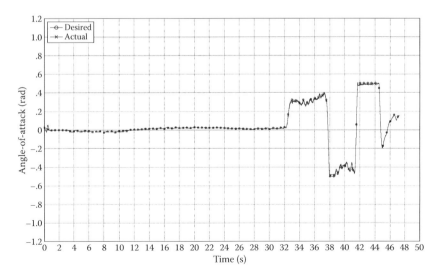

FIGURE 6.16
Angle-of-attack tracking performance

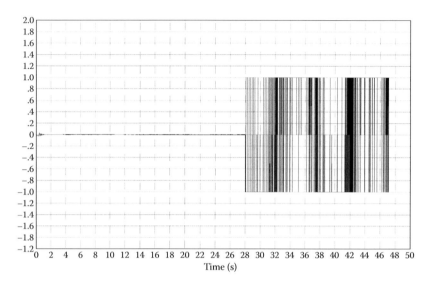

FIGURE 6.17
Attitude (normalized) commands.

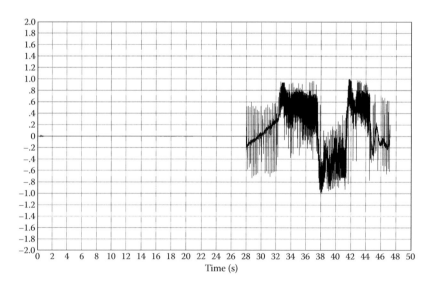

FIGURE 6.18
Attitude actuators.

6.7.6 Higher-Order SMC Quaternion Autopilot

In Section 6.7.1, we presented an inversion in the very simple case of a pitch motion and assuming flat nonrotating earth. While these assumptions are satisfactory for a missile with relatively slow velocities and relatively small ranges, they are not valid anymore when it comes to larger ranges and velocities. Given the insensitivity of the HOSM design to matched disturbances, the problem is not so much that terms f_γ, f_q, and f_α in Equations 6.56, 6.59, and 6.62, respectively, are going to be different and much more complex as we represent trajectory dynamics and attitude dynamics over oblate rotating earth instead of flat nonrotating earth in Sections 6.7.2 through 6.7.4.

The new challenges are as follows:

1. The calculation of prescribed attitude rates cannot be completed using simple pitch plane Equations 6.51 through 6.54. First, it is because attitude rates are defined in body axes while flight path angle and ground track angles are defined in the north-east-down (NED) local axis. As the vehicles move around the earth, the NED frame moves accordingly, and thus, it is necessary to add terms accounting for its rotation. The problem is compounded in oblate earth by the fact that down axis does not point toward the center of the earth but is simply locally perpendicular to the earth's geoid.

2. The challenge arises from the definition of angle of attack and sideslip angles as Euler angle rotations. As such, the sideslip angle definition becomes singular when the flight path angle is $\pm\dfrac{\pi}{2}$.

3. The third difficulty arises from the singularity and the nonlinearity of the relation between attitude rates and time derivatives of Euler angles:

$$
\begin{bmatrix} \dot{\phi} \\ \dot{\theta} \\ \dot{\psi} \end{bmatrix} =
\begin{bmatrix}
1 & \tan(\theta)\sin(\phi) & \tan(\theta)\cos(\phi) \\
0 & \cos(\phi) & \sin(\phi) \\
0 & \dfrac{\sin(\phi)}{\cos(\theta)} & \dfrac{\cos(\phi)}{\cos(\theta)}
\end{bmatrix}
\begin{bmatrix} p \\ q \\ r \end{bmatrix}.
\tag{6.65}
$$

The approach adopted is to calculate the quaternion from velocity axes to any suitable reference (.) system such as the earth-centered-inertial (ECI) NED. Assuming that ECI is the chosen reference, the transformation from velocity to ECI is constructed as follows: The *x*-axis is directed along the vehicle velocity vector. The *y*- and *z*-axes of the frame are calculated as

$$\mathbf{I}_y = \frac{\mathbf{I}_x \times \mathbf{rI}}{\left|\mathbf{I}_x \times \mathbf{rI}\right|}$$

$$\mathbf{I}_z = \frac{\mathbf{I}_x \times \mathbf{I}_y}{\left|\mathbf{I}_x \times \mathbf{I}_y\right|}.$$

(6.66)

The corresponding quaternion is given by $Q_{vel/(.)} = Q\left(\mathbf{T}^T_{vel/(.)}\right)$, where (.) represents the selected reference frame.

Posing $\mathbf{R} = \mathbf{T}^T_{vel/(.)}$, the formulation for calculating the quaternion from a rotation matrix more frequently used is

$$4q_0^2 = 1 + R_{1,1} + R_{2,2} + R_{3,3}$$

(6.67)

and

$$\begin{cases} q_1 = \dfrac{R_{3,2} - R_{2,3}}{4q_0} \\[2mm] q_2 = \dfrac{R_{1,3} - R_{3,1}}{4q_0} \\[2mm] q_3 = \dfrac{R_{2,1} - R_{1,2}}{4q_0}. \end{cases}$$

(6.68)

This can also be written as

$$Q = \begin{bmatrix} \cos(\phi/2) \\ \sin(\phi/2)e_1 \\ \sin(\phi/2)e_2 \\ \sin(\phi/2)e_3 \end{bmatrix}.$$

(6.69)

This formulation of the quaternion may unfortunately become singular when q_0 approaches zero. The rotation becomes a simple reflection and the vector $\mathbf{e}$ becomes undefined. The approach for overcoming this difficulty is to assume that the quaternion representing the rotation is now q_1, q_2, or q_3 and to calculate them by choosing one of the four possible formulations of Equations 6.67 and 6.70 as discussed in [21]:

$$4q_1^2 = 1 + R_{1,1} - R_{2,2} - R_{3,3}$$

$$4q_2^2 = 1 - R_{1,1} + R_{2,2} - R_{3,3}$$

(6.70)

$$4q_3^2 = 1 - R_{1,1} - R_{2,2} + R_{3,3}.$$

The other terms are calculated using

$$
\begin{cases}
4q_0 q_1 = R_{3,2} - R_{2,3} & 4q_1 q_2 = R_{2,1} - R_{1,2} \\
4q_0 q_2 = R_{1,3} - R_{3,1} & 4q_2 q_3 = R_{3,2} - R_{2,3} \\
4q_0 q_3 = R_{2,1} - R_{1,2} & 4q_1 q_3 = R_{3,1} - R_{1,3}.
\end{cases}
\tag{6.71}
$$

The algorithm evaluates concurrently the four formulations and selects the formulation that yields the larger value of $\cos(\phi/2)$. It is assumed here that the traditional formulation Equations 6.67 and 6.68 are used. It is reasonable to assume that the same quaternion formulation can be used throughout the entire flight.*

The reference quaternion representing the transformation from body frame to frame (.) is calculated as the quaternion product of the transformations by the quaternion velocity frame to frame (.) as

$$
Q_{body/(.)} = Q_{\eta/\xi} \otimes Q_{vel/(.)}
$$

$$
Q_{\eta/\xi} = Q(\mathbf{T}_{\eta/\xi})
\tag{6.72}
$$

where $\otimes$ is used to represent the quaternion product,

$$
\mathbf{T}_{\eta/\xi} =
\begin{pmatrix}
\cos(\eta) & 0 & \sin(\eta) \\
0 & 1 & 0 \\
-\sin(\eta) & 0 & \cos(\eta)
\end{pmatrix}
\begin{pmatrix}
\cos(\xi) & \sin(\xi) & 0 \\
\sin(\xi) & \cos(\xi) & 0 \\
0 & 0 & 1
\end{pmatrix}
\tag{6.73}
$$

where η and ξ represent the rotations around the y- and z-axes, respectively.

Prescribed roll, pitch, and yaw rates can be defined from the time derivative of the prescribed quaternion as

$$
\Omega^* =
\begin{bmatrix} p^* \\ q^* \\ r^* \end{bmatrix}
= 2
\begin{bmatrix}
-q_1 & q_0 & q_3 & -q_2 \\
-q_2 & -q_3 & q_0 & q_1 \\
-q_3 & q_2 & -q_1 & q_0
\end{bmatrix}
\begin{bmatrix} \dot{q}_0 \\ \dot{q}_1 \\ \dot{q}_2 \\ \dot{q}_3 \end{bmatrix} ;
\quad Q^*_{body/(.)} =
\begin{bmatrix} q_0 \\ q_1 \\ q_2 \\ q_3 \end{bmatrix}.
\tag{6.74}
$$

---

* The problem discussed here is only posed by the initiation of a quaternion using a rotation matrix. Once the initial quaternion has been calculated, it suffices to update it, and it does not matter anymore as $\cos(\phi/2)$ becomes zero thereafter. As $\cos(\phi/2) \to 0$, there are occurrences of a sign change of the three other components: while in general, the quaternion represents a rotation of a frame, around a vector when $\cos(\phi/2) = 0$, it represents a symmetry between initial and terminal frames in which case the rotation vector ceases to be unique.

Alternatively, we can calculate the prescribed attitude rate as

$$
\mathbf{\Omega}^* \times_{body} = \begin{bmatrix} 0 & -r^* & q^* \\ r^* & 0 & -p^* \\ -q^* & p^* & 0 \end{bmatrix} = \dot{\mathbf{T}}^*_{(.)/body} \, \mathbf{T}^*_{body/(.)}.
\tag{6.75}
$$

Introducing the angle error $\mathbf{\Delta\Omega I}$ as

$$
\mathbf{\Delta\Omega I} = \begin{bmatrix} \Delta\Omega I_p \\ \Delta\Omega I_q \\ \Delta\Omega I_r \end{bmatrix} = 2 \begin{bmatrix} -q_1 & q_0 & q_3 & -q_2 \\ -q_2 & -q_3 & q_0 & q_1 \\ -q_3 & q_2 & -q_1 & q_0 \end{bmatrix} (\mathbf{Q}^*_{body/(.)} - \mathbf{Q}_{body/(.)}); \quad \mathbf{Q}^*_{body/(.)} = \begin{bmatrix} q_0^* \\ q_1^* \\ q_2^* \\ q_3^* \end{bmatrix}.
\tag{6.76}
$$

Three proportional integral surfaces define the desired behavior of the error response as

$$
\begin{bmatrix} \sigma_p \\ \sigma_q \\ \sigma_r \end{bmatrix} = \begin{bmatrix} \omega_p \Delta\Omega I_p + \Delta p \\ \omega_q \Delta\Omega I_q + \Delta q \\ \omega_r \Delta\Omega I_r + \Delta q \end{bmatrix}; \quad \begin{bmatrix} \Delta p \\ \Delta q \\ \Delta q \end{bmatrix} = \mathbf{\Omega}^* - \mathbf{\Omega}
\tag{6.77}
$$

where $\mathbf{\Omega}$ represents the actual rotation rate in body axes.

Interestingly, when $\omega = \omega_p = \omega_q = \omega_r$, Equation 6.77 can be represented as

$$
\begin{bmatrix} \sigma_p \\ \sigma_q \\ \sigma_r \end{bmatrix} = 2 \begin{bmatrix} -q_1 & q_0 & q_3 & -q_2 \\ -q_2 & -q_3 & q_0 & q_1 \\ -q_3 & q_2 & -q_1 & q_0 \end{bmatrix} (\omega \Delta\mathbf{Q} + \Delta\dot{\mathbf{Q}}).
\tag{6.78}
$$

The term $(\omega\Delta\mathbf{Q} + \Delta\dot{\mathbf{Q}})$ represents a four-vector of sliding surfaces in the quaternion space. Equation 6.78 projects the four-vector of sliding surfaces in the quaternion space into a three-vector of sliding surfaces in body axes.

The control of pitch and yaw attitude motion during boost is achieved by means of continuous actuators. Corresponding control laws are designed as follows.

Handling the relative degree = 2 case, we proceed as in Section 6.7.4 and replace the supertwist design used in that section with an SSOSM controller as per Equations 6.29 and 6.30 as

$$U_{(.)} = \alpha_1 \left| \sigma'_{(.)} \right|^{2/3} \text{sign}(\sigma'_{(.)}) + \alpha_2 \int \left| \sigma'_{(.)} \right|^{1/3} \text{sign}(\sigma'_{(.)}) d\tau + z1_{(.)}; \quad (.) = q, r$$

$$\dot{z}_0 = v_0 + U_{(.)}; \quad v_0 = -3 \left| L \right|^{1/3} \left| z_0 - \sigma'_{(.)} \right|^{2/3} \text{sign}(z_0 - \sigma_{(.)}) + z_1$$

$$\dot{z}_1 = v_1; \quad v_1 = -2 \left| L \right|^{1/2} \text{sign}(z_1 - v_0) \left| z_1 - v_0 \right|^{1/2} + z_2 \tag{6.79}$$

$$\dot{z}_2 = v_2; \quad v_2 = -1.5 \left| L \right| \text{sign}(z_2 - v_1).$$

Since the SSOSM controller is designed to operate with systems of relative degree 1, we introduce, as in Section 6.7.4, auxiliary surfaces defined as

$$\sigma'_{(.)} = c\sigma_{(.)} + \dot{\sigma}_{(.)}. \tag{6.80}$$

The control of pitch and yaw attitude motion during KV autonomous flight and the control of roll motion during boost and KV flight are achieved by means of on–off PWM actuators. Corresponding control laws are designed as follows.

Handling the relative degree 2 case, the SOSM/NDSM controller is used. Here there is no need for a smooth controller, and it would not make sense to design a smooth controller using Equation 6.79 to then design a PWM implementation thereof. The PWM-SOSM/NDSM controller is designed directly as per Equation 6.16 as

$$\begin{cases} \dot{\chi}_{(.)} = \xi_{(.)} \left| \sigma_{(.)} \right|^{0.5} \text{sign}(\sigma_{(.)}) - \eta_{(.)} \left| J_{(.)} \right|^{0.5} \text{sign}(J_{(.)}); \quad (.) = p, q, r \\ J_{(.)} = \chi_{(.)} + \sigma_{(.)}, \quad u_{[.]} = \bar{\rho}_{(.)} \cdot \text{sign}(J_{(.)}); \quad [.] = \delta l, \delta m, \delta n. \end{cases} \tag{6.81}$$

6.7.7 Simulation Results for Integrated SOSM Guidance-Quaternion Autopilot

The quaternion autopilot (Equation 6.81) was simulated for the missile-target engagement kinematics (Equation 6.32) and missile dynamics (Equations 6.33 through 6.36). Figure 6.19 presents the four quaternion tracking errors, which are very small (less than 0.0008 most of the time). The spike at the end is caused by the switching from a continuous control as per Equation 6.79 to a PWM control law as defined by Equation 6.81. The switching of the control law and integrator reset actually only causes a minor and short transient error with a peak normalized amplitude that is equal to 0.015. The same can be said about the angle errors, which are less than 1 mrad most of the time and have a peak transient amplitude that is equal to 0.01 rad as shown Figure 6.20.

Angles plotted in Figure 6.20 are sometimes called rolling, pitching, and yawing angles; they should not be confused with ordinary roll, pitch, and yaw Euler angles. Actually, roll rate, pitch rate, and yaw rate errors are the time derivatives of rolling, pitching, and yawing angle errors; but the

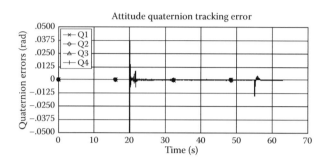

FIGURE 6.19
Quaternion tracking errors.

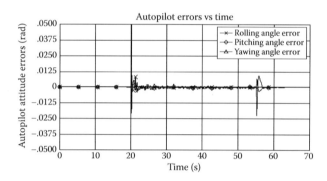

FIGURE 6.20
Autopilot angle errors.

reciprocal is not true; rolling, pitching, and yawing error angles differ from the integrals of roll, pitch, and yaw rate errors by the initial conditions of integration. Attitude rate errors are shown in Figure 6.21.

Roll rate, pitch rate, and yaw rate errors are smaller than 0.03 rad/s most of the time with the exception of the very brief (less than 1 s) transient when the switching from the continuous actuator and the on–off attitude actuator takes place. In this simulation, it was assumed for expediency's sake that the two autopilot modes were run back to back. A more detailed simulation would have included a brief coasting between the end of the booster and the beginning of the terminal flight.

As indicated before and as shown in the formulae presented before, the HOSM designs of different autopilot presented do not require values for interceptor control characteristics other than a very broad estimation of the upper bound of the disturbances. To demonstrate even further the inherent robustness of the design, we have applied multiplicative disturbances to the magnitude of the actuators. Applying this disturbance, as shown in Figure 6.22, it is possible for the maximum magnitude of an actuator to be as some point in time 1.3 times the nominal value and 50 ms after to be 0.8

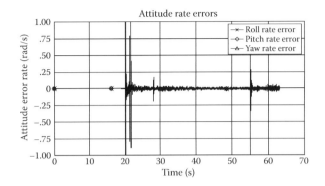

FIGURE 6.21
Attitude rate errors.

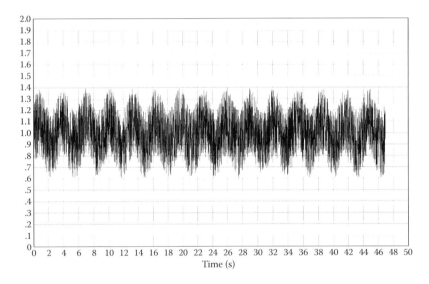

FIGURE 6.22
Multiplicative disturbance.

times this value. Evidently, such a rapid and large amplitude variation of the thrust level cannot be estimated in real time by a disturbance accommodating controller.

Likewise, the modeling of such effects would require very complex codes that account for the combination of the effects of the multiple simultaneous causes of such disturbances. In addition to the fast variation of the maximum thrust, a slow sinusoidal modulation has been introduced to account for the disturbances caused by slowly varying effects such as air density or velocity change.

6.8 Conclusions

The fundamentals of HOSM control and the most widely used HOSM/SOSM control/estimation algorithms were presented. Using a nonlinear uncertain model of a notional ballistic missile interceptor, we showed that the entire Guidance Navigation and Control Suite could be designed based on HOSM algorithms. Owing to the robustness of HOSM/SOSM, a flight path angle, pitch angle, and angle-of-attack SOSM autopilots were designed achieving excellent performance. Moreover, the control architecture is very simple and does not require gain scheduling. The design included a smooth guidance law with its embedded estimator of target acceleration, an angle-of-attack autopilot yielding continuous control law used during boost flight, a flight path angle autopilot yielding on–off commands of the divert actuators, and a pitch rate autopilot yielding on–off commands of the attitude actuators, both of them used during the autonomous KV flight. The endo-atmospheric interceptor was steered during boost phase by combined effects of aerodynamic lift and the orientation of the booster thrust, while during the autonomous KV flight, it was steered by combined effects of lift and divert actuators. This is a regime where the interactions of aerodynamics around the body and the firing of actuator exert on each other considerable disturbing effects, which are, for all matters and purposes, practically impossible to model and to estimate in real time. The proposed quaternion autopilot allows the direct controlling of attitude quaternion with respect to some reference frame, such as ECI, using body axis moments created by attitude actuators, whether continuous or on–off attitude actuators. The proposed autopilot design circumvents a number of difficulties such as the earth curvature, earth rotation, and nonlinearities and singularities of the relation between Euler angles and attitude rates, which need to be accounted for when conventional control designs are used. Theoretical and simulation results showed that HOSM/SOSM integrated guidance control is capable of remedying the difficult robustness challenge posed to missile system control in a simple and effective manner.

Nomenclature

$a_\alpha, a_\delta, a_\Delta$	Distance from the center of gravity along the longitudinal axis of the aerodynamic center and application points of attitude and divert actuators (in meters).
$d_\alpha, d_\delta, d_\Delta, \delta_\varepsilon$	Disturbance factors applied to aerodynamic lift, attitude and divert actuators, seeker measurement noise.

$D_\alpha, D_\delta, D_\Delta$	Disturbance terms are uniformly distributed within $\pm D$ intervals.
f_q, f_γ, f_α	Additive disturbances.
$\bar{C}$	Reference length (in meters).
α, γ, q	Angle of attack, flight path angles (in radians), and pitch rate (in radians per second).
C_{L_α}	Lift coefficient gradient with respect to angle of attack α.
V	Longitudinal velocity of a missile (in meters per second).
$Q_{a/b}$	Quaternion representing the transformation from frame a to frame b.
$T_{a/b}$	Transformation from frame a to frame b.
$V_\|, V_\perp$	Relative velocity components parallel and normal to line of sight (LOS; in meters per second)
$C_{M_q} \dfrac{\bar{C}}{V}$	Moment coefficient gradient with respect to pitch rate q (in seconds^{-1}).
$\varepsilon_\alpha, \varepsilon_\gamma, \varepsilon_q$	Angle of attack and flight path angle and pitch rate tracking errors (in radians and radians per second).
I_{yy}	Moment of inertia around the pitch axis (in kilogram square meter).
$\Gamma^*, \Gamma, \Gamma_{max}$	Commanded, actual, and maximum interceptor normal acceleration (in meters per square second).
$\hat{\Gamma}_T$	Estimated target acceleration vector (in meters per square second).
$\Gamma_\|, \Gamma_\perp$	Target acceleration components along the LOS and normal to it.
g	Acceleration of earth gravity (in meters per square second).
m	Missile mass (in kilograms).
δ, Δ	Normalized attitude and divert actuator forces.
ς	Continuous attitude control (in radians).
r	Target-interceptor range along the LOS (in meters).
$T_\varsigma, T_\delta, T_\Delta$	Sustainer, attitude, and divert thrust (in newtons).
$Z_\varsigma, Z_\alpha, Z_\delta, Z_\Delta$	Trajectory stability derivative with respect to ς, α, δ, and Δ, respectively (in seconds^{-1}).
$M_\varsigma, M_\alpha, M_\delta, M_\Delta$	Pitch rate stability derivative with respect to ς, α, δ, and Δ, respectively (in seconds^{-2}).
$T_\varsigma, T_{max\delta}, T_{max}$	Maximum sustainer attitude and divert actuator (in newtons).
$u_\delta, u_\Delta, u_\varsigma$	Attitude, divert actuator, and continuous normalized attitude command.
ρ	Air specific mass (in kilograms per cubic meter).
$\lambda, \hat{\lambda}$	LOS angle and bore sight angle (in radians).
ω_λ	Rotation rate of the LOS (in radians per second).

$\tau_\delta,\ \tau_\Delta,\ \tau_\varsigma$	Time constants of attitude on–off actuators, on–off divert actuators, and continuous attitude actuators (in seconds).
$(.)^*$	Commanded, reference value of variable (.).
$(.)_\delta;\ (.)_\Delta$	Value, variable associated to attitude and divert on–off attitude actuators.
$\varepsilon_{(.)}\ e_{(.)}$	Angular and linear tracking error with respect to variable (.).
$(.)'$	Variable associated with seeker bore sight steering.
$(.)''$	Variable associated with guidance.
$(\underline{.})$	Filtered variable.

References

1. Kennedy, K., Walker, B., and Mikelsen, C., AIT Real Gas Divert Jet Interactions; Summary of Technology, Publication of the AIAA, Reston, VA, Paper AIAA 98-5188, August 1998, pp. 225–234.
2. Edwards, C. and Spurgeon, S., *Sliding Mode Control: Theory and Applications*, Taylor & Francis, Bristol, 1998.
3. Idan, M. and Shima, T., Integrated sliding mode autopilot-guidance for dual control missiles, *AIAA Journal of Guidance Control and Dynamics*, 30(4), July–August 2007, 1081–1089.
4. Moon, J., Kim, K., and Kim, Y., Design of missile guidance law via variable structure control, *Journal on Guidance, Control, and Dynamics*, 24(4), 2001, 659–664.
5. Shkolnikov, I., Shtessel, Y., and Lianos, D., Integrated Guidance-Control System of a Homing Interceptor: Sliding Mode Approach, *Proceedings of 2001 AIAA Guidance, Navigation, and Control Conference*, Publication of the AIAA, Reston, VA, Paper AIAA-2001-4218, August 2001.
6. Thukral, A. and Innocenti, M., Sliding mode missile pitch autopilot synthesis for high angle of attack maneuvering, *IEEE Transactions on Control Systems Technology*, 6(3), 1998, 359–371.
7. Babu, K., Sarma, I., and Swamy, K., Switched bias proportional navigation for homing guidance against highly maneuvering target, *Journal of Guidance, Control and Dynamics*, 17(6), 1994, 1357–1363.
8. Zhou, D., Mu, C., and Xu, W., Adaptive sliding-mode guidance of a homing missile, *Journal of Guidance, Control and Dynamics*, 22(4), 1999, 589–594.
9. Yeh, F.-K., Chien, H.-H., and Fu, L.-C., Optimal midcourse guidance sliding-mode control for missiles with TVC, *IEEE Transactions on Aerospace and Electronic Systems*, 39(3), 2003, 824–837.
10. Levant, A., Higher order sliding modes, differentiation and output-feedback control, *International Journal of Control*, 76(9/10), 2003, 924–941.
11. Shtessel, Y., Shkolnikov, I., and Levant, A., Smooth second order sliding modes: missile guidance application, *Automatica*, 43(8), 2007, 1470–1476.

12. Shtessel, Y., Shkolnikov, I., and Brown, M., An asymptotic second-order smooth sliding mode control, *Asian Journal of Control*, 4(5), 2003, 498–504.
13. Shtessel, Y., Tournes, C., and Shkolnikov, I., Guidance and Autopilot for Missiles Steered by Aerodynamic Lift and Divert Trusters Using Second Order Sliding Modes, *Proceedings of the Conference on Guidance, Navigation and Control*, Publication of the AIAA, Reston, VA, Paper AIAA-2006-6784, 8, August 2006, pp. 5250–5292.
14. Levant, A., Sliding order and sliding accuracy in sliding mode control, *International Journal of Control*, 58(6), 1993, 1246–1263.
15. Shtessel, Y. B. and Tournes, C., Integrated higher order sliding mode controls and autopilots for dual control missiles, *Journal of Guidance Control and Dynamics*, 32(1), 2009, 79–94.
16. Tournes, C. and Shtessel, Y., Integrated Guidance and Autopilot for Dual Controlled Missiles Using Higher Order Sliding Mode Controllers and Observers, *Proceedings of the Conference on Guidance, Navigation and Control*, Publication of the AIAA, Reston, VA, Paper AIAA-2008-7433, August 2008.
17. Tournes, C. and Shtessel, Y. B., Integrated Autopilot and Guidance for Dual Control Missiles Using Higher Sliding Mode Controls and Observers, *World Congress of International Federation of Automatic Control*, Publication of the International Federation of Automatic Control, Omnipress, Seoul, Korea, 2008.
18. Tournes, C., Shtessel, Y. B., and Shkolnikov, I., Autopilot for missiles steered by aerodynamic lift and divert thrusters, *AIAA Journal of Guidance Control and Dynamics*, 29(3, May–June issue), 2006, 617–625.
19. Wise, K. A. and Broy, D. J., Agile missile dynamics and control, *AIAA Journal of Guidance, Control and Dynamics*, 21(3), 1998, 441–449.
20. Zarchan, P., Tactical and Strategic Missile Guidance, *Progress in Astronautics and Aeronautics*, 176, Publication of the AIAA, Reston, VA, 1998.
21. Stevens, B. L. and Lewis, F. L., *Aircraft Control and Simulation*, 1992, pp. 42–43, John Wiley and Sons, New York. *Guidance, Control and Dynamics*, 21(3), 1998, 441–449.
22. Bacciotti, A. and Rosier, L., Lyapunov functions and stability in control theory, *Lecture Notes in Control and Information Sciences*, 267, Springer-Verlag, New York, 2001.
23. Filippov, A. F., *Differential Equations with Discontinuous Right-Hand Side*, Kluwer, Dordrecht, Netherlands, 1988.
24. Krupp, D., Shkolnikov, I. A., and Shtessel, Y. B., 2-Sliding Mode Control for Nonlinear Plants with Parametric and Dynamic Uncertainties, *Proceedings of the Conference on Guidance, Navigation, and Control*, Publication of the AIAA, Reston, VA, Paper AIAA-2000-3965, August 2000.
25. Shkolnikov, I., Shtessel, Y., and Lianos, D., The effect of sliding mode observers in the homing guidance loop, *ImechE Journal on Aerospace Engineering, Part G*, 219(2), 2005, 103–111.
26. Khalil, H., *Nonlinear Systems*, Prentice-Hall, Upper Saddle River, New Jersey, 3rd edition, 2002.
27. Sweriduk, G., Ohlmeyer, E., Malyevac, D., and Menon, P., Integrated guidance and control of moving-mass actuated kinetic warheads, *Journal of Guidance, Control, and Dynamics*, 27(1), 2004, 118–126.
28. Xin, M., Balakrishnan, S. N., and Ohlmeyer, E., Integrated guidance and control of missiles with θ-D method, *IEEE Transactions on Control Systems Technology*, 4(6), 2006, 981–992.

Appendix: Evaluation of the Lipschitz Constant for Outer (Guidance) Loop

It is assumed that the target acceleration transversal to the LOS $\Gamma_\perp(t)$ is differentiable. Then the function $g(V_\parallel(t), V_\perp(t), r(t), \Gamma_\perp(t))$ is also differentiable, and the function $\dot{g}(V_\parallel(t), V_\perp(t), r(t), \Gamma_\perp(t))$ is continuous everywhere except for $r = 0$. This singularity point occurs when intercept by impact happens. However, technically, the intercept by impact ("hit-to-kill") happens when $r \neq 0$ but belongs to the interval $r \in [r_{min}, r_{max}] = [0.1, 0.25]$ m [2, 3, 11]. This fact is due to a certain size of the ballistic target and a particular intercept value of $r^0 \in [0.1, 0.25]$ m, named "zero intercept," and depends on this size. Therefore, the function $g(V_\parallel(t), V_\perp(t), r(t), \Gamma_\perp(t))$ is differentiable, and the function $\dot{g}(V_\parallel(t), V_\perp(t), r(t), \Gamma_\perp(t))$ is continuous everywhere until hit-to-kill "zero intercept" happens. Its Lipschitz constant is estimated in the following theorem.

Theorem 2

Assume that $|\Gamma| \leq \Gamma_{max}$, $|\dot{\Gamma}_\perp| \leq \dot{\Gamma}_\perp^{LIM}$, $|\Gamma_\perp| \leq \Gamma_\perp^{LIM}$, $|\Gamma_\parallel| \leq \Gamma_\parallel^{LIM}$, $|V_\perp(t)| \leq V_\perp^{LIM}$, $V_\parallel(0) = M \ll 0$, and $M \leq V_\parallel(t) \leq 0$, in a reasonable flight domain. Then the Lipschitz constant L for $\dot{g}(V_\parallel(t), V_\perp(t), r(t), \Gamma_\perp(t))$ can be estimated as

$$L \approx \dot{\Gamma}_\perp^{LIM} + \frac{V_\perp^{LIM}}{(r^0)^2}\left[(V_\perp^{LIM})^2 + 2M^2\right]. \qquad (6.A1)$$

∎

Proof

Using Equation 6.31 and taking into account inequalities $|\sin(\lambda - \gamma_M)| < c_1 < 1$ and $|\cos(\lambda - \gamma_M)| < c_2 < 1$, the following inequality holds:

$$\left|\dot{g}(V_\parallel(t), V_\perp(t), r(t), \Gamma_\perp(t))\right| \leq \dot{\Gamma}_\perp^{LIM} + \frac{\left|(\dot{V}_\parallel V_\perp + V_\parallel \dot{V}_\perp)r - V_\parallel^2 V_\perp\right|}{r^2} + \frac{c_0\left|(\dot{V}_\parallel r - V_\parallel^2)\right|}{2r\sqrt{r}} \leq \dot{\Gamma}_\perp^{LIM}$$

$$+ \frac{1}{(r^0)^2}\left[(V_\perp^{LIM})^3 + 2M^2 V_\perp^{LIM} + \frac{c_0\sqrt{r^0}}{2}(V_\perp^{LIM} + M^2)\right] + \frac{1}{r^0}[(\Gamma_\perp^{LIM} + c_1\Gamma_{max})V_\perp^{LIM}$$

$$+ \left|M\right|(\Gamma_\perp^{LIM} + c_2\Gamma_{max}) + \frac{c_0\sqrt{r^0}}{2}(\Gamma_\perp^{LIM} + c_1\Gamma_{max})] = L.$$

Since $0 < r^0 \ll 1$, then $L \approx \dot{\Gamma}_\perp^{LIM} + \frac{V_\perp^{LIM}}{(r^0)^2}\left[(V_\perp^{LIM})^2 + 2M^2\right]$ and the theorem is proven.

∎

7

Neoclassical Missile Guidance

Pini Gurfil

CONTENTS

7.1 Introduction

The challenging problem of missile guidance has been treated using several fundamental methodologies. The classical approach is to apply the missile maneuver acceleration proportionally to the measured line-of-sight (LOS) rate. The resulting guidance law is the well-known proportional navigation (PN). The modern approach to missile guidance is based upon optimal control theory (one-sided optimization) and differential games (two-sided optimization).

PN is the method most commonly used for guidance of homing missiles. A vast amount of literature exists on the subject (see, e.g., the works of Shneydor[1] and Zarchan[2] and the references therein). Modern guidance laws have also been thoroughly analyzed.[3–5] PN is known to yield reasonable miss distance when applied against nonmaneuvering or moderately maneuvering targets, whereas modern guidance laws can theoretically achieve zero miss distance (ZMD) against highly maneuvering targets. This merit is obtained at the expense of additional information, required for the implementation of these guidance laws. In particular, an estimation of time-to-go and target maneuver is required. The latter requirement is problematic owing to the inherent time delay of the estimator,[6–8] causing the modern guidance law performance to deteriorate. Moreover, modern guidance laws are often quite complicated; closed-form solutions exist only when system dynamics are neglected or approximated to first-order[9] or second-order[10] transfer functions. This complexity demands a considerable real-time computational capability. Furthermore, due to the fact that modern guidance laws result in an inverse of the system dynamics, their robustness has been doubted.[11]

In this chapter, we suggest an alternative approach, wherein the guidance law relies on LOS rate measurement *only*, similarly to the classical PN, yet its performance is similar to modern guidance laws, in the sense that ZMD can be obtained against highly maneuvering targets. This approach is therefore termed *neoclassical guidance*.[28–32] The main goal of this chapter is to present a new guidance law based upon LOS rate measurement only, whose main features are as follows:

1. In the case where LOS rate measurement is not corrupted by noise, the new guidance law yields ZMD for any flight time, against targets performing an arbitrary (bounded) deterministic maneuver, random maneuver, or deterministic maneuver with a random starting time.

2. The new guidance law yields ZMD for all flight times in the case of stochastic inputs, such as fading noise and passive- and active-receiver noise.

3. If the LOS rate measurement *is* corrupted by noise, a straightforward modification will give near-ZMD performance.

4. The maneuver acceleration required to achieve ZMD performance remains within reasonable limits, such that the overall maneuver effort is smaller than that required by PN guidance (PNG).

An important caveat is that the said guidance method preforms best under the assumptions of linearized dynamics and minimum phase transfer functions.

In the derivation of our neoclassical guidance law, called ZMD-PNG, we rely on the basic kinematic scenario used for the formulation of the PNG interception problem. Although in the general case, PNG is a nonlinear control problem, in order to apply known techniques of analysis and design, the system equations are linearized, yielding an equivalent linear time-varying system. The linearization is valid when it is assumed that the missile and the target approach the *collision course*. It is known that the linearized model faithfully represents the guidance dynamics[2,12,13] and that the miss distance associated with the linear approximation is very close to that obtained from the nonlinear model.[2,13]

The most popular tool for the analysis of the linear PNG loop is the method of adjoints.[2,14,15] This technique is based on the adjoint system impulse response and can be used to analyze miss distance caused by arbitrary inputs to the PNG system. Although this method renders analytical expressions for the miss distance as a function of flight time, it is very difficult to get closed-form solutions for high-order systems or for cases where the effective PN constant is not an integer. Thus, the use of this method is mainly numerical, that is, simulations of the adjoint loop are carried out in order to analyze miss distance. Except for simple cases,[2,15] no direct design information is available.

In this chapter, the adjoint formulae are utilized to the derivation of the ZMD-PNG law. We first examine the miss distance formulae for three main cases: deterministic target maneuvers, stochastic inputs (such as fading noise, passive- and active-receiver noise, and random target maneuver), and deterministic target maneuvers with random starting times. The key observation is that when the dynamics of the guidance loop, given by some transfer function, is positive real, and the PNG coefficient is larger than some threshold value, ZMD is obtained for all cases mentioned. This, of course, requires lead compensation, which can be achieved by augmenting the guidance commands by proportional-derivative (PD) controller. To prevent noise amplification problems, one can use lead-lag compensation. The design considerations delineated in this chapter are illustrated in simulations, which verify that ZMD-PNG gives small miss distance against highly maneuvering targets—even when the LOS rate measurement is noisy. However, a few caveats are in order: ZMD-PNG was proven useful for minimum-phase systems only. There are no analytical methods for applying it to nonminimum-phase systems. In addition, it was developed using linear models; hence, care should be taken in nonlinear applications. There are cases wherein ZMD-PNG has been shown to exhibit sensitivity to noise and optimal avoidance maneuvers. Finally, the term "ZMD" refers to the ideal case only. In practical applications, the miss distance cannot actually be "zero."

7.2 Miss Distance in PNG

The general formulation of a three-dimensional PN interception problem is rather complicated. However, assuming that the lateral and longitudinal maneuver planes are decoupled by means of roll control, one can deal with the equivalent two-dimensional problem in quite a realistic manner. We shall further assume that the geometry is two-dimensional. In addition to this basic assumption, we shall also assume that the gravitational component of the total missile lateral acceleration is compensated by means of g-bias. These assumptions enable the formulation of a general planar interception missile-target geometry as depicted in Figure 7.1. The figure describes a missile employing PN to intercept a maneuvering target.

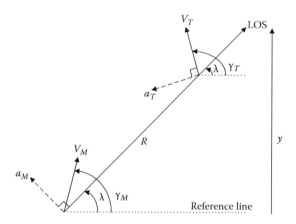

FIGURE 7.1
Two-dimensional geometry of a missile-target interception problem.

Based on Figure 7.1, a linearized model of the guidance dynamics can be developed. Such a model is widely used in the analysis of PNG.[1,2,14,16] A block diagram describing the linear model is given in Figure 7.2. In this linear time-varying system, missile acceleration a_M is subtracted from target acceleration a_T to form a relative acceleration $\ddot{y}$. A double integration yields the relative vertical position y (see Figure 7.1), which at the end of the engagement is the miss distance $y(t_f)$. By assuming that the closing velocity V_C is constant, the relative range is given by

$$R = V_C \cdot \tau \qquad (7.1)$$

where τ is the time-to-go, defined as

$$\tau \triangleq t_f - t. \qquad (7.2)$$

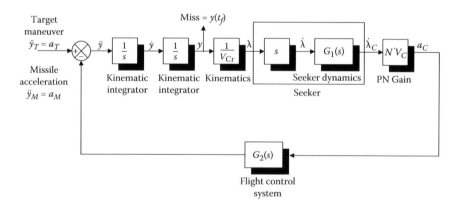

FIGURE 7.2
Linearized PNG model block diagram including the kinematics, flight control system, and seeker dynamics.

Dividing the relative vertical position y by the range given in Equation 7.1 yields the geometric LOS angle λ. The missile seeker is represented in Figure 7.2 as an ideal differentiator with an additional transfer function $G_1(s)$, representing the LOS rate measurement and noise filtering dynamics. The seeker generates a LOS rate command $\dot{\lambda}_C$, which is multiplied by the PN gain $N' \cdot V_C$ to form a commanded missile maneuver acceleration a_C, with N' being the effective PN constant. The flight control system, whose dynamics are represented by the transfer function $G_2(s)$, attempts to adequately maneuver the missile to follow the desired acceleration command.

A common use of the model depicted in Figure 7.2 is miss distance analysis. In particular, the method of adjoints is utilized.[1,2,14,15] The adjoint technique is based on the system impulse response and can be used to analyze miss distance as a function of flight time, provided that the system is linear. This method is utilized for the analysis of miss distance due to deterministic disturbances,[2] stochastic inputs,[14] and deterministic target maneuvers with random starting times.[15] The purpose of the subsequent discussion is to prove that there exists a class of PNG-based systems that yield ZMD for any type of input (deterministic, stochastic, random) and any given flight time.

7.2.1 Deterministic Disturbances

In the deterministic case, the miss distance is given by

$$y(t_f) = \mathcal{L}^{-1}\left\{Q(s) \cdot y_T(s)\right\} \tag{7.3}$$

where

$$Q(s) \triangleq \exp\left(N' \int_{\infty}^{s} H(\sigma)\,d\sigma\right) \tag{7.4}$$

$$G(s) \triangleq G_1(s)G_2(s) \tag{7.5}$$

$$H(s) \triangleq \frac{G(s)}{s} \tag{7.6}$$

and

$$y_T(s) = \mathcal{L}\left\{y_T(t)\right\}. \tag{7.7}$$

In the above expressions, $\mathcal{L}$ denotes the Laplace transform, and $y_T(t)$ denotes the deterministic system input, which can be either an initial condition or a deterministic target maneuver (see Figure 7.2). $G(s)$ represents the LOS rate measurement and flight control dynamics of the PNG loop (see Figure 7.2) and is assumed to be asymptotically stable with $G(0) = 1$.

7.2.2 Stochastic Inputs

Stochastic inputs are divided into two subcategories: noise inputs, such as fading noise, passive- and active-receiver noise (mainly in radar-guided missiles), and glint noise (mainly in radar-guided missiles and, to much smaller extent, in electro-optical missiles as well), and random target maneuvers, such as the random telegraph maneuver.[14,17] The expressions for the root-mean-square (RMS) miss distance in these cases, with $Q(s)$ as in Equation 7.4, are well known.[2,5]

RMS miss due to fading noise, which is a range-independent LOS angular noise, is given by

$$\frac{E[y^2(t_f)]\big|_{Fading}}{\Phi_{FN}} = \int_0^{t_f} \left\{ \mathcal{L}^{-1}\left[\frac{dQ(s)}{ds}\right] \right\}^2 dt \tag{7.8}$$

where Φ_{FN} is the power spectral density (PSD) of the fading noise (in square radians per hertz). The RMS miss due to a passive-receiver noise is similarly given by

$$\frac{E[y^2(t_f)]\big|_{Passive}}{\Phi_{PN}} = \int_0^{t_f} \left\{ \mathcal{L}^{-1}\left[\frac{d^2Q(s)}{ds^2}\right] \right\}^2 dt \tag{7.9}$$

where Φ_{PN} is the PSD of the passive-receiver noise (in square radians per hertz). The RMS miss due to an active-receiver noise is

$$\frac{E[y^2(t_f)]\big|_{Active}}{\Phi_{AN}} = \int_0^{t_f} \left\{ \mathcal{L}^{-1}\left[\frac{d^3Q(s)}{ds^3}\right] \right\}^2 dt \tag{7.10}$$

where Φ_{AN} is the PSD of the active-receiver noise (in square radians per hertz). Finally, the RMS miss due to glint noise is expressed as

$$\frac{E[y^2(t_f)]\big|_{Glint}}{\Phi_{GN}} = \int_0^{t_f} \left\{ \mathcal{L}^{-1}\left[1 - Q(s)\right] \right\}^2 dt \tag{7.11}$$

where Φ_{GN} is the PSD of the glint noise (in square meters per hertz).

A well-known example for a random target maneuver is the random tele-graph. We remind the reader that a random telegraph maneuver represents a policy, starting at time zero, in which the target executes either a maximum positive or negative acceleration $\pm a_T$ such that the number of sign changes per second follows a Poisson distribution and the average number of sign changes is ν per second. By demanding the equivalence of second-order miss distance statistics, the random sequence can be represented as a white noise passing through a shaping filter.[18] Thus, the following expression for the RMS miss distance is obtained[14,18]:

$$\frac{E[y^2(t_f)]\big|_{R.T.}}{\Phi_{RT}} = \int_0^{t_f} \left\{ \mathcal{L}^{-1}\left[\frac{1}{s/(2\nu)+1} \cdot \frac{Q(s)}{s^2} \right] \right\}^2 dt \qquad (7.12)$$

where $\Phi_{RT} = (a_T)^2_{max}/\nu$ is the PSD of the white noise, passing through the shaping filter $P(s) = 1/[s/(2\nu) + 1]$.

7.2.3 Deterministic Target Maneuvers with Random Starting Times

The target might initiate a maneuver at some random time during flight. It is assumed that the probability distribution function of the maneuver starting time is known. For instance, assume that the target performs a constant maneuver a_T whose starting time is uniformly distributed over the flight time. By demanding equivalence of second-order miss distance statistics, this maneuver can be mod-eled as a white noise, with PSD $\Phi_S = a_T^2/t_f$ passing through the shaping filter $1/s$ (see Zarchan[2]). In this case, the RMS miss distance is given by

$$\frac{E[y^2(t_f)]\big|_S}{\Phi_S} = \int_0^{t_f} \left\{ \mathcal{L}^{-1}\left[\frac{Q(s)}{s^3} \right] \right\}^2 dt. \qquad (7.13)$$

Another possibility is that the target performs a sinusoidal maneuver, whose starting time is uniformly distributed, that is, the phase of the maneu-ver, denoted φ, is a random variable:

$$a_T(t) = a_T \sin(\omega_T t + \varphi). \qquad (7.14)$$

The random-phase sinusoidal maneuver can be represented as a white noise with PSD $\Phi_{SIN} = a_T^2/t_f$ passing through the shaping filter $P(s) = 1/[\omega_T(s/\omega_T)^2 + \omega_T]$. The RMS miss distance is given by[14]

$$\frac{E[y^2(t_f)]\big|_{SIN.}}{\Phi_{SIN}} = \int_0^{t_f} \left\{ \mathcal{L}^{-1}\left[\frac{1/\omega_T}{(s/\omega_T)^2+1} \cdot \frac{Q(s)}{s^2} \right] \right\}^2 dt. \qquad (7.15)$$

In the general case, any target maneuver with random starting time can be represented as a white noise with PSD Φ_{in} passing through a shaping filter $P(s)$. Thus, the RMS miss distance is given by[14]

$$\frac{E[y^2(t_f)]}{\Phi_{in}} = \int_0^{t_f} \left\{ \mathcal{L}^{-1} \left[P(s) \cdot \frac{Q(s)}{s^2} \right] \right\}^2 dt. \tag{7.16}$$

In the subsequent discussion, we shall address the following problem: Determine the set of all possible strictly proper transfer functions $H(s)$, defined in Equation 7.6, such that the miss distance becomes zero for all flight times and all possible deterministic, stochastic, and random inputs to the guidance systems.

That is, we wish to find the following class:

$$\mathbf{H} = \{H(s) : y(t_f) \equiv 0 \; \forall t_f\}. \tag{7.17}$$

Notice that due to the definition of $H(s)$, finding the class $\mathbf{H}$ immediately characterizes the set $\mathbf{G}$, where

$$\mathbf{G} = \{G(s) : y(t_f) \equiv 0 \; \forall t_f\}. \tag{7.18}$$

7.3 Class of All PNG-Based Systems Yielding ZMD

Exclude for the moment the miss distance due to glint (Equation 7.11). Notice that in all other cases (Equations 7.3 and 7.8 through 7.16), if $Q(s)$ was identically equal to zero, no miss distance would be obtained. Thus, to characterize the class of ZMD-PNG–based systems, we have to find when $Q(s)$ is equal to zero. To begin, we notice that Equation 7.4 can be rewritten into the following form:

$$Q(s) = \frac{e^{NF(s)}}{e^{NF(\infty)}} \tag{7.19}$$

where

$$F(x) \triangleq \left[\int H(\sigma) d\sigma \right]_{\sigma = x}. \tag{7.20}$$

Since $e^{NF(s)} \neq 0$, $Q(s) \to 0$ if and only if $e^{N'F(\infty)} \to \infty$, which requires $F(\infty) \to \infty$. Hence, it is required to determine $H(s)$ for which $F(\infty) \to \infty$. To do this, the following theorem is introduced (for the complete proof, compare with the work of Gurfil[29]).

Theorem 7.1

Consider a strictly proper rational function of the form

$$H(s) = \frac{b(s)}{a(s)} = \frac{b_1 s^{n-1} + b_2 s^{n-2} + \cdots + b_n}{s^n + a_1 s^{n-1} + \cdots + a_n}, \ b_1 \geq 0, a_n = 0 \qquad (7.21)$$

where $a(s)$ and $b(s)$ are coprime polynomials. Denote by r the relative order of $H(s)$, that is, $r = \deg[a(s)] - \deg[b(s)]$. Under these conditions,

$$F(\infty) = \lim_{\sigma \to \infty} \left[\int H(\sigma) d\sigma \right] \to \begin{cases} 0 & \text{iff} \quad r \geq 2 \\ \infty & \text{iff} \quad r = 1. \end{cases} \qquad (7.22)$$

∎

The relevance of Theorem 7.1 to the classification of all PNG-based systems rendering ZMD is clarified by means of the following corollary.

Corollary of Theorem 7.1

$G(s) \in \mathbf{G}$ if and only if $b_1 > 0$, that is, $G(s)$ is biproper (i.e., the degree of the numerator equals the degree of the denominator):

$$\mathbf{G} = \{G(s) : b_1 > 0\}. \qquad (7.23)$$

∎

The set $\mathbf{G}$ contains all the transfer function $G(s)$ that imposes ZMD to the PNG system depicted in Figure 7.2. It is important to note that Theorem 7.1 and its corollary provide necessary and sufficient conditions. Consequently, if $y(t_f) \equiv 0 \ \forall t_f$, then $G(s) \in \mathbf{G}$; conversely, if $G(s) \in \mathbf{G}$, then $y(t_f) \equiv 0 \ \forall t_f$.

The interpretation of Theorem 7.1 and its corollary should be as follows. If $G(s)$, the dynamics of the PNG loop, is biproper, that is, the degree of the numerator equals the degree of the denominator, the miss distance will be identically zero for all flight times and all deterministic or random target maneuvers. Furthermore, in the stochastic case, ZMD is obtained for the cases of fading and passive- and active-receiver noise inputs. Since, in general, $G(s)$ is a strictly proper transfer function, that is, the degree of the denominator is greater than the degree of the numerator, a guidance controller composed of lead networks, such as PD controllers, should be added. However, pure lead causes noise amplification problems. Thus, instead of pure lead compensation, lead-lag networks should be used. The PN-based guidance law with the appropriate compensation will be called hereafter as ZMD-PNG.

It should be noted that the case of glint noise is different from all other inputs. First, observe that in the stochastic case, the expressions for miss distance comprise derivatives of $Q(s)$, whereas in the case of glint (Equation 7.11), the miss distance is calculated by taking the inverse Laplace transform of $1 - Q(s)$. Consequently, we notice that if $Q(S) \equiv 0$, then for long flight times,

$$\left. \frac{E[y^2(t_f)]}{\Phi_{GN}} \right|_{Glint} = \int_0^\infty \mathcal{L}^{-1}(1) \, dt = \int_0^\infty \delta(t) \, dt = 1. \tag{7.24}$$

That is, the RMS miss distance due to glint will be from the order of magnitude of the PSD of the noise. This implies that the implementation of the ZMD-PNG law is more appropriate in low-glint systems, such as missiles with electro-optical seekers, rather than in missiles with radar seekers, where the glint is much more dominant.

The above discussion is valid under the assumptions that the missile maneuver acceleration is unlimited and that the LOS measurement is not corrupted by noise. In real systems, these assumptions must be alleviated. The purpose of the following sections is to generalize the ZMD-PNG formalism for real-life systems.

7.4 Case of Saturating Missile Acceleration

The design of a guidance law for a given missile is effected primarily from the a priori knowledge of the missile acceleration capability. When the missile-target maneuver ratio is known, a suitable extension of the ZMD-PNG method can be synthesized by implementing functional analysis tools, mainly the concept of bounded-input bounded-output stability.[20–22]

The linear model used in previous sections of this chapter is based implicitly on the assumption of nonlimited missile maneuverability. Actually, every real missile system is subject to maneuverability saturation due to aerodynamic or structural constraints. To complete the guidance model description in this case, a new variable, the required missile maneuver acceleration, a_R, is introduced. The nonlinear relationship between the actual and required missile maneuver is defined by

$$a_M = a_{M_{max}} \text{ sat} \left\{ \frac{a_R}{a_{M_{max}}} \right\} \tag{7.25}$$

where

$$\text{sat}(x) \triangleq \begin{cases} x & \text{if} \quad x \leq 1 \\ 1 & \text{if} \quad |x| > 1. \end{cases} \tag{7.26}$$

Equation 7.25 implicitly assumes that the limit on the missile maneuverability is of the aerodynamic type, owing to mechanical limits of control fin deflection or to hinge moment saturation. This limit is at the output of the guidance channel, as shown in Figure 7.3a. However, in many missiles, the limit is imposed on the acceleration command. This limit is usually designed to be sufficiently conservative, such that an aerodynamic saturation, mentioned earlier, is not reached. In this case, the limited acceleration command is denoted a_L, where

$$a_L = a_{C_{max}} \text{ sat} \left\{ \frac{a_C}{a_{C_{max}}} \right\}. \tag{7.27}$$

This limit is depicted in Figure 7.3b.

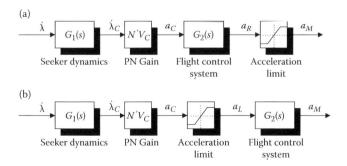

FIGURE 7.3
(a) Limited output acceleration, (b) limited acceleration command.

In a conventional PNG system, it is known[16,19] that an infinite missile acceleration is required near to intercept ($t \to t_f$). This means that saturation is always reached, and miss distance will be greater than predicted by linear analysis. Our goal in this section is to characterize a ZMD-PNG system in which saturation is *avoided*.

Consequently, it is necessary to find some bound μ on the required missile-target maneuver ratio, μ_r, defined as

$$\mu_r \triangleq \frac{\sup_{t \in [0, t_f]} |a_M(t)|}{\sup_{t \in [0, t_f]} |a_T(t)|}. \tag{7.28}$$

If μ is found to be smaller than the a priori known missile-target maneuver ratio μ_0, no saturation will occur. In the case of PNG, owing to the divergence of the state variables at the vicinity of the interception, we have

$$\lim_{t \to t_f} |a_M(t)| \to \infty \quad \forall a_T(t) \neq 0. \tag{7.29}$$

We shall prove that there is a way to modify the PNG law such that $\sup_{t \in [0, tf]} |a_M(t)| < \infty$. Moreover, we shall find a constant μ such that

$$\sup_{t \in [0, tf]} |a_M(t)| = \|a_M(t)\|_\infty \leq \mu \|a_T(t)\|_\infty = \mu \sup_{t \in [0, tf]} |a_T(t)|. \tag{7.30}$$

A proper assessment of μ, together with a proper design modification of the PNG, might yield the required nonsaturating guidance system.

Inequality 7.30 describes a bounded-input bounded-output stability problem in the functional space $L^\infty[0, t_f]$. This problem has been solved in the work of Gurfil et al.[32] and may be summarized in the design guidelines, sufficient to avoid output acceleration saturation:

> *Design Guideline 1.* Given flight control dynamics $G_1(s)$ and seeker dynamics $G_2(s)$, design a controller $K(s)$ such that $G(s) = K(s)G_1(s)G_2(s)$ is a phase lead network with a maximal phase lead not exceeding 180°.
>
> *Design Guideline 2.* Given the missile-target maneuver ratio μ_0, choose N' such that $N'/(N' - 2) \geq \mu_0$, or equivalently, $N' \geq \dfrac{2\mu_0}{\mu_0 - 1}$.

From Design Guideline 2, it is obvious that the smaller the value of μ_0 is, the larger the N' that should be chosen. When following the design guidelines, it is assured that saturation is prevented. However, we have still not provided an insight into the newly developed guidance law. The purpose of the next section is to provide a new outlook on ZMD-PNG and show how it can be formulated as an estimation problem.

7.5 ZMD Guidance as Estimation Problem

ZMD-PNG is merely PNG with an additional lead controller, designed to render the total dynamics of the guidance loop positive real. Owing to the inherent phase lead required, ZMD-PNG may significantly amplify the LOS rate measurement noise to infeasible levels. The purpose of this section is to establish a suitable design procedure of ZMD-PNG for the general stochastic guidance system where the LOS rate measurement is corrupted by noise.

Let us first summarize the results of the previous discussion in the following theorems:

Theorem 7.2

Let $\{PR\}$ be the class of positive real transfer functions, and denote $G(s) \triangleq G_1(s)G_2(s)$. If $\exists K(s)$ ("the guidance controller") such that $K(s)G(s)/s \in \{PR\}$, then $\exists a_M < \infty$ such that $y(t_f) = 0 \ \forall t_f, \ \forall |a_T| \leq a_{Tmax}$.

∎

Theorem 7.3

Let the missile maneuver acceleration a_M be limited, $|a_M| \leq a_{Mmax}$. Denote $\mu_0 = a_{Mmax}/a_{Tmax}$. If $\exists K(s)$ such that $K(s)G(s)/s \in \{PR\}$, and in addition, $N' \geq 2\mu_0/(\mu_0 - 1)$, then $y(t_f) = 0 \ \forall t_f, \ \forall |a_T| \leq a_{Tmax}$.

∎

The positive realness requirement necessitates the guidance controller to be a PD phase lead controller of the form

$$K(s) = \prod_{i=1}^{m} (\tau_{iZ}s + 1) \tag{7.31}$$

where m denotes the relative degree of $G(s)$.

In addition to the realization difficulty, the noise amplification of the PD guidance controller (Equation 7.31) may be prohibitively high. Thus, when the LOS rate measurement is corrupted by noise, the guidance controller must be modified into a lead-lag network so that a certain extent of filtering is used. This results in a suboptimal operation since the ZMD property does not hold when the positive realness condition is violated. However, when the filtering, or "roll-off," is performed at high frequencies, a near-ZMD performance can be achieved. The major question is, therefore, how to design an appropriate guidance controller for the stochastic case, which will provide the necessary noise filtering on one hand but will not increase the miss distance on the other hand. In order to quantify this trade-off, we first discuss the physical interpretation of the deterministic ZMD-PNG. This will provide us with the necessary tools needed in order to extend the ZMD-PNG law to the stochastic case.

The deterministic ZMD-PNG is implemented as described in Figure 7.4a. The kinematic LOS rate $\dot{\lambda}$ is measured by the seeker, whose output is the measured LOS rate, $\dot{\lambda}_m$. $\dot{\lambda}_m$ constitutes an input to the guidance controller $K(s)$. At the guidance controller output, we have

$$(\dot{\lambda}_m)_{eq} \triangleq \dot{\lambda}_m + \bar{\tau}_{Z_1}\ddot{\lambda}_m + \ldots + (\bar{\tau}_{Z_m})^m \overset{(m+1)}{\lambda_m} \qquad (7.32)$$

(a)

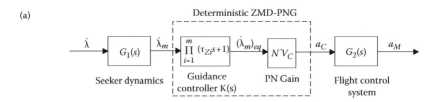

(b)

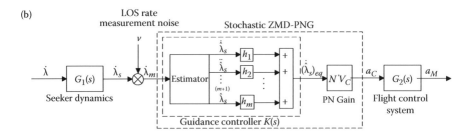

FIGURE 7.4
Implementation of ZMD-PNG. (a) Deterministic case. Measured LOS rate is fed to the PD guidance controller, which generates an equivalent LOS rate. (b) Stochastic case. High-order LOS rate derivative estimator and a weighting vector generate an estimated equivalent LOS rate.

where

$$\bar{\tau}_{Z_1} = \sum_{i \in C_1^m} \tau_{Z_i} = \sum_{i=1}^{m} \tau_{Z_i}$$

$$(\bar{\tau}_{Z_2})^2 = \sum_{i,j \in C_2^m} \tau_{Z_i} \cdot \tau_{Z_j}$$

$$(\bar{\tau}_{Z_3})^3 = \sum_{i,j,k \in C_3^m} \tau_{Z_i} \cdot \tau_{Z_j} \cdot \tau_{Z_k} \qquad (7.33)$$

$$\vdots$$

$$(\bar{\tau}_{Z_m})^m = \sum_{i,j,k,\ldots,m \in C_m^m} \tau_{Z_i} \cdot \tau_{Z_j} \cdot \tau_{Z_k} \cdot \ldots \cdot \tau_{Z_m} = \prod_{i=1}^{m} \tau_{Z_i}$$

and C_p^n is the usual combination space of choosing p terms out of n.

The variable $(\dot{\lambda}_m)_{eq}$ has LOS rate units. It represents the sum of the measured LOS rate, $\dot{\lambda}_m$, and higher-order derivatives of λ_m multiplied by the coefficients of the polynomial $K(s)$. Thus, $(\dot{\lambda}_m)_{eq}$ can be viewed as an *equivalent* LOS rate. The equivalent LOS rate is multiplied by the PN gain $N'V_C$ to form a commanded missile acceleration. In this sense, the ZMD-PNG law is a generalization of PNG; instead of applying an acceleration command that is merely proportional to the measured LOS rate, the acceleration command is proportional to an equivalent LOS rate.

The superior performance of ZMD-PNG is therefore a result of extracting additional information from the LOS rate measurement. In other words, a *prediction* procedure is employed to obtain the future state of the target. This can be immediately seen by recalling that[2]

$$\ddot{\lambda}(t) = \{[2 - N'G(s)] \cdot \dot{\lambda}(t) + a_T/V_C\}/t_{go} \qquad (7.34)$$

which implies

$$\ddot{\lambda}(t) \sim f(a_T/R). \qquad (7.35)$$

Hence, the use of $\ddot{\lambda}$ (and possibly higher-order derivatives, if needed) as an additional signal for generation of the commanded maneuver acceleration is equivalent, in a sense, to using information regarding both target maneuver and relative range (or time-to-go), as required by optimal guidance methods. However, ZMD-PNG extracts this information *indirectly* from the LOS rate measurement, whereas optimal guidance laws (OGLs) need *direct* estimation of target maneuver and either measurement or estimation of the relative range.

The measurement of the LOS rate is corrupted by noise, usually as a result of sensor imperfections (such as rate-gyro noise). A simple differentiation of the measured LOS rate to generate high-order derivatives is therefore infeasible. Rather, high-order derivatives should be *estimated* from the LOS rate measurement. To this end, consider the measurement equation

$$\dot{\lambda}_m = \dot{\lambda}_s + v \tag{7.36}$$

where v is a zero-mean Gaussian white noise, and $\dot{\lambda}_s$ is the noise-free LOS rate measurement.

It is required to recover high-order derivatives of $\dot{\lambda}_s$ out of the noisy measurement (Equation 7.36). The highest-order derivative required is determined by m, the relative order of $G(s)$. A diagram of the estimation process, which constitutes the stochastic ZMD-PNG, is depicted in Figure 7.4b. The input to the estimator is the noisy measurement $\dot{\lambda}_m$. The estimator provides the estimated high-order derivatives, $\dot{\hat{\lambda}}_s, \ddot{\hat{\lambda}}_s, ..., \overset{(m+1)}{\hat{\lambda}}_s$, which are then combined using some weighting vector **h** to yield the estimated equivalent LOS rate:

$$(\dot{\hat{\lambda}}_s)_{eq} = \dot{\hat{\lambda}}_s + h_1 \ddot{\hat{\lambda}}_s + ... + h_m \overset{(m+1)}{\hat{\lambda}}_s = [1\, h_1 \cdots h_m] \begin{bmatrix} \dot{\hat{\lambda}}_s \\ \ddot{\hat{\lambda}}_s \\ \vdots \\ \overset{(m+1)}{\hat{\lambda}}_s \end{bmatrix} \triangleq \mathbf{h}^T \hat{\mathbf{\Lambda}}. \tag{7.37}$$

Qualitatively speaking, since $\dot{\lambda}_m$ is the only measurement, the higher the order of the required LOS rate derivative is, the larger the estimation error will be. Fortunately, the relative order of the overall loop dynamics is often small; thus, practically speaking, there is usually no need to estimate LOS rate derivatives beyond the LOS jerk ($\dddot{\lambda}_s$). Nevertheless, we consider hereafter the general estimation procedure.

In order to estimate high-order LOS rate derivatives, we use a variant of the well-known exponentially correlated acceleration (ECA) approach, which is a kinematic-model estimator[23] first introduced by Singer[24] for estimation of the target maneuver. Several modifications will be performed to adapt this approach to the estimation of high-order LOS rate derivatives. The new technique will be referred to as *exponentially correlated high-order LOS rate derivative* (ECH) estimation. The application of the ECA approach to estimation of high-order LOS rate derivatives stems mainly from the firm relationship between the target acceleration and the high-order LOS rate derivatives, as pointed out in Equation 7.4, and the probabilistic interpretation of ECA, described as follows.

Let the highest-order LOS rate derivative, $\overset{(m+1)}{\lambda_s}$, be a zero-mean random variable uniformly distributed within the limits $\pm(\overset{(m+1)}{\lambda_s})_{max}$. The probability of attaining an extremum value is π_{max}. The variance of $\overset{(m+1)}{\lambda_s}$ is therefore given by

$$\sigma^2_{\overset{(m+1)}{\lambda_s}} = \left[\left(\overset{(m+1)}{\lambda_s}\right)_{max}\right]^2 [1 + 4\pi_{max}]/3. \tag{7.38}$$

Using a standard Wiener–Kolmogorov whitening procedure,[25] a shaping filter of the exponentially correlated $\overset{(m+1)}{\lambda_s}$ can be derived:

$$\overset{(m+1)}{\lambda_s}(t) = \frac{-\overset{(m)}{\lambda}(t)}{\tau_E} + \frac{w(t)}{\tau_E} \tag{7.39}$$

where τ_E is the random process correlation time, and $w(t)$ is a zero-mean Gaussian white noise with PSD:

$$\tilde{q} = 2\sigma^2_{\overset{(m+1)}{\lambda_s}}/\tau_E^3. \tag{7.40}$$

The dynamics of the ECH model can thus be written in the state space representation:

$$\dot{\Lambda}(t) = \begin{bmatrix} 0_{m\times 1} & I_m \\ 0_{1\times m} & -1/\tau_E \end{bmatrix} \Lambda(t) + \begin{bmatrix} 0_{m\times 1} \\ 1/\tau_E \end{bmatrix} w(t) = A\Lambda(t) + \mathbf{g}w(t) \tag{7.41}$$

where $\Lambda^T = [\dot{\lambda}_S, \ddot{\lambda}_S, \ldots, \overset{(m+1)}{\lambda_S}]$, I_m is an $m \times m$ identity matrix, $0_{k\times l}$ is a $k \times l$ zero matrix, and $w(t)$ is a zero-mean Gaussian white noise with PSD $\tilde{q}$,

$$E[w(t)] = 0$$

$$E[w(t)w(\theta)] = \tilde{q}\delta(t - \theta). \tag{7.42}$$

The initial conditions satisfy

$$E[\Lambda(0)] = \overline{\Lambda}(0)$$

$$\text{cov}[\Lambda(0)] = P_0. \tag{7.43}$$

The measurement equation (Equation 7.36) is rewritten as follows:

$$\dot{\lambda}_m = [1, 0, \ldots 0]\mathbf{\Lambda} + v(t) = \mathbf{c}^T\mathbf{\Lambda} + v(t) \tag{7.44}$$

where

$$E[v(t)] = 0 \tag{7.45}$$
$$E[v(t)v(\theta)] = r\delta(t - \theta)$$

and r is the PSD of the measurement noise. It is important to note that r as used in the estimator design need not necessarily be equal to the actual LOS rate measurement noise. Also, we assume that the initial conditions, measurement noise, and process noise are uncorrelated:

$$E[\mathbf{\Lambda}(0)w(t)] = E[\mathbf{\Lambda}(0)v(t)] = E[w(t)v(t)] = 0. \tag{7.46}$$

The ECH model formulated in Equations 7.41 through 7.46 can be used to obtain an estimated state vector $\hat{\mathbf{\Lambda}}$ using the well-known time-varying Kalman–Bucy filter (TVKF):

$$\dot{\hat{\mathbf{\Lambda}}}(t) = A\hat{\mathbf{\Lambda}}(t) + \mathbf{k}(t)[\dot{\lambda}_m - \mathbf{c}^T\hat{\mathbf{\Lambda}}(t)]$$
$$\mathbf{k}(t) = P(t)\mathbf{c}/r \tag{7.47}$$
$$\dot{P}(t) = AP(t) + P(t)A^T + \tilde{q}\mathbf{g}\mathbf{g}^T - P(t)\mathbf{c}\mathbf{c}^T P(t)/r$$

where

$$P(t) \triangleq E[(\mathbf{\Lambda} - \hat{\mathbf{\Lambda}})(\mathbf{\Lambda} - \hat{\mathbf{\Lambda}})^T)]. \tag{7.48}$$

Note that the ECH constitutes a kinematic-model estimator. Thus, $\exists P > 0$ such that $\lim_{t \to \infty} P(t) \to P$, and the steady-state Kalman–Bucy filter (SSKF) is given by

$$\dot{\hat{\mathbf{\Lambda}}}(t) = A\hat{\mathbf{\Lambda}}(t) + \mathbf{k}[\dot{\lambda}_m - \mathbf{c}^T\hat{\mathbf{\Lambda}}(t)]$$
$$\mathbf{k} = P\mathbf{c}/r \tag{7.49}$$
$$AP + PA^T + \tilde{q}\mathbf{g}\mathbf{g}^T - P\mathbf{c}\mathbf{c}^T P/r = 0.$$

It is important to note that in the specific application discussed here, there is no significant advantage of using the TVKF rather than the SSKF. We are merely interested in a structured, systematic, robust, and tunable method to roll off the noise obtained by a direct differentiation of λ_m. To this end, the performance of the SSKF is perfectly feasible. Moreover, since the analysis of ZMD-PNG and the associated proof-of-concept were carried out in the frequency domain, it is only natural to continue probing the problem in this manner.

With this notion in mind, we adopt the SSKF and obtain the vector of filter transfer functions between the $m + 1$ components of the vector $\hat{\mathbf{\Lambda}}$ and the input $\dot{\lambda}_m$,

$$
\mathbf{F}(s) = \begin{bmatrix} \dfrac{\dot{\hat{\lambda}}_s(s)}{\dot{\lambda}_m(s)} \\[2mm] \dfrac{\ddot{\hat{\lambda}}_s(s)}{\dot{\lambda}_m(s)} \\[2mm] \vdots \\[2mm] \dfrac{\overset{(m+1)}{\hat{\lambda}}_s(s)}{\dot{\lambda}_m(s)} \end{bmatrix} = C_F(sI - A_F)^{-1}\mathbf{b_F} \tag{7.50}
$$

where

$$
\begin{aligned}
C_F &= I_{m+1} \\
A_F &= A - \mathbf{k} \cdot \mathbf{c}^T \\
\mathbf{b_F} &= \mathbf{k}.
\end{aligned} \tag{7.51}
$$

Equation 7.50 presents a closed-form expression for the estimates $\dot{\hat{\lambda}}_s, \ddot{\hat{\lambda}}_s, \ldots, \overset{(m+1)}{\hat{\lambda}}_s$. However, in order to obtain the guidance controller whose output is $(\dot{\hat{\lambda}}_s)_{eq}$, these estimates should be combined using some weighting vector $\mathbf{h}$. Thus, in the stochastic case, we have

$$
K(s) = \frac{(\dot{\hat{\lambda}}_s)_{eq}(s)}{\dot{\lambda}_m(s)} = \mathbf{h}^T\mathbf{F}(s). \tag{7.52}
$$

By observation, the denominator of $K(s)$ is of order $m + 1$, and the numerator is of order m. Thus, the overall effect of the ECH filter, in terms of the overall dynamics, is adding a first-order lag. This, of course, results in miss distance increase due to the violation of the positive realness condition stated in Theorems 7.2 and 7.3. The purpose of the weight vector $\mathbf{h}$ is to partially recover the miss distance performance obtained prior to the filtering process. In practice, $\mathbf{h}$ quantifies the trade-off between miss distance and noise filtering. A method for selecting this vector is as follows. First, note that Equation 7.50 can be partitioned as

$$\mathbf{F}(s) = \frac{1}{d_{SSKF}(s)} \begin{bmatrix} n_1(s) \\ n_2(s) \\ \vdots \\ n_{m+1}(s) \end{bmatrix} \tag{7.53}$$

where $d_{SSKF}(s)$ is the filtering dynamics of the SSKF, and $n_i(s)$ is the numerator polynomial of the (*i*-1)-order LOS rate derivative estimator. In the deterministic case, we had $d_{SSKF}(s) \equiv 1$ since no filtering was used. To satisfy the positive realness condition in the deterministic case, formulated by Theorems 7.1 and 7.2, $\mathbf{h} = [1, h_1, \ldots h_m]^T$ would have been required to satisfy

$$[n_1(s) + h_1 n_2(s) + \ldots + h_m n_{m+1}(s)] \cdot G(s)/s \in \{PR\}. \tag{7.54}$$

The same procedure can be adopted in the stochastic case. That is, first we ignore the presence of the denominator polynomial and assume $d_{SSKF}(s) \equiv 1$. Next, we find $\mathbf{h}$ such that Equation 7.54 is satisfied. In a sense, this concept represents a variant of the separation principle, where the estimation and control designs are done separately and then interconnected. The overall system robustness reduction that stems from using the separation principle as a design methodology reflects itself in our case by the slight increase in miss distance compared with the deterministic case.

7.6 Illustrative Example

In the previous sections, a new guidance law, ZMD-PNG, was synthesized. The purpose of this section is to investigate the performance of ZMD-PNG when implemented in a real-life electro-optical missile. Furthermore, a detailed comparison between ZMD-PNG, OGL, and PNG will be given. The comparison includes both deterministic simulation runs and a Monte Carlo analysis.

The missile models used here are believed to constitute a faithful representation of a large family of electro-optical guided missiles. The real-life models include a detailed flight control system, which consists of an aerodynamic model, fin actuators, and sensors, and an electro-optical tracking loop, which includes a detailed model of the seeker and tracker. Both the flight control system and the tracking loop include nonlinear effects such as state saturation and field-of-view limits. Exogenous disturbances such as fin bias and measurement noise are modeled as well. For a detailed description of these models, the reader is referred to the work of Gurfil.[32]

7.6.1 Guidance Law Synthesis

In order to design a ZMD-PNG law, transfer functions of the flight control system, $G_2(s) = a_M(s)/a_C(s)$, and the tracking loop, $G_1(s) = \dot{\lambda}_m/\dot{\lambda}$, are required. These transfer functions can be found in two steps. First, the nonlinear terms are left out. Second, the resulting high-order linear models are reduced using a state truncation method, such as balanced realization. It is important to stress that this procedure is used for the guidance design *only*, not for the overall performance evaluation of the missile, where the complete, detailed nonlinear stochastic models are used.

We start with model reduction of the complex flight control system,[32] which has 9 zeros and 13 poles. Using balanced realization state truncation, and the parameter values given in the work of Gurfil,[32] the following reduced-order transfer function is obtained:

$$G_2(s) \approx \frac{-s/40.3 + 1}{(s/23.3 + 1)(s/1.93 + 1)}. \tag{7.55}$$

Obviously, $G_2(s)$ is nonminimum phase due to the fact that the missile is aft-steered. From the engineering standpoint, it is evident that the right half plane zero is "fast." Hence, an additional state truncation yields

$$G_2(s) \approx \frac{1}{0.56s + 1}. \tag{7.56}$$

This transfer function constitutes an adequate approximation to the overall flight control system dynamics in both the frequency and time domains. It is subsequently used for ZMD-PNG design.

The tracking loop overall transfer function, $G_1(s)$, is obtained in a similar manner. Using the numerical values in the work of Gurfil,[32] neglecting the FOV saturation and the pure tracking delay, we have

$$G_1(s) \approx \frac{1}{0.1s + 1}. \tag{7.57}$$

The overall transfer function of the guidance loop is therefore

$$G(s) = G_1(s)G_2(s) = \frac{1}{(0.1s + 1)(0.56s + 1)}. \tag{7.58}$$

The relative order of $G(s)$ is $m = 2$, so the highest order LOS rate derivative required is the LOS jerk, $\dddot{\lambda}_S$. Assume that the correlation time of $\dddot{\lambda}_S$ is equal to the total time constant of the reduced-order system (Equation 7.58), that is, $\iota_E - 0.66$. Applying, we write the ECH model for this case.

$$\dot{\mathbf{\Lambda}}(t) = \begin{bmatrix} 0 & 1 & 0 \\ 0 & 0 & 1 \\ 0 & 0 & -1.51 \end{bmatrix} \mathbf{\Lambda}(t) + \begin{bmatrix} 0 \\ 0 \\ 1.51 \end{bmatrix} w(t) \tag{7.59}$$

where $\mathbf{\Lambda}^T = [\lambda_s, \dot{\lambda}_s, \ddot{\lambda}_s]$.

The PSD of the process noise is evaluated using Equation 7.40. To this end, let

$$(\dddot{\lambda}_s)_{\max} = 750 \deg/s^3, \pi_{\max} = 0.5, \tag{7.60}$$

which yields

$$\tilde{q} = 45{,}000 \deg^2/s^9. \tag{7.61}$$

The measurement equation is given by

$$\dot{\lambda}_m = [1, 0, 0]\mathbf{\Lambda} + v(t) \tag{7.62}$$

with

$$r = 0.04 \deg^2/s^2 \tag{7.63}$$

Note that, generally speaking, the parameters in Equations 7.60 through 7.63 constitute tuning parameters that reflect the usual trade-offs associated with the Kalman filter design.

The SSKF estimator is initialized as follows:

$$\hat{\mathbf{\Lambda}}(0) = 0. \tag{7.64}$$

The resulting transfer function vector is

$$\mathbf{F}(s) = \begin{bmatrix} \dfrac{\dot{\hat{\lambda}}(s)}{\dot{\lambda}_m(s)} \\[2mm] \dfrac{\ddot{\hat{\lambda}}(s)}{\dot{\lambda}_m(s)} \\[2mm] \dfrac{\dddot{\hat{\lambda}}(s)}{\dot{\lambda}_m(s)} \end{bmatrix} = \dfrac{1}{0.0001s^3 + 0.038s^2 + 0.087s + 1} \begin{bmatrix} 0.0037s^2 + 0.087s + 1 \\ 0.0814s^2 + s \\ 0.876s^2 \end{bmatrix}. \tag{7.65}$$

The next step is to choose the weighting vector **h**. To this end, we follow the design guideline discussed above:

$$\mathbf{h} = [1, 0.3, 0.2]^T \tag{7.66}$$

which yields the guidance controller

$$K(s) = \mathbf{h}^T F(s) = \frac{0.203s^2 + 0.387s + 1}{0.0001s^3 + 0.038s^2 + 0.087s + 1}. \tag{7.67}$$

The final step is choosing N'. To this end, we utilize the design principle given in Theorem 7.2. Assuming that the missile-target maneuver ratio is $\mu_0 = 2$ requires $N' \geq 4$. We chose $N' = 5$ to account for possible uncertainties in μ_0 (i.e., the case where the actual missile-target maneuver ratio is smaller than 2). In summary, the ZMD-PNG command to the flight control system is

$$a_C = 5 \cdot V_C \cdot \frac{0.203s^2 + 0.387s + 1}{0.0001s^3 + 0.038s^2 + 0.087s + 1} \cdot \dot{\lambda}_m. \tag{7.68}$$

The performance of the guidance law (Equation 7.68) will be compared to OGL and the classical PNG. The commanded OGL maneuver acceleration is[9]

$$a_C = N'(\xi)[y + \dot{y}\hat{t}_{go} + 0.5\hat{a}_T\hat{t}_{go}^2 - a_M\tau_D^2(e^{-\xi} + \xi - 1)]/\hat{t}_{go}^2$$

$$N'(\xi) = \frac{6\xi^2(e^{-\xi} + \xi - 1)}{2\xi^3 - 6\xi^2 + 6\xi + \xi - 12\xi e^{-\xi} - 3e^{-2\xi}} \tag{7.69}$$

where $\hat{t}_{go}$ is the estimated time-to-go, $\hat{a}_T$ is the estimated target maneuver, τ_D is the so-called "design" time constant, and $\xi \triangleq \hat{t}_{go}/\tau_D$. Due to the fact that the most common OGL was originally conceived for first-order dynamics, there is a considerable mismatch between actual missile dynamics and the OGL. To deal with this mismatch, it was suggested[26] to use a "design" time constant, τ_D, which is about 1.5 times larger than the equivalent time constant of the system. In our case, the equivalent time constant is 0.66 s, so $\tau_D = 1$ s. It is also assumed that the time-to-go is estimated exactly, that is, $\hat{t}_{go} = t_{go}$. For the estimation of the target maneuver, the following model was adopted

$$\hat{a}_T = e^{-s\tau_a} \cdot a_T. \tag{7.70}$$

In Equation 7.70, it is assumed that the target maneuver estimator constitutes a pure time delay of τ_a seconds. We chose $\tau_a = 0.2$ s, which is a rather optimistic value, since often the estimation delay is even larger.

The PNG design is far simpler. The sole degree of freedom is N'. We chose $N' = 5$, so that with PNG, we have

$$a_C = 5 \cdot V_C \cdot \dot{\lambda}_m. \tag{7.71}$$

7.6.2 Engagement Scenario

The engagement scenario assumes a constant closing velocity of 1000 m/s and a missile velocity of 500 m/s. This means that the scenario is head-on (in terms of projection of the relative velocity on the LOS).

The target maneuver simulated to test the performance of the three guidance laws was a random phase sinusoidal maneuver. In this maneuver, the target performs a periodic maneuver perpendicular to its velocity vector; this type of an engagement simulates a missile attempting to intercept a ballistic missile on atmosphere reentry, which is one of the most challenging problems faced by modern homing missiles.[27] In this case, tested in the following simulations, the target performs a so-called barrel-roll maneuver. A projection of this maneuver on the plane yields a sinusoidal target acceleration time history. In addition, it was assumed that the starting time of the maneuver is unknown; it can be initiated at any given time point within the flight time of the interceptor. This represents the fact that the ballistic missile maneuver is initiated randomly due to atmospheric perturbations. Thus, the target maneuver can be described by

$$a_T(t) = a_{T_0} \sin(\omega_T t + \phi) \tag{7.72}$$

where a_{T_0} is the maneuver magnitude, ω_T is the frequency, and ϕ is the phase, assumed uniformly distributed between 0 and 2π, which is completely equivalent to a random maneuver initiation time. In this example, the numerical values chosen were $a_{T_0} = 10g$ and $\omega_T = 1.7$ rad/s. (This frequency is the optimal avoidance frequency for $N' = 5$; see the work of Shinar and Steinberg.[19])

7.6.3 Guidance Law Performance

The performance of ZMD-PNG, PNG, and OGL against the random phase sinusoidal target maneuver includes both deterministic simulation runs and a Monte Carlo analysis. In the deterministic examination, all the parameters are kept in their nominal values. The simulation was performed using the full nonlinear stochastic models described previously. For the deterministic case only, a constant flight time, $t_f = 5$ s, and a constant target maneuver phase, $\phi = 0$, were chosen. The time histories of the actual and commanded accelerations for $V_C = 1000$ m/s are depicted in Figure 7.5.

Evidently, both ZMD-PNG and OGL yield superior performance compared with PNG. Note that with PNG, the actual maneuver acceleration saturates

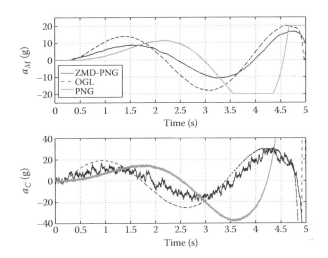

FIGURE 7.5
Comparison of actual and commanded missile acceleration shows that ZMD-PNG requires less maneuver effort than OGL and PNG.

1.5 s before impact, which seriously increases the miss distance. However, with ZMD-PNG, this saturation is avoided, as expected, and the miss distance is much smaller. Note also that ZMD-PNG requires a smaller maneuver effort than OGL because the OGL used here is actually suboptimal due to the high-order system dynamics.

As seen in Figure 7.5, the acceleration command generated by ZMD-PNG is somewhat noisy. The noise level reaches $\pm 1g$ at $\pm \sigma$, which is feasible, and has no substantial implications on the performance of the system. The three guidance laws examined yielded the following miss distance:

$$y(t_f)|_{\text{ZMD-PNG}} = -0.108 \text{ m}, \ y(t_f)|_{\text{OGL}} = 0.114 \text{ m}, \ y(t_f)|_{\text{PNG}} = 60.2 \text{ m}. \quad (7.73)$$

Thus, while ZMD-PNG and OGL render a similar miss distance, PNG induces a considerably larger miss. This implies that PNG cannot deal adequately with sinusoidal targets, as was also mentioned by Zarchan.[14]

We proceed with a thorough statistical examination of the guidance law performance using a Monte Carlo technique. In each simulation run, parameter values, as well as the seed used to generate the noise signals, are randomly selected according to prespecified probability distribution functions. After a large database of simulation runs has been created, the results, especially the miss distance, are statistically analyzed.

In this example, each simulation run used random parameter values and a random target maneuver phase. For each flight time, 300 simulation runs were performed. The procedure was repeated for flight times ranging from

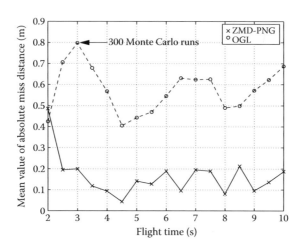

FIGURE 7.6
Mean values of the absolute miss distance with ZMD-PNG are smaller than with OGL.

2 to 10 s. In order to evaluate miss distance statistics, the mean value of the absolute miss distance, defined by

$$\bar{y}(t_f) = \frac{1}{300} \sum_{i=1}^{300} \left| y(t_f) \right|_i, \tag{7.74}$$

was examined. In addition, standard deviations of miss distance for each flight time were considered. The results are depicted in Figures 7.6 and 7.7.

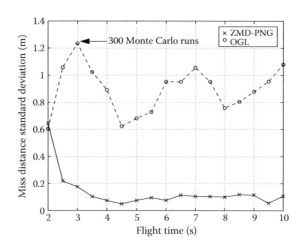

FIGURE 7.7
Standard deviations of the miss distance with ZMD-PNG are smaller than with OGL.

Figure 7.6 compares the mean value of absolute miss distance, and Figure 7.7 compares the standard deviation of miss distance. PNG was not considered here because the deterministic examination showed that it yielded miss distance that was larger than the miss obtained with ZMD-PNG and OGL by an order of magnitude.

Clearly, ZMD-PNG yields smaller mean and RMS miss than OGL. This is true for all flight times greater than 2 s. Moreover, with ZMD-PNG, the miss distance is more robust to variations in flight times and in missile parameters. These observations imply that the newly developed guidance law can deal adequately with highly maneuvering targets. This merit is achieved with neither estimating target maneuver nor estimating time-to-go.

7.7 Summary and Conclusions

In this chapter, a new guidance law was presented. The derivation of this law was composed of two main stages: first, linear PNG kinematics was adopted; then, the expressions for miss distance, derived by the method of adjoints, were analyzed. This analysis showed that ZMD could be obtained for any flight time, provided that the guidance transfer function is positive real and that the PNG coefficient is larger than some threshold value.

The design procedure of the new guidance law, ZMD-PNG, involves lead compensation. When the LOS rate measurement is noisy, lead-lag compensation may be used. Simulations of ZMD-PNG that compared its performance with PNG and an OGL have given rise to the following observations:

1) The overall maneuver effort required by the ZMD is smaller than that required by PNG.

2) The miss distance is considerably smaller than that obtained with either PNG or OGL. Actually, it is very close to zero.

Perhaps the main advantage of ZMD-PNG over modern guidance laws is that it does not require the estimation of target maneuver or time-to-go. It uses LOS rate measurements only.

The best performance of ZMD-PNG is expected in low-glint systems, such as missiles with electro-optical seekers. This is due to the inherent lead compensation involved, which renders the system more sensitive to glint effects. Thus, the main drawback of ZMD-PNG is its sensitivity to the electro-optical target tracking noise, which is equivalent to glint in radar seekers. Systems having considerable target tracking noise may not be a suitable platform for ZMD-PNG.

ZMD-PNG constitutes a simple improvement to the well-known PNG law. It guarantees that the saturation of maneuver acceleration is avoided.

Based on previous studies, it was shown that preventing saturation yields ZMD. ZMD-PNG is based upon the assumption that the target maneuver is bounded. It offers two design guidelines to follow: First, the total dynamics of the guidance system should be designed positive real; second, the effective PN constant should be chosen according to a simple function of the given missile-target maneuver ratio. ZMD-PNG exhibits a significant improvement compared with PNG. The main disadvantage of the proposed law is that it might increase noise sensitivity. However, this obstacle could be overcome by introducing some lag into the system.

ZMD-PNG applies an acceleration command proportionally to an equivalent LOS rate instead of merely the LOS rate. The equivalent LOS rate is a linear combination of the measured LOS rate and high-order derivatives of the LOS rate estimated from the LOS rate measurement using a Kalman–Bucy filter.

ZMD-PNG does not require either estimation of target maneuver or measurement of the relative range, as needed when implementing OGLs. ZMD-PNG yields small, near-zero, miss distances against highly maneuvering targets for a wide range of flight times. This conclusion was established based upon Monte Carlo analysis of miss distance. The acceleration commands generated by ZMD-PNG, although noisier than acceleration commands of OGL, are reasonable in magnitude and are actually smaller than acceleration commands of an OGL designed for a first-order system.

Acknowledgments

This work was carried out as part of the author's doctoral studies in the Faculty of Aerospace Engineering, Technion, Israel Institute of Technology. The author is in debt of gratitude to his PhD advisors Mario Jodorkovsky and Moshe Guelman.

Nomenclature

a	Lateral acceleration
L^p	Normed space
LOS	Line of sight
N'	Effective navigation constant
OGL	Optimal guidance law
PN	Proportional navigation
PNG	Proportional navigation guidance
PSD	Power spectral density

PR	Positive real
R	Missile-target relative range
r	Relative order of a rational function
t_f	Flight time
V	Velocity
V_C	Closing velocity
y	Relative vertical position
$y(t_f)$	Miss distance
ZMD	Zero miss distance
γ	Flight path angle
λ	Line-of-sight angle
μ	Bound on maneuver ratio
μ_0	Given maneuver ratio
μ_r	Required maneuver ratio
τ	Time-to-go
τ_1	Missile time constant
ζ	Damping coefficient
ω_n	Natural frequency

Subscripts:

$(\ )_T$	Target
$(\ )_M$	Missile
$(\ )_C$	Commanded value
$(\ )_f$	Final value
$(\ )_0$	Initial value
$\|\cdot\|_p$	p-norm

Superscripts:

$(\dot{\ })$	Time differentiation

References

1. Shneydor, N., *Missile Guidance and Pursuit*, Horwood, West Sussex, 1998, pp. 101–124; Chap. 5.
2. Zarchan, P., *Tactical and Strategic Missile Guidance*, Progress in Astronautics and Aeronautics, Vol. 124, AIAA, Washington, DC, 1990, pp. 37–110; Chaps. 3–5.
3. Rusnak, I., Meir., L, Optimal guidance for high order and acceleration constrained missile, *Journal of Guidance, Control and Dynamics*, 14(3), May 1991, 589–596.
4. Rusnak, I., Advanced guidance laws for acceleration constrained missile, randomly maneuvering target and noisy measurement, *Transactions of the IEEE Regional Conference on Aerospace Control Systems*, Westlake, CA, May 1993, pp. 223–232.

5. Gutman, S., On optimal guidance for homing missiles, *Journal of Guidance, Control and Dynamics*, 2(4), July 1979, 296–300.
6. Chang, W. T., Lin, S. A., Incremental maneuver estimation model for target tracking, *IEEE Transactions on Aerospace and Electronic Systems*, 28(2), April 1992, 439–451.
7. Hepner, S. A. R., Geering, H. P., Adaptive two-time-scale tracking filter for target acceleration estimation, *Journal of Guidance, Control, and Dynamics*, 14(3), May–June 1991, 581–588.
8. Moose, R. L., An adaptive state estimation solution to the maneuvering target problem, *IEEE Transactions on Aerospace and Electronic Systems*, AES-17(3), June 1975, 359–362.
9. Cottrel, R. G., Optimal intercept guidance for short-range tactical missiles, *AIAA Journal*, 9, July 1971, 1414–1415.
10. Holder, E. J., Sylvester, V. B., An analysis of modern versus classical homing guidance, *IEEE Transactions on Aerospace and Electronic Systems*, 26(4), July 1990, 599–605.
11. Weiss, H., Hexner, G., Modern guidance laws with model mismatch, in *Proceedings of the IFAC Symposium on Missile Guidance*, January 1998, Tel-Aviv, Israel, 10–21.
12. Guelman, M., A qualitative study of proportional navigation, *IEEE Transactions on Aerospace and Electronic Systems*, AES-7(4), July 1971, 637–643.
13. Gurfil, P., Jodorkovsky, M., Guelman, M., Finite time stability approach to proportional navigation systems analysis, *Journal of Guidance, Control and Dynamics*, 21(6), November 1998, 853–861.
14. Zarchan, P., Complete statistical analysis of nonlinear missile guidance systems—SLAM, *Journal of Guidance and Control*, 2(1), January 1979, 71–78.
15. Zarchan, P., Proportional navigation and weaving targets, *Journal of Guidance, Control and Dynamics*, 18(5), September–October 1995, 969–974.
16. Shinar, J., Divergence range of homing missiles, *Israel Journal of Technology*, 14, July 1976, 47–55.
17. Lipman, Y., Shinar, J., Oshman, Y., Stochastic analysis of the interception of maneuvering antisurface missiles, *Journal of Guidance, Control and Dynamics*, 20(4), July–August 1997, 707 714.
18. Fitzgerald, R. J., Shaping filters for disturbances with random starting times, *Journal of Guidance and Control*, 2(2), March–April 1979, 152–154.
19. Shinar, J., Steinberg, D., Analysis of optimal evasive maneuvers based on a linearized two-dimensional kinematic model, *Journal of Aircraft*, 14(8), August 1977, 795–802.
20. Sandberg, I. W., Some results on the theory of physical systems governed by nonlinear functional equations, *Bell Systems Technology Journal*, 44(5), 1965, 871–898.
21. Zames, G., On input–output stability of time-varying nonlinear feedback systems—Part II: Conditions involving circles in the frequency plane and sector nonlinearities, *IEEE Transactions on Automatic Control*, AC-11, 1966, 465–476.
22. Mossaheb, S., The circle criterion and the L^p stability of feedback systems, *SIAM Journal on Control and Optimization*, 20(1), January 1982, 144–151.
23. Bar-Shalom, Y., Li., X. R., *Estimation and Tracking: Principles, Techniques and Software*, Artech House, Boston, 1993, pp. 625–655; Chap. 9.
24. Singer, R. A., Estimating optimal tracking filter performance for manned maneuvering targets, *IEEE Transactions on Aerospace and Electronic Systems*, AES-6(4), July 1970.

25. Fitzgerald, R. J., Shaping filters for disturbances with random starting times, *Journal of Guidance and Control*, 2(2), March–April 1979, 152–154.

26. Weiss, H., Hexner, G., Modern Guidance Laws with Model Mismatch, in *Proceedings of the IFAC Symposium on Missile Guidance*, January 1998, Tel-Aviv, Israel.

27. Ohlmeyer, E., J., Root-mean-square miss distance of proportional navigation missile against sinusoidal target, *Journal of Guidance, Control, and Dynamics*, 19(3), May–June 1996, 563–568.

28. Gurfil, P., Jodorkovsky, M., Guelman, M., Design of non-saturating guidance systems, *Journal of Guidance, Control and Dynamics*, 23(4), July 2000, 693–700.

29. Gurfil, P., Jodorkovsky, M., Guelman, M., Neoclassical guidance for homing missiles, *Journal of Guidance, Control and Dynamics*, 24(3), May 2001.

30. Gurfil, P., Synthesis of zero miss distance guidance via solution of an optimal tuning problem, *Control Engineering Practice*, 9(10), October 2001, 1117–1130.

31. Gurfil, P., Robust guidance for electro-optical missiles, *IEEE Transactions on Aerospace and Electronic Systems*, 39(2), April 2003, 450–461.

32. Gurfil, P., Zero miss distance guidance based on line-of-sight rate measurement only, *Control Engineering Practice*, 11(7), July 2003, 819–832.

8

Differential Geometry Applied to Missile Guidance

Brian A. White and Antonios Tsourdos

CONTENTS

8.1 Introduction

Homing guidance is usually referred to as *two-point guidance,* as there are two reference points defining the engagement geometry: the missile and the target. There has been a lot of interest in the development of homing guidance techniques in the literature over the past 30 years or so. Some of the literature has focused on the development of geometric approaches to guidance algorithm design. The use of differential geometric concepts gives the approach a sound basis for more generalized guidance techniques that allow for curved as well as straight trajectories to be considered for both the target and the missile. Augmented proportional navigation (PN) includes the target acceleration as part of the algorithm but still assumes straight-line interception geometry.

8.2 Homing Guidance Acquisition Geometry

The geometry of homing guidance for a nonmaneuvering target and missile is shown in Figure 8.1.

Figure 8.1 shows the case of a target that is flying in a straight line at constant velocity v_t. The missile is flying at velocity v_m, also in a straight line. Both trajectories are assumed to intercept at the *impact point* at point I. The target and missile centers of gravity (c.g.) together with the impact point form a triangle, which will be called the *impact triangle*. Two sides of the triangle form the predicted straight-line trajectories of the missile and the target, while the third is formed by connecting the missile and target c.g. This establishes the line of sight from the missile homing head to the target, and this line is thus labeled the *sight line*. In order to establish the conditions for impact, consider a time T such that the target has traveled in a straight line and at constant velocity from its initial position in Figure 8.1 to the impact point. The length of this trajectory s_t will be

$$s_t = v_t T. \tag{8.1}$$

In order for the missile to arrive at the impact point at the same time as the target, it must travel a distance s_m in the same time T, that is,

$$s_m = v_m T. \tag{8.2}$$

The ratio of the trajectory lengths is then given by

$$\frac{s_m}{s_t} = \frac{v_m T}{v_t T}$$

$$= \frac{v_m}{v_t} \tag{8.3}$$

$$= \gamma.$$

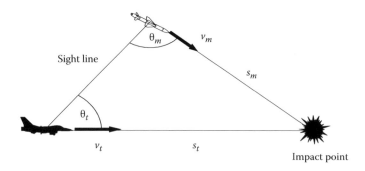

FIGURE 8.1
Homing geometry.

Equation 8.3 shows that in order to impact on a target flying at constant velocity in a straight line, the missile must maneuver until the trajectory lengths of the *impact triangle* are in the same ratio as the target and missile velocities. As the target velocity and heading are either unknown or estimated, and targets can maneuver, there must be an active control system to acquire and maintain this impact geometry. The geometry is not fixed, however, as only the trajectory lengths need matching with the missile and target velocities. The range of impact triangles possible is shown in Figure 8.2.

This shows the locus of possible impact triangles, where the missile position lies on a circle of radius s_m (the *impact circle*), and the missile velocity vector v_m lies along the radius of the impact circle. From this figure, it can be seen that an impact triangle can be produced to give a head-on collision (point A) or a tail-chase (point B) and any variant in between.

The ratio of velocities $\dfrac{v_m}{v_t} = \gamma$ is important here as, if the ratio falls below unity, the possible missile position is restricted to a position in front of the target. The impact circle for a range of velocity ratios of

$$\gamma = \{2 \quad 1.5 \quad 1 \quad 0.75\} \tag{8.4}$$

is shown in Figure 8.3.

The missile is moving left to right ahead of the target, at initial ranges of 1 km to 10 km, from the impact point, in steps of 1 km. The impact circle

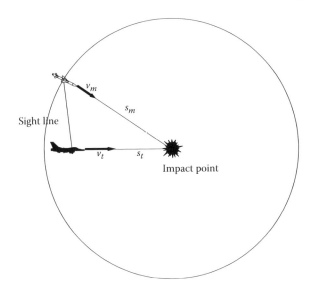

FIGURE 8.2
Range of impact triangle locus.

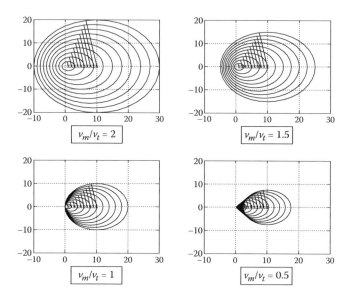

FIGURE 8.3
Impact circles for a range of velocity ratios.

for each of these points is then plotted. From this figure, it can be seen that a velocity ratio larger than 1 produces possible missile positions for impact both behind and in front of the target, as the circle encloses the target at the origin of the coordinate system. The impact circles for velocity ratios less than or equal to unity show that there is a region that cannot produce an impact triangle, as the circle does not enclose the target. This implies that for certain missile positions relative to the target, no intercept is possible. A simple example would be a tail-chase geometry, where the missile is to the left of the target. A velocity ratio of 2 or greater is desirable to enable the target to be engaged from any relative missile position. Current trends are to reduce the cost of missile systems, and so, the speed advantage required for acquisition under all conditions will be lost as missiles with speed ranges comparable with the target are designed. Such systems are usually area defense systems where the predominant geometry is head-on rather than tail-chase; hence, the missile will almost always have an impact geometry to guide onto.

The conditions for interception can be examined by exploring the geometry in Figure 8.1. As has been stated, the intercept geometry is not unique; the only condition that is required for a nonmaneuvering intercept is that the ratio of the trajectories is the same as the ratio of the missile and target velocities. Figure 8.4 shows the intercept geometry that is required.

The geometry is drawn using the sight line axes centered in the missile; hence, x is along the sight line, and y is normal to it, with r as the sight line

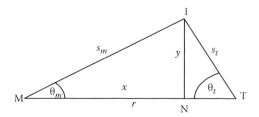

FIGURE 8.4
Intercept geometry.

range between the missile and the target. From the figure, using the sine rule, we have

$$\frac{s_m}{\sin(\theta_t)} = \frac{s_t}{\cos(\theta_m)} \tag{8.5}$$

or

$$\sin(\theta_t) = \frac{s_m}{s_t} \sin(\theta_m)$$
$$= \gamma \sin(\theta_m). \tag{8.6}$$

As the target angle to the sight line (θ_t) varies, the intercept point I will change. The locus of the intercept point can be determined by using Pythagoras on two triangles. The first is (M I N), made up of the missile position M, the intercept point I, and the intercept of the normal from the intercept point onto the sight line N. The second is triangle (T I N), replacing the missile position with the target position. Hence,

$$s_m^2 = x^2 + y^2$$
$$s_t^2 = (r - x)^2 + y^2. \tag{8.7}$$

Using Equation 8.4, this yields

$$x^2 + y^2 = \gamma^2 ((r - x)^2 + y^2)$$

$$x^2 + 2 \frac{\gamma^2 r}{(1 \quad \gamma^2)} x + y^2 = \frac{\gamma^2 r^2}{(1 - \gamma^2)}. \tag{8.8}$$

Completing the square yields

$$\left(x+\frac{\gamma^2 r}{(1-\gamma^2)}\right)^2 + y^2 = \frac{\gamma^2 r^2}{(1-\gamma^2)} + \frac{\gamma^4 r^2}{(1-\gamma^2)^2}$$

$$= \left(\frac{\gamma r}{(1-\gamma^2)}\right)^2 . \tag{8.9}$$

This equation represents a circle with radius r_l and center c_l with respect to the sight line axes, where

$$r_l = \frac{\gamma r}{(1-\gamma^2)}$$

$$c_l = -\frac{\gamma^2 r}{(1-\gamma^2)} . \tag{8.10}$$

This circle represents the locus of the intercept points and can be used to assess the guidance algorithm. Figure 8.5 shows the intercept locus for a range of γ given by Equation 8.4.

For $\gamma > 1$, the circles enclose the target, which implies that the missile can intercept the target for any target velocity direction. This is shown in Figure 8.6 for $\gamma = 2$.

For the case of $\gamma < 1$, the missile is enclosed within the circle, implying that it can only intercept targets that have a velocity vector that intercepts the

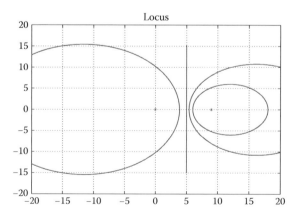

FIGURE 8.5
Intercept geometry locus.

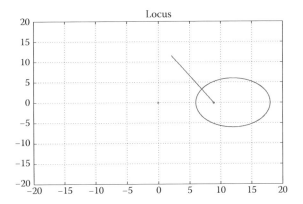

FIGURE 8.6
Intercept geometry.

locus. Figure 8.7 shows a case for γ = 0.75, where the target velocity vector does not intercept the locus and hence cannot be intercepted by the missile.

The locus also has another interpretation. Given that the target is traveling at constant velocity in a fixed direction, as shown in Figure 8.8, the locus represents the earliest intercept that the missile can achieve. This is based on the fact that the shortest distance between two points is a straight line. Hence, the straight-line intercept geometry in Figure 8.8 represents the earliest intercept geometry. This implies that if the target maneuvers away from a straight-line trajectory to get to the intercept with the circular locus, then the missile can achieve the intercept within the circle that contains the target if it travels on a straight-line trajectory. Conversely, if the missile deviates from a straight-line trajectory, then the intercept point must lie outside the intercept

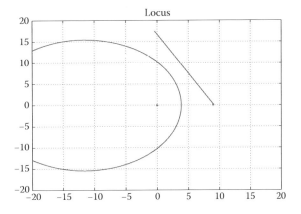

FIGURE 8.7
Nonintercept geometry.

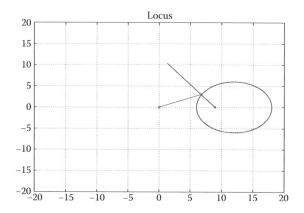

FIGURE 8.8
Earliest intercept geometry.

circle. This interpretation is useful in capturing the area in which the target can evade intercept and also the area in which the missile can successfully defend.

8.3 Differential Geometry Kinematics

Consider a two-dimensional engagement scenario, shown in Figure 8.9.

As the sensor that determines the relative motion and position of the target and missile is located in the nose of the missile for homing guidance, the sight line between the target and the missile is an important measure of the relative geometry. From Figure 8.9, we have

$$r = r_t - r_m. \tag{8.11}$$

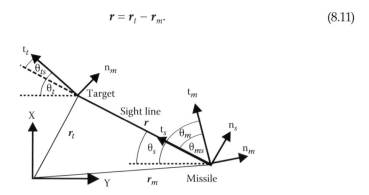

FIGURE 8.9
Guidance geometry.

Defining derivatives with respect to time $\dot{r}$ and with respect to arc length r' by

$$r' = \frac{d\mathbf{r}}{ds}$$

(8.12)

$$\dot{r} = \frac{d\mathbf{r}}{dt}$$

and differentiating Equation 8.11 with respect to time yields

$$\dot{r} = \dot{r}_t - \dot{r}_m$$ (8.13)

$$\dot{r}_t = r'_t \frac{ds_t}{dt}$$

(8.14)

$$\dot{r}_m = r'_m \frac{ds_m}{dt}.$$

If the assumption is made that both the missile and target velocities are constant, and by noting that the sight line range **r** vector can be expressed in terms of sight line coordinates defined by basis vectors $\mathbf{t}_s$ and $\mathbf{n}_s$, where $\mathbf{t}_s$ is the basis vector along the sight line and $\mathbf{n}_s$ is the basis vector normal to the sight line, as shown in Figure 8.9, Equation 8.13 can be written in the form

$$\dot{\mathbf{r}} = \dot{\mathbf{r}}_t - \dot{\mathbf{r}}_m$$

(8.15)

$$\dot{r}\mathbf{t}_s + r\dot{\theta}_s\mathbf{n}_s = V_t\mathbf{t}_t - V_m\mathbf{t}_m.$$

This equation represents the components of the target velocity relative to the missile. Components of the relative velocity along and normal to the sight line are given by projection onto the basis vectors $\mathbf{t}_s$ and $\mathbf{n}_s$. Hence,

$$\mathbf{t}'_s.\dot{\mathbf{r}} = \dot{r}$$
$$= V_t\mathbf{t}'_s.\mathbf{t}_t - V_m\mathbf{t}'_s.\mathbf{t}_m$$

(8.16)

$$\mathbf{n}'_s.\dot{\mathbf{r}} = r\dot{\theta}_s$$
$$= V_t\mathbf{n}'_o.\mathbf{t}_t - V_m\mathbf{n}'_c.\mathbf{t}_m.$$

(8.17)

Missile-to-target relative acceleration is given by differentiating Equation 8.15 and noting

$$\dot{\mathbf{t}}_s = \dot{\theta}_s \mathbf{n}_s$$
$$\dot{\mathbf{n}}_s = -\dot{\theta}_s \mathbf{t}_s \tag{8.18}$$

to give

$$(\ddot{r}\mathbf{t}_s + \dot{r}\dot{\theta}_s\mathbf{n}_s) + (\dot{r}\dot{\theta}_s\mathbf{n}_s + r\ddot{\theta}_s\mathbf{n}_s - r\dot{\theta}_s^2\mathbf{t}_s) = (V_t\dot{\mathbf{t}}_t - V_m\dot{\mathbf{t}}_m). \tag{8.19}$$

Hence, the Serret–Frenet equations can be rewritten in terms of a constant velocity trajectory in the form

$$\dot{\mathbf{t}} = \kappa V \mathbf{n}$$
$$= \dot{\theta} \mathbf{n} \tag{8.20}$$

$$\dot{\mathbf{n}} = -\kappa V \mathbf{t}$$
$$= -\dot{\theta} \mathbf{t} \tag{8.21}$$

where κ is the curvature of the missile trajectory, and $\dot{\theta}$ is the instantaneous rotation rate of the Serret–Frenet frame about the binormal vector $\mathbf{b}$. The normal vector $\mathbf{n}$ is a unit vector that defines the direction of the curvature of the trajectory (cf. Figure 8.9), and the binormal vector $\mathbf{b}$ is orthonormal to $\mathbf{t}$ and $\mathbf{n}$, forming a right-handed triplet $(\mathbf{t}, \mathbf{n}, \mathbf{b})$. Hence,

$$(\ddot{r} - r\dot{\theta}_s^2)\mathbf{t}_s + (r\ddot{\theta}_s + 2\dot{r}\dot{\theta}_s)\mathbf{n}_s = V_t^2\kappa_t\mathbf{n}_t - V_m^2\kappa_m\mathbf{n}_m. \tag{8.22}$$

Components along and normal to the sight line can be determined by projection onto the basis vectors $\mathbf{t}_s$ and $\mathbf{n}_s$. For a missile producing a lateral acceleration f_m, where

$$f_m = V_m^2\kappa_m$$
$$= V_m\dot{\theta}_m. \tag{8.23}$$

The acceleration components along and normal to the sight line can be determined as

$$(\ddot{r} - r\dot{\theta}_s^2) = f_t\mathbf{t}_s'.\mathbf{n}_t - f_m\mathbf{t}_s'.\mathbf{n}_m \tag{8.24}$$

$$(r\ddot{\theta}_s + 2\dot{r}\dot{\theta}_s) = f_t\mathbf{n}_s'.\mathbf{n}_t - f_m\mathbf{n}_s'.\mathbf{n}_m. \tag{8.25}$$

8.4 Geometric Guidance

The guidance algorithm that defines the missile lateral acceleration f_m determines the missile response to target motion. For fixed velocity, Equation 8.23 shows that the curvature κ_m is linearly related to the lateral acceleration f_m. For a geometric approach, the curvature κ is a more natural parameterization for the intercept trajectory. Two cases will be considered:

- Direct intercept missile with a non maneuvering target
- Maneuvering intercept missile with a maneuvering target

The second of these cases is developed in the paper of White et al. [1] and will be presented in a modified manner in this chapter. Although these cases are treated separately, they produce a unified guidance algorithm by controlling the sight line geometry. Consider each case in turn. The geometry of each intercept configuration is determined by the requirement to match the missile arc length s_m to the target arc length s_t. For constant velocity target and missile, these are related to the missile and target velocities such that

$$s_m = \gamma s_t$$

$$\gamma = \frac{v_m}{v_t}.$$

(8.26)

8.4.1 Direct Intercept Geometry of Nonmaneuvering Target

This geometry is the classic PN geometry case, where the missile and the target are assumed to have a constant velocity and, once on the intercept trajectory, fly in straight lines with no maneuver. The geometry of a nonmaneuvering target with a direct, straight-line-intercepting missile trajectory is shown in Figure 8.10.

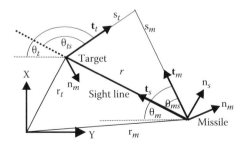

FIGURE 8.10
Guidance geometry: direct intercept of a nonmaneuvering target.

Note that the intercept triangle **TIM** is invariant in that as the missile and the target move along their respective straight-line trajectories, the **TIM** triangle does not change shape but shrinks as the missile approaches the intercept point. The missile-to sight line angle θ_{ms} and the target-to-sight line angle θ_{ts} remain constant over the whole engagement. From the intercept triangle in Figure 8.10, forming a vector sum, we have

$$s_m \mathbf{t}_m = r\mathbf{t}_s + s_t \mathbf{t}_t \tag{8.27}$$

and noting the matching condition in Equation 8.26 gives

$$\gamma s_t \mathbf{t}_m = r\mathbf{t}_s + s_t \mathbf{t}_t$$

$$\mathbf{t}_m = \frac{1}{\gamma}\left[\frac{r}{s_t}\mathbf{t}_s + \mathbf{t}_t\right]. \tag{8.28}$$

Equation 8.28 shows that the missile tangent vector $\mathbf{t}_m$ can be obtained from the target tangent vector $\mathbf{t}_t$ and the sight line tangent vector $\mathbf{t}_s$. As this represents the solution for all intercept triangles, the ratio r/s_t will be fixed for a particular geometry, regardless of the size of the impact triangle. It can be visualized as a vector addition and is shown in Figure 8.11.

This representation is in a nondimensional form and will thus represent the solution for all ranges between the missile and the target. The ratio r/s_t is fixed for the whole solution, and thus, as the range r decreases, so will the target arc length s_t. Given the geometry of the target basis vector $\mathbf{t}_t$ and the range basis vector $\mathbf{t}_s$, the direction of the missile basis vector $\mathbf{t}_m$ is fixed. In Equation 8.28, the ratio r/s_t can be obtained by the use of the cosine rule in Figure 8.11. From Figure 8.11, we have

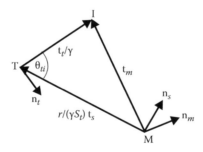

FIGURE 8.11
Guidance geometry: matching condition.

$$\frac{1}{\gamma^2} + \frac{1}{\gamma^2}\left(\frac{r}{s_t}\right)^2 - 2\frac{1}{\gamma^2}\left(\frac{r}{s_t}\right)\cos(\theta_{ti}) = 1$$

$$\left(\frac{r}{s_t}\right)^2 - 2\cos(\theta_{ti})\left(\frac{r}{s_t}\right) - (\gamma^2 - 1) = 0.$$

(8.29)

This quadratic in r/s_t can be solved explicitly to give

$$\left(\frac{r}{s_t}\right) = \cos(\theta_{ti}) \pm \sqrt{\cos^2(\theta_{ti}) + \gamma^2 - 1}.$$

(8.30)

Given that $\gamma > 1$ and that $r > 0$ and $s_t > 0$, the solution is

$$\left(\frac{r}{s_t}\right) = \cos(\theta_{ti}) + \sqrt{\cos^2(\theta_{ti}) + \gamma^2 - 1}$$

$$= \cos(\theta_{ti}) + \sqrt{\gamma^2 - \sin^2(\theta_{ti})}.$$

(8.31)

This solution can be used to find the required direction of the missile tangent vector $\hat{\mathbf{t}}_m$. Direct substitution of (r/s_t) into Equation 8.28 gives

$$\hat{\mathbf{t}}_m = \frac{1}{\gamma}\left[\frac{r}{s_t}\mathbf{t}_s + \mathbf{t}_t\right]$$

(8.32)

where $\hat{\mathbf{t}}_m$ is the required missile tangent intercept solution for the engagement.

8.4.2 Guidance Algorithm for Direct Intercept

From Section 8.4.1, the condition required for the intercept of a nonmaneuvering target is given by

$$\hat{\mathbf{t}}_m = \frac{1}{\gamma}\left[\frac{r}{s_t}\mathbf{t}_s + \mathbf{t}_t\right].$$

(8.33)

A geometric interpretation of Equation 8.33 is shown in Figure 8.11. An error vector can be defined as the angle θ_e, between $\hat{\mathbf{t}}_m$ and $\mathbf{t}_m$, where $\hat{\mathbf{t}}_m$ is the solution of Equation 8.33. The form and stability of the guidance algorithm can be determined by use of a simple Lyapunov function V, given by

$$V = \frac{1}{2}\theta_\varepsilon^2$$

(8.34)

where θ_ε is the angle between the missile tangent $\mathbf{t}_m$ and the required tangent vector $\hat{\mathbf{t}}_m$, given by

$$\theta_\varepsilon = \theta_m - \hat{\theta}_m. \tag{8.35}$$

The time derivative of the function V is given by

$$\frac{dV}{dt} = \dot{\theta}_\varepsilon \theta_\varepsilon. \tag{8.36}$$

Hence, for stability, we require

$$\dot{\theta}_\varepsilon \theta_\varepsilon < 0. \tag{8.37}$$

From a definition of θ_ε, we have

$$\dot{\theta}_\varepsilon = \dot{\theta}_m - \dot{\hat{\theta}}_m \tag{8.38}$$

where $\dot{\hat{\theta}}_m$ is the rate of change of the desired tangent vector $\hat{\mathbf{t}}_m$ as the geometry changes, and $\dot{\theta}_m$ is the rate of change of the missile tangent vector $\mathbf{t}_m$. The rate of change of the desired missile tangent vector $\hat{\mathbf{t}}_m$ is given by differentiation of Equation 8.33 to give

$$\dot{\hat{\mathbf{t}}}_m = \frac{1}{\gamma}\left[\left(\frac{r}{s_t}\right)\dot{\mathbf{t}}_s + \frac{d}{dt}\left(\frac{r}{s_t}\right)\mathbf{t}_s\right]. \tag{8.39}$$

As the target is not maneuvering, the rate of change of the target tangent vector $\mathbf{t}_t = 0$. The sight line vector $\mathbf{t}_s$ rate of change is given by Equation 8.18, and hence

$$\dot{\hat{\mathbf{t}}}_m = \frac{1}{\gamma}\left[\left(\frac{r}{s_t}\right)\dot{\theta}_s\mathbf{n}_s + \frac{d}{dt}\left(\frac{r}{s_t}\right)\mathbf{t}_s\right]. \tag{8.40}$$

This equation shows that the rate of change is made up of components along the sight line basis vectors $\mathbf{t}_s$ and $\mathbf{n}_s$. In fact, as the missile tangent vector $\hat{\mathbf{t}}_m$ is a unit vector, the resultant rate of change must be normal to the desired missile tangent vector along $\hat{\mathbf{n}}_m$. Hence,

$$\begin{aligned} \dot{\hat{\mathbf{t}}}_m &= \dot{\hat{\theta}}_m \hat{\mathbf{n}}_m \\ &= \frac{1}{\gamma}\left[\left(\frac{r}{s_t}\right)\dot{\theta}_s\mathbf{n}_s + \frac{d}{dt}\left(\frac{r}{s_t}\right)\mathbf{t}_s\right] \end{aligned} \tag{8.41}$$

where $\dot{\hat{\theta}}_m$ will have maximum and minimum values given by the maximum and minimum values of the components in Equation 8.41. The rate of change of r/s_t in Equation 8.41 can be obtained from the solution in Equation 8.31:

$$\frac{r}{s_t} = \cos(\theta_{ti}) + \sqrt{\gamma^2 - \sin(\theta_{ti})^2}. \tag{8.42}$$

Differentiating this equation with respect to time gives

$$\frac{d}{dt}\left(\frac{r}{s_t}\right) = \left(-\sin(\theta_{ti}) - \sin(\theta_{ti})\cos(\theta_{ti})\left(\gamma^2 - \sin^2(\theta_{ti})\right)^{-1/2}\right)\dot{\theta}_{ti}$$

$$= -\sin(\theta_{ti})\left[1 + \frac{\cos(\theta_{ti})}{(\gamma^2 - \sin^2(\theta_{ti}))^{1/2}}\right]\dot{\theta}_{ti}. \tag{8.43}$$

Also, from the definition of θ_{ti} and the fact that the target is nonmaneuvering, we have

$$\dot{\theta}_{ti} = \dot{\theta}_s + \dot{\theta}_t$$
$$= \dot{\theta}_s. \tag{8.44}$$

Hence,

$$\frac{d}{dt}\left(\frac{r}{s_t}\right) = -\sin(\theta_{ti})\left[1 + \frac{\cos(\theta_{ti})}{(\gamma^2 - \sin^2(\theta_{ti}))^{1/2}}\right]\dot{\theta}_s$$

$$= -\delta(\theta_{ti})\dot{\theta}_s \tag{8.45}$$

$$\delta(\theta_{ti}) = \sin(\theta_{ti})\left[1 + \frac{\cos(\theta_{ti})}{(\gamma^2 - \sin^2(\theta_{ti}))^{1/2}}\right].$$

Hence, Equation 8.41 can be written as

$$\dot{\hat{\mathbf{t}}}_m = \frac{1}{\gamma}\left[\left(\frac{r}{s_t}\right)\mathbf{n}_s - \delta(\theta_{ti})\mathbf{t}_s\right]\dot{\theta}_s$$

$$= \dot{\theta}_m \hat{\mathbf{n}}_m \tag{8.46}$$

or

$$\dot{\theta}_m = \frac{1}{\gamma}\left[\left(\frac{r}{s_t}\right)\mathbf{n}_s' - \delta(\theta_{ti})\mathbf{t}_s'\right]\hat{\mathbf{n}}_m\dot{\theta}_s. \tag{8.47}$$

To help in obtaining the maximum and minimum values, note that each component of $\dot{\hat{\theta}}_m$ in Equation 8.33 can be visualized by simple geometry. Figure 8.12 shows the geometric construction of $\hat{\mathbf{t}}_m$. The figure shows that there are maximum and minimum values for the rate of change $\dot{\theta}_m$.

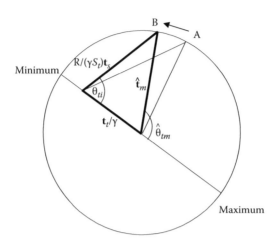

FIGURE 8.12
Maximum and minimum $\hat{\dot{\theta}}_m$.

From the figure, assuming that the sight line to target angular rate $\dot{\theta}_{ti}$ is given by the sight line rate $\dot{\theta}_s$, then the rate of change, at the maximum and minimum points, is given by

$$\hat{\dot{\theta}}_m = \frac{1}{\gamma}\frac{r}{s_t}\dot{\theta}_s. \tag{8.48}$$

This assumes that the rate of change of r/s_t at these points is zero due to the fact that the vector $\mathbf{t}_s$ is normal to the circle at these points. The maximum length of r/s_t is given by

$$\max\left(\frac{1}{\gamma}\frac{r}{s_t}\right) = 1 + \frac{1}{\gamma}$$

$$\max\left(\frac{r}{s_t}\right) = \gamma + 1. \tag{8.49}$$

Also from the figure, the minimum ratio is given by

$$\min\left(\frac{1}{\gamma}\frac{r}{s_t}\right) = 1 - \frac{1}{\gamma}$$

$$\min\left(\frac{r}{s_t}\right) = \gamma - 1. \tag{8.50}$$

For both the minimum and the maximum, the target and missile tangent vectors are parallel to the sight line. Hence,

$$(\gamma - 1) \leq \frac{r}{s_t} \leq (\gamma + 1). \tag{8.51}$$

Also, from the definition of θ_{ti} and the fact that the target is nonmaneuvering, we have

$$\dot{\theta}_{ti} = \dot{\theta}_s + \dot{\theta}_t$$
$$= \dot{\theta}_s. \tag{8.52}$$

Hence,

$$\max\left(\dot{\hat{\theta}}_m\right) = \max\left(\frac{1}{\gamma}\frac{r}{s_t}\right)\dot{\theta}_s$$
$$= \left(1 + \frac{1}{\gamma}\right)\dot{\theta}_s \tag{8.53}$$
$$\min\left(\dot{\hat{\theta}}_m\right) = \left(1 - \frac{1}{\gamma}\right)\dot{\theta}_s.$$

This property can be used to formulate a guidance law to guide the missile tangent vector $\mathbf{t}_m$ onto the desired tangent vector $\hat{\mathbf{t}}_m$. From the Lyapunov equation, we have,

$$\frac{dV}{dt} = \dot{\theta}_\varepsilon \theta_\varepsilon \tag{8.54}$$
$$= (\dot{\theta}_m - \dot{\hat{\theta}}_m)\theta_\varepsilon.$$

If the missile is controlled using the following guidance law:

$$\dot{\theta}_m = -\left(1 + \frac{1}{\gamma}\right)|\dot{\theta}_s|\theta_\varepsilon - K\theta_\varepsilon \tag{8.55}$$
$$K > 0$$

then we have

$$\frac{dV}{dt} = -\left(1 + \frac{1}{\gamma}\right)|\dot{\theta}_s|\theta_\varepsilon^2 - \dot{\hat{\theta}}_m\theta_\varepsilon^2 - K\theta_\varepsilon^2 \tag{8.56}$$

which is negative semidefinite. The missile tangent vector $\mathbf{t}_m$ is controlled by defining the curvature of the missile trajectory. This is achieved by applying lateral acceleration. The curvature of the trajectory and hence the tangent vector is controlled by the Serret–Frenet equations:

$$\dot{\mathbf{t}}_m = \kappa_m V_m \mathbf{n}_m \tag{8.57}$$

$$\dot{\mathbf{n}}_m = -\kappa_m V_m \mathbf{t}_m \tag{8.58}$$

or

$$\left(\dot{\mathbf{t}}_m \quad \dot{\mathbf{n}}_m \right) = \left(\mathbf{t}_m \quad \mathbf{n}_m \right) \begin{pmatrix} 0 & -\kappa_m \\ \kappa_m & 0 \end{pmatrix} V_m$$

$$= \left(\mathbf{t}_m \quad \mathbf{n}_m \right) \begin{pmatrix} 0 & -\dot{\theta}_m \\ \dot{\theta}_m & 0 \end{pmatrix}. \tag{8.59}$$

A physical interpretation of the Serret–Frenet equations can be obtained by examining Figure 8.10, where the missile target and normal vectors $\mathbf{t}_m$ and $\mathbf{n}_m$ are shown. A positive κ_m rotates the tangent vector $\mathbf{t}_m$ in the direction of the normal vector $\mathbf{n}_m$, with the speed of rotation $\dot{\theta}_m$ being given by the magnitude of the curvature. Substituting for the missile trajectory curvature gives

$$\kappa_m = \frac{\dot{\theta}_m}{V_m}$$

$$= \frac{1}{V_m}\left[\left(1+\frac{1}{\gamma}\right)\dot{\theta}_s - K\theta_\varepsilon\right]$$

$$K > 0.$$

8.4.3 Direct Intercept Engagement Simulation

An engagement was simulated for a 300 m/s target against two missile velocities with $\gamma = \{2 \quad 1.5\}$. The sensor is assumed to be able to measure

- Range
- Range rate
- Sight line angle
- Sight line rate

The target is flying on an initial course of 90° and at a range of 10 km on an initial bearing of 0° to the launch site of the missile. The missile Latax is capped at 40g.

For this case, the target acceleration is zero. Two cases are considered to show the global convergence of the solution. The first is an engagement where the initial missile bearing is set to 0°. This will set the missile to within 90° of the correct flight angle for interception. The second case considered is when the initial missile bearing is set to 180°, which will set the missile onto a bearing that initially is moving away from the target. This case is particularly testing, as a solution for PN is for the missile and target to be of an impact geometry with both traveling away from each other. The sight line rate for this case is zero, and PN will not demand any acceleration; hence, the geometry will be maintained. Because of the global convergence properties of the approach, this will not happen for this algorithm. The results for $\theta_m = 0°$ are shown in Figure 8.13a through d.

The target is acquired when flying at $\theta_t = 90°$, and the missile is launched at a range of 7 km at $\theta_m = 0°$. The guidance gain is set to $K = 1$ and shows good convergence, as seen by the convergence of the Lyapunov variable ε. The case for $\gamma = 1.5$ is shown in Figure 8.14.

The case for an initial missile angle of −90° for $\gamma = 2$ is shown in Figure 8.15 and for $\gamma = 1.5$ in Figure 8.16.

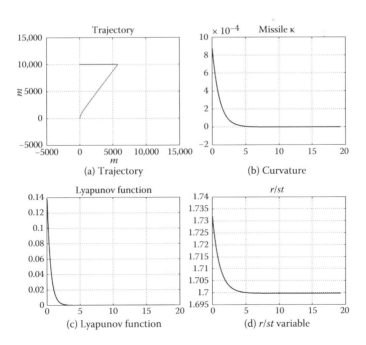

FIGURE 8.13
Nonmaneuvering target and missile with $\gamma = 2$ and $\theta_m = 0°$.

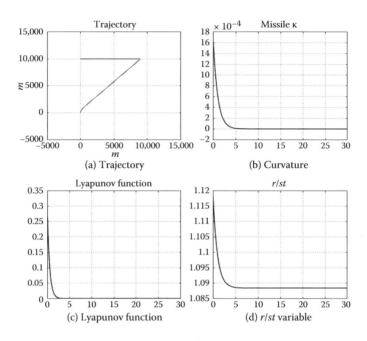

FIGURE 8.14
Nonmaneuvering target and missile with $\gamma = 1.5$ and $\theta_m = 0°$.

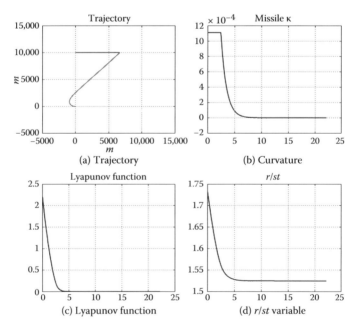

FIGURE 8.15
Nonmaneuvering target and missile with $\gamma = 2$ and $\theta_m = -90°$.

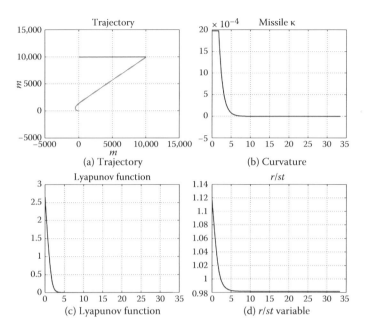

FIGURE 8.16
Nonmaneuvering target and missile with $\gamma = 1.5$ and $\theta_m = -90°$.

8.4.4 Maneuvering Intercept Geometry of Maneuvering Target

This approach can be extended to the second case: that of a constant maneuver target trajectory being intercepted by a constant maneuver missile trajectory. The geometry of a maneuvering target with a maneuver intercepting missile trajectory is shown in Figure 8.17.

The intercept point **I** can be determined by considering the target maneuver arc and the missile maneuver arc geometry. This is shown in Figure 8.18.

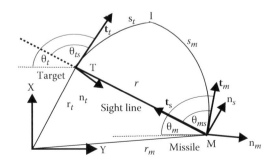

FIGURE 8.17
Guidance geometry: maneuver intercept of a maneuvering target.

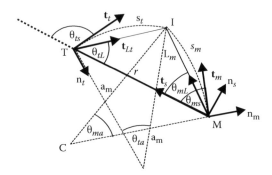

FIGURE 8.18
Maneuver intercept geometry of a maneuvering target.

Following the same approach as the nonmaneuvering target with direct intercept, the intercept triangle **TIM** is defined by the target chord vector $\mathbf{t}_{Lt}$ and the missile arc chord vector $\mathbf{t}_{Lm}$.

$$L_m \mathbf{t}_{Lm} = r\mathbf{t}_s + L_t \mathbf{t}_{Lt} \tag{8.60}$$

where the arc chord basis vectors $\mathbf{t}_{Lm}$ and $\mathbf{t}_{Lt}$ can be obtained from the missile and target basis vectors $\mathbf{t}_m$ and $\mathbf{t}_t$ by a rotation through $-\theta_{ta}/2$ and $-\theta_{ma}/2$, respectively. The arc lengths L_m and L_t are given by

$$L_m = \beta s_m$$

$$\beta = \frac{\sin(\theta_{ma}/2)}{\theta_{ma}/2}$$

$$\tag{8.61}$$

$$L_t = \alpha s_t$$

$$\alpha = \frac{\sin(\theta_{ta}/2)}{\theta_{ta}/2}.$$

Equation 8.60 can thus be written in the form

$$\mathbf{t}_{Lm} = \frac{1}{\gamma\beta}\left[\left(\frac{r}{s_t}\right)\mathbf{t}_s + \alpha\mathbf{t}_{Lt}\right]. \tag{8.62}$$

The matching condition for the maneuvering missile case can be visualized as a vector addition, as shown in Figure 8.19.

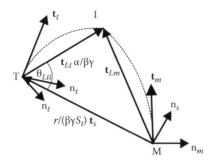

FIGURE 8.19
Guidance geometry: matching condition.

From this figure, the matching condition can be calculated by again applying the cosine rule to the intercept triangle **TIM**. Hence,

$$\left(\frac{r}{s_t}\right)^2 - 2\alpha\cos(\theta_{Lt})\left(\frac{r}{s_t}\right) - (\gamma^2\beta^2 - \alpha^2) = 0. \tag{8.63}$$

This equation has no explicit solution, unlike the case for the nonmaneuvering target and missile. A solution is possible, however, by iteration. The arc length s_m and the arc angle θ_{ma} are related by

$$\gamma r\left(\frac{s_t}{r}\right) = \frac{\theta_{ma}}{\kappa_m} \tag{8.64}$$

together with

$$\theta_{ta} = \frac{\kappa_t}{\gamma\kappa_m}\theta_{ma}. \tag{8.65}$$

An implicit solution to Equation 8.63 is possible by iteration, as κ_m and r/s_t can be treated as independent variables. Initialization is straightforward in the initial missile curvature of $\kappa_m = 0$ with r/s_t given by Equation 8.30 for the nonmaneuvering case. Hence, given an initial κ_m and r/s_t, the angle θ_{ma} can be determined using Equation 8.64, and the angle θ_{ta} can be determined using Equation 8.65. As both α and β can now be evaluated using Equation 8.61, a new solution for r/s_t can be calculated using

$$\left(\frac{r}{s_t}\right) = \alpha\cos(\theta_{ti}) + \sqrt{\beta^2\gamma^2 - \alpha^2\sin^2(\theta_{ti})}. \tag{8.66}$$

The new estimate of missile curvature κ_m can be obtained by substituting the new value for r/s_t in Equation 8.64. Hence,

$$
\begin{aligned}
\kappa_m &= \frac{\theta_{ma}}{\gamma r}\left(\frac{r}{s_t}\right) \\
&= \frac{\theta_{ma}}{\gamma r}\left(\alpha\cos(\theta_{ti}) + \sqrt{\beta^2\gamma^2 - \alpha^2\sin^2(\theta_{ti})}\right).
\end{aligned}
\tag{8.67}
$$

The new solutions for κ_m and r/s_t can then be used to iterate onto a further solution. This iteration continues until the values of κ_m and r/s_t converge onto a solution. The solution to this equation also shows that the missile curvature κ_m can vary from zero to a maximum. The maximum solution is given by the condition that

$$
\beta^2\gamma^2 - \alpha^2\sin^2(\theta_{ti}) \geq 0
\tag{8.68}
$$

in order to ensure that the solution is real. The maximum κ_m can be obtained by differentiating Equation 8.67 to give

$$
\begin{aligned}
\frac{\partial\kappa_m}{\partial\theta_{ma}} &= \frac{1}{\gamma r}\left(\frac{r}{s_t}\right) + \frac{\theta_{ma}}{\gamma r}\frac{\partial}{\partial\theta_{ma}}\left(\frac{r}{s_t}\right) \\
&= 0.
\end{aligned}
\tag{8.69}
$$

Now,

$$
\begin{aligned}
\frac{\partial}{\partial\theta_{ma}}\left(\frac{r}{s_t}\right) &= \frac{\partial\alpha}{\partial\theta_{ma}}\cos(\theta_{ti}) + \frac{\partial}{\partial\theta_{ma}}(\beta^2\gamma^2 - \alpha^2\sin^2(\theta_{ti}))^{1/2} \\
&= \beta\gamma^2(\beta^2\gamma^2 - \alpha^2\sin^2(\theta_{ti}))^{-1/2}
\end{aligned}
\tag{8.70}
$$

where

$$
\begin{aligned}
\frac{\partial\beta}{\partial\theta_{ma}} &= \frac{1}{\theta_{ma}}\left[\cos(\theta_{ma}/2) - \beta\right] \\
\frac{\partial\alpha}{\partial\theta_{ma}} &= \frac{\partial\theta_{ta}}{\partial\theta_{ma}}\frac{\partial\alpha}{\partial\theta_{ta}} \\
&= \frac{\partial\theta_{ta}}{\partial\theta_{ma}}\frac{1}{\theta_{ta}}(\cos(\theta_{ta}/2) - \alpha).
\end{aligned}
\tag{8.71}
$$

Hence,

$$
\frac{\partial \kappa_m}{\partial \theta_{ma}} = \frac{1}{\gamma r} \left(\cos(\theta_{ti}) + (\beta^2 \gamma^2 - \alpha^2 \sin^2(\theta_{ti}))^{1/2} \right)
$$
$$
+ \frac{\beta \gamma}{r} \frac{(\cos(\theta_{ma}/2) - \beta)}{(\beta^2 \gamma^2 - \alpha^2 \sin^2(\theta_{ti}))^{1/2}}.
$$

(8.72)

Setting this to zero and rearranging yields

$$
(\beta^2 \gamma^2 - \alpha^2 \sin^2(\theta_{ti})) + \cos(\theta_{ti})(\beta^2 \gamma^2 - \alpha^2 \sin^2(\theta_{ti}))^{1/2} + \gamma \beta (\cos(\theta_{ma}/2) - \beta) = 0
$$

(8.73)

This is also well behaved and can be solved iteratively using Newton's algorithm.

8.4.5 Guidance Algorithm for Maneuvering Intercept

From Section 8.4, the condition required for maneuvering intercept of a maneuvering target is given by

$$
\hat{\mathbf{t}}_{Lm} = \frac{1}{\gamma \beta} \left[\left(\frac{r}{s_t} \right) \mathbf{t}_s + \alpha \mathbf{t}_{Lt} \right]
$$

(8.74)

where

$$
\alpha = \frac{\sin(\theta_{ta}/2)}{\theta_{ta}/2}
$$

$$
\beta = \frac{\sin(\theta_{ma}/2)}{\theta_{ma}/2}
$$

(8.75)

$$
\theta_{ta} = \frac{r \kappa_t}{(r/s_t)}
$$

$$
\theta_{ma} = \gamma \frac{r \kappa_m}{(r/s_t)}.
$$

With the same Lyapunov function as the nonmaneuvering case in Section 8.4.2, we have, as before

$$
\frac{dV}{dt} = \dot{\theta}_\varepsilon \theta_\varepsilon
$$

(8.76)

with

$$
\dot{\theta}_\varepsilon = \dot{\theta}_m - \dot{\hat{\theta}}_m.
$$

(8.77)

To calculate $\dot{\hat{\theta}}_m$, the rate of change of the desired missile tangent chord vector $\hat{\mathbf{t}}_{Lm}$ is required. Thus,

$$\dot{\hat{\mathbf{t}}}_{Lm} = \frac{1}{\gamma \beta} \left[\frac{d}{dt}\left(\frac{r}{s_t}\right)\mathbf{t}_s + \left(\frac{r}{s_t}\right)\dot{\theta}_s \mathbf{n}_s \right]$$
$$- \frac{1}{\gamma \beta^2}\left[\left(\frac{r}{s_t}\right)\mathbf{t}_s + \mathbf{t}_t\right]\dot{\beta}$$
$$+ \frac{d}{dt}\left(\alpha \mathbf{t}_{tL}\right). \tag{8.78}$$

This equation can be written in the form

$$\dot{\hat{\mathbf{t}}}_m = \frac{1}{\gamma}\hat{\mathbf{n}}_m \dot{\hat{\theta}}_m \tag{8.79}$$

where

$$\hat{\mathbf{n}}_m = \left[\mathbf{t}_s \quad \left(\tilde{\alpha}\mathbf{t}_{tL} - \frac{1}{2}\mathbf{n}_{tL}\right) - \frac{\tilde{\beta}}{\beta}\left[\left(\frac{r}{s_t}\right)\mathbf{t}_s + \mathbf{t}_t\right] \quad \left(\frac{r}{s_t}\right)\mathbf{n}_s \right] \tag{8.80}$$

and

$$\dot{\hat{\theta}}_m = \begin{bmatrix} \dfrac{d}{dt}\left(\dfrac{r}{s_t}\right) \\[2mm] \dot{\theta}_{ta} \\[2mm] \dot{\theta}_{ma} \\[2mm] \dot{\theta}_s \end{bmatrix}. \tag{8.81}$$

The solution to Equations 8.80 and 8.81 will result in a vector that points in the direction of the missile normal vector $\mathbf{n}_m$ with magnitude given by $\dot{\hat{\theta}}_m$. Hence, if the same guidance law as the direct intercept case is used, where

$$\kappa_m = \frac{\dot{\theta}_m}{V_m}$$
$$= \dot{\hat{\theta}}_m \operatorname{sign}(\theta_\varepsilon) - K\theta_\varepsilon \tag{8.82}$$

then the Lyapunov function will again be negative definite, as is given by

$$\frac{dV}{dt} = -K\theta_\varepsilon^2. \tag{8.83}$$

8.4.6 Maneuver Intercept Engagement Simulation

An engagement was simulated for a 300 m/s target against two missile velocities with $\gamma = \{2 \;\; 1.5\}$ with the same sensor suite as for the nonmaneuvering case. The target is flying on an initial course of 90° and at a range of 10 km on an initial bearing of 0° to the launch site of the missile. The missile Latax is capped at 40g.

For this case, the target acceleration is ±2g. Again, two cases are considered to show the global convergence of the solution. The first is an engagement where the initial missile acceleration is set to 2g and the second to −2g. In both cases, the initial missile bearing is set to 0°. The results for target acceleration of −2g and missile acceleration of 2g are shown in Figure 8.20a through d.

The target is acquired when flying at $\theta_t = 90°$, and the missile is launched at a range of 7 km at $\theta_m = 0°$. The guidance gain is set to $K = 1$ and shows good convergence, as seen by the convergence of the Lyapunov variable ε. The case for $\gamma = 1.5$ is shown in Figure 8.21.

The second case for target acceleration of 2g and for an initial missile acceleration of 2g for $\gamma = 2$ is shown in Figure 8.22, and the one for $\gamma = 1.5$ is shown in Figure 8.23.

Again, convergence is shown, and the target is intercepted.

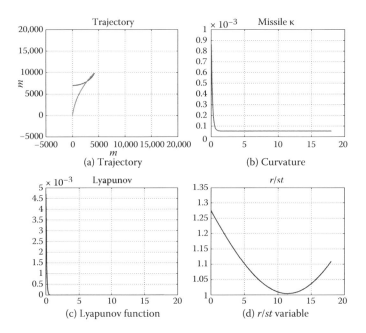

FIGURE 8.20
Maneuvering target (−2g) and maneuvering missile (2g) with $\gamma = 1.5$ and $\theta_m - 0°$.

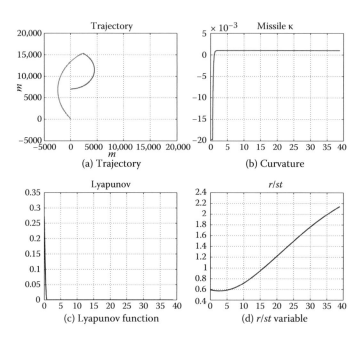

FIGURE 8.21
Maneuvering target ($-2g$) and maneuvering missile ($2g$) with $\gamma = 2$ and $\theta_m = 0°$.

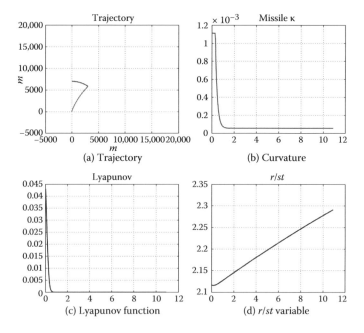

FIGURE 8.22
Maneuvering target (g) and maneuvering missile ($2g$) with $\gamma = 2$ and $\theta_m = 0°$.

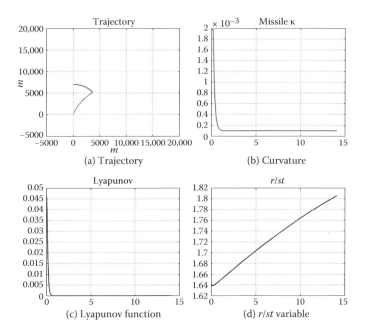

FIGURE 8.23
Maneuvering target (2g) and maneuvering missile (2g) with $\gamma = 1.5$ and $\theta_m = 0°$.

8.5 Geometry Control

The geometry of the engagement can be controlled to some extent by examination of the impact triangle. The figure is reproduced in Figure 8.24.

The intercept point **I** is stationary for the missile and target velocity vectors lying along the impact triangle sides. To see this, consider the velocity of the impact point v_I with respect to the target **T**. We have

$$v_I \mathbf{t_t} = \mathbf{v_t t_t} + \dot{\mathbf{s}_t} \mathbf{t_t}. \tag{8.84}$$

The rate of change of the impact point must lie along the target velocity vector for the case of a nonmaneuvering target, as the impact point must lie along the target trajectory. Hence,

$$v_I = v_t + \dot{s}_t. \tag{8.85}$$

Now from Section 8.4.1, intercept conditions must obey Equation 8.31, reproduced here.

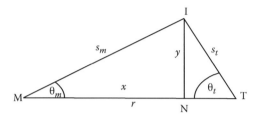

FIGURE 8.24
Intercept geometry.

$$\frac{r}{s_t} = \cos(\theta_t) + \sqrt{\gamma^2 - \sin^2(\theta_t)}$$

$$r = \left[\cos(\theta_t) + \sqrt{\gamma^2 - \sin^2(\theta_t)}\right]s_t \qquad (8.86)$$

Differentiating this equation with respect to time gives

$$\dot{r} = \left[\cos(\theta_t) + \sqrt{\gamma^2 - \sin^2(\theta_t)}\right]\dot{s}_t$$

$$= -\left[\sin(\theta_t) + \frac{\sin(\theta_t)\cos(\theta_t)}{\sqrt{\gamma^2 - \sin^2(\theta_t)}}\right]\dot{\theta}_t s_t. \qquad (8.87)$$

Now,

$$\dot{r} = -\left(v_t\cos(\theta_t) + v_m\cos(\theta_m)\right)$$

$$= -\left(\cos(\theta_t) + \sqrt{\gamma^2 - \sin^2(\theta_t)}\right)v_t \qquad (8.88)$$

and for a nonmaneuvering target,

$$\dot{\theta}_s = \dot{\theta}_t$$

$$= \frac{\left[\gamma\sin(\theta_m) - \sin(\theta_t)\right]}{r}v_t. \qquad (8.89)$$

Substituting for $\dot{r}$ and $\dot{\theta}_s$ yields

$$\left[\cos(\theta_t) + \sqrt{\gamma^2 - \sin^2(\theta_t)}\right]\dot{s}_t = -\left[\cos(\theta_t) + \sqrt{\gamma^2 - \sin^2(\theta_t)}\right]v_t$$

$$= +\left[\frac{\cos(\theta_t) + \sqrt{\gamma^2 - \sin^2(\theta_t)}}{\sqrt{\gamma^2 - \sin^2(\theta_t)}}\right]\left[\gamma\sin(\theta_m) - \sin(\theta_t)\right]\sin(\theta_t)s_t v_t \qquad (8.90)$$

$$\dot{s}_t = \left[\frac{\gamma\sin(\theta_m) - \sin(\theta_t)}{\sqrt{\gamma^2 - \sin^2(\theta_t)}}\sin(\theta_t)s_t - 1\right]v_t.$$

Substituting into Equation 8.85 yields

$$v_I = \left[\frac{\gamma\sin(\theta_m) - \sin(\theta_t)}{\sqrt{\gamma^2 - \sin^2(\theta_t)}}\sin(\theta_t)s_t\right]v_t. \qquad (8.91)$$

For impact, the sight line rate $\dot{\theta}_s = 0$, or

$$\gamma\sin(\theta_m) - \sin(\theta_t) = 0. \qquad (8.92)$$

Hence, for impact, we have

$$v_I = 0 \qquad (8.93)$$

and the impact point is stationary in space.

In order to change the position of the impact point, the missile-to-sight line angle θ_m can be set to give a positive or negative rate of change of the impact point along the target tangent vector t_t.

References

1. B. A. White, R. Zbokowski, and A. Tsourdos. Direct intercept guidance using differential geometry concepts. *IEEE Transactions on Aerospace and Electronic Systems*, 43(3), July 2007.
2. F. P. Adler. Missile guidance by three dimensional proportional navigation. *Journal of Applied Physics*, 27:500–507, 1956.
3. R. K. Aggarwal. Optimal Missile Guidance for Weaving Targets. In 35th *IEEE Conference on Decision and Control*, pp. 2775–2779, 1996.
4. G. M. Anderson. Comparison of optimal control and differential game intercept missile guidance laws. *AIAA Journal of Guidance and Control*, 4(2):109–115, 1981.

5. S. N. Balakrishnan, D. T. Stansbery, J. H. Evers, and J. R. Cloutier. Analytical Guidance Laws and Integrated Guidance/Autopilot for Homing Missiles. In *IEEE International Conference on Control Applications*, pp. 27–32, 1993.

6. G. W. Cherry. A General Explicit, Optimizing Guidance Law for Rocket-Propellant Spacecraft. In *AIAA/ION Astrodynamics, Guidance and Control Conference*, 1964. Paper 64-638.

7. Y. C. Chiou and C. Y. Kuo. Geometric approach to three dimensional missile guidance problem. *Journal of Guidance, Control, and Dynamics*, 21(2):335–341, 1998.

8. H. Cho, C. K. Ryoo, and M. J. Tahk. Implementation of optimal guidance laws using predicted missile velocity. *Journal of Guidance, Control, and Dynamics*, 22(4):579–588, 1999.

9. J. R. Cloutier, J. H. Evers, and J. J. Feeley. Assessment of air-to-air missile guidance and control technology. *IEEE Control Systems Magazine*, 9(6):27–34, 1989.

10. J. E. Cochran Jr., T. S. No, and D. G. Thaxton. Analytical solutions to a guidance problem. *Journal of Guidance, Control, and Dynamics*, 14(1):117–122, 1991.

11. R. G. Cottrell. Optimal intercept guidance for short range tactical missiles. *AIAA Journal*, 9(7):1414–1415, 1971.

12. R. G. Cottrell, T. L. Vincent, and S. H. Sadati. Minimizing interceptor size using neural networks for terminal guidance law synthesis. *Journal of Guidance, Control, and Dynamics*, 19(3):557–502, 1996.

13. C. D. Yang, E. B. Yeh, and J. H. Chen. Generalized guidance law of homing missiles. *IEEE Transactions on Aerospace and Electronic Systems*, 25, 898–902, March 1989.

14. E. A. Dijksman. *Motion Geometry of Mechanisms*. Cambridge University Press, Cambridge, 1976.

15. J. E. Steck and S. N. Balakrishnan. Use of Hopfield neural networks in optimal guidance. *IEEE Transactions on Aerospace and Electronic Systems*, 30(1):287–293, 1994.

16. A. Gray. *Modern Differential Geometry of Curves and Surfaces with Mathematica*. CRC Press, Boca Raton, 2nd edition, 1998.

17. M. Guelman. A qualitative study of proportional navigation. *IEEE Transactions on Aerospace and Electronic Systems*, 7:637–643, July 1971.

18. S. Gutman. On optimal guidance for homing missiles. *AIAA Journal of Guidance and Control*, 3(4):296–300, 1979.

19. E. J. Song, H. Lee, and M. J. Tahk. On-line suboptimal midcourse guidance using neural networks. In *The Society of Instrument and Control Engineers Annual Conference*, Tottori, Japan, pp. 1313–1318, 1996.

20. C. Y. Kuo and Y. C. Chiou. Geometric analysis of missile guidance command. *IEE Proceedings: Control Theory and Applications*, 147(2):205–211, 2000.

21. C. Y. Kuo, D. Soetanto, and Y. C. Chiou. Geometric analysis of flight control command for tactical missile guidance. *IEEE Transactions on Control Systems Technology*, 9(2):234–243, 2001.

22. H. Lee, Y. I. Lee, E. J. Song, B. C. Sun, and M. J. Tahk. Missile guidance using neural networks. *Control Engineering Practice*, 5(6):753–762, 1997.

23. K. Y. Lian, L. C. Fu, D. M. Chuang, and T. S. Kuo. Nonlinear autopilot and guidance for a highly maneuverable missile. In *American Control Conference*. IEEE, Baltimore, pp. 2293–2297, 1994.

24. G. Lightbody and G. W. Irwin. Neural model reference adaptive control and application to a BTT-CLOS guidance system. In *IEEE International Conference on Neural Networks*. IEEE, Florida, pp. 2429–2435, 1994.

25. C.-F. Lin. *Modern Navigation Guidance and Control Processing*. Prentice-Hall, New Jersey, 1991.

26. C. M. Lin and Y. J. Mon. Fuzzy-logic-based guidance law design for missile systems. In *IEEE International Conference on Control Applications*. IEEE, Toronto, Canada, pp. 421–426, 1999.

27. J. M. Lin and T. S. Chio. Guidance system design by LEQG/DC method. *Journal of Control Systems and Technology*, 4(1):1–11, 1996.

28. J. M. Lin and S. W. Lee. Bank-to-turn optimal guidance with linear exponential quadratic Gaussian performance criterion. *Journal of Guidance, Control, and Dynamics*, 14(5):951–958, 1995.

29. M. M. Lipschutz. *Differential Geometry*. Schaum's Outline Series. McGraw-Hill, New York, 1969.

30. Yang S. M. Analysis of optimal midcourse guidance law. *IEEE Transactions on Aerospace and Electronic Systems*, 32(1):419–425, 1996.

31. P. K. Menon and E. J. Ohlmeyer. Integrated design of agile guidance and control systems. In *7th IEEE Mediterranean Conference on Control and Automation*, Haifa, Israel, pp. 1469–1494, 1999.

32. S. K. Mishra, I. G. Sarma, and K. N. Swamg. Performance evaluation of two fuzzy-logic-based homing guidance schemes. *Journal of Guidance, Control, and Dynamics*, 17(6):1381–1391, 1993.

33. O. Ariff, R. Zbikowski, A. Tsourdos, and B. A. White. Differential geometric guidance based on the involute of the target's trajectory. *Journal of Guidance, Control, and Dynamics*, 28(5):990–996, 2005.

34. B. O'Neill. *Elementary Differential Geometry*. Academic Press, San Diego, 2nd edition, 1997.

35. N. F. Palumbo and T. D. Jackson. Integrated missile guidance and control: A state dependent Riccati differential equation approach. In *IEEE International Conference on Control Applications*. IEEE, Hawaii, pp. 243–248, 1999.

36. I. Rusnak. Guidance law based on an exponential cost criterion for high order missile and maneuvering target. In *American Control Conference*, IEEE, Chicago, pp. 2386–2390, 1992.

37. I. Rusnak. Advanced guidance laws for acceleration-constrained missile, randomly maneuvering target and noisy measurements. *IEEE Transactions on Aerospace and Electronic Systems*, 32(1):456–464, 1996.

38. D. J. Salmond. Foundations of modern missile guidance. *Journal of Defence Science*, 1(2):171–180, 1996.

39. D. Serakos and C.-F. Lin. Linearized kappa guidance. *Journal of Guidance Control and Dynamics*, 18(5):975–980, 1995.

40. D. Serakos and C.-F. Lin. Three dimensional mid-course guidance state equations. In *Proceedings of the 1999 American Control Conference*. IEEE, San Diego, CA, vol. 6, pp. 3738–3742, 1999.

41. N. A. Shneydor. *Missile Guidance and Pursuit. Kinematics, Dynamics and Control*. Horwood Publishing, Chichester, 1998.

42. J. Z. Ben-Asher and I. Yaesh. *Advances in Missile Guidance Theory*. American Institute of Aeronautics and Astronautics, Reston, 1998.

43. P. Zarchan. Tactical and strategic missile guidance. *Progress in Astronautics and Aeronautics*, vol. 124, AIAA, 1990.

9

Differential Game-Based Interceptor Missile Guidance

Josef Shinar and Tal Shima

CONTENTS

9.1 Introduction

9.1.1 Vector Equations for Interceptor Missile Guidance

The large family of guided missiles can be divided into two main groups: those that are aimed to hit static or slow-moving surface targets (buildings, tanks, ships, etc.) and those whose targets are airborne (and probably maneuvering) objects. Missiles of the second category are called interceptor missiles. The motion of an interceptor missile, as well as that of its target, takes place in 3-D space. The relative motion of the interceptor missile with respect to the target is also 3-D. In the equations of 3-D relative motion, bold letters indicate vectors.

The line-of-sight vector **R**, connecting the interceptor missile and the target, is

$$\mathbf{R} = \mathbf{R}_T - \mathbf{R}_M = \dot{R} \cdot \vec{1}_R. \tag{9.1}$$

The relative velocity vector $\mathbf{V}_R = \mathbf{V}_T - \mathbf{V}_M$ can be decomposed into two components, one along the line-of-sight vector and the other normal to it:

$$\mathbf{V}_R = \mathbf{V}_T - \mathbf{V}_M = \dot{R} \cdot \vec{1}_R + R \cdot \dot{\vec{1}}_R = \dot{R} \cdot \vec{1}_R + \Delta \mathbf{V}_n. \tag{9.2}$$

The normal component is proportional to the range R (the length of the line-of-sight vector) and changes the direction of **R** at the rate of Ω_R:

$$\Delta \mathbf{V}_n = R \cdot \dot{\vec{1}}_R = R \cdot \Omega_R \cdot \vec{1}_n. \tag{9.3}$$

The line-of-sight angular velocity vector (line-of-sight rate) is

$$\mathbf{\Omega}_R = (\mathbf{R} \times \mathbf{V}_R)/R^2 = \Omega_R \cdot \vec{1}_{\Omega_R}. \tag{9.4}$$

The three unit vectors $[\vec{1}_R, \vec{1}_n, \vec{1}_{\Omega_R}]$ define a Cartesian (right-handed) line-of-sight coordinate system. Between these unit vectors, the following relationships exist:

$$\vec{1}_R = \vec{1}_n \times \vec{1}_{\Omega_R}; \quad \vec{1}_n = \vec{1}_{\Omega_R} \times \vec{1}_R; \quad \vec{1}_{\Omega_R} = \vec{1}_R \times \vec{1}_n. \tag{9.5}$$

Moreover,

$$\dot{\vec{1}}_R = \Omega_R \cdot \vec{1}_n; \quad \dot{\vec{1}}_n = -\Omega_R \cdot \vec{1}_R. \tag{9.6}$$

In an interception, the range rate $\dot{R}$ (obtained as a scalar product of the vectors $\mathbf{V}_R$ and $\vec{1}_R$) is generally negative. Its magnitude is called the closing velocity V_c:

$$V_c = -\dot{R}; \; V_R^2 = V_c^2 + (\Delta V_n)^2. \tag{9.7}$$

The relative motion takes place in a *plane*, defined by the vectors $\mathbf{R}$ and $\mathbf{V}_R$. This plane, however, is *not a fixed* one in general, because the vector $\mathbf{R}$ also changes its direction if $\Omega_R \neq 0$.

The vector of relative acceleration $\mathbf{a}_R = \mathbf{a}_T - \mathbf{a}_M$ is

$$\mathbf{a}_R = \frac{d}{dt}(\dot{R} \cdot \vec{1}_R + R \cdot \Omega_R \cdot \vec{1}_n) = \ddot{R} \cdot \vec{1}_R + \dot{R} \cdot \Omega_R \cdot \vec{1}_n + \dot{R} \cdot \Omega_R \cdot \vec{1}_n + R \cdot \dot{\Omega}_R \cdot \vec{1}_n - R \cdot \Omega_R^2 \cdot \vec{1}_R,$$

which can be rewritten as

$$\mathbf{a}_R = (\ddot{R} - R \cdot \Omega_R^2) \cdot \vec{1}_R + (2\dot{R} \cdot \Omega_R + R \cdot \dot{\Omega}_R) \cdot \vec{1}_n. \tag{9.8}$$

If $\mathbf{a}_R = 0$, the direction of the relative velocity vector remains constant, and the plane of relative motion does not rotate. In this case, two scalar differential equations are obtained:

$$\ddot{R} - R \cdot \Omega_R^2 = 0 \tag{9.9}$$

$$R \cdot \dot{\Omega}_R + 2\dot{R} \cdot \Omega_R = 0. \tag{9.10}$$

Equation 9.10 can be directly integrated, yielding

$$R^2 \cdot \Omega_R = R_0^2 \cdot \Omega_{R_0} = \text{constant}. \tag{9.11}$$

If the components of the acceleration vectors $\mathbf{a}_T$ and $\mathbf{a}_M$ are given, the relative motion in an interception scenario can be constructed based on the solution of the nonlinear differential equation (Equation 9.8). This equation clearly illustrates that the interception scenario is governed by nonlinear dynamics in 3-D space. A nonlinear planar analysis is valid only if $\mathbf{a}_R = 0$.

Both $\mathbf{a}_T$ and $\mathbf{a}_M$ are generated by respective acceleration commands in the respective (interceptor and target) body frames. The relationship between the actual and commanded accelerations is described by some transfer functions in the respective body coordinates. The guidance law of the interceptor is the realization of the strategy $U(\mathbf{X}, t)$ that maps the available information to the acceleration command a_M^c, where X is the state vector. The acceleration command of the target a_T^c is determined by another strategy $V(X, t)$.

9.1.2 Pursuit–Evasion Game Formulation

An (aerial) interception scenario belongs to the family of so-called *pursuit-evasion* problems. The objective of the interceptor missile (called in the sequel the *pursuer*) is to destroy the target (called in the sequel the *evader*). Target destruction can be achieved either by a direct hit or (if a hit cannot be achieved) by detonating an explosive warhead in its vicinity. Therefore, the natural cost function of the interception is the distance of the closest approach (called in the sequel the *miss distance*) to be minimized by the *pursuer*.

The evader's acceleration strategy can be either known or unknown to the pursuer. If (and only if) the evader's acceleration strategy or its future actual acceleration profile is known to the pursuer, the interception can be formulated as a one-sided optimal control problem. Otherwise, the evader's trajectory is not predictable, and the optimal control formulation is conceptually inappropriate. In such a case, assuming that the evader's acceleration bounds are known, a *robust* control formulation, requiring successful interception against any feasible (or admissible) target maneuver, can be used.

Since in many aerial interception scenarios the evader's acceleration is independently controlled, another relevant formulation of the problem is in the context of *zero-sum differential games*. In such a game, the pursuer and the evader wish to optimize (minimize/maximize) the same cost function by simultaneously determining their respective optimal strategies. If the processes of *minmax* and *maxmin* lead to the same solution, the game has a *saddle point*, and the respective optimized cost is the *value* of the game. In such a case, the solution of the game is a triplet, composed of the *optimal strategies* of the pursuer and the evader and the *value* of the game, all expressed as a function of the state variables.

Based on such a game solution, the interceptor's guidance law (the realization of the *optimal pursuer strategy*) and the best evasive maneuver (the realization of the *optimal evader strategy*) can be found. If both players use their optimal strategies, the outcome of the interception (the *guaranteed miss distance*) will be the *value* of the game. The pursuer cannot achieve a smaller miss distance, and the evader cannot generate a larger one, as long as the opponent uses its optimal strategy.

The formulation of an aerial interception as a zero-sum differential game was first suggested by Isaacs [1], and since then, it has been used in a great number of research papers and publications. Due to the nonlinear nature of the scenario, only very few reduced dimensional pursuit–evasion games, based on oversimplified assumptions, could be solved.

9.1.3 Modeling Assumptions

In order to obtain some kind of generalized (hopefully closed-form) solutions, in all analytical studies, a set of simplifying models relating to the scenario and to the interceptor's and target's dynamics were used. When the analytical solution is obtained, it becomes necessary to verify the validity of each simplifying assumption in the context of the solution. In the sequel, the most common assumptions are reviewed.

A great part of interception analysis has been carried out in a deterministic mind set, assuming that all state variables and parameters are known to both participants. This means that all state variables of the problem can be (and are) measured with high accuracy. This *perfect information* assumption is unfortunately not valid. Some variables, such as acceleration of the opponent, are not measurable, so they have to be reconstructed from measured data by an observer. Moreover, all measurements are imprecise. This fact is expressed by saying that an actual measurement is the sum of the actual value plus an additive error, modeled as a noise of a given family. Having a large sequence of measurements, the noise can be filtered, and the unmeasured state variables can be obtained from an estimator. Nevertheless, the outcome of a realistic noise-corrupted scenario will not be the same as the outcome of a perfect information analysis.

Another frequently used assumption is that the flying vehicles can be represented by their center of gravity, where the mass is concentrated. Such an assumption, neglecting the angular motions, called the *point-mass* approximation, is very useful for trajectory computations and for miss distances that are either negligibly small or very large. If the miss distance is of the order of the interceptor and/or the target dimensions, a lethality analysis with more details is needed.

In many studies, interceptor and target velocities are assumed to be constant or known as a function of time. In the case of a maneuvering aerial vehicle, this assumption is simply not physical due to the maneuver-dependent induced aerodynamic drag force. The different velocity profiles lead to different flight times and different miss distances.

The maneuvering dynamics of a flying vehicle have in fact a rather complex (not necessarily linear) structure, while in many studies, ideal (instantaneous) dynamics or first-order linear dynamics are assumed. While the assumption of ideal interceptor dynamics can lead to totally unrealistic results, the representation of first-order dynamics preserves, at least qualitatively, correct behavior. In any case, the value of the *equivalent* time constant has to be selected carefully for approximating the true dynamics.

In a great number of studies, the interception is confined to a plane, mostly for the sake of simplicity. Even if guidance laws developed using planar models can work in 3-D space, such an analysis may neglect some inherently 3-D features of the scenario.

9.1.4 Linearized Interception Model

In spite of adopting some, or even all, of the above-mentioned simplifying assumptions, interception kinematics remains generally nonlinear. There is no need to emphasize the difficulties of analyzing nonlinear problems, particularly when optimization is involved. Therefore, much effort has been devoted to creating linear interception models in order to obtain closed-form optimal solutions. The linearization is based on assuming that the relative interception trajectory is sufficiently close to the initial *collision course* trajectory, to be used as a reference.

The notion of a *collision course* comes from an ancient naval background for intercepting a vessel by another. It relates therefore to a planar constant speed scenario. The *collision* plane is defined by the line-of-sight vector $\mathbf{R}$ and the velocity vector of the evader (target) $\mathbf{V_E}$. Assuming that the target moves on a straight line and the pursuer (interceptor) velocity is larger ($V_P > V_E$), there exists a unique direction in the *interception* plane for the pursuer to reach the evader in a finite time, as illustrated in Figure 9.1.

Assuming constant speeds ($\dot{V}_P = \dot{V}_E = 0$), the two conditions for collision can be written as

$$V_P \sin \phi_P - V_E \sin \phi_E = 0 \qquad (9.12)$$

$$V_P \cos \phi_P - V_E \cos \phi_E = -\dot{R} = V_{c_0} > 0, \qquad (9.13)$$

where ϕ_P and ϕ_E are the respective aspect angles. Equation 9.12 is the scalar expression of the vector product between $\mathbf{R}$ and $\mathbf{V_R}$ in the collision plane, indicating (see Equation 9.4) that $\Omega_R = 0$, that is, the initial line-of-sight angle λ_0 remains fixed. Equation 9.12 determines the required direction of the interceptor missile with respect to the nonrotating line of sight:

$$(\phi_P)_{col} = \sin^{-1}[V_E(\sin \phi_E)/V_P]. \qquad (9.14)$$

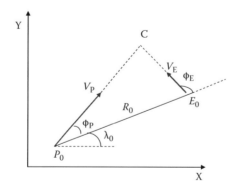

FIGURE 9.1
Collision course geometry.

This direction, called the *collision course*, is constant ($\dot\phi_P = 0$) as long as the target does not maneuver ($\dot\phi_E = 0$) and the velocities are fixed. Equation 9.14 can have a unique solution only if $V_P > V_E$, as in a missile/aircraft interception. If $V_P < V_E$, as in the case of antiballistic missile defense, there may be either two solutions or none.

If at the start of the interception engagement the difference $|(\phi_P)_0 - (\phi_P)_{col}|$ is small, and during the interception the differences $|\phi_E(t) - (\phi_E)_0|$ and $|\phi_P(t) - (\phi_P)_{col}|$ also remain small, one can write two sets of identical *linear* equations of motion normal to the reference line of sight in two perpendicular planes. By setting $\lambda_0 = 0$, the line-of-sight angle $\lambda(t)$ also remains small. It means that a valid trajectory linearization, based on the smallness of angular deviations from the collision course geometry, also leads to decoupling of the original 3-D motion in two planar motions in perpendicular planes [2]. The validity of the linearization is preserved, even if the velocities V_M and V_T are not constant but known as a function of time. In an aerial interception scenario, where both the interceptor and the target are equally affected by gravity, the respective term can be left out of the equations, and the direction of the two perpendicular planes is immaterial. This decoupling property is the reason for which the large majority of the interceptor missiles are designed in a cruciform configuration, having two identical guidance channels acting in perpendicular planes. In the sequel, this chapter concentrates on linearized planar interception models.

In such a linearized planar interception model, the X-axis of the (inertial) coordinate system is aligned with the initial line of sight, that is, $\Rightarrow R_0 = x_0$.

Based on Equation 9.13, the relative motion in the X direction becomes predictable as a function of time:

$$x(t) = R(t) = R_0 - V_c t = V_c(t_f - t), \tag{9.15}$$

where $t_f = R_0/V_c$ is the predicted final time of the engagement (collision). The state variable of interest is the relative position y between the interceptor missile and the target normal to the initial reference line of sight as seen in Figure 9.2:

$$y(t) = y_E(t) - y_P(t). \tag{9.16}$$

The basic equations of motion normal to the initial line of sight and the respective initial conditions are

$$\dot y(t) = V_E \sin\phi_E(t) - V_P \sin\phi_P(t); \quad y(0) = 0 \tag{9.17}$$

$$\ddot y(t) = (a_E)_n - (a_P)_n; \quad \dot y(0) = V_E \sin\phi_{E_0} - V_P \sin\phi_{P_0}, \tag{9.18}$$

where $(a_E)_n = a_E \cos\phi_E(t)$ and $(a_P)_n = a_P \cos\phi_P(t)$ are the acceleration components normal to the initial line of sight. The relationship between the actual

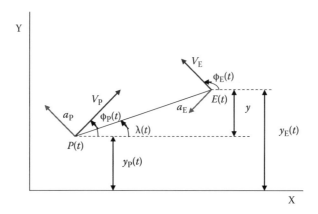

FIGURE 9.2
Planar interception geometry.

accelerations and the respective acceleration commands $(a_E^c)_n$ and $(a_P^c)_n$ is in general determined by a transfer function. Thus, the equations of a linearized planar pursuit–evasion game can be written in the following form:

$$\dot{X} = AX + Bu + Cv, \tag{9.19}$$

where $X \in R^n$ is the state vector $\{X^T = (y, \dot{y}, (a_P)_n, (a_E)_n, \ldots)\}$, and u and v are the normalized control variables $\{(a_P^c)_n = u(a_P)_{max}; (a_E^c)_n = v(a_E)_{max}\}$ satisfying the constraints

$$|u| = 1; \; |v| = 1 \tag{9.20}$$

while A is an $n \times n$ matrix, and B and C are n-dimensional vectors. This notation assumes that the maneuvering dynamics of neither the pursuer nor the evader are ideal, so in this case, both lateral accelerations are state variables.

9.2 Generalized Solution of Linear Differential Games

9.2.1 Generalized Game Formulation

As already mentioned in Section 9.1.2, an aerial interception engagement of an independently controlled maneuverable evader has to be formulated as a zero-sum differential game of pursuit–evasion. It is a two-person zero-sum game that in many ways can be considered as a "two-sided" optimal control problem, but its solution is generally more complicated. Although the necessary conditions of game optimality look similar to those of a two-sided optimal control problem, the sufficiency conditions are different and are more difficult to verify.

A two-person zero-sum differential game is defined by the following elements:

a. Admissible game space
b. Differential equations describing the game dynamics
c. The appropriate initial conditions
d. Admissible control sets of the *players*
e. Conditions of game termination
f. The cost function of the game J
g. The *players'* roles (who is the *minimizer/maximizer?*)
h. The information pattern of the game

For a linearized interception model, the first elements (a–d) are defined in the previous section by Equations 9.16 through 9.20, and the interception terminates at a fixed time t_f, are defined in Equation 9.15.

The natural cost function of an interception engagement is the miss distance

$$J = |\mathbf{DX}(t_f)| = |y(t_f)|, \tag{9.21}$$

where

$$\mathbf{D} = (1, 0, \dots 0). \tag{9.22}$$

In an interception, the roles of the players are obvious: the pursuing interceptor wants to minimize the miss distance, and the evading target wants to maximize it. In this chapter, similar to many other studies, *perfect information* is assumed, which means that both players have perfect knowledge of the state variables and the parameters of the engagement.

The solution of a two-person zero-sum differential game consists of four elements: the *optimal strategies* of the two players and (possibly) two outcomes, namely, the *upper value* and the *lower value* of the game. The players' *strategies* (U for the *minimizer* and V for the *maximizer*) are mappings from the sets of information available for each player to the respective set of admissible controls. Since the players optimize the cost function independently, it is important whether the minimization or the maximization occurs first. The upper value of the game J_{up} is defined as

$$J_{up} - \min_{} \max_{} [J(U, V)] = \max_{} [J(U^*, V)] \tag{9.23}$$

while the lower value of the game J_{low} is

$$J_{low} = \max \min \{J(U,V)\} = \min \{J(U,V^*)\}, \tag{9.24}$$

where U^* and V^* are the respective *optimal strategies*. These two outcomes are generally different. Obviously

$$J_{up} \geq J_{low}. \tag{9.25}$$

If both outcomes are equal (min max = max min), one says that the game has a value J^*, and the respective optimal strategies U^* and V^* are *saddle point strategies* (in other words, the game has a *saddle point*). If the information available for each player is the state vector of the game, the realizations of the optimal strategies are feedback controls. One has to note that satisfaction of the necessary conditions of game optimality provides only *candidate* optimal strategies. One way to verify the fulfillment of the sufficiency conditions is to fill the entire game space with candidate optimal trajectories.

Perfect information pursuit–evasion games with separated dynamics admit a saddle point and have a value. The game solution provides the optimal guidance law of the interceptor missile (*optimal pursuer strategy*), the optimal missile avoidance strategy (*optimal evader strategy*), and the respective guaranteed outcome (*value*) of the game.

A useful methodology that facilitates the solution of linear games is presented in the next subsection.

9.2.2 Terminal Projection Transformation

The vector differential Equation 9.19 can be reduced to a scalar one by using the transformation [3, 4] that can be called a *terminal projection*:

$$Z(t) = \mathbf{D}\boldsymbol{\Phi}(t_f, t)\, \mathbf{X}(t), \tag{9.26}$$

where $\boldsymbol{\Phi}(t_f, t)$ is the transition matrix of the original homogeneous system $\dot{\mathbf{X}} = \mathbf{A}\mathbf{X}$ and $\mathbf{D} = (1\ 0\ 0\ 0\ \ldots)$. The new state variable, denoted by $Z(t)$, is the *zero-effort miss distance*, the miss distance that is created if none of the players use any control until the final time of the interception. The notion of the zero-effort miss distance has a central role in modern missile guidance theory.

The cost function of the interception game can be written as

$$J = |Z(t_f)|. \tag{9.27}$$

The time derivative of $Z(t)$ becomes, using the well-known property of the transition matrix $\dot{\boldsymbol{\Phi}}(t_f, t) = -\boldsymbol{\Phi}(t_f, t)\mathbf{A}$,

$$\dot{Z}(t) = \tilde{B}(t_f, t)u(t) + \tilde{C}(t_f, t)v(t), \tag{9.28}$$

where

$$\tilde{B}(t_f,\ t) = \mathbf{D}\Phi(t_f,\ t)\mathbf{B}; \quad \tilde{C}(t_f,\ t) = \mathbf{D}\Phi(t_f,\ t)\mathbf{C}. \tag{9.29}$$

Integrating Equation 9.28 yields

$$Z(t_f) = Z(t) + \int_t^{t_f} \{\tilde{B}(t_f,\ t)u(t) + \tilde{C}(t_f,\ t)v(t)\}\, dt. \tag{9.30}$$

Since the dynamics of $Z(t)$ is directly controlled by $u(t)$ and $v(t)$, the realizations of the *candidate optimal strategies* of the players (the bounded *normalized optimal control functions*) are

$$u^*(t) = -\text{sign}\{\tilde{B}(t_f,\ t)Z(t)\} \tag{9.31}$$

$$v^*(t) = \text{sign}\{\tilde{C}(t_f,\ t)Z(t)\}. \tag{9.32}$$

Substituting Equations 9.30 and 9.31 into Equation 9.29 yields

$$Z(t_f) = Z(t) - \text{sign}\{Z(t)\} \int_t^{t_f} \left\{ \left|\tilde{B}(t_f,\ t)\right| - \left|\tilde{C}(t_f,\ t)\right| \right\} dt. \tag{9.33}$$

Assuming that $Z(t)$ does not change sign, a candidate optimal trajectory that terminates with the miss distance $Z(t_f)$ can be constructed by backward integration using Equation 9.33, and one can verify whether the family of such (*regular*) trajectories fills the entire game space. Regions that are left empty by such constructions are *singular*, and within them, another pair of optimal strategies has to be found. This procedure will be carried out in the sequel using different dynamic game models.

9.2.3 Hard or Soft Control Constraints

In reality, every interceptor missile and airborne target has inherent physical limitations on the maximal value of the admissible lateral accelerations. Such a saturation phenomenon creates inherently nonlinear dynamics with the difficulties of obtaining closed-form solutions. There are two well-known approaches to circumvent the effects of nonlinearities. The first one, as was demonstrated previously, consists of limiting the acceleration commands to the value of the maximal value of the admissible lateral accelerations, as indicated by Equation 9.20. In this case, the realizations of the candidate optimal strategies can become discontinuous (of the "bang-bang" type) as seen in Equations 9.31 and 9.32. Such "hard" control constraints guarantee that the actual lateral accelerations respect the admissible physical limits (at least as long as the physical limits are nondecreasing). The resulting chattering control creates an unnecessary excessive control effort and may also create other inconveniencies of implementation. Examples of pursuit–evasion games with bounded control are given in Section 9.3.

The other approach is motivated for obtaining a smooth control realization. It ignores the "hard" control constraints, but in order to avoid unnecessary excessive control efforts, the controls are included as a weighted penalty in the cost function. Such an approach creates a linear quadratic (LQ) problem formulation, which can be considered as a "soft" control constraint. The resulting control solution (as long as conjugate points are avoided) will be smooth, and saturation of the acceleration can be avoided by proper tuning of the penalty weighting coefficients. Examples of pursuit–evasion games with LQ formulation are given in Section 9.4.

9.3 Bounded Control Linear Differential Games

In this section, several dynamic models of planar linearized pursuit–evasion games with bounded control are presented. The presentations use the scalar (reduced) state variable $\mathbf{Z}$, the zero-effort miss distance, while the independent variable is the time-to-go defined by

$$t_{go} = t_f - t. \tag{9.34}$$

9.3.1 Ideal Pursuer and Evader Dynamics

This is the simplest game model, denoted as DGI (Differential Game/Ideal), where the directly controlled normalized lateral accelerations are the control variables ($u = (a_P)_n/(a_P)_{max}$; $v = (a_E)_n/(a_E)_{max}$) and the state vector has only two components:

$$\mathbf{X}^T = (y, \dot{y}). \tag{9.35}$$

The components of the transition matrix $\Phi(t_f, t)$ involved in this case are

$$\varphi_{11} = 1; \quad \varphi_{12} = t_{go}, \tag{9.36}$$

yielding the following expression for the zero-effort miss distance:

$$Z(t_{go}) = y + \dot{y}t_{go}. \tag{9.37}$$

The respective components of the vectors **B** and **C** are

$$b_1 = 0; \quad b_2 = -(a_P)_{\max}; \quad c_1 = 0; \quad c_2 = (a_E)_{\max} \tag{9.38}$$

and consequently

$$\tilde{B}(t_f, t) = -t_{go}(a_P)_{\max}; \quad \tilde{C}(t_f, t) = t_{go}(a_E)_{\max}. \tag{9.39}$$

By denoting the *pursuer/evader* maneuver ratio as

$$\mu = \frac{(a_P)_{\max}}{(a_E)_{\max}} \tag{9.40}$$

one obtains for such a game

$$J^* = Z(t_{go} = 0) = Z(t) - \frac{1}{2}t_{go}^2(\mu - 1)(a_E)_{\max}\,\mathrm{sign}\{Z(t)\}. \tag{9.41}$$

From Equation 9.41, it is clear that the game solution is determined by the value of μ. If $\mu < 1$, the zero miss distance cannot be achieved from any initial condition. Optimal trajectories originate from the t_{go}-axis (serving as a *dispersal line*) and fill the entire game space (Figure 9.3).

If $\mu = 1$, the optimal trajectories are lines that are parallel to the t_{go}-axis, filling the game space, as seen in Figure 9.4.

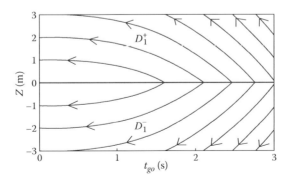

FIGURE 9.3
DGL/I game space decomposition for $\mu < 1$.

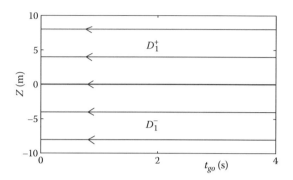

FIGURE 9.4
DGL/I game space decomposition for $\mu = 1$.

If $\mu > 1$, the zero miss distance is guaranteed from a part of the game space (*capture* zone). The boundaries of this region are two symmetrical parabolas determined by the equation

$$\pm Z^*(t_{go}) = \frac{1}{2} t_{go}^2 (\mu - 1)(a_E)_{max}, \tag{9.42}$$

as shown in Figure 9.5.

Within this *singular* region, denoted as D_0, the optimal strategies are *arbitrary*, and the value of the game is constant (zero). Outside it, there is the *regular* region, denoted as D_1, where the optimal strategies are given by Equations 9.31 and 9.32, while the value of the game depends on the initial conditions according to Equation 9.33. This game solution was published in [5]. The implementation of an interceptor guidance law based on this game

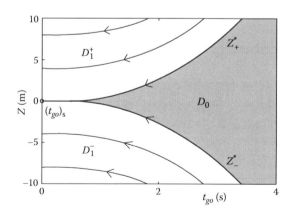

FIGURE 9.5
DGL/I game space decomposition for $\mu > 1$.

solution, denoted DGL/I, is not unique due to the existence of the singular region for $\mu > 1$, which is the case of practical interest. One option is to use the "bang-bang" guidance law of Equation 9.31 everywhere. Another option suggested in [5] is to use in the capture zone a linear guidance law in such a way that on the boundaries of the region, the maximal admissible acceleration is reached. This guidance law turns out to be proportional navigation (PN) because if linearization is valid, then the line-of-sight angle $\lambda(t) \ll 1$ can be written as

$$\lambda(t) \cong \tan \lambda(t) = y(t)/x(t). \tag{9.43}$$

Using this approximation, the line-of-sight rate becomes

$$\dot{\lambda} = \frac{d}{dt}\left(\frac{y}{x}\right) = \frac{\dot{y}x - y\dot{x}}{x^2} = \frac{\dot{y}}{x} - \frac{\dot{x}}{x^2}y. \tag{9.44}$$

Since for the linearized geometry $x(t) = V_c(t_{go})$,

$$\dot{\lambda} = \frac{\dot{y}}{V_c t_{go}} + \frac{y}{V_c t_{go}^2} = \frac{1}{V_c t_{go}^2}[y + \dot{y}t_{go}]. \tag{9.45}$$

The linear approximation of the miss distance is $y(t_f)$. Therefore, the expression for the linearized predicted zero-effort miss distance for PN becomes

$$Z_{PN}(t) = y + \dot{y}t_{go}, \tag{9.46}$$

leading us to conclude that

$$\dot{\lambda}(t) = \frac{1}{V_c t_{go}^2} Z_{PN}(t). \tag{9.47}$$

The classical form of PN is

$$(a_P)_n = N'V_c\dot{\lambda}. \tag{9.48}$$

In order to obtain maximal admissible acceleration on the boundaries of the capture zone, the effective navigation ratio N' depends on μ. By using Equation 9.42 with Equations 9.46 through 9.48, the appropriate value of N' is

$$N' - 2\mu/(\mu \quad 1). \tag{9.49}$$

By accepting this suggestion, the interceptor guidance law can be expressed in the entire reduced game space (Z, t_{go}) by using the saturation operator:

$$(a_P)_n = (a_P)_{max} \; \text{sat} \left\{ \frac{2\mu}{\mu-1} \frac{Z_{PN}(t_{go})}{t_{go}^2 \, (a_P)_{max}} \right\}. \tag{9.50}$$

The model of ideal dynamics, being far from reality, is unable to provide a reliable element in guided missile design. For such a purpose, more realistic models are needed.

9.3.2 Ideal Evader and First-Order Pursuer Dynamics

This model is motivated by acknowledging the strong effect of interceptor dynamics on the homing performance and approximates it by a first-order transfer function. Assuming ideal evader dynamics (although this is not possible in reality) provide the "worst case" for the pursuer and thus allows one to be on the "safe side" for guided missile design.

Using such a model, the state vector of the game is

$$\mathbf{X}^T = (y, \dot{y}, (a_P)_n). \tag{9.51}$$

The components of the transition matrix $\Phi(t_f, t)$ involved in this case are

$$\varphi_{11} = 1; \; \varphi_{12} = t_{go}; \; \varphi_{13} = -\tau_P^2 [e^{-\theta} + \theta - 1], \tag{9.52}$$

where

$$\theta = t_{go}/\tau_P \tag{9.53}$$

is the *normalized* time-to-go. The expression for the zero-effort miss distance becomes

$$Z(t_{go}) = y + \dot{y} t_{go} - (a_P)_n \tau_P^2 (e^{-\theta} + \theta - 1). \tag{9.54}$$

The respective components of the vectors $\mathbf{B}$ and $\mathbf{C}$ are

$$b_1 = 0; \;\; b_2 = 0; \;\; b_3 = \frac{(a_P)_{max}}{\tau_P}; \;\; c_1 = 0; \;\; c_2 = (a_E)_{max}; \;\; c_3 = 0, \tag{9.55}$$

and consequently

$$\tilde{B}(t_f, t) = -(a_P)_{max} \tau_P [e^{-\theta} + \theta - 1]; \;\; \tilde{C}(t_f, t) = t_{go} (a_E)_{max}. \tag{9.56}$$

By using θ and the normalized zero-effort miss distance defined by

$$z = Z/\tau_P^2 (a_E)_{max}, \qquad (9.57)$$

the expression 9.30 becomes

$$z(0) = z(\theta) - \text{sign}\{z(\theta)\} \int_0^\theta h(\theta)\,d\theta, \qquad (9.58)$$

where $h(\theta)$ is defined as

$$h(\theta) = \frac{\left\{ \left| \tilde{B}(t_f, t) \right| - \left| \tilde{C}(t_f, t) \right| \right\}}{\tau_P a_E^{max}} \qquad (9.59)$$

and for this dynamic model

$$\int_0^\theta h(\theta)\,d\theta = \int_0^\theta \{\mu[e^{-\theta} + \theta - 1] - \theta\}\,d\theta. \qquad (9.60)$$

One can see that for $\mu > 1$, this integral has a minimum obtained at $\theta = \theta_s$, where θ_s is the nonzero solution of the equation

$$\mu[e^{-\theta} + \theta - 1] - \theta = 0. \qquad (9.61)$$

For small values of θ ($0 < \theta_s$), the integrand is negative, which means that the zero miss distance [$z(0) = 0$] can never be achieved.

The two *limiting* trajectories (Z_+^*, Z_-^*) that satisfy the condition that $z(\theta)$ does not change sign reach the θ-axis tangentially at $\theta = \theta_s$. The normalized game space is decomposed into a singular region D_0, which is between these trajectories for $\theta > \theta_s$, and the regular region D_1. In D_1, the optimal *strategies* are given by Equations 9.31 and 9.32, while the value of the game depends on the initial conditions.

In the singular region, the optimal strategies are *arbitrary*, and the value of the game is a nonzero constant J_s. This *singular* game *value*, which is the smallest miss distance that an optimally playing pursuer can achieve against an optimally playing evader, depends on the physical parameter μ. Once θ_s is found from the solution of Equation 9.61, J_s can be computed from Equation 9.33 by setting $z(\theta_s) = 0$ followed by direct integration between θ_s and zero. The larger the value of μ, the smaller the value of θ_s and consequently the smaller the value of J_s. For a sufficiently large value of μ, the guaranteed miss distance J_s is very small. Every trajectory starting in D_0 must go through the *throat* [$z(\theta_s) = 0$]. This is a *dispersal point* for the evader to decide on the maneuver direction (Figure 9.6).

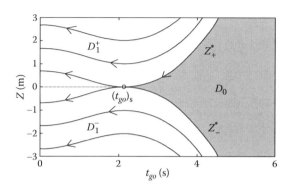

FIGURE 9.6
DGL/0 game space decomposition for $\mu > 1$.

The interceptor guidance law based on this game solution is denoted DGL/0. Its implementation (similarly to DGL/I) is not unique due to the existence of the singular region. One option is (similarly to DGL/I) to use the "bang-bang" guidance law of Equation 9.31 everywhere. In [6], it is suggested to use in D_0 a linear guidance law in such a way that on the boundaries of the region, the maximal admissible acceleration is reached. Another interesting option is to use in D_0 an LQ game solution that guarantees reaching the throat with minimal control effort.

The interceptor guidance law DGL/0 has an important advantage. Its implementation requires only (see Equations 9.45 and 9.54) the knowledge of the line-of-sight rate and its own acceleration but not the target acceleration. Although it cannot guarantee zero miss distance even in an ideal situation, the guaranteed miss distance can be made negligibly small by using sufficiently high maneuverability advantage of the interceptor.

In the practically unimportant case of $\mu < 1$, there is no singular region, and the game space decomposition looks similar to Figure 9.3.

9.3.3 First-Order Evader and Pursuer Dynamics

If there is sufficient information on the *evader* dynamics, approximating it by a first-order transfer function provides a more realistic and balanced game model than was used first in [7]. In this game, the state vector is

$$\mathbf{X}^{\mathrm{T}} = [y, \dot{y}, (a_E)_n, (a_P)_n]. \tag{9.62}$$

The components of the transition matrix $\Phi(t_f, t)$ involved in this case are

$$\varphi_{11} = 1; \quad \varphi_{12} = t_{go}; \quad \varphi_{13} = \tau_E^2[e^{-\theta/\varepsilon} + \theta/\varepsilon - 1]; \quad \varphi_{14} = -\tau_P^2[e^{-\theta} + \theta - 1], \tag{9.63}$$

where

$$\varepsilon = \tau_E/\tau_P. \tag{9.64}$$

The expression for the zero-effort miss distance is

$$Z(t_{go}) = y + \dot{y}t_{go} + (a_E)_n \tau_E^2 [e^{-\theta/\varepsilon} + \theta/\varepsilon - 1] - (a_P)_n \tau_P^2 (e^{-\theta} + \theta - 1) \tag{9.65}$$

and the respective components of the vectors **B** and **C** are

$$b_1 = 0; \quad b_2 = 0; \quad b_3 = 0; \quad b_4 = \frac{(a_P)_{max}}{\tau_P}; \quad c_1 = 0; \quad c_2 = 0; \quad c_3 = \frac{(a_E)_{max}}{\tau_E}; \quad c_4 = 0, \tag{9.66}$$

leading to

$$\tilde{B}(t_f, t) = -(a_P)_{max} \tau_P [e^{-\theta} + \theta - 1]; \quad \tilde{C}(t_f, t) = (a_E)_{max} \tau_E [e^{-\theta/\varepsilon} + \theta/\varepsilon - 1]. \tag{9.67}$$

For this game, the integral of $h(\theta)$ becomes

$$\int_0^\theta h(\theta) = \int_0^\theta \left\{ \mu(e^{-\theta} + \theta - 1) - \varepsilon \left(e^{-\theta/\varepsilon} + \theta/\varepsilon - 1 \right) \right\}. \tag{9.68}$$

Depending on the values of the physical parameters μ and ε, $h(\theta)$ can be either positive or negative. If $\mu > 1$ and $\mu\varepsilon < 1$, the function has a minimum at $\theta = \theta_s$, where θ_s is the nonzero solution of the equation

$$\mu[e^{-\theta} + \theta - 1] - \varepsilon[e^{-\theta/\varepsilon} + \theta/\varepsilon - 1] = 0 \tag{9.69}$$

and the game space decomposition is similar to the one of DGL/0 for $\mu > 1$. For $\mu > 1$ and $\mu\varepsilon \geq 1$, the only solution of Equation 9.69 is $\theta = 0$, $h(\theta)$ is always positive, and the game space decomposition is similar to the one of DGL/I with $\mu > 1$. In this case, from any initial condition inside the singular region D_0, *point capture* $[z(0) = 0]$ is guaranteed by using arbitrary strategies. For the case of $\mu < 1$ and $\mu\varepsilon \geq 1$, the singular region (where the zero miss distance can be achieved) is closed, as seen in Figure 9.7.

In Table 9.1, the conditions for various forms of game solution structures established in [8] are summarized.

Due to the existence of the singular region, the implementation of the interceptor guidance law based on this game solution, denoted as DGL/1, is also not unique, and options similar to those of DGL/I and DGL/0 can be adopted.

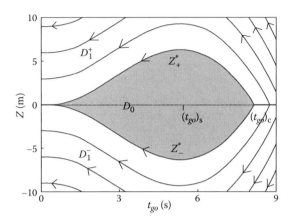

FIGURE 9.7
DGL/1 game space decomposition for $\mu < 1$ and $\mu\varepsilon \geq 1$.

TABLE 9.1

Conditions for Various Forms of Game Solution Structures

	$\mu < 1$	$\mu = 1$	$\mu > 1$
$\mu\varepsilon < 1$	Figure 9.3	Figure 9.3	Figure 9.6
$\mu\varepsilon = 1$	Figure 9.3	Figure 9.4	Figure 9.5
$\mu\varepsilon > 1$	Figure 9.7	Figure 9.5	Figure 9.5

It should also be noted that the implementation of DGL/1 requires the knowledge of the current evader maneuver as a component of the state vector, which cannot be measured from another platform. It has to be reconstructed from available measurements by an observer in a noise-free case or by an estimator if the available measurements are corrupted by noise.

9.3.4 Dual Maneuver Devices

The first-order dynamic model used in the previous subsections is the simplest approximation to introduce the *nonideal* behavior of a control system. In reality, the behavior of an airborne control system is more complex. It depends mainly on the location of the aerodynamic control surfaces. The force created by the deflection of a forward control surface (*canard*) is almost immediate, while the contribution of the other parts of the missile becomes active slightly later (due to the change of the angle of attack). On the other hand, using aerodynamic control surfaces located at the rear section of the missile creates a *nonminimum phase* effect because the initial direct lift is in the opposite direction compared with the total lift created by the required angle of attack. Therefore, it can be shown [9] that canard control provides better homing performance than tail control. However, if the missile is to perform sharp initial turns, tail control may be selected because such controls saturate at higher angles of attack. By using both canard and tail controls, a reasonable design compromise can be obtained that provides overall better performance. Higher pitching moments can be created, and there is an option to create very fast direct lift (although of limited magnitude) for terminal corrections. The additional degree of freedom offered by the dual control system requires special consideration in the guidance and control design. In several studies, different blending strategies between the canard and tail have been suggested. In a recent paper [10], a guidance strategy tailored for a missile with dual controls has been proposed.

The closed-loop maneuvering dynamics of an interceptor missile with dual control can be approximated by two first-order biproper transfer functions:

$$\frac{a_{P_c}}{a_c^c} = d_c + \frac{1 - d_c}{1 + s\tau_p}; \quad 1 > d_c > 0 \tag{9.70}$$

$$\frac{a_{P_t}}{a_t^c} = d_t + \frac{1 - d_t}{1 + s\tau_p}; \quad -1 < d_t < 0, \tag{9.71}$$

where a_{P_c} and a_{P_t} are the missile acceleration components (including the contribution of the airframe) due to the control actions of the canard and tail, respectively, and a_c^c and a_t^c are the commanded accelerations of the canard and tail channels. The constants d_c and d_t are the direct lift coefficients, and the second terms on the right-hand sides of Equations 9.70 and 9.71 represent the response of the missile airframe to the command. Note that the transfer function (Equation 9.70) is the *minimum phase* and that of Equation 9.71 is the *nonminimum phase*. Since we assume that the system is linear, the total acceleration of the interceptor missile, denoted as a_P, is

$$a_P = a_{P_c} + a_{P_t} = d_c u_c + d_t u_t + a_P^b, \tag{9.72}$$

where a_P^b is the specific force acting on the missile airframe, excluding the direct lift contributions of the control surfaces.

The maneuvering dynamics of the target is approximated by a first-order (strictly proper) transfer function

$$\frac{a_E}{a_E^c} = \frac{1}{1 + s\tau_E}. \tag{9.73}$$

In order to avoid control saturation, it is assumed that

$$\left|a_c^c\right| \le k_c \cdot a_P^{\max}; \quad 0 \le k_c \le 1 \tag{9.74}$$

$$\left|a_t^c\right| \le k_t \cdot a_P^{\max}; \quad 0 \le k_t \le 1. \tag{9.75}$$

The overall acceleration command should not exceed the maximum maneuvering capability of the airframe. Therefore, k_c and k_t must satisfy

$$k_c + k_t = 1. \tag{9.76}$$

The ratio between k_c and k_t determines the relative effectiveness of the canard and tail control channels. In the case of canard control, only $k_c = 1$ and $k_t = 0$, and for tail control, only $k_c = 0$ and $k_t = 1$. We assume that the control of the target is bounded by

$$\left|a_E^c\right| \le a_E^{\max}. \tag{9.77}$$

The state vector in the equations of relative motion normal to the initial line of sight is

$$\mathbf{X}^T = \begin{pmatrix} y & \dot{y} & a_E & a_P^B \end{pmatrix}. \tag{9.78}$$

The corresponding equations of motion are

$$\begin{aligned} \dot{x}_1 &= x_2 \\ \dot{x}_2 &= x_3 - x_4 - d_c a_c^c - d_t a_t^c \\ \dot{x}_3 &= [a_E^c - x_3]/\tau_E \\ \dot{x}_4 &= [(1 - d_c)a_c^c + (1 - d_t)a_t^c - x_4]/\tau_P. \end{aligned} \tag{9.79}$$

The zero-effort miss distance, based on Equation 9.26, is similar to Equation 9.65:

$$Z(t_{go}) = y + \dot{y}t_{go} + (a_E)\tau_E^2 \psi(\theta/\varepsilon) - (a_P^b)\tau_P^2 \psi(\theta) \tag{9.80}$$

where $\psi(\theta) = [e^{-\theta} + \theta - 1]$ with the only difference being that $(a_P)_n$ is replaced by (a_P^b).

The candidate optimal strategies for this problem are obtained in a similar way as for the previous models:

$$u_c^* = k_c \cdot \text{sign}[Z(t_f)] \tag{9.81}$$

$$u_t^* = k_t \cdot \text{sign}[Z(t_f)] \cdot \text{sign}[f(d_t, \theta)], \tag{9.82}$$

where

$$f(\delta, \zeta) = \delta \cdot \zeta + (1 - \delta)\psi(\zeta) \tag{9.83}$$

$$v^* = \text{sign}[Z(t_f)]. \tag{9.84}$$

We also define

$$t_{go}^t = \underset{t_{go}>0}{\arg}[f(d_t, t_{go}/\tau_P) = 0]. \tag{9.85}$$

Based on these candidate optimal strategies, the game space decomposition can be made. In Figure 9.8, a game space for $\mu > 1$ with one singular region D_0 is plotted. In this singular region, the optimal strategies are arbitrary, and the value of the game is zero. Note that the region D_1, where the optimal strategies are Equations 9.81 through 9.84 and the value of the game is a function of the initial conditions, is divided into two subregions denoted as D_{1-} and D_{1+} by a transition surface located at $t_{go} = t_{go}^t$. When reaching this transition surface, the tail control changes its sign due to its nonminimum phase feature.

For the case where the pursuer has a disadvantage in maneuverability, that is, $\mu < 1$, the decomposition is different. In such a case, a game space,

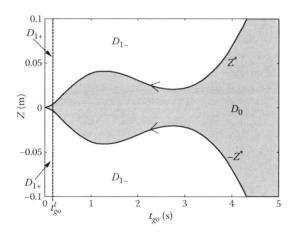

FIGURE 9.8

Game space with one singular region ($\mu > 1$)

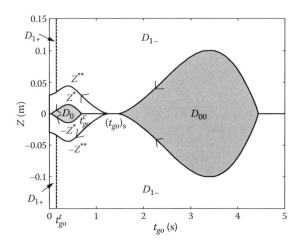

FIGURE 9.9
Game space with two closed singular regions ($\mu < 1$).

composed of two singular regions denoted D_0 and D_{00}, is obtained as shown in Figure 9.9. In D_{00}, the optimal strategies are arbitrary, and the value of the game is a nonzero constant. Nonetheless, there always exists another (small) closed singular region (D_0) from which the zero miss distance can be guaranteed for the interceptor. The existence of such a capture zone is the consequence of the direct lift (with ideal dynamics) of the interceptor. More details on this problem can be found in [10].

9.3.5 Time-Varying Parameters

As mentioned in Section 9.1.3 on modeling assumptions, in most cases, the speeds of the interceptor missile and the target are not constant. In the examples given, until now, a constant speed assumption was adopted, and the longitudinal accelerations were neglected. The longitudinal accelerations may have a component normal to the line of sight and affect the homing process. Moreover, in an interception performed in the vertical plane, such as in ballistic missile defense, the maneuvering capabilities of the interceptor and the target also vary with altitude. Assuming that the time dependency of these values is known as a function of the time-to-go in a given engagement, the resulting pursuit–evasion game can still be solved using a linear time-varying model of the problem, as proposed in [11].

The state vector in this problem also includes the aspect angles ϕ_E and ϕ_P:

$$\mathbf{X} = [x_1, x_2, x_3, x_4, x_5, x_6]^T = [y, \dot{y}, a_E, a_P, \phi_E, \phi_P]^T . \tag{9.86}$$

Nevertheless, these aspect angles are assumed to be small, and the approximations $\cos(\phi_i) \approx 1$ and $\sin(\phi_i) \approx \phi_i$ ($i = P, E$) are uniformly valid, suitable for linear analysis. Computation of the interception's final time for a given initial range x_0 of the endgame, needed for using the time-to-go, is obtained by

$$t_f = \arg\left\{ x_f = x_0 - \int_{t_0}^{t_f} [V_E(t) + V_P(t)]\,dt = 0 \right\}. \qquad (9.87)$$

From the known velocity profiles $V_E(t)$ and $V_P(t)$, the respective longitudinal accelerations $a_{xE}(t)$ and $a_{xP}(t)$ can be computed and substituted into the equations of motion, which become

$$\dot{X} = A(t)X + B(t)u + C(t)v; \quad X(0) = X_0 \qquad (9.88)$$

with

$$A(t) = \begin{bmatrix} 0 & 1 & 0 & 0 & 0 & 0 \\ 0 & 0 & -1 & 1 & -a_{xE}(t) & a_{xP}(t) \\ 0 & 0 & -1/\tau_E & 0 & 0 & 0 \\ 0 & 0 & 0 & -1/\tau_P & 0 & 0 \\ 0 & 0 & 1/V_E(t) & 0 & 0 & 0 \\ 0 & 0 & 0 & 1/V_P(t) & 0 & 0 \end{bmatrix} \qquad (9.89)$$

$$B(t) = \begin{bmatrix} 0 & 0 & 0 & a_P^{\max}(t)/\tau_E & 0 & 0 \end{bmatrix}^T \qquad (9.90)$$

$$C(t) = \begin{bmatrix} 0 & 0 & a_E^{\max}(t)/\tau_P & 0 & 0 & 0 \end{bmatrix}^T \qquad (9.91)$$

and the normalized controls

$$u = a_P^c / a_P^{\max}(t); \quad |u| \le 1 \qquad (9.92)$$

$$v = a_E^c / a_E^{\max}(t); \quad |v| \le 1. \qquad (9.93)$$

The zero-effort miss distance of this time variable problem is more complex than for the case of constant parameters discussed in Section 9.3.4;

$$z(t) = x_1(t) + x_2(t)t_{go}$$

$$+ x_3(t)\tau_E^2[\psi(\theta/\varepsilon) + III_E(t_f, t)] - x_4(t)\tau_P^2[\psi(\theta) + III_P(t_f, t)] \qquad (9.94)$$

$$+ x_5(t)[IV_E(t_f, t) - V_E(t)t_{go}] - x_6(t)[IV_P(t_f, t) - V_P(t)t_{go}],$$

where

$$III_i(t_f, t) = \int_t^{t_f} \frac{II_i(\zeta, t)}{\tau_i} d\zeta; \quad IV_i(t_f, t) = \int_t^{t_f} V_i(\zeta) d\zeta; \quad i = P, E \tag{9.95}$$

and

$$I_i(t_f, t) = \int_t^{t_f} \frac{e^{-(\zeta - t/\tau_P)}}{V_i(\zeta)} d\zeta; \quad II_i(t_f, t) = \int_t^{t_f} \frac{I_i(\zeta, t) a_{xi}(\zeta)}{\tau_i} d\zeta; \quad i = P, E. \tag{9.96}$$

Comparing Equation 9.94 with Equation 9.65, one can see the corrective terms $III_i(t_f, t)$ multiplying the lateral accelerations, as well as the additional terms due to the larger state vector.

This linear pursuit–evasion game with time-varying speeds and control bounds is solved similarly as the game with constant parameters in Section 9.3.4. The decomposition of the game space is similar to those shown in Figures 9.3 through 9.7. The only difference is that the maneuverability ratio μ is not constant. Therefore, the rules of Table 9.1 have to be rewritten for this more general case. The game space decomposition depends on the behavior of the candidate optimal trajectory derivative defined by

$$\frac{dZ^*}{dt_{go}} \triangleq \Gamma(t_{go}) \operatorname{sign} Z^* \tag{9.97}$$

and can be written explicitly as

$$\Gamma(t_{go}) = a_P^{max}(t)[\psi(t_{go}/\tau_P) + III_P(t_{go})]\tau_P - a_E^{max}(t)[\psi(t_{go}/\tau_E) + III_E(t_{go})]\tau_E. \tag{9.98}$$

These general rules can be summarized as follows:

If $\Gamma(t_{go})$ does not change sign and $\Gamma(t_{go}) < 0 \ \forall \ t_{go} \in (0, t_f] \Rightarrow$ Figure 9.3.
If $\Gamma(t_{go})$ does not change sign and $\Gamma(t_{go}) = 0 \ \forall \ t_{go} \in (0, t_f] \Rightarrow$ Figure 9.4.
If $\Gamma(t_{go})$ does not change sign and $\Gamma(t_{go}) > 0 \ \forall \ t_{go} \in (0, t_f] \Rightarrow$ Figure 9.5.
If $\Gamma(t_{go})$ changes sign once and $\Gamma(t_f) > 0$ while $\Gamma(0) < 0 \Rightarrow$ Figure 9.6.
If $\Gamma(t_{go})$ changes sign once and $\Gamma(t_f) > 0$ while $\Gamma(0) > 0 \Rightarrow$ Figure 9.7.

Although cases where $\Gamma(t_{go})$ changes sign more than once can be imagined, their occurrence (as well as of $\Gamma(t_{go}) = 0 \ \forall \ t_{go} \in (0, t_f]$) is rather unlikely. In all cases, the condition for the existence of a capture zone (a region where the zero miss distance is guaranteed) is determined by the limit of $\Gamma(0_+) > 0$.

A simplified example with linearly varying speed and maximum accelera-tion can be found in [8].

9.4 LQ Differential Games

9.4.1 General Solution of LQ Differential Games

In this section, the interception problem is solved using the framework of LQ differential games (LQDGs). The framework of an LQDG in its general formulation

$$J = \mathbf{X}_f^T \mathbf{Q}_f \mathbf{X}_f + \int_{t_0}^{t_f} [\mathbf{X}^T(t)\mathbf{Q}(t)\mathbf{X}(t) + \mathbf{u}^T(t)\mathbf{S}_u(t)\mathbf{u}(t) - \mathbf{v}^T(t)\mathbf{S}_v(t)\mathbf{v}(t)]dt \quad (9.99)$$

allows investigation of a large variety of problems. Each component of the terminal state can have a different weighting. The first term in the integral can serve for trajectory shaping, and the penalty terms can deal with high dimensional controls.

This section concentrates on the interception problem and uses a simpler approach. The terminal part of the cost considers the miss distance only; tra-jectory shaping is neglected, and the control variables are scalars. The anal-ysis uses the dynamic model of planar linearized pursuit–evasion games (Equation 9.19) as well as the relevant zero-effort miss distance transforma-tion (Equation 9.26). In the LQDG formulation, there are no constraints on the controls, but they are included as a weighted penalty in the cost function [11] in the form of

$$J = \frac{b}{2}Z^2(t_f) + \frac{1}{2}\int_{t_0}^{t_f}[u^2(t) - \gamma^2 v^2(t)]dt. \quad (9.100)$$

In this section, a slightly different, but essentially identical, cost function is used through dividing Equation 9.100 by b:

$$J = \frac{1}{2}Z^2(t_f) + \frac{1}{2}\int_{t_0}^{t_f}[\alpha u^2(t) - \beta v^2(t)]dt. \quad (9.101)$$

The cost is to be minimized by the pursuer and maximized by the evader subject to the dynamic Equation 9.19.

The Hamiltonian of the game is

$$H = 0.5(\alpha u^2 - \beta v^2) + \lambda_Z[\tilde{B}(t_f, t)u + \tilde{C}(t_f, t)v]. \tag{9.102}$$

The adjoint equation is

$$\dot{\lambda}_Z = 0; \quad \lambda_Z(t_f) = Z(t_f), \tag{9.103}$$

which yields for continuous λ_Z

$$\lambda_Z(t) = Z(t_f). \tag{9.104}$$

Since the controls are not constrained, the candidate optimal strategies are obtained from

$$\frac{\partial H}{\partial u} = 0; \quad \frac{\partial^2 H}{\partial u^2} > 0 \Rightarrow u^*(t) = -\frac{\tilde{B}(t_f, t)}{\alpha} Z(t_f)$$

$$\frac{\partial H}{\partial v} = 0; \quad \frac{\partial^2 H}{\partial v^2} < 0 \Rightarrow v^*(t) = \frac{\tilde{C}(t_f, t)}{\beta} Z(t_f). \tag{9.105}$$

Substituting Equation 9.105 into Equation 9.28 and integrating from an arbitrary t to t_f, we get an expression for $Z(t_f)$ that is of the form

$$Z(t_f) = Z(t) - Z(t_f)F_{\alpha\beta}(t_f, t), \tag{9.106}$$

which leads to

$$Z(t_f) = Z(t)/[1 - F_{\alpha\beta}(t_f, t)], \tag{9.107}$$

where $F_{\alpha\beta}$ is

$$F_{\alpha\beta}(t_f, t) = \int_t^{t_f} [-\tilde{B}^2(t_f, \tau)/\alpha + \tilde{C}^2(t_f, \tau)/\beta] d\tau. \tag{9.108}$$

Substituting Equations 9.106 and 9.107 into Equations 9.104 and 9.105, the candidate optimal strategies take the state feedback forms

$$u^*(t) = \frac{-\tilde{B}(t_f, t)}{\alpha[1 - F_{\alpha\beta}(t_f, t)]} Z(t) \tag{9.109}$$

$$v^*(t) = \frac{\tilde{C}(t_f, t)}{\beta[1 - F_{\alpha\beta}(t_f, t)]} Z(t). \tag{9.110}$$

These feedback controls indicate that the LQDG solution exists (i.e., there is no conjugate point) if $F_{\alpha\beta} < 1$. In a recent paper [12], it has been shown that for a game with constant parameters, this condition exists if the following relationship is satisfied:

$$\alpha/\beta = \sigma \le \min\ [\mu^2,\ \mu^2\varepsilon^2] \triangleq \sigma^0. \tag{9.111}$$

Based on the definition of $\sigma \triangleq k\sigma^0$ ($0 < k \le 1$) and Equation 9.111, the candidate optimal strategies (Equations 9.109 and 9.110) become

$$u^*(t) = \frac{-\tilde{B}(t_f,\ t)}{\alpha + \displaystyle\int_t^{t_f} [\tilde{B}^2(t_f,\ \tau)]d\tau - \sigma \int_t^{t_f} [\tilde{C}^2(t_f,\ \tau)]d\tau}\ Z(t) \tag{9.112}$$

$$v^*(t) = \frac{\tilde{C}(t_f,\ t)}{\beta + \dfrac{1}{\sigma}\displaystyle\int_t^{t_f} [\tilde{B}^2(t_f,\ \tau)]d\tau - \int_t^{t_f} [\tilde{C}^2(t_f,\ \tau)]d\tau}\ Z(t). \tag{9.113}$$

It can be easily verified that candidate optimal trajectories generated by these strategies fill completely the reduced $(Z,\ t_{go})$ game space. Therefore, they are optimal strategies of an LQDG.

By letting the penalty coefficients α and β tend toward zero (called in the literature *cheap control*) but keeping their ratio σ constant, satisfying Equation 9.111, a negligibly small miss distance can be achieved by the guidance law denoted LQDG/0:

$$u^*_{\alpha\beta=0} = \frac{-\tilde{B}(t_f,\ t)}{\displaystyle\int_t^{t_f} [\tilde{B}^2(t_f,\ \tau)]d\tau - \sigma \int_t^{t_f} [\tilde{C}^2(t_f,\ \tau)]d\tau}\ Z(t). \tag{9.114}$$

The corresponding optimal evader strategy is

$$v^*_{\alpha\beta=0}(t) = \frac{\tilde{C}(t_f,\ t)}{\dfrac{1}{\sigma}\displaystyle\int_t^{t_f} [\tilde{B}^2(t_f,\ \tau)]d\tau - \int_t^{t_f} [\tilde{C}^2(t_f,\ \tau)]d\tau}\ Z(t). \tag{9.115}$$

In order to avoid control saturation (i.e., respecting the physical constraint) using Equation 9.112 or 9.114 during interception, the value of μ has to be sufficiently high.

In the following subsections, examples of LQDGs with different dynamic models are presented.

9.4.2 Ideal Pursuer and Evader Dynamics

For this simplest dynamic model, $Z(t) = Z_{PN}(t)$, and

$$\tilde{B}(t_f,\ t) = -t_{go} a_P^{max}\ ;\ \tilde{C}(t_f,\ t) = t_{go} a_E^{max}. \tag{9.116}$$

In this case, the value of ε is not determined. Therefore, the condition 9.111 for not having a conjugate point becomes simply $\sigma \leq \mu^2$. If this condition is satisfied, the guidance law (the *optimal pursuer strategy*) can be written, assuming cheap control, in a well-known form of PN (Equations 9.55 and 9.56):

$$(a_P)_n = a_P^{max} u = \frac{N'}{t_{go}^2} Z_{PN}(t) \tag{9.117}$$

with

$$N' = \frac{3}{(1 - \sigma/\mu^2)}. \tag{9.118}$$

The solution of this game was presented first in [13].

9.4.3 Ideal Evader and First-Order Pursuer Dynamics

For this model, $Z(t)$ is given by Equation 9.63 and the expressions of $\tilde{B}(t_f,\ t)$ and $\tilde{C}(t_f,\ t)$ by Equation 9.67. Since in this model, the parameter ε, defined by Equation 9.64, is identically zero, Equation 9.111 predicts that for any given set of parameters $(\alpha,\ \beta,\ \mu)$, there exists a game duration (t_f) for which a conjugate point cannot be avoided. Nevertheless, for a given interception scheme where the values of t_f and μ are determined, the weighting parameters $(\alpha,\ \beta)$ can be chosen to avoid a conjugate point.

The guidance law based on this game solution can be written in the format of Equation 9.117 with

$$N' = \frac{6\theta^2 \psi(\theta)}{6\alpha / [\tau_P^3 (a_P^{max})^2] + \{2[1 - (\sigma/\mu^2)]\theta^3 + 3 - 6\theta^2 + 6\theta - 12\theta e^{-\theta} - 3e^{-2\theta}\}}. \tag{9.119}$$

If a cheap control approach is used for obtaining negligibly small miss distances and assuming that the game duration is such that there is no conjugate point, Equation 9.119 is reduced to

$$N' = \frac{6\theta^2 (e^{-\theta} + \theta - 1)}{2[1 - (\sigma/\mu^2)]\theta^3 + 3 - 6\theta^2 + 6\theta - 12\theta e^{-\theta} - 3e^{-2\theta}}. \tag{9.120}$$

9.4.4 First-Order Evader and Pursuer Dynamics

For this model, $Z(t)$ is given by Equation 9.65 and the expressions of $\tilde{B}(t_f, t)$ and $\tilde{C}(t_f, t)$ by Equation 9.67. In this case, by respecting the condition 9.111, a conjugate point can be always avoided, and a cheap control solution can be very attractive. The guidance gain N' becomes more complicated by adding another term in the denominators of Equations 9.119 and 9.120. For the cheap control case

$$N' = \frac{6\theta^2(e^{-\theta} + \theta - 1)}{\{2[1-(\sigma/\mu^2)]\theta^3 + 3 - 6\theta^2 + 6\theta - 12\theta e^{-\theta} - 3e^{-2\theta}\} + (\sigma/\mu^2)g(\theta,\varepsilon)}, \quad (9.121)$$

where

$$g(\theta,\varepsilon) = [-3\varepsilon^3 - 6\varepsilon^2\theta + 6\varepsilon\theta^2 + 12\varepsilon^2\theta e^{-\theta/\varepsilon} + 3\varepsilon^3 e^{-2\theta/\varepsilon}]. \quad (9.122)$$

In order to avoid control saturation (i.e., respecting the physical constraint) using the guidance laws indicated in this section during the interception, the value of μ has to be sufficiently high. For the model of the present subsection, the limit value of μ depends on ε and λ. Assuming $\lambda = \lambda^0$ ($k = 1$), for $\varepsilon \leq 1$ this limit is

$$\mu \geq \frac{3}{\varepsilon + \varepsilon^2(3-\varepsilon)} \quad (9.123)$$

and for $\varepsilon > 1$ it is

$$\mu \geq \frac{3}{1 + 3\varepsilon - \varepsilon^2}. \quad (9.124)$$

If saturation of the commanded acceleration is admissible, that is, one applies as the guidance law a modified version of Equation 9.112 or for cheap controls Equation 9.114,

$$\tilde{u}^*_{\alpha\beta=0} = \mathrm{sat}\left\{\frac{\mu(e^{-\theta} + \theta - 1)}{H_1(\theta) - \sigma H_2(\theta)} z(\theta)\right\}, \quad (9.125)$$

where $H_1(\theta) = \int_t^{t_f} [\tilde{B}^2(t_f, \tau)]d\tau$ and $H_2(\theta) = \int_t^{t_f} [\tilde{C}^2(t_f, \tau)]d\tau$ the limit value of μ that guarantees a maximal capture zone can be smaller than the values required by Equations 9.123 and 9.124. For $k \to 1$, this value approaches the value required for DGL/1. In this case, the capture zones of both guidance laws (namely, DGL/1 and the saturated version of LQDG/0 with $k \to 1$) are almost identical, but the control effort of the modified LQDG guidance law is lower. Both guidance laws also have similar performances with noisy measurements.

9.4.5 Dual Maneuver Devices

The quadratic cost function chosen for the derivation of an LQ guidance law for a dual control missile is

$$J = \frac{1}{2} Z^2(t_f) + \frac{1}{2} \int_{t_0}^{t_f} [\alpha_c u_c^2(t) + \alpha_t u_t^2(t) - \beta v^2(t)] dt, \tag{9.126}$$

where the weights α_c, α_t, and β are all positive. Note that letting $\alpha_c = 0$ corresponds to a guidance law for a tail-controlled missile, while letting $\alpha_t = 0$ corresponds to a canard control.

The zero-effort miss for this problem is given in Equation 9.80. Following the procedure outlined in Section 9.4.1, the following optimal strategies are obtained:

$$u_i^*(t) = \frac{N_i'}{a_{P_{max}} t_{go}^2} Z(t); \quad i = c, t \tag{9.127}$$

$$v^*(t) = \frac{N_v'}{a_{E_{max}} t_{go}^2} Z(t), \tag{9.128}$$

where

$$N_c' = \frac{\theta^2 f(d_c, \theta)}{\dfrac{\alpha_c}{\tau_P^3 (a_P^{max})^2} + \displaystyle\int_0^\theta \left[f^2(d_c, \zeta) + \frac{\alpha_c}{\alpha_t} f^2(d_t, \zeta) \right] d\zeta - \frac{\alpha_c}{\beta \mu^2} \displaystyle\int_0^{\theta/\varepsilon} \psi^2(\zeta) d\zeta} \tag{9.129}$$

$$N_t' = \frac{\theta^2 f(d_t, \theta)}{\dfrac{\alpha_t}{\tau_P^3 (a_P^{max})^2} + \displaystyle\int_0^\theta \left[\frac{\alpha_t}{\alpha_c} f^2(d_c, \zeta) + f^2(d_t, \zeta) \right] d\zeta - \frac{\varepsilon^3 \alpha_t}{\beta \mu^2} \displaystyle\int_0^{\theta/\varepsilon} \psi^2(\zeta) d\zeta} \tag{9.130}$$

$$N_v' = \frac{\varepsilon \theta^2 \psi(\theta/\varepsilon)}{\dfrac{\beta}{\tau_P^3 (a_P^{max})^2} + \mu^2 \displaystyle\int_0^\theta \left[\frac{\beta}{\alpha_c} f^2(d_c, \zeta) + \frac{\beta}{\alpha_t} f^2(d_t, \zeta) \right] d\zeta - \varepsilon^3 \displaystyle\int_0^{\theta/\varepsilon} \psi^2(\zeta) d\zeta} \tag{9.131}$$

and $f(\delta, \zeta)$ was defined by Equation 9.83 as $f(\delta, \zeta) = \delta \cdot \zeta + (1 - \delta)\psi(\zeta)$.

By adopting the cheap control approach leading to a perfect interception, one sets $\alpha_c = \alpha_t = \beta = 0$, but keeping the ratios $\alpha_c/\alpha_t = \kappa_{ct}$, $\alpha_c/\beta = \sigma_c$, and $\alpha_t/\beta = \sigma_t$ fixed, Equations 9.121 through 9.123 become

$$N'_c = \frac{\theta^2 f(d_c, \theta)}{\int_0^\theta [f^2(d_c, \zeta) + \kappa_{ct} f^2(d_t, \zeta)] d\zeta - \frac{\varepsilon^3 \sigma_c}{\mu^2} \int_0^{\theta/\varepsilon} \psi^2(\zeta) d\zeta} \tag{9.132}$$

$$N'_t = \frac{\theta^2 f(d_t, \theta)}{\int_0^\theta \left[\frac{1}{\kappa_{ct}} f^2(d_c, \zeta) + f^2(d_t, \zeta)\right] d\zeta - \frac{\varepsilon^3 \sigma_t}{\mu^2} \int_0^{\theta/\varepsilon} \psi^2(\zeta) d\zeta} \tag{9.133}$$

$$N'_v = \frac{\varepsilon \theta^2 \psi(\theta/\varepsilon)}{\mu^2 \int_0^\theta \left[\frac{1}{\sigma_c} f^2(d_c, \zeta) + \frac{1}{\lambda_t} f^2(d_t, \zeta)\right] d\zeta - \varepsilon^3 \int_0^{\theta/\varepsilon} \psi^2(\zeta) d\zeta}. \tag{9.134}$$

From Equations 9.124 through 9.126, it is clear that perfect interception (i.e., zero miss distance) can be achieved, avoiding a conjugate point if the ratios σ_c and σ_t are selected so that for the existing physical parameters d_c, d_t, and ε and for $0 \le \theta_0 = t_f/\tau_P$

$$\mu^2 \int_0^\theta \left[\frac{1}{\sigma_c} f^2(d_c, \zeta) + \frac{1}{\sigma_t} f^2(d_t, \zeta)\right] d\zeta > \varepsilon^3 \int_0^{\theta/\varepsilon} \psi^2(\zeta) d\zeta. \tag{9.135}$$

More details on this problem can be found in [14].

9.4.6 Time-Varying Parameters

As in the case of bounded control linear differential games, solutions based on the assumption of constant speeds can be extended to more realistic scenarios, where the speeds and the maneuvering capabilities of both players are known as the function of time (or time-to-go) along the predicted nominal interception trajectory. A detailed discussion on this topic, with players that have first-order dynamics, is presented in [15]. Here only the main results are outlined.

The model with players of first-order dynamics is presented by Equations 9.86 through 9.98. The solution of an LQDG with time-varying parameters can be easily constructed using these equations together with the general solution of the LQDG presented in Equations 9.104 through 9.110 and 9.112 through 9.115. The only difference to keep in mind relates to the limit value of the ratio (α/β), denoted by σ^0, which provides the sufficiency condition for avoiding a conjugate point. Since in the time-varying case, μ is not constant, Equation 9.111 has to be replaced by another definition of σ^0:

$$\sigma^0 \triangleq \inf_{t \in [0, t_f)} \frac{\displaystyle\int_t^{t_f} [\tilde{B}^2(t_f, \tau)]d\tau}{\displaystyle\int_t^{t_f} [\tilde{C}^2(t_f, \tau)]d\tau}. \tag{9.136}$$

By respecting the condition $\alpha/\beta = \sigma \leq \sigma^0$, a guidance law based on the cheap control solution expressed by Equation 9.114 can be implemented.

9.5 Conclusion

In this chapter, an overview of interceptor guidance laws, derived using the formulation of perfect information zero-sum differential games, was presented. Closed-form solutions, implementable online in an airborne interceptor, were obtained by using planar linearized models.

Two families of interceptor guidance laws, both using the zero-effort miss distance concept, were obtained by assuming two different cost functions and control constraints. The first family, assuming bounded controls, was characterized by discontinuous controls of the "bang-bang" type. The second family, based on an LQ formulation, provided a smooth control realization. However, when using this formulation, the cost that was optimized was not only the miss distance; similar performance as the first family could be achieved by applying a cheap control approach. A special effort was made for casting both guidance law families in a similar framework, enabling direct comparison.

This chapter summarized differential game-based guidance laws, known to the authors, starting from classical work like [13] and also including recent papers that introduced biproper dynamics and time-varying models.

The main advantage of the differential game formulation is its robustness with respect to the entire class of admissible future target maneuvers. Not including the contribution of the target acceleration in the zero-effort miss distance calculation is equivalent to assuming ideal target dynamics, which is a pessimistic (worst case) assumption, leading to a nonzero guaranteed miss distance.

For nonideal target dynamics, in order to achieve zero miss distances, the current target acceleration (which is one of the state variables) has to be known. Other modern guidance laws, based on optimal control theory, assume (sometimes unjustified) knowledge of future target acceleration.

Simulation studies demonstrated that a pair of planar linear guidance laws in perpendicular planes can be implemented successfully in typical 3-D interception scenarios.

The basic limitation of these guidance laws is that they all are based on the assumption of perfect information. As mentioned in Section 9.1.3, this assumption, which is a necessity for obtaining closed-form deterministic solutions, is not valid in real interception scenarios. The existence of noisy measurements requires incorporating a state estimator in the guidance loop.

It should be obvious that the homing performance in a realistic noise-corrupted scenario cannot be as good as the one predicted by a perfect information analysis. Instead of the deterministic figure of the guaranteed miss distance, the outcome of a realistic noise-corrupted scenario has to be presented in stochastic terms by the probability distribution of miss distances. The complex subject of how to design a state estimator for interceptor guidance and how to incorporate it in the guidance and control system is the topic of a separate chapter.

References

1. Isaacs, R., *Differential Games*, Wiley, New York, 1965.
2. Adler, F. P., Missile guidance by three-dimensional proportional navigation, *J. Appl. Phys.*, 27, 1956, 500.
3. Bryson, A. E. and Ho, Y. C., *Applied Optimal Control*, Blaisdell Publishing Company, New York, 1969, p. 154.
4. Krasovskii, N. N. and Subbotin, A. I., *Game Theoretical Control Problems*, Springer, New York, 1988.
5. Gutman, S. and Leitmann, G., Optimal strategies in the neighborhood of a collision course, *AIAA J.*, 14(9), 1976, 1210.
6. Gutman, S., On optimal guidance for homing missiles, *J. Guidance Control*, 3(4), 1979, 296.
7. Shinar, J., Solution techniques for realistic pursuit–evasion games, in *Advances in Control and Dynamic Systems*, vol. 17, Academic Press, New York, 1981, p. 63.
8. Shima, T. and Shinar, J., Time varying pursuit evasion game models with bounded controls, *J. Guidance Control Dynamics*, 25(3), 2002, 425.
9. Gutman, S., The superiority of canards in homing missiles, *IEEE Trans. Aerosp. Electron Syst.*, 39(3), 2003, 740.
10. Shima, T. and Golan, O., Bounded differential game guidance law for a dual controlled missile, *IEEE Trans. Control Syst. Tech.*, 14(4), 2006, 719.
11. Ben-Asher, J. and Yaesh, I., *Advances in Missile Guidance Theory*, Progress in Astronautics and Aeronautics, vol. 180, AIAA, Washington, DC, 1998, p. 89.

12. Turetsky, V. and Shinar, J., Missile guidance laws based on pursuit–evasion game formulations, *Automatica*, 39(4), 2003, 607.

13. Ho, Y. C., Bryson, A. E., and Baron, S., Differential games and optimal pursuit evasion strategies, *IEEE Trans. Auto. Control*, 10(4), 1965, 385.

14. Shima, T. and Golan, O., Linear quadratic differential game guidance law for dual control missiles, *IEEE Trans. Aerosp., Electron. Syst.*, 41(3), 2007, 834.

15. Turetsky, V. and Glizer, V. Y. Continuous feedback control strategy with maximal capture zone in a class of pursuit games, *Int. Game Theory Rev.*, 7(1), 2005, 1.

10

Optimal Guidance Laws with Impact Angle Control

Chang-Kyung Ryoo, Min-Jea Tahk, and Hangju Cho

CONTENTS

Impact angle control guidance laws have a variety of applications for missiles and unmanned aerial vehicles. For antiship or antitank missiles, terminal impact angle control could maximize the effectiveness of warheads by attacking the weakest part of a target. For inertially or global positioning system-guided missiles, vertical approach to the target can reduce the miss distance by nullifying the effect of navigation error in the vertical channel. Impact angle control may find useful applications for a ground target not only to ensure that the warhead does not ricochet off the target but also to achieve maximum penetration performance. Impact angle control laws can also be used as a convenient tool for waypoint guidance or path planning of unmanned aerial vehicles. With these diverse application areas, there have been extensive studies on the issue. Among them, the optimal guidance laws (OGLs) with impact angle control, derived in the framework of optimal control theory, draw special attention for their adaptability to various engagement circumstances.

In an attempt to replace the well-known proportional navigation (PN) guidance law that is widely used in various forms in practical homing missile applications, OGLs have been studied extensively for many years. It has been known that PN guidance with the navigation constant N is nothing but a special case of OGLs that minimizes the energy cost weighted by the inverse of $(N - 3)$th power of time-to-go [1]. It is worth noting at this point that the minimization of the energy cost, which is typically given by the integral of the square of the acceleration command, is closely related to that of aerodynamic drag [2] or the mean power consumption of the actuators and consequently to increased effective range of engagement or smaller actuation systems. Many different formulations of the optimal guidance problem are possible, and the OGL satisfying the specified impact angle as well as the zero miss distance can certainly be obtained by using optimal control theory.

In this chapter, we show how to derive optimal impact angle control guidance laws by using the well-known result of the linear quadratic (LQ) optimal control theory. While we assume in the derivation that the speed of the missile is constant and the target is stationary or moving at constant speed, the resultant guidance laws can be applied for a varying-speed missile to intercept a maneuvering target and yield moderate performance.

An optimal control problem for a linear system and its solution are outlined in Section 10.1 for easy reference. In Section 10.2, the optimal impact angle control law for a lag-free missile is derived, and its properties are discussed. A time-to-go computation method based on curved path length over missile speed is provided. The performance of the OGL for a maneuvering target will be also discussed in this section. In Section 10.3, an optimal impact angle control guidance law for a first-order missile will be derived. The optimal impact angle control laws can also be applied nicely to the energy-optimal waypoint guidance problem. This issue will be addressed in Section 10.4. The proposed approach can be used not only for waypoint guidance itself but also for real-time generation of the energy optimal trajectory passing through multiple waypoints.

10.1 LQ Optimal Control Problem and Its Solution

In this section, we present a well-known LQ optimal control problem, the form of which is most frequently encountered when some optimal guidance algorithm is sought for an advanced missile system, and we summarize its solution for the use in the subsequent sections. Consider the following LQ optimal control problem, which has wide applications in various fields: Find an optimal control function $u(t)$ that minimizes the cost

$$J = \frac{1}{2} x(t_f)^T S_f x(t_f) + \frac{1}{2} \int_{t_0}^{t_f} u^T R u \, dt \tag{10.1}$$

subject to

$$\dot{x} = Ax + Bu, \ x(t_0) = x_0 \tag{10.2}$$

with terminal constraint

$$Dx(t_f) = E \tag{10.3}$$

where $S_f \geq 0$ and $R > 0$. The solution u^* to this LQ optimal control problem is found to be [4, 5]

$$u^* = -R^{-1}B^T [(S - FG^{-1}F^T)x + FG^{-1}E]$$
$$= R^{-1}B^T FG^{-1}(F^T x - E) - R^{-1}B^T Sx \tag{10.4}$$

where

$$\bar{S} = S - FG^{-1}F^T$$
$$\dot{S} = -A^T S - SA + SBR^{-1}B^T S, \ S(t_f, t_f) = S_f$$
$$\dot{F} = -(A^T - SBR^{-1}B^T)F, \ F(t_f, t_f) = D^T$$
$$\dot{G} = F^T BR^{-1}B^T F, \ G(t_f, t_f) = 0. \tag{10.5}$$

Simpler Case—Hard Terminal Constraint Only

The most widely used LQ optimal control formulation in guidance problems is to minimize overall control effort with hard terminal constraints. In this case, the above optimal control problem is reduced to the following: Find $u(t)$ that minimizes

$$J = \frac{1}{2} \int_{t_0}^{t_f} u^T R u \, dt \tag{10.6}$$

subject to

$$\dot{x} = Ax + Bu,\ x(t_0) = x_0 \qquad (10.7)$$

with the terminal constraint

$$Dx(t_f) = E. \qquad (10.8)$$

The solution to this LQ optimal control problem is now given in a simpler form by

$$u^* = R^{-1}B^T FG^{-1}(F^T x - E) \qquad (10.9)$$

where

$$\dot{F} = -A^T F,\ F(t_f, t_f) = D^T \qquad (10.10)$$

$$\dot{G} = F^T BR^{-1}B^T F,\ G(t_f, t_f) = 0. \qquad (10.11)$$

10.2 Optimal Impact Angle Control Law for Lag-Free Missile (OGL)

The PN guidance law with the navigation constant N' can be obtained as a solution to the LQ optimal control problem minimizing the time-varying weighted energy cost $\int t_{go}^{-(N'-3)} u^2$ subject to zero miss distance [1]. (N' here is called the navigation constant. The PN law with $N' = 3$ is energy optimal, but $N' > 3$ is recommended in practice to provide the missile with sufficient capability to deal with disturbances and uncertainties, such as initial position/heading error and abrupt target maneuver [6].)

In this section, we address an optimal guidance problem with the terminal constraints of zero miss distance and specified impact angle. We follow the same line of thought in the problem formulation as in the derivation of PN with various navigation constants as a family of optimal guidance laws; in this way, we could see how the well-known PN guidance law with a certain navigation constant would change when an impact angle constraint is required additionally. In other words, the resulting guidance law could be regarded as a generalized PN that fulfills the additional impact angle constraint. We assume a lag-free missile and consider the same cost of time-varying weighted energy in the formulation. Although the OGL is obtained for a stationary target, we will see that the proposed law can also be applied to intercept a maneuvering target as in the case of the PN law.

10.2.1 Derivation of OGL

Consider the homing guidance geometry for a stationary target shown in Figure 10.1. Here, V, θ, and θ_f denote the missile velocity, the flight path angle, and the desired impact angle, respectively. The acceleration command applied normal to the velocity vector is denoted by u and the line-of-sight (LOS) angle by σ. Other variables in Figure 10.1 are self-explanatory.

The equations of motion for the homing problem are given by

$$\dot{z}(t) = V(t)\sin\theta(t), \; z(0) = 0$$
$$V(t)\dot{\theta}(t) = u(t), \; \theta(0) = \theta_0.$$
(10.12)

Under the assumption that V is constant and θ is small, we can linearize Equation 10.12 as

$$\dot{z}(t) \approx V\theta(t)$$
$$V\dot{\theta}(t) = u(t).$$
(10.13)

Let

$$v(t) = V\theta(t).$$
(10.14)

Then, we obtain the linear differential equation

$$\dot{x} = Ax + Bu, \; x(0) = x_0$$
(10.15)

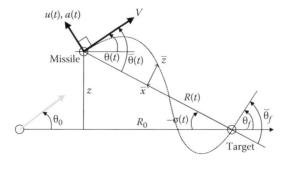

FIGURE 10.1
Homing guidance geometry

where

$$x = [z \quad v]^T, \ x_0 = [0 \quad v_0]^T, \ v_0 = V\theta_0 \qquad (10.16)$$

and

$$A = \begin{bmatrix} 0 & 1 \\ 0 & 0 \end{bmatrix}, \ B = \begin{bmatrix} 0 \\ 1 \end{bmatrix}. \qquad (10.17)$$

Now we consider the same LQ optimal control problem: Find u that minimizes the cost

$$J = \frac{1}{2} \int_0^{t_f} R(t)u^2(t)dt \qquad (10.18)$$

subject to Equation 10.15 with Equation 10.17 and the terminal constraints

$$Dx(t_f) = E \qquad (10.19)$$

where

$$D = \begin{bmatrix} 1 & 0 \\ 0 & 1 \end{bmatrix}, \ E = \begin{bmatrix} 0 \\ v_f \end{bmatrix}, \ v_f = V\theta_f \qquad (10.20)$$

and

$$R(t) = \frac{1}{(t_f - t)^N}, \ N \geq 0. \qquad (10.21)$$

Note that R is identical to the time-varying weighting function used in [1] where $N + 3$ turns out to be the navigation constant N' of the PN guidance law. If we choose $N = 0$, we get pure energy optimal guidance command. Using the time-to-go dependent weighting function given by Equation 10.21, we can shape the guidance command profile. For $N > 0$, the cost of control becomes radically expensive as $t \rightarrow t_f$ so that the command eventually reduces to 0 at $t = t_f$. If $N < 0$ is chosen, the command increases without bound or blows up.

Now we derive the state-feedback solution by using the result summarized in the previous section. First, to compute F in Equation 10.9, set

$$F \triangleq \begin{bmatrix} f_{11} & f_{12} \\ f_{21} & f_{22} \end{bmatrix}. \qquad (10.22)$$

Then we have four differential equations for f_{ij}:

$$\dot{f}_{11} = 0, \quad f_{11}(t_f) = 1$$
$$\dot{f}_{12} = 0, \quad f_{12}(t_f) = 0$$
$$\dot{f}_{21} = -f_{11}, \quad f_{21}(t_f) = 0$$
$$\dot{f}_{22} = -f_{12}, \quad f_{22}(t_f) = 1.$$

(10.23)

The solutions of the above equations are easily obtained to be

$$F = \begin{bmatrix} 1 & 0 \\ t_{go} & 1 \end{bmatrix},$$

(10.24)

where t_{go} denotes the time-to-go and is defined by

$$t_{go} = t_f - t.$$

(10.25)

Now, the right-hand side of Equation 10.11 is given by

$$F^T B R^{-1} B^T F = \frac{1}{t_{go}^{-N}} \begin{bmatrix} t_{go}^2 & -t_{go} \\ -t_{go} & 1 \end{bmatrix}.$$

(10.26)

Integrating Equation 10.26, we have

$$G = \begin{bmatrix} -\dfrac{1}{N+3} t_{go}^{N+3} & -\dfrac{1}{N+2} t_{go}^{N+2} \\ -\dfrac{1}{N+2} t_{go}^{N+2} & -\dfrac{1}{N+1} t_{go}^{N+1} \end{bmatrix}.$$

(10.27)

By substituting Equations 10.24 and 10.27 into Equation 10.9, the state-feedback closed-form impact angle control optimal guidance law for a lag-free missile, which we call OGL [7] for brevity, is obtained as

$$u^* = -\frac{V}{t_{go}^2} \left[\frac{N_z}{V} z(t) + N_\theta t_{go} \theta(t) + N_f t_{go} \theta_f \right]$$

(10.28)

where

$$N_z = (N+2)(N+3)$$
$$N_\theta = 2(N+2)$$
$$N_f = (N+1)(N+2).$$

(10.29)

It is interesting to note that the above guidance gains are not linear to N. Examples of the guidance gain set $\{N_z, N_0, N_f\}$ for some integer values of N are shown in Table 10.1. Note that the following relationship between gains holds:

$$N_z = N_0 + N_f. \tag{10.30}$$

For $N = 0$, Equation 10.28 becomes the pure energy optimal impact angle control guidance law [8]. Note that if we use the relationship in Equation 10.13, that is, $\dot{z} = V\theta$, and write $\dot{z}_f = V\theta_f$, then we get

$$u^* = \frac{1}{t_{go}^2}[N_z(-z - t_{go}\dot{z}) - N_f t_{go}(\dot{z}_f - \dot{z})]. \tag{10.31}$$

This guidance law is the same as the GENEX, the OGL proposed in [9]. If we choose $N = 0$ and N_z, $\theta_f = 0$, then we get

$$u^* = -\frac{1}{t_{go}^2}[6z(t) + 4t_{go}v(t)] \tag{10.32}$$

which represents the optimal rendezvous solution discussed in [4] and [10].

The time history of the optimal control can be computed by substituting Equation 10.28 for u in Equation 10.13 and integrating and is found to be

$$u^*(t) = t_{go}^N(C_S + C_R t_{go}) \tag{10.33}$$

where

$$C_S = \frac{VN_f}{t_f^N}\left[\frac{N+3}{V}z_0 + t_f\theta_0 + (N+2)t_f\theta_f\right]$$

$$C_R = -\frac{VN_z}{t_f^{N+1}}\left[\frac{N+2}{V}z_0 + t_f\theta_0 + (N+1)t_f\theta_f\right]. \tag{10.34}$$

TABLE 10.1

Examples of Gain Set of OGL

N	N_z	N_0	N_f
0	6	4	2
1	12	6	6
2	20	8	12

We thus see from Equation 10.33 that the optimal control is represented by an $(N + 1)$th-order polynomial function of t_{go}. For $N = 0$, the guidance command is a linear function of time and converges to a finite nonzero value of C_S as $t \to t_f$ ($t_{go} \to 0$). Note that the magnitude of guidance commands with $N > 0$ always decreases to 0 as t goes to t_f. The maximum magnitude of the guidance command occurs at $t = 0$ or $t = t_f$ for $N = 0$, while it always occurs at $t = 0$ for $N > 0$. Command convergence to 0 for $N > 0$ is a valuable property of OGL; it minimizes the possibility of command saturation in the terminal homing phase, which might occur in situations not considered during the course of optimal guidance derivation, such as sudden target maneuver and large delay in missile maneuver. In other words, this property could give some margin for guidance command to handle external disturbances, model uncertainties, and command limits. If the missile speed decreases considerably in the final stage of engagement due to aerodynamic drag, we may have to use a larger gain set to avoid severe performance degradation. On the other hand, a smaller gain set is preferred to reduce sensor noise sensitivity. In most cases, the best guidance gain is chosen via nonlinear simulations taking account of all adversary effects, nonlinearity, and uncertainties.

Figures 10.2 and 10.3 represent the normalized command histories of OGL for $\theta_f = 0$ and $\theta_0 = 0$, respectively. As discussed above, the guidance command of OGL approaches nonzero values for $N = 0$ (energy optimal case) and always converges to 0 for $N = 1$ and 2.

Note that if we take t ($\geq t_0$), $z(t)$, and $\theta(t)$ as initial values of the problem, then Equation 10.34 becomes

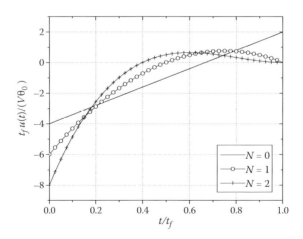

FIGURE 10.2
Normalized guidance command histories of OGL for $\theta_f = 0$.

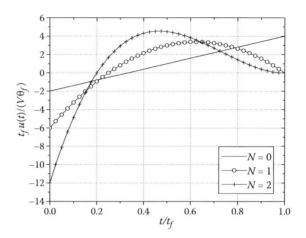

FIGURE 10.3
Normalized guidance command histories of OGL for $\theta_0 = 0$.

$$C_S = \frac{VN_f}{t_{go}^N}\left[\frac{N+3}{V}z(t) + t_{go}\theta(t) + (N+2)t_{go}\theta_f\right]$$

$$C_R = -\frac{VN_z}{t_{go}^{N+1}}\left[\frac{N+2}{V}z(t) + t_{go}\theta(t) + (N+1)t_{go}\theta_f\right].$$

(10.35)

If $z(t)$ and $\theta(t)$ are the values on the optimal path from t_0 to t_f, then the optimal path newly obtained at t should coincide with that computed with initial conditions at t_0 (this is called the principle of optimality [3]). Hence, we can deduce that as long as $z(t)$ and $\theta(t)$ stay on the optimal path, C_S and C_R in Equation 10.35 remain the same constant values. We could thus check the degree of deviation from optimality by monitoring the changes in C_s and C_R during the engagement. (Deviation from the optimal value is inevitable due to disturbances and uncertainties such as missile speed variation, time-to-go calculation error, and time lag of the missile response.)

By substituting Equation 10.33 into Equation 10.13 and integrating it, we observe that $\theta(t)$ and $z(t)$ are $(N+2)$th- and $(N+3)$th-order polynomial functions, respectively:

$$\theta(t) = -\frac{1}{V}\left(\frac{1}{N+2}C_R t_{go}^{N+2} + \frac{1}{N+1}C_S t_{go}^{N+1}\right) - \theta_f$$

$$z(t) = \frac{1}{(N+2)(N+3)}C_R t_{go}^{N+3} + \frac{1}{(N+1)(N+2)}C_S t_{go}^{N+2} + V\theta_f t_{go}$$

(10.36)

$$= \frac{1}{N_z}C_R t_{go}^{N+3} + \frac{1}{N_f}C_S t_{go}^{N+2} + V\theta_f t_{go}.$$

Adjoint Analysis of OGL for First-Order Lag System

Together with the command limit, the time lag of missile response to command (hereafter, it is called missile autopilot lag) is one of the most dominant error sources when using the guidance laws derived under the assumption of a lag-free missile. Indeed, if OGL is applied to a missile model with autopilot lag of an arbitrary order, both the miss distance and the impact angle error are inevitable. Here, we investigate the effect of the missile autopilot lag on the terminal guidance errors using the adjoint simulation method [11].

Figure 10.4 shows the linear homing loop using OGL for a first-order autopilot lag system. While PN guidance loop considers only miss distance as an output, the impact angle control problem has two terminal guidance errors under consideration: miss distance and impact angle error. Hence, the adjoint simulations should be performed twice, once for miss distance and another time for impact angle error. Figures 10.5 and 10.6 show adjoint loops

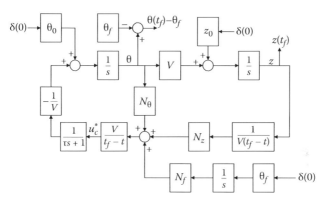

FIGURE 10.4
Linear homing loop of OGL for a first-order lag system.

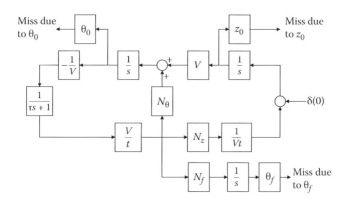

FIGURE 10.5
Adjoint loop of OGL for $z(t_f)$.

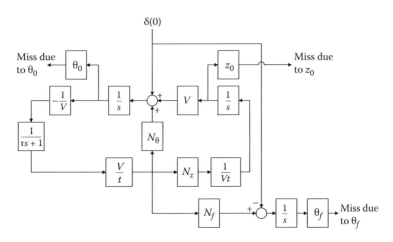

FIGURE 10.6
Adjoint loop of OGL for $\theta(t_f) - \theta_f$.

for miss distance and impact angle error, respectively. These adjoint loops are obtained from the linear homing loop in Figure 10.4 by using the technique in [11].

In Figures 10.7 and 10.8, the miss distance due to initial heading angle is shown to be less sensitive to the change of N than that due to impact angle requirements. On the other hand, the impact angle error due to either initial heading or the required impact angles greatly varies according to the values of N, as shown in Figures 10.9 and 10.10. For PN guidance law, the miss distance due to initial heading angles increases as N decreases [11], but the miss distance and the impact angle error tend to decrease for OGL as N decreases. Both terminal errors can be made sufficiently small if t_f is greater than 15 times τ.

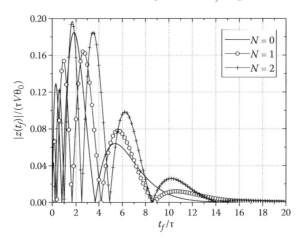

FIGURE 10.7
Normalized miss distance of OGL induced by θ_0.

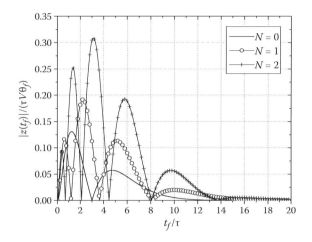

FIGURE 10.8
Normalized miss distance of OGL induced by θ_f.

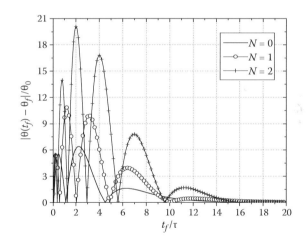

FIGURE 10.9
Normalized impact angle error of OGL induced by θ_0.

This implies that the target acquisition by a missile seeker should occur sufficiently far from the target to guarantee small terminal errors.

10.2.2 Time-to-Go Calculation for Impact Angle Control Laws

In general, time-to-go explicitly appears in the closed-form optimal guidance laws like OGL, but it cannot be directly measured by any sensor or device. Therefore, one should estimate the time-to-go for the implementation of OGLs, and accurate estimation of time-to-go is frequently very important

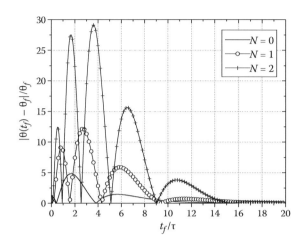

FIGURE 10.10
Normalized impact angle error of OGL induced by θ_f.

because poor estimate not only degrades the guidance performance but also makes the overall missile trajectory deviate far from the optimal one [12].

The most widely used and the simplest time-to-go calculation method is the range over the missile velocity:

$$t_{go} = \frac{R}{V}. \tag{10.37}$$

It is readily seen that Equation 10.37 is quite effective in many situations where the missile trajectory is almost straight or slightly curved. For the impact angle control laws, however, this method may not be adequate because the trajectory generated by the impact angle control laws is generally much curved. We therefore need a better time-to-go calculation method for OGL.

In this section, we devise a time-to-go calculation method for OGL where the path curvature is calculated by the approximated closed-form trajectory solution of OGL. It is based on the fact that the trajectory under OGL can be represented, in the ideal case, by some polynomial function of time-to-go, as seen in Equation 10.36, or of range-to-go, as discussed below.

As depicted in Figure 10.1, let $\bar{x}$ and $\bar{z}$ be the coordinates of the missile's future trajectory in the LOS frame at t, and $\bar{\theta}$ and $\bar{\theta}_f$ be defined as $\theta + \sigma$ and $\theta_f + \sigma$, respectively. Suppose that $\bar{z}$ is expressed as the $(N + 3)$th-order polynomials of $\bar{x}$:

$$\bar{z}(\bar{x}) = a_{N+3}\bar{x}^{N+3} + a_{N+2}\bar{x}^{N+2} + \ldots + a_1\bar{x} + a_0, \quad N = 0, 1, 2, \ldots. \tag{10.38}$$

This approximation is quite reasonable for small $\bar{\theta}$ because t_{go} in Equation 10.36 can be replaced by $\bar{x}/V$. From the first equation of Equation 10.13

$$\bar{\theta}(\bar{x}) = \frac{\dot{\bar{z}}}{V} = \frac{\dot{\bar{x}}}{V}\frac{d\bar{z}}{d\bar{x}}$$

$$\approx -\left[(N+3)a_{N+3}\bar{x}^{N+2} + (N+2)a_{N+2}\bar{x}^{N+1} + \dots a_2\bar{x} + a_1\right](\because \dot{\bar{x}} \approx -V). \tag{10.39}$$

$N + 4$ boundary conditions are required to determine the coefficients in Equations 10.38 and 10.39. For $N = 0$, there are four boundary conditions that involve the position and the flight path angle at t and at t_f:

$$\bar{z} = 0 \quad \text{at} \quad \bar{x} = R$$

$$\bar{\theta} = \bar{\theta}(t) \quad \text{at} \quad \bar{x} = R$$

$$\bar{z} = 0 \quad \text{at} \quad \bar{x} = 0 \tag{10.40}$$

$$\bar{\theta} = \bar{\theta}_f \quad \text{at} \quad \bar{x} = 0.$$

For $N \geq 1$, we need N additional conditions, which follow from the zero acceleration requirement at $t = t_f$ (or $\bar{x} = 0$), as implied by Equation 10.33:

$$\left.\frac{d^{N-1}u^*(t)}{(dt)^{N-1}}\right]_{t=t_f} = 0 \quad \Rightarrow \quad \left.\frac{d^N\bar{\theta}}{(d\bar{x})^N}\right]_{\bar{x}=0} \approx 0. \tag{10.41}$$

Resultant coefficient sets for several N's are shown in Table 10.2. Note that the trajectory approximation discussed here is available only for integer values of N.

TABLE 10.2

Coefficient Sets of Trajectory Polynomial

N	a_0	a_1	a_2	a_3	a_4	a_5
0	0	$-\bar{\theta}_f$	$\dfrac{\bar{\theta} + 2\bar{\theta}_f}{R}$	$-\dfrac{\bar{\theta} + \bar{\theta}_f}{R^2}$	-	-
1	0	$-\bar{\theta}_f$	0	$\dfrac{\bar{\theta} + 3\bar{\theta}_f}{R^2}$	$-\dfrac{\bar{\theta} + 2\bar{\theta}_f}{R^3}$	-
2	0	$-\bar{\theta}_f$	0	0	$\dfrac{\bar{\theta} + 4\bar{\theta}_f}{R^3}$	$-\dfrac{\bar{\theta} + 3\bar{\theta}_f}{R^4}$

Once the future missile trajectory is obtained, t_{go} is simply calculated by the remaining length of the curved path to the impact point over the missile speed V:

$$
\begin{aligned}
t_{go} &= \frac{1}{V} \int_0^R \sqrt{1 + \left(\frac{d\bar{z}}{d\bar{x}} \right)^2} \, d\bar{x} \\
&\approx \frac{R}{V} (1 + k).
\end{aligned}
\tag{10.42}
$$

The time-to-go given by Equation 10.42 can be interpreted as the range over missile velocity compensated by the factor k that accounts for the length increase due to the path curvature. With the approximation given by

$$
\sqrt{1 + \left(\frac{d\bar{z}}{d\bar{x}} \right)^2} \approx 1 + \frac{1}{2} \left(\frac{d\bar{z}}{d\bar{x}} \right)^2,
\tag{10.43}
$$

the typical values of k for each N are found to be

$$
k = \begin{cases}
\left(2\bar{\theta}^2(t) - \bar{\theta}_f \bar{\theta}(t) + 2\bar{\theta}_f^2 \right) / 30 & \text{for} \quad N = 0 \\
\left(3\bar{\theta}^2(t) - 2\bar{\theta}_f \bar{\theta}(t) + 12\bar{\theta}_f^2 \right) / 70 & \text{for} \quad N = 1 \\
\left(4\bar{\theta}^2(t) - 3\bar{\theta}_f \bar{\theta}(t) + 36\bar{\theta}_f^2 \right) / 126 & \text{for} \quad N = 2.
\end{cases}
\tag{10.44}
$$

The proposed time-to-go calculation method does not require any special information beyond usual measurement requirements for guidance law implementation. We note that the proposed method could produce large time-to-go calculation error in the beginning of the flight when $|\theta_0| = \pi/2$, where the approximation $|d\bar{z}/d\bar{x}| \ll 1$ is not quite valid. However, this time-to-go calculation error becomes negligible as the missile approaches the target since $\bar{\theta}$ and $\bar{\theta}_f$ are going to be nullified regardless of the values of θ_f.

10.2.3 Implementing OGL: First Variant

The OGL given by Equation 10.28 yields the bad performance for some applications where the small angle approximation for linearization is not quite valid. For the simplicity of the discussion, let the time-to-go be calculated by the simplest way, that is, range over missile speed. Then the OGL of Equation 10.28 can be rewritten as

$$u^* = -\frac{V^2}{R}\left[N_z\frac{z(t)}{R} + N_\theta\theta(t) + N_f\theta_f\right].$$ (10.45)

For $N > 0$, u^* is supposed to go to 0 as the missile approaches the target. We see from Equations 10.30 and 10.45 that this zero command at the impact instant is achieved with $\theta \to \theta_f$ and $z/R \to -\theta_f$ as $t \to t_f$. Now note that for a larger impact angle, for instance, one near 90°, $|z|$ is almost the same as R during the final phase of homing, which in turn implies that z/R approaches not $-\pi/2$ but -1 as $t \to t_f$. Thus, the guidance command in this case does not vanish as the missile approaches the target and might yield a considerable amount so that huge guidance error could occur. This problem can be avoided by selecting the guidance coordinate frame in such a way that θ_f is small; however, such a choice of guidance frame could develop large θ at other time regions of engagement, violating the small θ assumption and yielding nonoptimal behavior there. Nevertheless, large deviation of θ in the final phase of engagement from the assumption underlying our optimal guidance formulation is far worse, in terms of guidance error, than large digress in the earlier phase.

There is other way to resolve the problem where the performance of the guidance law (Equation 10.28) is made insensitive to the engagement geometry (e.g., large impact angle) by using $\sigma(t)$ instead of $z(t)$ as a guidance variable in the law. Note that when the sight line angle $\sigma(t)$ is small, it can be approximated by

$$\sigma(t) = -\sin^{-1}\frac{z}{R} \approx -\frac{z}{Vt_{go}}$$ (10.46)

and

$$\dot{z} = V\sigma - Vt_{go}\dot{\sigma} \quad\Rightarrow\quad \theta = \sigma - t_{go}\dot{\sigma}.$$ (10.47)

Equation 10.28 is now rewritten as

$$u^*(t) = -\frac{V}{t_{go}}[-N_z\sigma(t) + N_\theta\theta(t) + N_f\theta_f].$$ (10.48)

For convenience, OGL given by Equation 10.48 is called the first variant of OGL. Note that all variables in Equation 10.48 are angular variables, and the equation does not have the problem discussed above, even in the case of a large θ_f. In this sense, the first variant of OGL is more robust to engagement scenarios and has wider applications.

Nonlinear Simulations

To examine the performance of the first variant of OGL and the time-to-go calculation method proposed in the previous section, we perform nonlinear simulations. Here, we are to compare the first variant of OGL with another well-known impact angle control law, the biased PN guidance (BPNG) proposed in [13], which is simple to implement and does not use time-to-go.

BPNG in [13] is given by

$$u_{BPN}(t) = N'V(\dot{\sigma}(t) - \dot{\sigma}_b(t)) \qquad (10.49)$$

where N' is a navigation constant, $\dot{\sigma}(t)$ is the LOS rate, and $\dot{\sigma}_b$ is the time-varying bias term to control the impact angle. $\dot{\sigma}_b$ is defined by

$$\dot{\sigma}_b(t) = \frac{\eta V(\sigma_d - \sigma(t))}{N'R(t)\cos(\theta(t) - \sigma(t))}. \qquad (10.50)$$

Here, η is an arbitrary positive constant, and σ_d is the desired LOS angle at the time of impact for which we use θ_f in our simulation study. It is noted from Equation 10.50 that BPNG is singular when $\theta(t) = \sigma(t)$, and thus, the capture region and launch envelope of BPNG are more limited than the first variant of OGL. The most attractive feature of BPNG is that it does not require the knowledge of time-to-go. We choose $N = 4.0$ and $\eta = 1.3$ for BPNG as used in [13] for demonstration.

Two distinct cases of nonlinear simulations are considered—case 1: $\theta_0 = \theta_f = 90°$ and case 2: $\theta_0 = 45°$, $\theta_f = -45°$. Initial conditions are shown in Table 10.3. It is assumed that the speed of the missile is constant, there is no command limit, and the target is fixed.

Figures 10.11 and 10.12 show that the missile under OGL travels a longer path for a larger value of N, and the time of flight is increased accordingly, as shown in Figures 10.13 and 10.14. It is also shown in Figures 10.13 and 10.14 that, for $N = 1$ and 2, the guidance command is large in the beginning of the flight and reduces to 0 as the missile approaches the target. However, for $N = 0$, the guidance command reaches a finite value, as expected in Equation 10.33, that depends on θ_0, r_0, and θ_f. Although the guidance law with $N = 0$ represents the optimum energy trajectory with the smallest flight time, this

TABLE 10.3

Initial Conditions for Nonlinear Simulations

Parameters	Values
Missile position (x_0, z_0)	0 m, 0 m
Missile velocity (V)	200 m/sec
Target position (x_t, z_t)	0 m, 4000 m

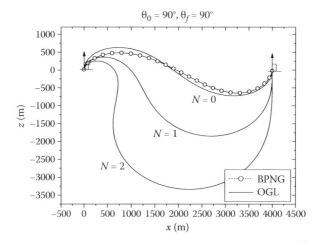

FIGURE 10.11
Trajectories produced by OGL for case 1.

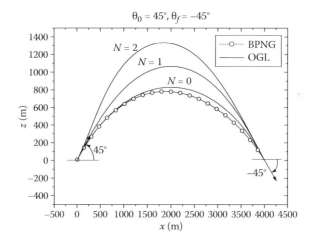

FIGURE 10.12
Trajectories produced by OGL for case 2.

property of nonzero terminal acceleration implies that it may produce larger miss distances when there are disturbances, acceleration command limits, and some uncertainties not considered beforehand.

We observe from Figures 10.13 and 10.14 that the guidance command of BPNG tends to blow up as the missile approaches the target. Thus, considerable terminal miss distance and impact angle error could result for BPNG if some command limit is imposed. It is also interesting to note that BPNG

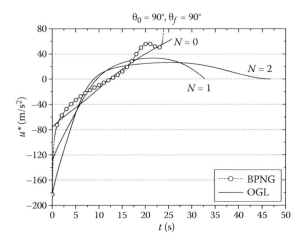

FIGURE 10.13
Guidance command histories of OGL for case 1.

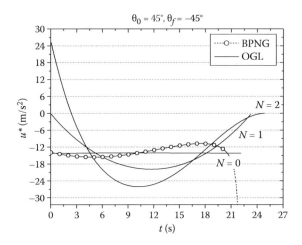

FIGURE 10.14
Guidance command histories of OGL for case 2.

command profile is quite similar to that of OGL with $N = -0.5$, though the latter is not shown here.

Figures 10.15 and 10.16 show the time histories of the estimated total time of flight $t_f(t)$, which is computed by the method proposed in Section 10.2.2. Note that quite a large t_f estimation error is produced for case 1 in the beginning of the flight, which occurs because small value approximation of $|d\bar{z}/d\bar{x}|$ is not valid. For $|\theta_0| \leq 60°$, the proposed time-to-go method provides good estimates for all N, as shown in Table 10.4. Here, the performance measure

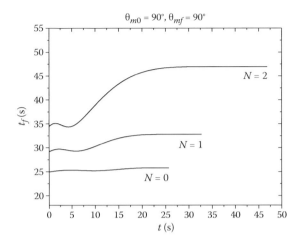

FIGURE 10.15
Estimated time-of-flight histories of OGL for case 1.

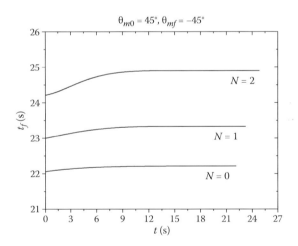

FIGURE 10.16
Estimated time-of-flight histories of OGL for case 2.

$\Delta t_{fmax}(t)/t_f(t_f)$ is the normalized maximum estimation error of t_f, where $t_f(t) = t_{go} + t$ and $\Delta t_{fmax} = \max_{0 \le t \le t_f}\{t_f(t) - t_f(t_f)\}$. Table 10.4 shows that the proposed method provides far better estimates of the time-to-go than the conventional method of $t_{go} = R/V$. The negative values in this table mean the underestimates of the time-to-go, which cause the guidance command to get larger than needed. No error sources, such as the system lag or the command saturation, are taken into account in these simulations, and therefore, no terminal errors (miss distances and impact angle errors) have been developed.

TABLE 10.4

Time-to-Go Calculation Errors, $\Delta t_{fmax}/t_f(\%)$, for Lag-Free Autopilot

θ_0 (degrees)	θ_f (degrees)	R/V			Proposed Method		
		$N = 0$	$N = 1$	$N = 2$	$N = 0$	$N = 1$	$N = 2$
90	90	−21.30	−39.21	−59.25	−3.28	−11.13	−26.52
−90	90	−31.86	−46.37	−63.66	−8.94	−18.32	−34.41
60	60	−10.24	−19.00	−29.44	−0.72	−2.44	−5.93
−60	60	−16.20	−23.63	−32.85	−2.07	−4.29	−8.20
30	30	−2.69	−5.00	−7.87	−0.05	−0.16	−0.40
−30	30	−4.43	−6.46	−9.05	−0.14	−0.30	−0.57

10.2.4 OGL for Moving Target: Second Variant

Recall that OGL given by Equation 10.28 or its first variant given by Equation 10.48 is a guidance law for a stationary target and does not give a satisfactory performance when the target is moving. We thus consider in this section an optimal guidance problem for a moving target in the same framework as in the previous sections.

Let the target travel with a constant velocity V_T in the direction represented by the constant angle θ_T (Figure 10.17).

Let θ_{MT} be the angle that satisfies

$$V \sin\theta_{MT} = V_T \sin\theta_T. \tag{10.51}$$

Note that if the missile maintains its flight direction $\theta(t)$ to be equal to θ_{MT} throughout the engagement and if $z(0) = 0$, then the trajectories of the missile and the target together with the initial sight line form a collision triangle.

Now let $\theta = \theta_{MT} + \Delta\theta$. Then

$$\begin{aligned}
\dot{z} &= V \sin\theta - V_T \sin\theta_T \\
&= V \sin(\theta_{MT} + \Delta\theta) - V_T \sin\theta_T \\
&= V \sin\theta_{MT} \cos\Delta\theta + V \cos\theta_{MT} \sin\Delta\theta - V_T \sin\theta_T.
\end{aligned} \tag{10.52}$$

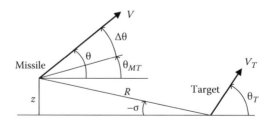

FIGURE 10.17
Interception geometry for a moving target.

Assuming $\Delta\theta$ is small and letting $\bar{V} = V\cos\theta_{MT}$, we have

$$\dot{z} \approx V\Delta\theta\cos\theta_{MT} = \bar{V}\Delta\theta. \tag{10.53}$$

Also, if we let $\bar{u} = u\cos\theta_{MT}$, then

$$\bar{V}\Delta\dot{\theta} = V\Delta\dot{\theta}\cos\theta_{MT} = u\cos\theta_{MT} = \bar{u}. \tag{10.54}$$

Thus, we have the following set of equations:

$$\dot{z} = \bar{V}\Delta\theta$$
$$\bar{V}\Delta\dot{\theta} = \bar{u}. \tag{10.55}$$

Remark. $V\sin\theta_{MT}$ given in Equation 10.51 is the missile velocity component to counter the target movement and to keep the missile near the target centered reference line, and $\Delta\theta$ can be considered to be the effective control used to form the missile trajectory as required around the reference line with the impact angle θ_f. Thus, if the reference line is chosen in such a way that θ_{MT} is not far from θ_f, then $\Delta\theta$ can be assumed to be small since θ should approach θ_f as time goes on when the applied guidance law works. Choice of such a reference line is always possible; take for example the base of zero-effort collision triangle defined in [14] for the case of impact angle constraint.

Note that Equation 10.55 is exactly the same set of equations as Equation 10.13 except for different symbols of variables. Thus, we can formulate the same optimal impact angle control guidance problem as in Section 10.2.1 with the terminal constraints $z(t_f) = 0$ and $\Delta\theta(t_f) = \theta(t_f) - \theta_{MT} = \theta_f - \theta_{MT}$ and obtain in a straightforward manner the solution given by

$$\bar{u}^* = -\frac{\bar{V}}{t_{go}^2}\left[\frac{N_z}{\bar{V}}z + N_\theta t_{go}\Delta\theta + N_f t_{go}(\theta_f - \theta_{MT})\right] \tag{10.56}$$

or

$$u^* = -\frac{V}{t_{go}^2}\left[\frac{N_z z}{V\cos\theta_{MT}} + N_\theta t_{go}(\theta - \theta_{MT}) + N_f t_{go}(\theta_f - \theta_{MT})\right]$$
$$= -\frac{V}{t_{go}}\left[N_z\left(\frac{z}{Vt_{go}\cos\theta_{MT}} - \theta_{MT}\right) + N_\theta\theta + N_f\theta_f\right]. \tag{10.57}$$

To get the second equality, we have used the relationship between the guidance gains given in Equation 10.30. Note that if the target is stationary, that is, $V_T = 0$, then $\theta_{MT} = 0$ so that Equation 10.57 reduces to Equation 10.28.

If the target is indeed flying straight with constant velocity, we can implement Equation 10.57 with θ_{MT} computed by using the information from a target tracker, even before the missile launch. In practice, however, we need to consider the case of target maneuvering and thus constantly estimate somehow θ_{MT} during the engagement. An alternative way is to use other variables that can reflect the effect of varying θ_{MT} or, equivalently, of the target maneuver.

Note that $z \approx -R\sigma$ for small σ so that $\dot{z} \approx -\dot{R}\sigma - R\dot{\sigma}$. Hence, from Equation 10.53, we have

$$V \cos\theta_{MT}\Delta\theta \approx -\dot{R}\sigma - R\dot{\sigma}. \tag{10.58}$$

Thus

$$\Delta\theta \approx \frac{-\dot{R}\sigma - R\dot{\sigma}}{V \cos\theta_{MT}} \tag{10.59}$$

or

$$\theta_{MT} \approx \theta + \frac{\dot{R}\sigma + R\dot{\sigma}}{V \cos\theta_{MT}}. \tag{10.60}$$

Substituting Equation 10.60 into Equation 10.57, we have

$$
\begin{aligned}
u^* &\approx -\frac{V}{t_{go}}\left[N_z\left(\frac{-R\sigma}{Vt_{go}\cos\theta_{MT}} - \theta - \frac{\dot{R}\sigma + R\dot{\sigma}}{V\cos\theta_{MT}} \right) + N_\theta\theta + N_f\theta_f \right] \\
&= \frac{N_z}{\cos\theta_{MT}}\left(\frac{R\sigma}{t_{go}^2} + \frac{\dot{R}\sigma + R\dot{\sigma}}{t_{go}} \right) + \frac{N_f V}{t_{go}}(\theta - \theta_f) \\
&\approx -\frac{N_z\dot{R}}{\cos\theta_{MT}}\dot{\sigma} + \frac{N_f V}{t_{go}}(\theta - \theta_f)
\end{aligned} \tag{10.61}
$$

where we have used the relationship in Equation 10.30 and $R \approx -\dot{R}t_{go}$ to get the second and third equalities.

Now that

$$\cos\theta_{MT} = \sqrt{1 - \frac{V_T^2}{V^2}(1 - \cos^2\theta_T)} \tag{10.62}$$

from Equation 10.51, $\cos\theta_{MT} \approx 1$ when $V \gg V_T$, which is the case for most realistic engagement scenarios. Thus, we now have an impact angle control guidance law for a moving target given by

$$u^* \approx -N_z \dot{R}\dot{\sigma} + \frac{N_f V}{t_{go}}(\theta - \theta_f). \tag{10.63}$$

We note that Equation 10.63 is a biased true PN guidance law [15].

In implemental aspect, Equation 10.63 is more viable than Equation 10.57; $\dot{\sigma}$ and $\dot{R}$ can be obtained directly from a typical microwave seeker and θ from an on-board inertial sensor of the missile, and t_{go} is the only variable that the guidance computer has to estimate. The situation, however, is somewhat different when a passive sensor such as an infrared seeker is employed; in this case, the seeker does not provide the information of $\dot{R}$, and we need to approximate $\dot{R}$ using some easily obtainable physical variables.

Note that

$$\begin{aligned}
\dot{R} &= -V\cos(\theta - \sigma) + V_T \cos(\theta_T - \sigma) \\
&= -V\cos(\theta_{MT} + \Delta\theta - \sigma) + V_T \cos(\theta_T - \sigma) \\
&\approx -V\cos\theta_{MT} + V_T \cos\theta_T \\
&= -V\cos\theta_{MT}\left(1 - \frac{V_T \cos\theta_T}{V\cos\theta_{MT}}\right)
\end{aligned} \tag{10.64}$$

when σ and $\Delta\theta$ are small.

Using Equations 10.61 and 10.64, we have

$$u^* \approx N_z V\left(1 - \frac{V_T \cos\theta_T}{V\cos\theta_{MT}}\right)\dot{\sigma} + \frac{N_f V}{t_{go}}(\theta - \theta_f) \tag{10.65}$$

which in turn reduces, when $V \gg V_T$, to an alternative form of Equation 10.63:

$$u^* \approx N_z V\dot{\sigma} + \frac{N_f V}{t_{go}}(\theta - \theta_f). \tag{10.66}$$

The impact angle control guidance laws (Equations 10.63 and 10.66) have been obtained from Equation 10.61 under the assumption that the target speed is slow enough when compared with the missile speed—as a matter of fact, Equations 10.63 and 10.66 can be derived by manipulating carefully the OGL (Equation 10.48) for a stationary target.

Comparing Equation 10.63 or 10.66 with Equation 10.61, we also immediately note that the only difference between these guidance laws lies in the gains of $\dot{\sigma}$ and expect that the impact angle control guidance law of the form of Equation 10.63 or 10.66 will give moderate guidance performance against a maneuvering target as well as a constant-speed nonmaneuvering target. Finally, we note that Equation 10.63 or 10.66 is now independent of the choice of the reference line as is Equation 10.48.

Figures 10.18 through 10.21 show the results of the nonlinear engagement simulation for intercepting a constant-speed nonmaneuvering target with

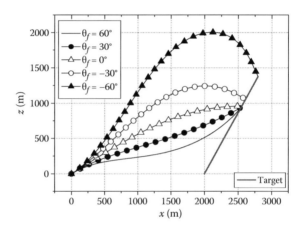

FIGURE 10.18
Missile/target trajectories (OGL for constant-velocity target).

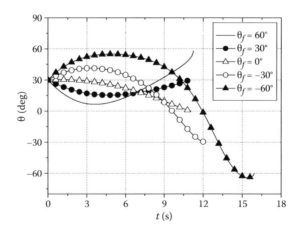

FIGURE 10.19
Missile flight path angle histories (OGL for constant-velocity target).

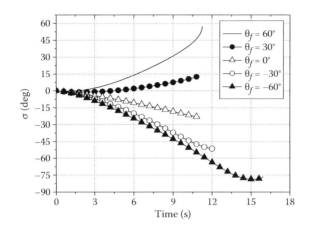

FIGURE 10.20
LOS angle histories (OGL for constant-velocity target).

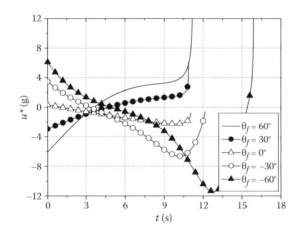

FIGURE 10.21
Missile guidance command histories (OGL for constant-velocity target).

the various impact angles. The second variant of OGL given by Equation 10.66 with $N = 0$ has been applied. Here, we used the time-to-go computation method given by Equation 10.42 with Equation 10.44 in Section 10.2.2. The speed of the missile is 250 m/s, and the target travels at the constant speed of 100 m/s in the direction of 120°. From Figure 10.19, we observe that all the impact angle requirements are satisfied. We also note from Figure 10.20 that the LOS angles are not coincident with the flight path angles at the impact instant; this is because the target is moving.

Figures 10.22 through 10.25 show the results of the nonlinear simulation for a maneuvering target with −2g. Other simulation conditions are the same as in the previous constant-velocity-target scenario. All the specified impact angles have been realized, as seen in Figure 10.23. It is also interesting to note from Figure 10.25 that the guidance commands abruptly change and tend to blow up in the later phase due to target maneuver that the guidance law does not take account of.

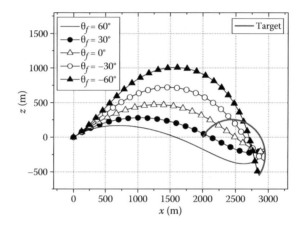

FIGURE 10.22
Missile/target flight trajectories (OGL for maneuvering target).

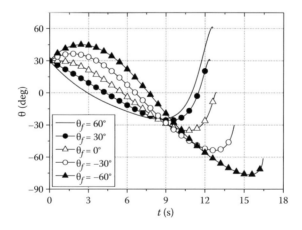

FIGURE 10.23
Missile flight path angle histories (OGL for maneuvering target).

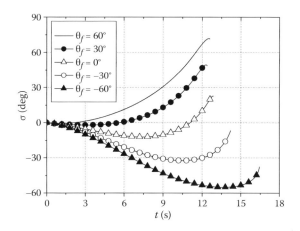

FIGURE 10.24
LOS angle histories (OGL for maneuvering target).

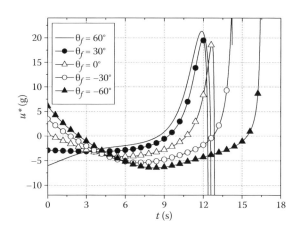

FIGURE 10.25
Missile guidance command histories (OGL for maneuvering target).

10.3 OGL for First-Order Lag Missile (OGL/1)

In the real world, the missile response is not perfect and should be modeled by a higher-order dynamics together with various nonlinear components in order to represent its behavior with high degree of fidelity. However, we rarely consider all the complexities of a missile model when developing an OGL simply because it is impossible or too complicated to get a closed-form solution and because the consideration of the first-order missile model provides in many cases sufficient improvement of performance over the guidance laws based on the assumption of perfect autopilot or lag-free missile response. In this section, we will derive the optimal impact angle control guidance law for a first-order missile model and discuss its performance.

Assume that the missile can be approximated by a first-order lag system with time constant τ, that is,

$$\frac{a(s)}{u(s)} = \frac{1}{\tau s + 1}. \tag{10.67}$$

Here, a is the achieved acceleration in response to the guidance command u and normal to the missile velocity as shown in Figure 10.1.

The equations of motion for a homing problem to intercept a stationary target are given by

$$\dot{z}(t) = V(t)\sin\theta(t), \quad z(0) = 0$$
$$V(t)\dot{\theta}(t) = a(t), \quad \theta(0) = \theta_0. \tag{10.68}$$

Under the assumption that V is constant and θ is small, we linearize Equation 10.68 as

$$\dot{z}(t) \approx V\theta(t)$$
$$\dot{v}(t) = a(t) \tag{10.69}$$

where

$$\dot{v}(t) \approx V\theta(t). \tag{10.70}$$

By augmenting Equation 10.67 to Equation 10.69, we obtain the linear differential equation

$$\dot{\xi} = A\xi + Bu, \quad \xi(0) = \xi_0 \tag{10.71}$$

where

$$\xi = [z \quad v \quad a]^T, \quad \xi_0 = [0 \quad V\theta_0 \quad 0]^T \tag{10.72}$$

and

$$A = \begin{bmatrix} 0 & 1 & 0 \\ 0 & 0 & 1 \\ 0 & 0 & -1/\tau \end{bmatrix}, \quad B = \begin{bmatrix} 0 \\ 0 \\ 1/\tau \end{bmatrix}. \tag{10.73}$$

Now we consider the LQ optimal control problem: find u that minimizes the cost

$$J = \frac{1}{2}\int_0^{t_f} R(\tau)u^2(\tau)d\tau \tag{10.74}$$

subject to Equation 10.71 and the terminal constraints

$$D\xi(t_f) = E \tag{10.75}$$

where

$$R = 1/t_{go}^N, \; N \geq 0 \tag{10.76}$$

and

$$D = \begin{bmatrix} 1 & 0 & 0 \\ 0 & 1 & 0 \end{bmatrix}, E = \begin{bmatrix} 0 \\ V\theta_f \end{bmatrix}. \tag{10.77}$$

The state-feedback solution (OGL/1) to this optimal control problem can also be computed by the same method as used in the previous section and is given below:

$$u^* = \frac{t_{go}^N}{\tau(g_{11}g_{22} - g_{12}g_{21})}\Big[(A_1 f_{11} + A_2 f_{12})z(t) + (A_1 f_{21} + A_2 f_{22})V\theta(t)$$

$$+ (A_1 f_{31} + A_2 f_{32})a(t) - A_2 V\theta_f\Big] \tag{10.78}$$

where

$$A_1 = f_{31}g_{22} - f_{32}g_{21}$$

$$A_2 = f_{32}g_{11} - f_{31}g_{12} \tag{10.79}$$

and

$$f_{11} = f_{22} = 1, \; f_{12} = 0$$

$$f_{21} = t_{go}$$

$$f_{31} = \tau t_{go} + \tau^2 e^{-t_{go}/\tau} - \tau^2$$

$$f_{32} = \tau\left(1 - e^{-t_{go}/\tau}\right) \tag{10.80}$$

$$g_{11} = -\frac{\tau^2}{N+1}t_{go}^{N+1} + \frac{2\tau}{N+2}t_{go}^{N+2} - \frac{1}{N+3}t_{go}^{N+3}$$

$$+2\tau e^{-t_{go}/\tau}\left[\sum_{i=0}^{N+1}\frac{(N+1)!}{(N+1-i)!}\tau^{i+1}t_{go}^{N+1-i} - \sum_{i=0}^{N}\frac{N!}{(N-i)!}\tau^{i+2}t_{go}^{N-i}\right] \tag{10.81}$$

$$+\tau^2 e^{-2t_{go}/\tau}\sum_{i=0}^{N}\frac{N!}{(N-i)!}\left(\frac{\tau}{2}\right)^{i+1}t_{go}^{N-i} - \tau^{N+3}N!\left[\left(\frac{1}{2}\right)^{N+1} + 2N\right]$$

$$g_{12} = g_{21} = \frac{\tau}{N+1}t_{go}^{N+1} - \frac{1}{N+2}t_{go}^{N+2}$$

$$+e^{-t_{go}/\tau}\left[2\sum_{i=0}^{N}\frac{N!}{(N-i)!}\tau^{i+2}t_{go}^{N-i} - \sum_{i=0}^{N+1}\frac{(N+1)!}{(N+1-i)!}\tau^{i+1}t_{go}^{N+1-i}\right] \tag{10.82}$$

$$-\tau e^{-2t_{go}/\tau}\sum_{i=0}^{N}\frac{N!}{(N-i)!}\left(\frac{\tau}{2}\right)^{i+1}t_{go}^{N-i} + \tau^{N+2}N!\left[\left(\frac{1}{2}\right)^{N+1} + N - 1\right]$$

$$g_{22} = -\frac{1}{N+1}t_{go}^{N+1} - \tau^{N+1}N!\left[\left(\frac{1}{2}\right)^{N+1} - 2\right]$$

$$\tag{10.83}$$

$$-2e^{-t_{go}/\tau}\sum_{i=0}^{N}\frac{N!}{(N-i)!}\tau^{i+1}t_{go}^{N-i} + e^{-2t_{go}/\tau}\sum_{i=0}^{N}\frac{N!}{(N-i)!}\left(\frac{\tau}{2}\right)^{i+1}t_{go}^{N-i}.$$

We can easily show that OGL/1 becomes OGL as $\tau \to 0$. Using Equation 10.46, we obtain a variant of OGL/1 given by

$$u^* = \frac{Vt_{go}^N}{\tau(g_{11}g_{22} - g_{12}g_{21})}\left[(A_1f_{11} + A_2f_{12})t_{go}\sigma(t) + (A_1f_{21} + A_2f_{22})\theta(t)\right.$$

$$\tag{10.84}$$

$$\left. + \frac{A_1f_{31} + A_2f_{32}}{V}a(t) - A_2\theta_f\right].$$

If the missile indeed has the first-order response lag with known time constant τ, OGL/1 gives the best performance, as shown in Figures 10.26 and 10.27. (Here, $\tau = 1$ s is chosen.) As $t_{go} \to 0$ (or $t \to t_f$), the guidance commands converge to 0 regardless of the choice of N and the initial conditions; this

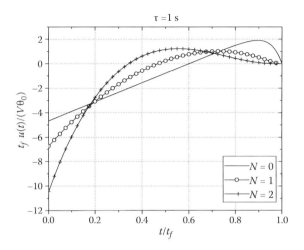

FIGURE 10.26
Normalized guidance command histories of OGL/1 for $\theta_f = 0$.

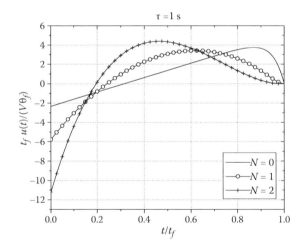

FIGURE 10.27
Normalized guidance command histories of OGL/1 for $\theta_0 = 0$.

result is worthwhile to note since OGL with $N = 0$ yields nonzero guidance command at the time of impact (Figures 10.2 and 10.3). We also observe that similar to PN law, larger initial guidance command is produced when larger N is selected.

If the missile responds exactly as the first-order lag model does, zero miss distance and zero impact angle error will be achieved by OGL/1. In practice, however, missile dynamics will differ from a simple first-order system, and some miss distance and impact angle error will occur. Nevertheless, if a missile has a large time lag in its response, OGL/1 will yield much smaller terminal guidance errors than OGL.

10.4 Energy Optimal Waypoint Guidance

In this section, an energy optimal waypoint guidance problem is addressed as an application of the impact angle control guidance laws. Consider a typical antiship missile (ASM) performing sea skimming at a prescribed altitude in order to enhance survivability. Careful path planning for the midcourse guidance of the ASM is then important to avoid obstacles and threats, such as islands and air defense systems of hostile forces; to minimize the chance of engaging in friendly forces; and so on. As depicted in Figure 10.28, the path planning of such ASMs can be easily accomplished by assigning appropriate waypoints on the way to the target. The midcourse guidance of an ASM is then to make ASM fly through those multiple waypoints in the given order and in an optimal fashion. Typical missile guidance laws can be used to guide the missile from one waypoint to another, regarding the second waypoint as a target. However, unlike the case of homing to a target, it is important at the time of hitting a waypoint to make the missile heading optimal for the guidance to the next waypoints. Again, the impact angle control guidance plays an important role here.

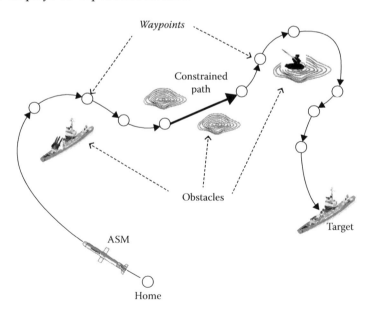

FIGURE 10.28
Path planning example of ASM missions.

In this section, a planar waypoint guidance method for ASMs or unmanned aerial vehicles (UAVs) is proposed, where the pure energy optimal guidance law, OGL with $N = 0$, is used in between the two waypoints. We will look at the energy optimal trajectory optimization problem with waypoint constraints and show that it can be reduced to an unconstrained optimization problem of finding the optimal boundary conditions at waypoints; the boundary conditions thus obtained are then achieved by the energy OGL. The proposed method generates the energy optimal trajectory by straight-forward computation. Indeed, if the ASM is lag-free, optimal boundary conditions become waypoint passing angles, which can be simply determined from a set of linear algebraic equations. Since there are no time-consuming numerical optimizations in this approach, the energy optimal trajectory passing through all the waypoints can be easily computed in real time.

10.4.1 Equivalence of Optimal Control Problems

Consider N waypoints that will be visited by the ASM in a given order, as shown in Figure 10.29. We define the flight segment between the $(i - 1)$th waypoint and the ith waypoint as the ith segment. The position of the ith waypoint is denoted by (x_i, z_i), and the passing angle at that waypoint is given by θ_i. The position and the flight path angle of the ASM at t are denoted by $(x(t), z(t))$, and $\theta(t)$, respectively.

Now, consider the following energy optimal control problem.

OCP-1: Find $u(t)$ that minimizes

$$J = \int_0^{t_f} u^2(t)\,dt = \sum_{i=1}^{N} \int_{t_{i-1}}^{t_i} u^2(t)\,dt \tag{10.85}$$

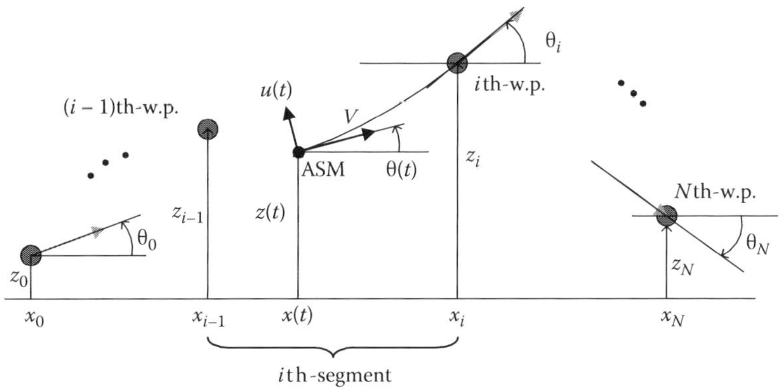

FIGURE 10.29
Geometry of planar waypoint guidance.

subject to

$$\dot{x} = V \cos\theta$$

$$\dot{z} = V \sin\theta \qquad (10.86)$$

$$\dot{\theta} = u/V$$

with the constraints

$$x(t_i) = x_i \text{ and } z(t) = z_i, \text{ for } i = 0,1,\ldots,N, \qquad (10.87)$$

where t_i is defined as the time when the *ASM reaches the i*th waypoint, and $t_0 = 0$ *and* $t_N = t_f$.

In OCP-1, different optimal control will be obtained depending on the values of t_i if t_i is fixed. Hence, t_i in OCP-1 must not be fixed except $i = 0$ and $i = N$ in order to obtain the optimal control that globally minimizes the cost.

Let the cost of each segment in Equation 10.85 be defined by

$$J_i = \int_{t_{i-1}}^{t_i} u^2(t)\,dt \quad \text{for} \quad i = 1,2,\ldots,N, \qquad (10.88)$$

and $u^*(t)$ be the solution to OCP-1. The minimum cost J^* is then

$$J^* = \sum_{i=1}^{N} J_i^* \qquad (10.89)$$

where

$$J_i^* = \int_{t_{i-1}}^{t_i} [u^*(t)]^2\,dt \quad \text{for} \quad i = 1,2,\ldots,N. \qquad (10.90)$$

Consider the following optimal control problem restricted to the *i*th segment.

OCP-2: For the given time interval $[t_{i-1}, t_i]$, find $\bar{u}(t)$ that minimizes

$$\bar{J}_i = \int_{t_{i-1}}^{t_i} \bar{u}^2(t)\,dt \qquad (10.91)$$

subject to Equation 10.86 with the initial condition

$$x(t_{i-1}) = x_{i-1}, \; z(t_{i-1}) = z_{i-1}, \text{ and } \theta(t_{i-1}) = \theta_{i-1} \qquad (10.92)$$

and the terminal constraints

$$x(t_i) = x_i,\ z(t_i) = z_i,\ \text{and}\ \theta(t_i) = \theta_i. \tag{10.93}$$

Let $\bar{u}^*(t)$ be the solution to OCP-2; the minimum cost $\bar{J}_i^*$ is then obtained as

$$\bar{J}_i^* = \int_{t_{i-1}}^{t_i} [\bar{u}^*(t)]^2\,dt. \tag{10.94}$$

Note that OCP-2 is another formulation of the energy-optimal guidance problem with terminal constraint on the impact angle.

We define the sets of the waypoint passing times and angles as follows:

$$T = \{t_i,\ i = 0,1,\ldots,N\} \tag{10.95}$$

$$\Theta = \{\theta_i,\ i = 0,1,\ldots,N\}. \tag{10.96}$$

Now suppose that the solution to OCP-1 is known. The optimal waypoint passing angle at the ith waypoint is denoted by $\theta_i^* = \theta(t_i^*)$, where t_i^* implies the optimal waypoint passing time. Let

$$T^* = \{t_i^*,\ i = 0,1,\ldots,N\} \tag{10.97}$$

$$\Theta^* = \{\theta_i^*,\ i = 0,1,\ldots,N\}. \tag{10.98}$$

The following theorem provides the relationship between OCP-1 and OCP-2.

Theorem 10.1

If $\Theta = \Theta^*$ and $T = T^*$, then $\bar{J}_i^* = J_i^*$ for $i = 1,\ldots,N$. Moreover, $\bar{u}^*(t) = u^*(t)$ for $t \in [t_{i-1}^*, t_i^*]$. ∎

Proof

For the Nth segment, from the principle of optimality [3], J_N^* with $u^*(t)$ is the minimum cost for OCP-1. From the assumption, the initial conditions and terminal constraints of OCP-2 are identical to the boundary conditions of OCP-1 for the Nth segment. If $\bar{u}^*(t) \neq u^*(t)$, there exists another optimal control for OCP-2. This contradicts that $\bar{u}^*(t)$ is the optimal control for OCP-2. Hence, $\bar{u}^*(t) = u^*(t)$ for $t \in [t_{N-1}^*, t_N^*]$ and $\bar{J}_N^* = J_N^*$. For $[t_{N-2}^*, t_N^*]$, also from the principle of optimality, we see that $J_{N-1}^* + J_N^*$ is the minimum cost. Therefore, $\bar{J}_{N-1}^* + \bar{J}_N^* \leq \bar{J}_{N-1}^* + \bar{J}_N^*$. Since $\bar{J}_N^* = J_N^*$, we have $J_{N-1}^* \leq \bar{J}_{N-1}^*$. This contradicts $\bar{J}_{N-1}^*$ being the minimum cost of OCP-2 if $\theta_i = \theta_i^*$ and $t_i = t_i^*$ for $i = N-2, N-1$. Hence, $J_{N-1}^* = \bar{J}_{N-1}^*$ and $\bar{u}^*(t) = u^*(t)$ for $t \in [t_{N-2}^*, t_{N-1}^*]$. By repeating the procedure up to the first segment, we prove the theorem. ∎

Theorem 10.1 implies that the optimal trajectory passing through all the waypoints can be obtained as a family of the independent OCP-2's solutions for each segment if Θ^* and T^* are known. We do not consider here the case where specific waypoint passing angles are prescribed; in this case, it can easily be seen that Theorem 10.1 still holds.

Now suppose that $\bar{u}_i^*(t)$, the optimal control of OCP-2, is given in a state-feedback form:

$$\bar{u}_i^*(t) := \Gamma(t, x(t), z(t), \Theta(t); \theta_i) \quad \text{for} \quad t \in [t_{i-1}, t_i].$$

(10.99)

Under the assumption that such a state-feedback law exists, we consider the following parameter optimization problem:

POP-1: For the given t_i, find θ_i for $i = 0,1,\ldots,N$, which minimizes

$$
\begin{aligned}
\tilde{J} &= \sum_{i=1}^{N} \int_{t_{i-1}}^{t_i} [\bar{u}_i^*(t)]^2 \, dt \\
&= \sum_{i=1}^{N} \int_{t_{i-1}}^{t_i} [\Gamma(t, x(t), z(t), \Theta(t); \theta_i)]^2 \, dt
\end{aligned}
$$

(10.100)

subject to Equation 10.86.

POP-1 is just the problem of finding the optimal waypoint passing angles when the state-feedback optimal control law of OCP-2 is applied for each flight segment. The following theorem states the relationship between POP-1 and OCP-1.

Theorem 10.2

For a given T^*, let $\tilde{\Theta}^* = \{\tilde{\theta}_i^*, i = 0,1,\cdots,N\}$ be the solution to POP-1, and the minimum cost $\tilde{J}^*$ be given by

$$\tilde{J}^* = \sum_{i=1}^{N} \int_{t_{i-1}^*}^{t_i^*} [\Gamma(t, x(t), z(t), \Theta(t); \tilde{\theta}_i^*)]^2 \, dt.$$

(10.101)

Then, $\tilde{\Theta}^* = \Theta^*$ and $\tilde{J}^* = J^*$.   ■

Proof

Suppose that $\tilde{\Theta}^* \neq \Theta^*$. Then

$$\tilde{J}^* < \sum_{i=1}^{N} \int_{t_{i-1}^*}^{t_i^*} [\Gamma(t, x(t), z(t), \theta(t); \theta_i^*)]^2 \, dt. \tag{10.102}$$

Recall that Γ is the state-feedback optimal control law obtained from OCP-2. By Theorem 10.1, if Θ^* is used for OCP-2, we have

$$\sum_{i=1}^{N} \int_{t_{i-1}^*}^{t_i^*} [\Gamma(t, x(t), z(t), \theta(t); \theta_i^*)]^2 \, dt$$

$$= \sum_{i=1}^{N} \bar{J}_i^* = J^*. \tag{10.103}$$

The inequality given by Equation 10.102 can be satisfied only by violating the fact that J^* is the minimum cost for OCP-1. Hence, by contradiction, the theorem is proven. ∎

Theorem 10.2 states that, for the given T^*, Θ^* can be obtained as the solution to POP-1 without solving OCP-1 if the closed-form state-feedback OGL for OCP-2 exists. Once Θ^* is obtained, the energy optimal trajectory for OCP-1 can be produced simply by applying Γ, the state-feedback control law for OCP-2, to each flight segment. Thus, OCP-1 now reduces to the problem of finding Θ^*, that is, to the POP-1. Note that POP-1 is a parameter optimization problem and can be solved with significantly less numerical effort compared with OCP-1.

Theorems 10.1 and 10.2 are easily expanded to the case for missiles with first-order lag autopilot where lateral acceleration is explicitly included in the missile kinematics [16].

10.4.2 Waypoint Guidance Scheme Based on OGL

Pure Energy Optimal Guidance between Waypoints for Lag-Free System
Note that OCP-2 is nothing but a special case of the optimal control problem discussed in Section 10.2.1. In fact, the closed-form state-feedback solution Γ in Equation 10.99 is given by the pure energy optimal impact angle control guidance law, OGL (Equation 10.28) with $N = 0$, that is,

$$\bar{u}^*(t) = -\frac{V}{t_{go}^2}\left[\frac{6z(t)}{V} + 4t_{go}\theta(t) + 2t_{go}\theta_f\right] \tag{10.104}$$

or

$$\bar{u}^*(t) = \frac{V}{t_{go}}[6\sigma(t) - 4\theta(t) - 2\theta_f].$$

(10.105)

From Equation 10.33, the open-loop time history of the energy optimal control is

$$\bar{u}^*(t) = C_R t_{go} + C_S$$

(10.106)

where C_R and C_S are constants, although they can be expressed in terms of the state variables as

$$C_R = -\frac{6V}{t_{go}^3}\left[\frac{2z(t)}{V} + t_{go}\theta(t) + t_{go}\theta_f\right]$$

$$C_S = \frac{2V}{t_{go}^2}\left[\frac{3z(t)}{V} + t_{go}\theta(t) + 2t_{go}\theta_f\right].$$

(10.107)

At $t = 0$, we have $z(0) = 0$, $\theta(0) = \theta_0$, and $t_{go} = t_f$. Hence, the constants C_R and C_S are

$$C_R = -\frac{6V}{t_f^2}(\theta_0 + \theta_f), \quad C_S = \frac{2V}{t_f}(\theta_0 + 2\theta_f).$$

(10.108)

The optimal cost $\bar{J}^*$ defined in Equation 10.94 is then computed as

$$\bar{J}^* = \frac{1}{2}\int_0^{t_f}\left[\bar{u}^*(t)\right]^2 dt = \frac{1}{2}\int_0^{t_f}(C_R t_{go} + C_s)^2 dt$$

$$= \frac{2V^2}{t_f}(\theta_0^2 + \theta_0\theta_f + \theta_f^2).$$

(10.109)

It should be noted here that $\bar{J}^*$ is represented by a quadratic function of the initial launch angle and the terminal impact angle. By extending the result (Equation 10.109) to the multiple waypoints case, we can obtain an analytical expression of the cost function for POP-1 in terms of flight path angles at waypoints.

Calculation of Θ^* for POP-1

Let σ_i be the slant angle of the line connecting the $(i-1)$th-waypoint and the ith waypoint as depicted in Figure 10.30. Since θ_0 and θ_f in Equation 10.109 are defined with respect to the line between the initial point and the final

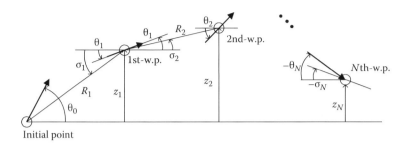

FIGURE 10.30
Definitions of σ_i and R_i.

point of the flight segment, we obtain the analytic expression of the cost function for POP-1:

$$
\tilde{J} = \sum_{i=1}^{N} \int_{t_{i-1}}^{t_i} [\bar{u}^*(t)]^2 \, dt
$$

$$
= 4V^2 \sum_{i=1}^{N} \frac{1}{\Delta t_i} [(\theta_{i-1} - \sigma_i)^2 + (\theta_{i-1} - \sigma_i)(\theta_i - \sigma_i) + (\theta_i - \sigma_i)^2]
$$

(10.110)

where $\Delta t_i = t_i - t_{i-1}$. For simplicity, we approximate Δt_i by the minimum flight time for each segment as

$$
\Delta t_i \approx R_i / V
$$

(10.111)

where R_i is the distance between the $(i-1)$th and ith waypoints. Then, Equation 10.110 is rewritten as

$$
\tilde{J} \approx 4V^3 \sum_{i=1}^{N} \frac{1}{R_i} [(\theta_{i-1} - \sigma_i)^2 + (\theta_{i-1} - \sigma_i)(\theta_i - \sigma_i) + (\theta_i - \sigma_i)^2].
$$

(10.112)

Here, σ_i and R_i are fixed and easily calculated from the waypoint positions.
Note that the energy cost is approximated by a quadratic function of the elements in Θ, the waypoint passing angles. Thus, the necessary condition to minimize Equation 10.112 yields a simple linear algebraic equation of Θ. The optimal trajectory can also be independently obtained for each leg defined by a set of consecutive waypoints. Depending on whether the initial and/or final waypoint is fixed (prescribed) as illustrated in Figure 10.31, each leg belongs to one of the following four cases.

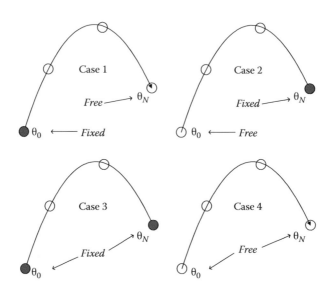

FIGURE 10.31
Types of waypoint clusters.

Case 1: Fixed θ_0 and free θ_N.

In this case, the unknown parameter vector is defined by

$$\Theta^* = [\theta_1^* \quad \theta_2^* \quad \cdots \quad \theta_{N-1}^* \quad \theta_N^*]^T \tag{10.113}$$

and Θ^* should satisfy the following necessary condition:

$$\left[\frac{\partial \tilde{J}}{\partial \theta_i}\right]_{\theta_i = \theta_i^*} = 0, \, i = 1, 2, ..., N \tag{10.114}$$

or

$$0 = \begin{bmatrix} 2\theta_1^*\left(\dfrac{1}{R_1} + \dfrac{1}{R_2}\right) + \dfrac{\theta_2^*}{R_2} + \dfrac{\theta_0}{R_1} - \dfrac{3\sigma_1}{R_1} - \dfrac{3\sigma_2}{R_2} \\[3mm] \dfrac{\theta_1^*}{R_2} + 2\left(\dfrac{1}{R_2} + \dfrac{1}{R_3}\right)\theta_2^* + \dfrac{\theta_3^*}{R_3} - \dfrac{3\sigma_2}{R_2} - \dfrac{3\sigma_3}{R_3} \\[3mm] \vdots \\[3mm] \dfrac{\theta_{N-1}^*}{R_{N-1}} + 2\left(\dfrac{1}{R_{N-1}} + \dfrac{1}{R_N}\right)\theta_{N-1}^* + \dfrac{\theta_N^*}{R_N} - \dfrac{3\sigma_{N-1}}{R_{N-1}} - \dfrac{3\sigma_N}{R_N} \\[3mm] \dfrac{\theta_{N-1}^*}{R_N} + \dfrac{2\theta_N^*}{R_N} - \dfrac{3\sigma_N}{R_N} \end{bmatrix}. \tag{10.115}$$

From Equation 10.115, we have

$$\Theta^* = R_A^{-1}\sigma_R \tag{10.116}$$

where

$$
R_A =
\begin{bmatrix}
2\left(\dfrac{1}{R_1}+\dfrac{1}{R_2}\right) & \dfrac{1}{R_2} & 0 & 0\ldots 0 & 0 \\[2ex]
\dfrac{1}{R_2} & 2\left(\dfrac{1}{R_2}+\dfrac{1}{R_3}\right) & \dfrac{1}{R_3} & 0\ldots 0 & 0 \\[2ex]
\vdots & \vdots & \vdots & \vdots & \vdots \\[2ex]
0 & 0 & \dfrac{1}{R_{N-1}} & 2\left(\dfrac{1}{R_{N-1}}+\dfrac{1}{R_N}\right) & \dfrac{1}{R_N} \\[2ex]
0 & 0 & 0 & \dfrac{1}{R_N} & \dfrac{2}{R_N}
\end{bmatrix}
\tag{10.117}
$$

and

$$
\sigma_R = 3\left[-\dfrac{\theta_0}{3R_1}+\dfrac{\sigma_1}{R_1}+\dfrac{\sigma_2}{R_2} \quad \dfrac{\sigma_2}{R_2}+\dfrac{\sigma_3}{R_3} \quad \cdots \quad \dfrac{\sigma_{N-1}}{R_{N-1}}+\dfrac{\sigma_N}{R_N} \quad \dfrac{\sigma_N}{R_N} \right]^T .
\tag{10.118}
$$

Case 2: Free θ_0 and fixed θ_N.

Parameter vector: $\quad \Theta^* = [\theta_0^* \quad \theta_1^* \quad \cdots \quad \theta_{N-2}^* \quad \theta_{N-1}^*]^T .$ (10.119)

Necessary condition: $\quad \left[\dfrac{\partial \tilde{J}}{\partial \theta_i}\right]_{\theta_i=\theta_i^*} = 0, \quad \text{for} \quad i = 0,1,\ldots,N-1.$ (10.120)

In a manner similar to case 1, we have

$$
R_A = \begin{bmatrix}
\dfrac{2}{R_1} & \dfrac{1}{R_1} & 0 & 0...0 & 0 \\[3mm]
\dfrac{1}{R_1} & 2\left(\dfrac{1}{R_1}+\dfrac{1}{R_2}\right) & \dfrac{1}{R_2} & 0...0 & 0 \\[3mm]
\vdots & \vdots & \vdots & \vdots & \vdots \\[3mm]
0 & 0 & \dfrac{1}{R_{N-2}} & 2\left(\dfrac{1}{R_{N-2}}+\dfrac{1}{R_{N-1}}\right) & \dfrac{1}{R_{N-1}} \\[3mm]
0 & 0 & 0 & \dfrac{1}{R_{N-1}} & 2\left(\dfrac{1}{R_{N-1}}+\dfrac{1}{R_N}\right)
\end{bmatrix}
$$

$$(10.121)$$

and

$$
\sigma_R = 3\left[\dfrac{\sigma_1}{R_1} \quad \dfrac{\sigma_1}{R_1}+\dfrac{\sigma_2}{R_2} \quad \cdots \quad \dfrac{\sigma_{N-2}}{R_{N-2}}+\dfrac{\sigma_{N-1}}{R_{N-1}} \quad \dfrac{\sigma_{N-1}}{R_{N-1}}+\dfrac{\sigma_N}{R_N}-\dfrac{\theta_N}{3R_N}\right]^T.
$$

$$(10.122)$$

Case 3: Fixed θ_0 and fixed θ_N.

Parameter vector: $\Theta^* = [\theta_1^* \quad \theta_2^* \quad \cdots \quad \theta_{N-2}^* \quad \theta_{N-1}^*]^T.$ (10.123)

Necessary condition: $\left[\dfrac{\partial \tilde{J}}{\partial \theta_i}\right]_{\theta_i=\theta_i^*} = 0, \quad \text{for} \quad i = 1, 2, \ldots, N-1.$ (10.124)

$$
R_A = \begin{bmatrix}
2\left(\dfrac{1}{R_1}+\dfrac{1}{R_2}\right) & \dfrac{1}{R_2} & 0 & 0...0 & 0 \\[3mm]
\dfrac{1}{R_2} & 2\left(\dfrac{1}{R_2}+\dfrac{1}{R_3}\right) & \dfrac{1}{R_3} & 0...0 & 0 \\[3mm]
\vdots & \vdots & \vdots & \vdots & \vdots \\[3mm]
0 & 0 & \dfrac{1}{R_{N-2}} & 2\left(\dfrac{1}{R_{N-2}}+\dfrac{1}{R_{N-1}}\right) & \dfrac{1}{R_{N-1}} \\[3mm]
0 & 0 & 0 & \dfrac{1}{R_{N-1}} & 2\left(\dfrac{1}{R_{N-1}}+\dfrac{1}{R_N}\right)
\end{bmatrix}
$$

$$(10.125)$$

and

$$\sigma_R = 3\left[-\frac{\theta_0}{3R_1} + \frac{\sigma_1}{R_1} + \frac{\sigma_2}{R_2} \quad \frac{\sigma_2}{R_2} + \frac{\sigma_3}{R_3} \quad \cdots \quad \frac{\sigma_{N-2}}{R_{N-2}} + \frac{\sigma_{N-1}}{R_{N-1}} \quad \frac{\sigma_{N-1}}{R_{N-1}} + \frac{\sigma_N}{R_N} - \frac{\theta_N}{3R_N}\right]^T.$$

(10.126)

Case 4: Free θ_0 and free θ_N.

Parameter vector: $\quad \Theta^* = [\theta_0^* \quad \theta_1^* \quad \cdots \quad \theta_{N-1}^* \quad \theta_N^*]^T.$ (10.127)

Necessary condition: $\quad \left[\dfrac{\partial \tilde{J}}{\partial \theta_i}\right]_{\theta_i = \theta_i^*} = 0, \quad$ for $\quad i = 0, 1, \ldots, N.$ (10.128)

$$R_A = \begin{bmatrix} \dfrac{2}{R_1} & \dfrac{1}{R_1} & 0 & 0\ldots0 & 0 \\[2mm] \dfrac{1}{R_1} & 2\left(\dfrac{1}{R_1} + \dfrac{1}{R_2}\right) & \dfrac{1}{R_2} & 0\ldots0 & 0 \\[2mm] \vdots & \vdots & \vdots & \vdots & \vdots \\[2mm] 0 & 0 & \dfrac{1}{R_{N-1}} & 2\left(\dfrac{1}{R_{N-1}} + \dfrac{1}{R_N}\right) & \dfrac{1}{R_N} \\[2mm] 0 & 0 & 0 & \dfrac{1}{R_N} & \dfrac{2}{R_N} \end{bmatrix}$$

(10.129)

and

$$\sigma_R = 3\left[\frac{\sigma_1}{R_1} \quad \frac{\sigma_1}{R_1} + \frac{\sigma_2}{R_2} \quad \cdots \quad \frac{\sigma_{N-1}}{R_{N-1}} + \frac{\sigma_N}{R_N} \quad \frac{\sigma_N}{R_N}\right]^T.$$

(10.130)

If the prescribed waypoints require a sharp turn, the approximation of the flight time given by Equation 10.111 may not be appropriate, and the error in the optimal solution calculated from the linear algebraic equations can be large. As illustrated in Figure 10.32, adding new waypoints in the course of the turn can alleviate this problem. If wind is encountered, the trajectory produced by the proposed waypoint guidance synthesis may deviate from the desired path, which is different from the behavior we expect in the path regulation methods. By assigning more waypoints along the trajectory, the wind effect could be overcome.

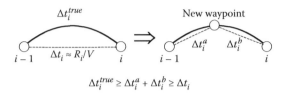

$$\Delta t_i^{true} \geq \Delta t_i^a + \Delta t_i^b \geq \Delta t_i$$

FIGURE 10.32
Addition of a new waypoint for more accurate flight time calculation.

The optimal waypoint guidance scheme is summarized in Figure 10.33. First, the scheme checks whether the waypoints in the remaining leg of the flight have been changed. If there is some waypoint change, the optimal waypoint passing angles are recalculated by using the linear algebraic equations. OGL then produces the guidance command for each flight segment with optimal flight passing angles at waypoints. The proposed guidance scheme produces the energy optimal trajectory in real time without any in-flight numerical optimization.

The proposed method does not require additional hardware to implement. Since the position of each waypoint is fixed and known before or in flight, a navigation system to measure the current state of the ASM is sufficient to implement the proposed scheme.

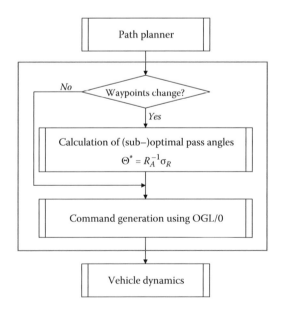

FIGURE 10.33
Energy optimal waypoint guidance scheme.

10.4.3 Numerical Examples

We compare a couple of trajectory solutions for a lag-free ASM: the numerical solution of OCP-1 and the solution of POP-1 using OGL. We investigate the performance of the proposed methods by carrying out nonlinear simulations for two different mission scenarios. As shown in Figure 10.34, nine waypoints on a plane are considered. The last waypoint (WP9) coincides with the initial point, and thus, the entire flight path is divided into nine flight segments. In Scenario 1, there is no waypoint with the prescribed passing angle, while in Scenario 2, two waypoints have been prescribed the passing angles: 45° at WP3 and –180° at WP7. The speed of the ASM is 100 m/s and remains constant during the flight. These mission scenarios look unrealistic for an ASM, because the missile returns to the launch site again, but could serve as excellent examples to demonstrate the applicability to all possible cases of the prescribed boundary conditions.

To solve OCP-1 directly by a numerical method, we adopt the use of the input parameter optimization technique [17]. In this technique, a dynamic optimal control problem is converted to a static parameter optimization problem by parameterizing control input (this technique has recently been widely used for trajectory optimization owing to its robustness to the initial guess). Ten parameterized control inputs and the flight time for each segment are considered as the parameters to be optimized. That is, the parameter set is given by

$$X = \bigcup_{i=1}^{N} X_i \tag{10.131}$$

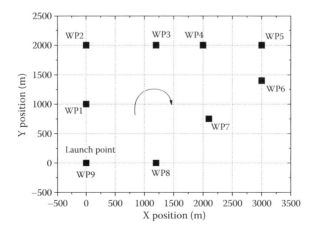

FIGURE 10.34
Distribution of waypoints for simulations.

where $N = 9$ and

$$X_i = \{u_{i1}, u_{i2}, ..., u_{i10}, t_i\} \text{ for the } i\text{th segment.} \qquad (10.132)$$

The total number of parameters to be optimized is 99 for OCP-1. After parameterization of the control input, some typical parameter optimization method, such as sequential quadratic programming (SQP) [18], can be directly employed to solve the problem. The PC used in this example for optimization has a 3.0 GHz Intel CPU with 512 MB RAM. All programs are coded by C++.

In POP-1 with OGL as the guidance law for each flight segment, the linear algebraic equations to obtain Θ^* are solved. Since Scenario 1 does not have any prescribed waypoint condition, it corresponds to Case 4 in Figure 10.31. In Scenario 2, the entire flight path is divided into three distinct legs: Case 2 from WP1 to WP3, Case 3 from WP3 to WP7, and Case 1 from WP7 to WP9.

For Scenario 1, both methods produce almost the same optimal waypoint passing angles as shown in Table 10.5. The minimum cost is achieved by OCP-1. The cost difference between OCP-1 and POP-1 using OGL is less than 0.6%. From Figures 10.35 and 10.36, we observe that trajectories and guidance command profiles obtained from both optimization methods are similar. Numerical optimization for OCP-1 takes 278 s, while the proposed method (POP-1) requires only several milliseconds.

For Scenario 2, it takes 615 s to obtain the optimal solution to OCP-1. Again, the proposed method requires only a few milliseconds to obtain near-optimal results, shown in Table 10.6. In this scenario, the cost of the

TABLE 10.5

J^*, Θ^* of Scenario 1 for Lag-Free ASM

	OCP-1	POP-1(OGL)
J^*	4907.62	4937.00
θ_0^* (degrees)	84.35	84.17
θ_1^*	101.17	101.67
θ_2^*	50.96	49.15
θ_3^*	−14.65	−14.28
θ_4^*	14.85	14.84
θ_5^*	−48.31	−48.92
θ_6^*	−123.21	−122.36
θ_7^*	−143.92	−144.00
θ_8^*	−155.42	−155.92
θ_9^*	167.70	167.96

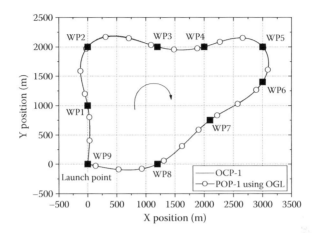

FIGURE 10.35
Comparison of the optimized trajectories of Scenario 1.

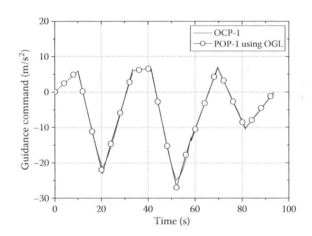

FIGURE 10.36
Comparison of optimal guidance commands of Scenario 1.

proposed method differs from OCP-1 by only 0.3%. Figures 10.37 and 10.38 show again that trajectories and command profiles obtained from both optimization methods are almost the same. It is worth noting from Figure 10.38 that there are guidance command discontinuities at the prescribed waypoints.

TABLE 10.6

J^*, Θ^* of Scenario 2 for a Lag-Free ASM

	OCP-1	POP-1(OGL)
J^*	9635.03	9664.06
θ_0^* (degrees)	81.46	82.08
θ_1^*	106.91	105.85
θ_2^*	30.78	34.54
θ_3^*	45 (given)	
θ_4^*	−2.20	−1.92
θ_5^*	−46.90	−47.62
θ_6^*	−116.79	−116.47
θ_7^*	−180 (given)	
θ_8^*	−144.32	−145.53
θ_9^*	162.04	162.77

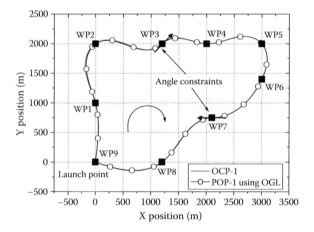

FIGURE 10.37
Comparison of the optimized trajectories of Scenario 2.

Note also that the numerical solutions for OCP-1 have been obtained without consideration of any specified waypoint passing time t_i. Hence, OCP-1's trajectories for Scenarios 1 and 2 are globally minimizing the energy. On the other hand, the trajectories based on POP-1 using OGL may vary according to how we assign a passing time to each waypoint. Interestingly enough, the numerical simulation results show that OGL with Θ^* obtained by POP-1 also produces the globally energy-optimized trajectory.

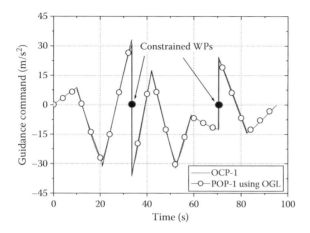

FIGURE 10.38
Comparison of optimal guidance commands of Scenario 2.

10.5 Summary

In this chapter, the optimal impact angle control guidance laws for a stationary target as well as for a constantly moving target have been derived, and their properties have been investigated. We have made a great deal of effort to examine OGL, the optimal impact angle control guidance law for a lag-free missile. With the consideration of the cost weighted by $1/t_{go}^N$, OGL with different gain sets has been obtained and offers the freedom to choose depending on the circumstances, just as we choose different navigation constants in PN guidance law. OGL requires good estimates of time-to-go for the best performance. It has been shown in Section 10.2.2 that the time-to-go computation based on the curved path length over the missile speed yields a far better estimate than the usual method of range over missile speed. OGL can be expressed as a biased PN law, as we have explored in Section 10.2.3, where a moving target is considered; it is composed of the conventional PN term to intercept the target and the bias term to satisfy the impact angle constraint. In particular, one of the second variants (Equation 10.66) of OGL has been shown to give good performance for a maneuvering target as well as a constant-speed nonmaneuvering target when combined with the time-to-go calculation method in Section 10.2.4. In Section 10.3, the same formulation and methodology have been applied to extend the result to the case of missiles modeled by a first-order lag system.

Although OGL has been devised initially to intercept a target in the desired direction, it can also constitute a very efficient waypoint guidance scheme for ASMs or UAVs. By applying OGL to guide a vehicle from one waypoint to the next, and by assigning the flight path angle—or the impact angle—at each waypoint in a proper (optimal) manner, the entire trajectory passing through all the waypoints becomes energy optimal. The optimal flight path angles at waypoints are given by a simple algebraic equation.

Not shown in this chapter, we note that OGL for a maneuvering target could produce huge impact angle error if a very large impact angle is assigned and that its launch boundary for successful intercept could be quite limited if the widely used time-to-go computation of range over closing velocity (or missile speed) is used. One of the main contributors to these problems is apparently the inaccurate time-to-go computation, which does not take account of target moving. Further study is needed to address these issues.

List of Symbols

a	Acceleration of a missile
a_t	Acceleration of a target
u	Guidance command of a missile
θ	Flight path angle of a missile
θ_t	Flight path angle of a target
V	Velocity of a missile
V_t	Velocity of a target
σ	Sight line angle
$\dot{\sigma}$	Line-of-sight angular rate (sight line angular rate)
R	Closing range
$\dot{R}$	Range rate
J	Performance cost
N'	Navigation constant
t	Time
t_{go}	Time-to-go
t_f	Time of flight
τ	Time constant of a first-order lag system

References

1. E. Kreindler, Optimality of proprtional navigation, *AIAA Journal*, 11(6), 878–880, 1973.
2. H. Cho, C. K. Ryoo, and M. J. Tahk, Implementation of optimal guidance laws using predicted velocity profiles, *Journal of Guidance, Control, and Dynamics*, 22(4), 579–588, 1999.
3. D. E. Kirk, *Optimal Control Theory—An Introduction*, Prentice-Hall, Englewood Cliffs, NJ, 1970.

4. A. E. Bryson, Jr. and Y.-C. Ho, *Applied Optimal Control*, John Wiley & Sons, NY, pp. 158–167, 1975.

5. D. G. Hull, *Optimal Control Theory for Applications*, Springer-Verlag, NY, 2003.

6. C.-F. Lin, *Modern Navigation, Guidance, and Control Processing*, Prentice Hall Inc., NJ, 1991.

7. C. K. Ryoo, H. Cho, and M. J. Tahk, Time-to-go weighted optimal guidance law with impact angle constraints, *IEEE Transactions on Control Systems Technology*, 14(3), 483–492, 2006.

8. C. K. Ryoo, H. Cho, and M. J. Tahk, Optimal guidance laws with terminal impact angle constraint, *Journal of Guidance, Control, and Dynamics*, 28(4), 724–732, 2005.

9. E. J. Ohlmeyer and C. A. Phillips, Generalized vector explicit guidance, *Journal of Guidance, Control, and Dynamics*, 29(2), 261–268, 2006.

10. J. Z. Ben-Asher and I. Yaesh, *Advances in Missile Guidance Theory*, Vol. 180, Progress in Astronautics and Aeronautics, AIAA Inc., VA, 1998.

11. P. Zarchan, *Tactical and Strategic Missile Guidance*, 4th ed., AIAA Inc., VA, pp. 11–15, 2003.

12. M. J. Tahk, C. K. Ryoo, and H. Cho, Recursive time-to-go estimation for homing guidance missiles, *IEEE Transactions on Aerospace and Electronic Systems*, 38(1), 13–24, 2002.

13. B. S. Kim, J. G. Lee, and H. S. Han, Biased PNG law for impact with angular constraint, *IEEE Transactions on Aerospace and Electronic Systems*, 34(1), 277–288, 1998.

14. H. Cho, C. K. Ryoo, A, Tsourdos, and B. White, Optimal Guidance Solution for Impact Angle Control Based on Linearization About Collision Triangle, *The 18th IFAC World Congress, 2011*, Milano, Italy, pp. 3909–3914, Aug. 28–Sep. 2, 2011.

15. M. Guelman, The closed-form solution of the true proportional navigation, *IEEE Transactions on Aerospace and Electronic Systems*, AES-12(4), 472–482, 1976.

16. C. K. Ryoo, H. S. Shin, and M. J. Tahk, Energy optimal waypoint guidance synthesis for anti-ship missiles, *IEEE Transactions on Aerospace and Electronic Systems*, 46(1), 80–95, 2010.

17. D. G. Hull, Conversion of optimal control problems into parameter optimization problems, *Journal of Guidance, Control, and Dynamics*, 20(1), 57–60, 1997.

18. C. Lawrence, K. L. Zhou, and A. L. Tits, *User's Guide for CFSQP Version 2.5: A C Code for Solving (Large Scale) Constrained Nonlinear (Minmax) Optimization Problems, Generating Iterates Satisfying All Inequality Constraints*, Institute for Systems Research, University of Maryland, 1997.

11

Integrated Design of Estimator and Guidance Law*

Josef Shinar and Tal Shima

CONTENTS

---

* The mathematical models and the respective guidance laws based on linear differential game theory can be found in Chapter 9 of this book.

11.1 Introduction

11.1.1 Role of Estimator in Guidance System

Almost all the guidance laws that are implemented in homing missile hardware, as well as in most attempts to develop new guidance laws, have been based on assuming *perfect* (instantaneous, accurate, and disturbance-free) knowledge of all state variables. In many modern guidance laws, the knowledge of target acceleration is necessary for achieving improved homing performance. In reality, however, all the measurements acquired by the sensors of a guided missile are noise corrupted. To use these measurements as a basis of a feedback control, the noisy signals have to be filtered. Moreover, not all components of the state vector are measured or even measurable. For example, relative acceleration from a moving platform, or the acceleration of the other object, cannot be measured. Such state variables have to be reconstructed in the ideal case, where the available measurements are noise-free, by an observer. In reality, such reconstruction has to be made by using the available noise-corrupted measurements by an estimator. Thus, in a guidance system, the estimator performs a dual role, the role of a filter and the role of an observer. Fulfillment of each role requires a different dynamical behavior. Smoothing out the erroneous effects of the measurement noise needs slow dynamics. At the other end, the reconstruction of nonmeasured state variables, which are required for an efficient feedback control, should be performed as quickly as possible. In designing the estimator, a compromise is needed.

In realistic interception scenarios with noise-corrupted measurements, the estimator has become an indispensable element of the guidance system because the homing performance of an interceptor missile has been limited by the estimation accuracy.

Most missile guidance laws were derived using a linearized kinematical model. It is well known that for a linear process with zero-mean Gaussian white noise, the Kalman filter [1] is the optimal estimator in the sense of minimum variance. This optimality depends on the assumption that the filter design is based on the correct dynamic system model, which includes also the (deterministic) input. Model uncertainties are represented within the filter as a *process noise*, assumed to be also zero-mean Gaussian and white. Unknown inputs also can be considered as a stochastic process and are approximated [2] by the output of a (linear) *shaping filter* driven by Gaussian white noise. The above-mentioned optimality of the Kalman filter is achieved by taking into account the spectral densities of the measurement noise and the process noise. The estimation process has inherent dynamics, creating a nonzero time delay of the information on the estimated variables. Such a delay leads to the deterioration of the control performance, especially in the presence of disturbances or fast input changes.

11.1.2 Certainty Equivalence Validity

11.1.2.1 Background

The homing guidance of an interceptor missile is an optimal control problem (with the objective to minimize the miss distance) using a sequence of noise-corrupted measurements. Thus, it has to be considered as a stochastic optimal control problem. In order to reduce the complexity for solving such problems, two important properties for linear systems were formulated. The first one is called the *separation* property, and it says that control and the estimation logic can be derived separately. Another closely related (but not identical) property is the *certainty equivalence*, which states that the optimal control function of the stochastic optimal control problem is the same as the related *deterministic* optimal control problem, with the only difference being that the state variables are replaced by their estimated values. The validity of *certainty equivalence* has been rigorously proven a long time ago for linear quadratic problems with white Gaussian noise [3] and extended later to include also the cases with non-Gaussian and colored noise [4] as well as nonquadratic cost [5].

Realistic interceptor guidance problems are characterized (in addition to noise-corrupted measurements) by bounded controls and saturated state variables, as well as non-Gaussian random disturbances. The validity of the *separation* and *certainty equivalence* properties has never been proven for this class of problems. In spite of that, in the 50-year-long history of guided missiles, it has been of common practice to design the estimators and missile guidance laws independently. The estimators were simple Wiener or Kalman filters, and the guidance laws were derived using simplified (linearized and planar) deterministic models. In most cases, such a comfortable design approach has been acceptable because, due to the substantial maneuverability advantage of guided missiles over their manned aircraft targets, it succeeded in satisfying the performance requirements.

11.1.2.2 Illustrative Example

In a more stressing scenario, such as ballistic missile defense, the usefulness of relying on the *separation* and *certainty equivalence* properties becomes
strongly questionable, as demonstrated by the following example [6]. This is
a planar interception endgame scenario with the parameters given in Table
11.1. The guidance law used by the interceptor (pursuer) is derived from the
solution of a perfect-information linear pursuit–evasion game with bounded
controls, called DGL/1 [7], and the estimator was a Kalman filter augmented
with a *shaping filter* using an exponentially correlated acceleration (ECA)
model [8]. Such a shaping filter has first-order dynamics with two tuning
parameters, the correlation time of the maneuver τ_s and the (assumed) level
of the process noise, expressed by its standard deviation $\sigma_s = a_E^{max}/C_s$. In [6],
the parameters of the shaping filter were $\tau_s = 1.5$ s and $C_s = 2$; here, a similar
example with $\tau_s = 0.4$ s and $C_s = 1$ is added.

In a *perfect-information* (noise-free) planar scenario and without an estimator in the guidance loop, the interception parameters of Table 11.1 guaranteed zero miss distance against any admissible target maneuver. Due to the
noisy measurements and the presence of an estimator in the guidance loop,
the outcome is very different.

The simulation results depicted in Figure 11.1 show the average miss distance of 100 Monte Carlo runs in a typical endgame scenario with "head on"
initial conditions against a randomly maneuvering target. The commanded
lateral acceleration of the target is of the "bang-bang" type, and the direction
change can take place anywhere within the endgame's short duration. The
average miss distance is plotted as the function of $(t_{go})_{sw}$, the time-to-go of the
direction change. These results indicate that if the direction change is commanded early enough, the resulting miss distance is very small. However,
if the change in command takes place close to the final time, large miss distances are created.

TABLE 11.1

Horizontal Endgame Parameters

Parameter	Value
Interceptor (pursuer) velocity	$V_P = 2300$ m/s
Target (evader) velocity	$V_E = 2700$ m/s
Interceptor lateral acceleration limit	$a_P^{max} = 20$ g
Target lateral acceleration limit	$a_E^{max} = 10$ g
Time constant of the interceptor	$\tau_P = 0.2$ s
Time constant of the target	$\tau_E = 0.2$ s
Initial endgame range	$R_0 = 20$ km
Duration of endgame engagement	$t_f = 4$ s
Measurement noise	$\sigma_{ang} = 0.1$ mrad
Sampling rate	$f = 100$ Hz

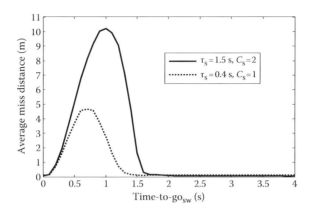

FIGURE 11.1
Homing performance of DGL/1 against randomly switched "bang-bang" command.

The main reason for the degraded homing performance is the inherent delay introduced by the convergence time of the target maneuver estimation process. The guidance law DGL/1 can correct the error created by the delay only if the change of the acceleration command occurs in the early part of the endgame. In this case, the estimated acceleration converges to its true value, and sufficient time remains until intercept. The filter design minimizes the variance of the estimation error, and the guidance law receives almost correct values of the *zero-effort miss distance* soon enough for achieving good precision [95% of the miss distances are less than 20 cm for $(t_{go})_{sw}$ beyond a critical value]. If the change of the acceleration command occurs later, the combination of such an estimator with the deterministic game optimal guidance law (OGL) fails to provide satisfactory results because of the delay. As will be shown in the sequel, there are other guidance laws that provide better performance, demonstrating that *certainty equivalence* property is not valid in this case.

The comparison of the two cases in Figure 11.1 indicates that the value of the delay can be reduced if the estimator bandwidth is increased by selecting different tuning parameters of the shaping filter. Using the faster shaping filter, the estimation delay becomes shorter; the average "peak" miss distance is reduced from more than 10 m to only 4.6 m. This is achieved at the expense of less efficient filtering, which leads to larger residual estimation errors and increased miss distances for changes of acceleration commands occurring in the early part of the endgame. In Figure 11.2, the cumulative probability distributions of the miss distances obtained for the acceleration command changes occurring with $(t_{go})_{sw} > 1.6$ s are compared for both estimators. For an improved homing performance, the error due to the estimation delay, as well as the variance of the converged estimation error, has to be reduced.

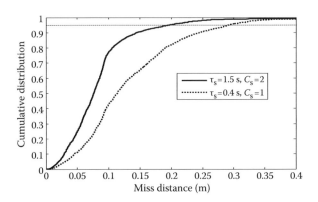

FIGURE 11.2
Cumulative probability distribution for $(t_{go})_{sw} > 1.6$ s.

The shaping filter model used in [6] is probably not ideal for this type of random bang-bang maneuver. By using another shaping filter that assumes a randomly starting (RST) maneuver [2] (a shaping filter that has only a single tuning parameter, denoted by C_r), the qualitative behavior is very similar. By enlarging the estimator bandwidth, the "peak" miss distance created by a maneuver switch executed near the intercept becomes smaller, but the miss distances obtained by earlier maneuver switch (denoted as the "plateau") are slightly larger. Comparison of the "peak" versus "plateau" relationship for both shaping filters, as plotted in Figure 11.3, shows that for the present example, the ECA provides better results than the RST.

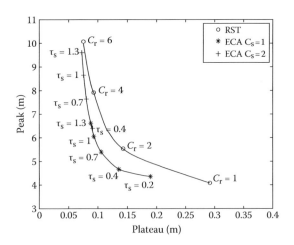

FIGURE 11.3
Comparison of "peak" and "plateau" relationship.

11.1.3 Estimation in Interception Endgames

The estimation error consists of two components. The first one is dynamic in nature and is expressed by the delay for converging to an eventually new value of the estimated state variable. The second component is of stochastic nature and expressed by the variance of the converged estimate. The design of a Kalman filter minimizes this second component. The reason for this approach has been that, in general, the control processes are of long duration, and abrupt variations of the state variables are not expected. In such cases, the estimation delay is not critical.

In an interception endgame, the situation is different. The endgame is of short duration, and estimation errors occurring near the intercept are crucial. In order to satisfy the requirement for a small miss distance, the estimation process has to become faster as the final time of the endgame approaches. Within the modern guidance laws, the knowledge of the time-to-go is an essential element, and a considerable effort is invested to obtain it accurately. However, the currently used estimation processes completely ignore this fast-changing variable, although it is available in the guidance system. For effective terminal guidance, this inherent lack of information of the classical Kalman filters has to be corrected.

The earlier-mentioned requirements to reduce both the estimation delay and the variance of the converged estimation error are contradictory. The convergence time associated with identifying a rapid target maneuver change is composed of the maneuver detection time and the estimator's response time. Short detection time comes at the price of high false-alarm rate, while short response time requires large bandwidth, generating large estimation errors. Good filtering, providing a small estimation error variance, requires narrow bandwidth, leading to a slow response. This controversy raises the following question: *can a single estimator satisfy the contradictory requirements of homing accuracy?* In the absence of available theory, the answer was sought in extensive Monte Carlo simulations [6]. These studies lead us to conclude that no estimator is globally optimal for all guidance laws/interception scenarios, and no *unique* "optimal" combination of guidance law and (Kalman filter type) estimator could be found for all feasible target maneuvers. Consequently, the answer to the above-raised question is negative, and new approaches have to be developed.

In the search for a suitable optimal estimator for the task of intercepting randomly maneuvering targets, a conceptual difficulty has been encountered. Since target maneuver dynamics is not ideal, the target acceleration is a state variable, a part of the interception model. As stated earlier, for linear systems with zero-mean, white and Gaussian measurement, and process noises, the Kalman filter [1], based on the correct model of the system dynamics, is the optimal minimum variance estimator. The measurement noise used in interception simulations has indeed such characteristics, but the representation of a random target maneuver as the output of a shaping filter driven by a zero-mean, white Gaussian noise [2] is only an approximation. Moreover, each type of target maneuver requires a different shaping filter approximation.

The disturbance inputs are random acceleration commands and can be discontinuous, representing a random jump process. They are bounded and certainly neither white nor Gaussian. In several recent papers [9, 10], it was shown that in such cases, the optimal estimator is of infinite dimension. Thus, every computationally feasible (finite dimensional) estimator can be, at best, only a suboptimal approximation, and the search for a single feasible optimal estimator associated with interceptor guidance is not a well-posed problem. Similarly, it should be of no surprise that the *certainty equivalence* and *separation* properties, both involving the concept of optimality, are not valid for the interception of randomly maneuvering targets. In investigating the interception of randomly maneuvering targets, the estimation and the guidance problems cannot be separately treated.

11.1.4 Partial Separation Property

For cases where the *certainty equivalence* property cannot be proven, a *partial separation* property was asserted [11], stating that the estimator can be designed independently of the controller, but the derivation of the optimal control function has to be based on the conditional probability density function (conditioned on the measurement history) of the estimated state variables. Unfortunately, there have been very few past works that used a rigorous practical approach for implementing this idea. In Section 11.4, several recent studies, based on direct application of the "partial" separation property, are reviewed. In Section 11.5, a heuristic approach is attempted, based on the insight generated by extensive simulation results and on control engineering intuition.

11.2 Estimation Process

11.2.1 Modeling Considerations

As mentioned earlier, in a guidance system, the estimator performs a dual role, the role of a filter and the role of an observer. Fulfillment of each role requires a different dynamical behavior. The observer design has to be based on the knowledge of the system model. Given an earlier state of the system, a new state is computed by propagating the dynamic equations of the system ahead of time. Comparison of the output of this new state with an *ideal* (noise-free) measurement of the output at the appropriate time creates the *reconstruction error*. For a stable observer, this error converges to zero asymptotically. In order to achieve a fast and reliable reconstruction, the gain of the observer can be selected as high as possible, subject to the acceptable stability margin. If the system model is perfectly known, but the measurement of the output is noise corrupted, the difference (called the *innovation*) is due to the *measurement noise*. In order to minimize the variance of the *estimation error*, the gain has to be inversely proportional to the covariance of the measurement noise. The larger this covariance is, the smaller the estimator gain is and the slower its dynamics is. The uncertainty of the dynamic system model can be expressed as a *process noise*. In computing the estimated state, the estimator makes a compromise between two uncertain values, the propagated uncertain model and the noise-corrupted measurement. This compromise is determined by the estimator gain, which depends on the covariances of the measurement noise and the process noise. Unknown inputs can be considered as stochastic processes and are approximated [2] by the output of a (linear) *shaping filter* driven by Gaussian white noise.

A very important source of model uncertainty is the unknown value of the system input. As mentioned earlier, if the input is a random variable, it is generally modeled by a linear shaping filter driven by white Gaussian noise. In order to obtain a second-order statistical similarity between the actual random input and output of the shaping filter, it is necessary that the spectral density of the white Gaussian noise will be at the level that provides identical autocorrelation functions for both processes. In the case of an interceptor missile homing on a randomly maneuvering target, this approach was adapted [2], but one has to remember that this is only an approximation. By this approach, the equations of the shaping filter become a part of the system model. Assuming that the dynamic model of the target is known, the random element is the commanded target acceleration.

There are two types of target maneuvers discussed frequently in the literature as examples of random target maneuvers: (1) periodical maneuvers with random phase (a projection of a "barrel roll" type maneuver on a plane) and (2) "bang-bang" type maneuvers with random switch. In addition to the random elements, the structure of the maneuver itself contains unknown parameters, such as the frequency and the amplitude in the periodical case or the direction and the magnitude of the acceleration in a "bang-bang" type maneuver. In order to properly discover the correct dynamic model of the interception scenario, first the maneuver structure has to be identified. The search for the necessary unknown parameters of the maneuver has to be addressed subsequently.

11.2.2 Model Identification

One way to overcome the problem of the unknown target maneuver model is to create a finite set of hypotheses corresponding to feasible maneuvers. This approach assumes that the actual target maneuver will be near to one of these hypotheses. The estimator constructed using such an approach is called a multiple-model adaptive estimator (MMAE) [12]. The MMAE is composed of a set of elemental estimators (Kalman filters), each one corresponding to one of the possible hypotheses. The entire set (bank) of the Kalman filters runs in parallel, using the same measurement sequence. Using the innovation process of each elemental estimator, the *a posteriori* probability for the correctness of its hypothesis can be recursively computed. The estimated state vector can now be computed by fusing the state vectors of the different Kalman filters according to some criterion. The two commonly used criteria are the minimum mean square error (MMSE) [13] and the maximum *a posteriori* probability (MAP) [14]. In the MMSE method, the estimated state vector is the weighted average of the outputs from all the elemental filters, based on their *a posteriori* probabilities. In the MAP method, the estimated state vector of the filter with the maximum *a posteriori* probability is used.

In the past, MMAEs were seldom applied to interception scenarios because of the excessive computational load associated with the use of a large number of Kalman filters representing the possible evasion strategies [15]. In a later paper [16], an innovative approach was introduced. The estimator includes several models of the target maneuver with identical dynamics but different controls. This feature enables using many elemental filters in the MMAE with a highly reduced computational load. This reduction has no effect on the estimation accuracy. Thus, given a limited computational capacity, more models can be used, and a substantial improvement can be achieved in the homing accuracy.

11.3 Delayed-Information Differential Games

11.3.1 Deterministic Estimation Models

As mentioned earlier, for good homing performance, the error due to the estimation delay, as well as the variance of the converged estimation error, has to be reduced. Since the design of the Kalman filter provides minimum variance, the aim of some investigations was to compensate the effect of the estimation delay. One approach in this direction is to include the estimation delay in a deterministic model of the interception problem by neglecting the stochastic features of the estimation process due to the noise-corrupted measurements. Based on such approximation, the interception scenario of a maneuvering target is reformulated as a delayed-information pursuit–evasion game with bounded controls. In this formulation, the evader has *perfect information* on all the state variables, as well as on the estimation delay of the pursuer. The game solution is based on an intuitive approach, inspired by the idea of *reachable sets* [17]. At every point of the time, the reachable set of the evader is created based on the information available to the pursuer. The objective is to reach the center of the convex hull of this reachable set. The interceptor guidance law is determined by the optimal pursuer strategy of this game. In this chapter, several deterministic guidance laws, based on different perfect-information models, are described. In principle, each of those models can be parameterized by the estimation delay, and for each case, another delay-compensating guidance law has to be derived.

11.3.2 Delay-Compensating Guidance Law

11.3.2.1 Analytical Solution

The delay-compensating guidance law, which was first published in the literature [18, 19], is denoted DGL/C. It considered a linearized planar constant-speed game model with first-order dynamics of both players (like DGL/1) and a fixed delay (Δt_{est}) in estimating the evader acceleration. Here, only the essential steps of the game solution, as well as the respective simulation results, are presented and discussed. The interested reader can find the mathematical details in the references.

The corresponding perfect-information guidance law (DGL/1) is based on using the zero-effort miss distance as the state variable of the game. It is defined by

$$Z(t_{go}) = y + \dot{y}t_{go} - a_P \tau_P^2 \psi(t_{go}/\tau_P) + a_E \tau_E^2 \psi(t_{go}/\tau_E) \tag{11.1}$$

where y and $\dot{y}$ are the relative separation and its time derivative, respectively, a_E and a_P are the respective accelerations of the evader and the pursuer, all normal to the initial line of sight, τ_E and τ_P are the respective time constants, and $\psi(.)$ is the function

$$\psi(\zeta) = e^{-\zeta} + \zeta - 1 \tag{11.2}$$

The state vector of the motion normal to the initial line of sight is defined as

$$X^T = (x_1, x_2, x_3, x_4) = (y, \dot{y}, a_E, a_P) \tag{11.3}$$

and the game dynamics is described by the following set of differential equations and the respective initial conditions:

$$\dot{x}_1 = x_2; \qquad\qquad x_1(0) = 0 \tag{11.4}$$

$$\dot{x}_2 = x_3 - x_4; \qquad\qquad x_2(0) = x_2^0 \neq 0 \tag{11.5}$$

$$\dot{x}_3 = (a_E^c - x_3)/\tau_E; \quad x_3(0) = 0 \tag{11.6}$$

$$\dot{x}_4 = (a_P^c - x_4)/\tau_P; \quad x_4(0) = 0 \tag{11.7}$$

where a_E^c and a_P^c are the commanded lateral accelerations of the evader and the pursuer, respectively:

$$a_E^c = a_E^{max} v(t) \quad |\mathbf{v}| \le 1 \tag{11.8}$$

$$a_P^c = a_P^{max} u(t) \quad |\mathbf{u}| \le 1 \tag{11.9}$$

The nonzero initial condition x_2^0 represents the respective initial velocity components not aligned with the initial (reference) line of sight. These components are small compared with the components along the line of sight. The final time of the interception t_f is given, allowing us to define the time-to-go by

$$t_{go} = t_f - t \tag{11.10}$$

The natural cost function of the game is the miss distance, defined as

$$J = |x_1(t_f)| = |Z(0)|. \tag{11.11}$$

The objective of the interceptor missile (the pursuer of the game) is the following.

Given the dynamic system (Equations 11.4 through 11.7) with the set of initial conditions and the constraints 11.8 and 11.9, minimize the cost function (Equation 3.11) against any admissible control $v(t)$ of the evader, subject to the following set of available measurements:

$$h_i(t) = x_i(t); \; i = 1, 2, 4 \; ; \; h_3(t) = x_3(t - \Delta t_{est}). \tag{11.12}$$

The basic idea of the delayed-information game solution is to replace $Z(t_{go})$, defined in Equation 11.1, with a new state variable $Z^c(t_{go})$, which is the position of the *uncertainty set* (reachable set) center created by the information delay Δt_{est}. By denoting the first three terms of Equation 11.1 as $Z^0(t_{go})$, the new variable $Z^c(t_{go})$ can be written as

$$Z^c(t_{go}) = Z^0(t_{go}) + \Delta Z_E^c(t_{go}) \tag{11.13}$$

where $\Delta Z_E^c(t_{go})$ is the position of the center of the *acceleration uncertainty* segment. Due to the delayed measurement $a_E(t - \Delta t_{est})$, the uncertain value of $a_E(t) = x_3(t)$ is bounded by

$$[x_3(t)]_{\min} \le x_3(t) \le [x_3(t)]_{\max} \tag{11.14}$$

where the extreme values $[x_3(t)]_{\min}$ and $[x_3(t)]_{\max}$ are computed by integrating Equation 11.6 with $a_E^c = -a_E^{\max}$ and $a_E^c = a_E^{\max}$, respectively:

$$[x_3(t)]_{\min} = x_3(t - \Delta t_{est})e^{-\Delta test/\tau} - a^{\max}(1 - e^{-\Delta test/\tau}) \tag{11.15}$$

$$[x_3(t)]_{\min} = x_3(t - \Delta t_{est})e^{-\Delta test/\tau} + a^{\max}(1 - e^{-\Delta test/\tau}) \tag{11.16}$$

The center of the *acceleration uncertainty* segment is

$$[x_3(t)]^c = \{[x_3(t)]_{\max} + [x_3(t)]_{\min}\}/2 = a(t - \Delta t_{est})e^{-\Delta test/\tau} \tag{11.17}$$

and as a consequence, in the new variable $Z^c(t_{go})$, defined in Equation 11.13, the term effected by the estimation delay is

$$\Delta Z^c(t) = \tau^2 \psi(\tau_{go}/\tau)a(t - \Delta t_{est})e^{-\Delta test/\tau}. \tag{11.18}$$

Comparing Equation 11.18 with the last term of Equation 11.1, one can conclude that the contribution of the evader acceleration to $Z^c(t_{go})$ is discounted, due to the delay Δt_{est} in its estimation, exponentially by its ratio to the time constant τ_E. For zero estimation delay, one obtains the zero effort miss of Equation 11.1, leading to the guidance law DGL/1, while for a very large estimation delay, the term in Equation 11.18 becomes negligibly small, leading to a guidance law that does not consider evader acceleration (DGL/0).

Since the rigorous mathematical solution is rather complex [18], here only the main results are presented. The first step toward the solution is transforming the original imperfect (delayed) information game to a perfect-information game with delayed control of the evader. The necessary conditions of optimality provide the candidate optimal strategies of the game:

$$u^* = v^* = \text{sign} \{Z^c(t_{go})\} \tag{11.19}$$

leading us to determine the optimal game dynamics as

$$dZ^c/dt_{go} = \Gamma^c(t_{go}, \Delta t_{est}) \, \text{sign} \{Z^c(t_{go})\} \tag{11.20}$$

with

$$\Gamma^c(t_{go}, \Delta t_{est}) = a_P^{\max} \tau_P \psi(t_{go}/\tau_P)$$

$$- a_E^{\max}[\tau_E \psi(t_{go}/\tau_E)e^{-\Delta test/\tau_E} + t_{go}(1 - e^{-\Delta test/\tau_E})] \tag{11.21}$$

By integrating Equation 11.21 for any value of $Z^c(0)$, the candidate optimal trajectories of the game are obtained. It is easy to see that for small positive values of t_{go}, $\Gamma^c(t_{go}, \Delta t_{est})$ is negative, but if $\mu = a_P^{max}/a_E^{max}$ is sufficiently large, it becomes positive (for any value of $\varepsilon = \tau_E/\tau_P > 0$) as t_{go} increases. As a consequence, the decomposition of the game space (Z^c, t_{go}) has the structure shown in Figure 11.4 as an example, using the endgame parameters of Table 11.1.

The two *limiting* trajectories (Z_+^{c*}, Z_-^{c*}) that satisfy the condition that $Z^c(t_{go})$ does not change sign reach the t_{go} axis tangentially at $t_{go} = (t_{go}^c)_s$. This value of $(t_{go}^c)_s$ is the nonzero solution of the equation

$$\Gamma^c(t_{go}, \Delta t_{est}) = 0. \tag{11.22}$$

The (Z^c, t_{go}) game space is decomposed into a *singular* region $\mathcal{D}_0$, which is between these trajectories for $t_{go} > (t_{go}^c)_s$ and the *regular* region $\mathcal{D}_1$. In $\mathcal{D}_1$, the *optimal strategies* are given by Equation 11.19, while the nonzero *value* of the game depends on the initial conditions.

In the *singular* region, the *optimal strategies* are arbitrary, all optimal trajectories must go through the point $(0, (t_{go}^c)_s)$ and therefore the *value* of the game in this entire region is a nonzero constant J_s^c. This *singular* game *value*, which is the smallest miss distance that an optimally playing *pursuer* can achieve against an optimally playing *evader*, depends on Δt_{est} and the physical parameters of the game [the *pursuer/evader* maneuverability ratio $\mu = (a_P)_{max}/(a_E)_{max}$ and the *evader/pursuer* time constant ratio $\varepsilon = \tau_E/\tau_P$]. Once $(t_{go}^c)_s$ is found from the solution of Equation 11.22, J_s^c can be computed from Equation 11.20 by setting $Z^c(t_{go}^c)_s = 0$ and by direct integration between $(t_{go}^c)_s$ and zero.

If $\mu > 1$ and $\mu\varepsilon \geq 1$, in the perfect-information game ($\Delta t_{est} = 0$), the only solution of Equation 11.22 is $(t_{go}^c)_s = 0$, and as a consequence, also $J_s^c = 0$. For $\Delta t_{est} > 0$, such a situation is excluded, and the game space decomposition

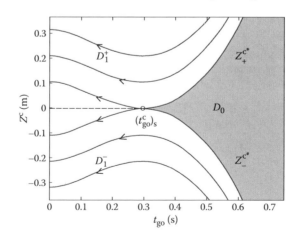

FIGURE 11.4
Decomposition of the (Z^c, t_{go}) game space.

structure shown in Figure 11.4 remains valid for any combination of $\mu > 1$ and ε. It means that *capture* (i.e., zero miss distance) cannot be achieved due to the existence of the information delay. This important conclusion can be reconfirmed by comparing the analytical solution of this game expressed by $(t^c_{go})_s$ and J^c_s with the similar expressions $[(t_{go})_s$ and $J_s]$ in the solution of the perfect-information game with the guidance law DGL/1. An example of these expressions for the endgame parameters given in Table 11.1 is presented in Figure 11.5 as functions of the estimation delay.

Implementation of the delay-compensating guidance law DGL/C is not straightforward. The guidance loop must include an estimator, and the appropriate value of the information delay (assumed to be constant) is not obvious. Depending on the actual model of the measurement noise, the information delay may vary with the range (time-to-go). The numerical value of the information delay that provides the best guaranteed homing performance has to be found, for any given estimator, by an offline search.

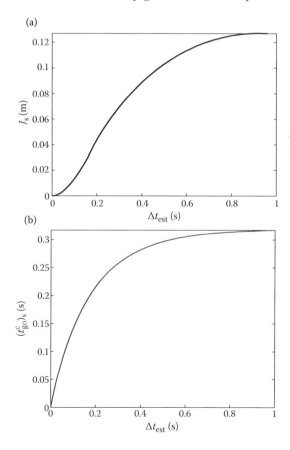

FIGURE 11.5
Values of $(t^c_{go})_s$ and J^c_s as functions of Δt_{est}. (Example for the case of Table 11.1.)

11.3.2.2 Simulation Results

The analytical results of the previous subsection had to be tested by simulations that include noise-corrupted measurements and an estimator in the guidance loop. For an endgame example, based on the data of Table 11.1, a large set of Monte Carlo simulations was carried out against a target performing a "bang-bang" type maneuver with randomly switched acceleration command. The estimator in the guidance loop was a Kalman filter augmented with a *shaping filter* using an ECA model [8] with different tuning parameters. For each estimator, defined by its tuning parameters, different values of the (*a priori* unknown) fixed estimation delay were tested. In order to obtain reliable statistics, in each endgame scenario (defined by the estimator, the assumed fixed estimation delay, and the timing of the switch for the maneuver command), 100 randomly selected measurement noise samples were used.

The Monte Carlo simulations provided the following results. As expected, using DGL/C instead of DGL/1 resulted in reducing the value of the

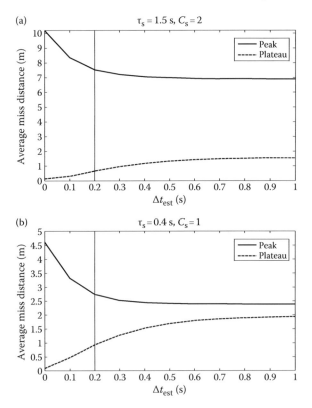

FIGURE 11.6
(a) Characteristic miss distances as functions of the estimation delay with a slow estimator ($\tau_s = 1.5$ s, $C_s = 2$). (b) Characteristic miss distances as functions of the estimation delay with a fast estimator ($\tau_s = 0.4$ s, $C_s = 1$).

maximum miss distance ("peak") obtained by a late maneuver change but increased the miss distance variance obtained by earlier maneuver changes. By increasing the estimation delay used in the guidance law, the reduction of the "peak," as well as the miss distances for early maneuver changes ("plateau"), became larger. While the miss distance of the "plateau" is monotonically increasing with the assumed estimation delay, the reduction of the "peak" reaches saturation. The maximum amount of achievable reduction in the "peak" depends on the estimator parameters.

The larger the estimator bandwidth (the smaller the tuning parameters) becomes, the more reduction of the "peak" can be achieved. The results are illustrated in Figures 11.6a and b; the reduction of the "peak" together with the average value of the "plateau" are shown as the function of the assumed estimation

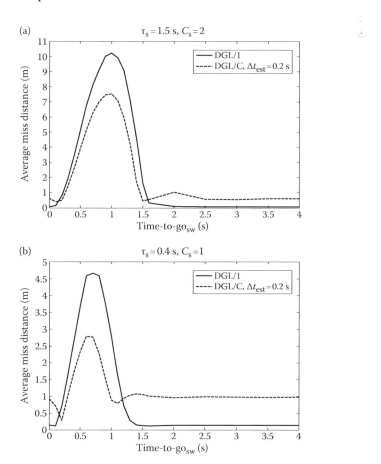

FIGURE 11.7
(a) Homing performance of DGL/1 and DGL/C against randomly switched "bang-bang" command with a slow estimator ($\tau_s = 1.5$ s, $C_s = 2$). (b) Homing performance of DGL/1 and DGL/C against randomly switched "bang-bang" command with a fast estimator ($\tau_s = 0.4$ s, $C_s = 1$).

delay for the two estimators used in Figure 11.4. These results clearly indicate that in spite of the monotonic behavior, assuming excessive values of estimation delay is not useful. It seems that an assumed fixed estimation delay of the order of the (assumed) evader first-order time constant is a reasonable compromise.

In Figures 11.7a and b, the homing performances of DGL/C and DGL/1 against randomly switched "bang-bang" type commanded evader acceleration are compared using a slow and a faster ECA estimator. These figures clearly indicate the benefit of the deterministic delayed estimation model, assuming a constant information delay in the target acceleration, as well as its limitations. The guaranteed (worst-case) miss distance is substantially reduced, but very small miss distances cannot be achieved.

11.3.3 Refined Deterministic Estimation Models

The main reason for the limitations of the proposed deterministic estimation model is being only a rough approximation of the real estimation process. The stochastic features of the estimation process are neglected. The information of all the estimated state variables, not only the target acceleration, is delayed. Although the effect of these other delays is less important, they should not be fully neglected. The assumption of a constant information delay is also not accurate. The delay induced by the estimator dynamics, determined by the noise model and its power density, is generally range dependent.

Several recent works proposed new methods to improve homing performance, using either refined deterministic estimation models or a more elaborate computation. In [20], a more sophisticated computing method is used. Although it assumes that the information delay affects only the estimated target acceleration, it takes into account the (assumed-to-be-perfect) measurements of the other state variables during the estimation delay. This method leads to an impressive reduction of the *target acceleration uncertainty* segment and consequently a new deterministic pursuit–evasion game. The guidance law based on this game solution yields an additional impressive reduction of the guaranteed miss distance compared with DGL/C. Unfortunately, simulations incorporating a realistic noise model and an estimator in the loop show much less improvement in homing performance. In another study [21], the time-varying nature of the information delay induced by the estimation was assumed, showing a marginal improvement compared with DGL/C.

A recent work [22] considered, in addition to the information delay in the estimated target acceleration, a smaller fixed delay in the relative velocity normal to the line of sight. Based on this model, a new deterministic pursuit–evasion game was formulated and solved. Simulation results, using the guidance law derived from the new game solution (denoted as DGL/CC) with realistic noise model and an estimator, showed an improvement in the reduction of the guaranteed miss distance compared with DGL/C against "bang-bang" maneuvers with random switch. This improvement came at the expense of increased average miss distances against target maneuver switches at the early part of the endgame.

11.3.4 Section Summary

In this section, several deterministic estimation models were reviewed. The stochastic features of the estimation process, as well as the eventual randomness of the target behavior, are neglected in the deterministic models. Each of them led to formulation of a different deterministic pursuit–evasion game model for the interception endgame of a maneuverable target. The game solution yielded the guidance law (the optimal pursuer strategy), the worst admissible target maneuver (the optimal evader strategy), and the guaranteed miss distance (the value of the game). The optimality of the game solution from the interceptor missile point of view is the reduction of the guaranteed (worst-case) miss distance compared with other guidance laws. This reduction came in each case at the expense of increasing miss distances against nonoptimal target maneuvers. For this reason, the practical value of such guidance laws is not obvious.

11.4 Estimation-Dependent Guidance Laws

The attempts reviewed in the previous section were motivated by the principle of *partial separation* property asserted by Witsenhausen [11], stating that the estimator can be designed independently of the controller, but the derivation of the optimal control function has to consider the results of the estimation process. As mentioned earlier, there have been very few past works that used a rigorous practical approach for implementing this idea. In this section, a direct and rigorous approach is taken by deriving the optimal control functions on the basis of the conditional probability density function (conditioned on the measurement history) of the estimated state variables, taking explicitly into account the saturated acceleration command of the interceptor. In the next subsection, linear and nonlinear estimation-dependent guidance laws are presented for the case of Gaussian noise models, with the Kalman filter used for the estimation. In the following subsection, the assumption of Gaussian noise was abandoned, resulting in the use of a *particle filter* and an estimation-dependent guidance law, based on an extension of the reachable-set concept.

11.4.1 Stochastic Optimal Control Guidance Laws

In this subsection, linear and nonlinear estimation-dependent one-sided optimal control-based guidance laws are presented for the case where the noise is Gaussian. These problems have been solved in [23] and [24], respectively. Here, only the essential elements of the derivation are given. For further details, the reader is referred to the above references.

11.4.1.1 Problem Formulation

Consider a first-order missile (pursuer) with a time constant τ_P and bound a_P^{max} on the acceleration command. Its bounded control can be described by the following standard saturation function:

$$sat(u) = \begin{cases} U_P & u > U_P \\ u & -U_P \leq u \leq +U_P, \\ -U_P & u < -U_P \end{cases} \qquad (11.23)$$

where in the investigated problem, $U_P = a_P^{max}$.

The target (evader) is assumed to perform a random maneuver with known statistics, and consequently, the respective shaping filter can be used for the derivation of the Kalman filter [25].

The state vector for this problem is defined by

$$x^T = (y \quad \dot{y} \quad a_E \quad a_P), \qquad (11.24)$$

where y is the relative displacement between the target and the interceptor in the Y direction. The corresponding state equation is

$$\dot{x} = Ax + B sat(u) + Gw \quad ; \quad x(0) = x_0, \qquad (11.25)$$

where

$$A = \begin{bmatrix} 0 & 1 & 0 & 0 \\ 0 & 0 & 1 & -1 \\ 0 & 0 & 0 & 0 \\ 0 & 0 & 0 & -1/\tau_P \end{bmatrix} ; B = \begin{bmatrix} 0 \\ 0 \\ 0 \\ 1/\tau_P \end{bmatrix} ; G = \begin{bmatrix} 0 & 0 \\ 0 & 0 \\ \sqrt{q} & 0 \\ 0 & \sigma_P \end{bmatrix} , \qquad (11.26)$$

and $w = [w_E, w_P]$ is a two-dimensional white Gaussian noise representing the target and the pursuing missile acceleration uncertainty. Their power spectral densities are q and σ_P^2, respectively.

It is assumed that the range to the target can be accurately measured and that the LOS angle is measured with white Gaussian noise ν with a standard deviation of σ_ϕ:

$$Y = cx + \nu, \qquad (11.27)$$

where

$$c = [1/r \ 0 \ 0 \ 0]. \qquad (11.28)$$

Moreover, a simplified glint model is assumed. Hence, the measurement statistics has a lower bound at a critical range r_c:

$$V = \begin{cases} \sigma_\phi^2 & r > r_c \\ \sigma_\phi^2 \cdot (r_c/r)^2 & r \le r_c \end{cases}. \tag{11.29}$$

The cost function to be minimized is

$$J = E\left[\mathbf{x}_f^T \mathbf{S}_f \mathbf{x}_f + \int_0^{t_f} Ru^2\, dt \right], \tag{11.30}$$

where

$$\mathbf{S}_f = \begin{bmatrix} s & 0 & 0 & 0 \\ 0 & 0 & 0 & 0 \\ 0 & 0 & 0 & 0 \\ 0 & 0 & 0 & 0 \end{bmatrix}. \tag{11.31}$$

11.4.1.2 Estimator

Based on the one-way separation discussed earlier, the estimator is derived independently from the control law. The estimated state vector is denoted $\hat{\mathbf{x}}$. We denote the estimation error as e, and it satisfies

$$e = \hat{\mathbf{x}} - \mathbf{x}. \tag{11.32}$$

Assuming that $\hat{\mathbf{x}}_0$, $\omega_T(t)$, $\omega_M(t)$, and $v(t)$ are mutually uncorrelated, we obtain the following filter equations:

$$\dot{\hat{\mathbf{x}}} = \mathbf{A}\hat{\mathbf{x}} + \mathbf{B}\mathrm{sat}(u) + \mathbf{K}_f(Y - c\hat{\mathbf{x}}) \tag{11.33}$$

$$\mathbf{K}_f = \mathbf{P}_{ee}\mathbf{c}^T\mathbf{V}^{-1} \tag{11.34}$$

$$\dot{\mathbf{P}}_{ee} = \mathbf{A}\mathbf{P}_{ee} + \mathbf{P}_{ee}\mathbf{A}^T - \mathbf{P}_{ee}\mathbf{c}^T\mathbf{V}^{-1}\mathbf{c}\mathbf{P}_{ee} + \mathbf{G}\mathbf{G}^T \;;\; \mathbf{P}_{ee}(0) = \mathbf{P}_{ee}^0 \tag{11.35}$$

where $\mathbf{P}_{ee}$ is the covariance matrix of the estimation error and $\mathbf{P}_{ee}^0$ is given. The estimator has the following initial statistics: $E[\hat{\mathbf{x}}(0)] = \hat{\mathbf{x}}_0$, $E[\hat{\mathbf{x}}(0)\hat{\mathbf{x}}^T(0)] = \mathbf{P}_{\hat{x}\hat{x}}^0$.

For the stochastic guidance law derivation presented next, we will use the expected value of the estimated state vector $\hat{\mathbf{x}}$ as well as the covariance matrix $\mathbf{P}_{ee}$.

11.4.1.3 Linear Stochastic Guidance Law

Let us first obtain the linear guidance law, denoted as SOGL, in the form of

$$u - \mathbf{k}_c\, \hat{\mathbf{x}}. \tag{11.36}$$

We then approximate the nonlinear saturation function sat(u) with a random input describing function L [26] and obtain

$$\dot{\hat{x}} = (A + LBk_c)\hat{x} + K_f(Y - c\hat{x}) \tag{11.37}$$

$$\dot{P}_{\hat{x}\hat{x}} = (A + LBk_c)P_{\hat{x}\hat{x}} + P_{\hat{x}\hat{x}}(A + LBk_c)^T + K_f V K_f^T; P_{\hat{x}\hat{x}}(0) = P_{\hat{x}\hat{x}}^0 \tag{11.38}$$

while the cost function to be minimized is given by

$$J = E[x_f^T S_f x_f + \int_0^{t_f} (x^T Q x + Ru^2) dt]$$

$$\tag{11.39}$$

$$= tr[(P_{\hat{x}\hat{x}}^f + P_{ee}^f)S_f] + \int_0^{t_f} \{tr[(P_{\hat{x}\hat{x}} + P_{ee})Q] + Rk_c P_{\hat{x}\hat{x}} k_c^T\} dt.$$

Notice that P_{ee} does not depend on k_c, and therefore, in the previous section, the estimator was derived independently from the guidance law. The solution of this guidance problem constitutes a two-point boundary value problem. The details of the solution appear in [23]. The important aspect is that the solution yields $k_c(t)$ that is dependent on $P_{ee}(t)$, that is, an estimation-dependent guidance law. Specifically, the guidance law is

$$u = \frac{N'(t_{go})}{t_{go}^2} \hat{Z}(t_{go}) \tag{11.40}$$

where $\hat{Z}(t_{go})$ is the estimated zero-effort miss, expressed in this problem as (using a shaping filter as an integrator driven by white noise)

$$\hat{Z} = \hat{y} + \dot{\hat{y}} t_{go} - \hat{a}_M[\exp(-\theta) + \theta - 1] + \hat{a}_T t_{go}^2/2. \tag{11.41}$$

Note that this form of the ZEM is equivalent to assuming in the perfect information derivation known constant target acceleration. $N'(t_{go})$ is dependent on $P_{ee}(t)$, satisfying $\lim_{t_{go}\to\infty} N'(t_{go}) = 3$. Note that this guidance law degenerates to the classical OGL of Cottrell [27], with the appropriate navigation gain $N'(t_{go})$, only if the acceleration bound is not active. In such a case, the problem is LQG, and the certainty equivalence principle is valid.

In Figures 11.8 and 11.9, the effective navigation gain of SOGL is plotted for an example scenario. The results are plotted for different saturation limits and noise levels. It is evident that the effective navigation gain of SOGL

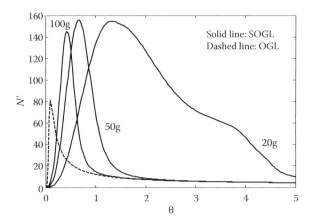

FIGURE 11.8
Example of effective navigation gain for different acceleration limits.

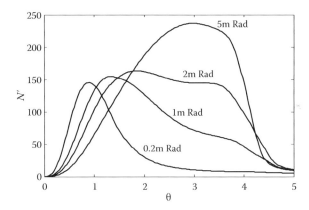

FIGURE 11.9
Example of effective navigation gain for different measurement noise levels.

spreads the control effort over the entire scenario. It is also evident that as the saturation limit is raised, the maximum value of N' occurs later in the scenario, since the need to perform hard maneuvers early in the engagement is reduced. Note that N' obtained by solving the OGL case, calculated with the same finite terminal weight ($s = 10^9$) as used in the SOGL laws, coincides with the results of SOGL only for an extremely large acceleration limit. Also observe that as the noise level is raised, the maximum value of N' increases and occurs earlier in the scenario, since the need to perform maneuvers early in the engagement is increased.

Remember that the actual gain is defined as $N'(t_{go})/t_{go}^2$. It is clear that although SOGL enforces a larger navigation gain $N'(t_{go})$, it actually provides significantly lower maximum gains than OGL. This is achieved since SOGL considers from the outset the bounded acceleration and thus uses a higher navigation gain earlier in the scenario (larger t_{go}).

The homing performance compared with OGL was investigated in [23]. It was shown that as long as the saturation limit is active (i.e., the interceptor does not have a large maneuver advantage over the target), the new estimation-dependent guidance law provides superior performance. If the interceptor has a large maneuver capability (i.e., the saturation influence is reduced), the guidance laws have similar homing performance, as in the limit case of no saturation, and SOGL degenerates to OGL.

11.4.1.4 Nonlinear Stochastic Guidance Law

The nonlinear stochastic guidance law for this problem was obtained in [24]. Here too, since the certainty equivalence principle is not valid, the resulting OGL depends on the conditional probability density function of the estimated states. The general cost function chosen in this case was

$$J = E\left\{\phi\left[x(t_f)\right] + \frac{1}{2}\int_0^{t_f} Ru^2 dt\right\}. \tag{11.42}$$

Note that this function is not necessarily quadratic. Selecting $\phi[x(t_f)] = s_f y^2(t_f)$, we obtain the same cost function as in Equation 11.30, enabling us to compare the results also with the LQG OGL.

In order to obtain the optimal control command, the resulting Hamilton–Jacobi–Bellman equation

$$0 = \frac{\partial v}{\partial t} + \min_u\left\{l(x,u) + \frac{\partial v}{\partial x} f(x,u)\right\} + \frac{1}{2}\frac{\partial^2 v}{\partial x^2} g^2 \tag{11.43}$$

needs to be solved numerically, where $v(x, t)$ is the optimal cost to go from the initial state x at time t. The solution was obtained using the Markov chain approximation method [28]. The essence of the method is in approximating the stochastic differential equation of the problem by a discrete time and space Markov chain. The Markov chain is obtained by the finite differencing equation (Equation 11.43) and by identifying the coefficients in the finite difference equation with the transition probabilities of the Markov chain.

The solution of this equation for the fourth-order state vector x is computationally intensive and is currently impractical when using a desktop computer. In order to substantially reduce the amount of calculations required to obtain the numerical solution, the problem can be reduced to a scalar one by using the zero-effort miss distance concept.

In this approach, the target maneuver was modeled as a first-order Gauss–Markov process:

$$\dot{a}_E = -a_E / \tau_E + w_E \tag{11.44}$$

where w_E is a white process noise with spectral density Q. As a consequence, the expression of the zero-effort miss distance will be similar as in Equation 11.1, rather than Equation 11.41, yielding

$$\dot{Z} = -\text{sat}(a_c)\tau_P \psi\left(t_{go}/\tau_P\right) + w\tau_E^2 \psi\left(t_{go}/\tau_E\right). \tag{11.45}$$

Using this approach, we obtain the nonlinear optimal control law in the form

$$u = k(t_{go}, \hat{Z}) \tag{11.46}$$

where $\hat{Z}$ is the output of the estimation process discussed in Section 11.4.1.2.

An example of the numerically obtained function $k(t_{go}, \hat{Z})$ is plotted in Figures 11.10 and 11.11 for $t_{go} = 0.25$s and $t_{go} = 0.5$s, respectively. This non-linear guidance law generates larger acceleration commands earlier in the scenario than the classical linear OGL. Further, lowering the bound on the missile acceleration causes the guidance law to reach saturation for lower values of the estimated zero-effort miss. In fact, for the acceleration bound of 150 m/s², the guidance law is essentially a sign function for the entire inves-tigated time interval.

It is evident that this guidance law is slightly nonlinear, generating larger acceleration commands than a purely linear guidance law. Note that for the

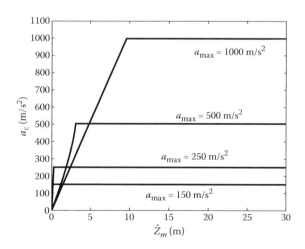

FIGURE 11.10
Function $k(t_{go}, \hat{Z}_m)$ at $t_{go} = 0.25$s.

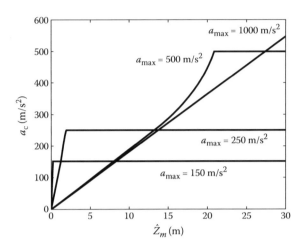

FIGURE 11.11
Function $k(t_{go}, \hat{Z}_m)$ at $t_{go} = 0.5s$.

acceleration limit of 1000 m/s², the command is linear and is identical to the OGL gain.

The homing performance compared with OGL was investigated in [24]. It was shown that, similarly to SOGL presented earlier, as long as the saturation limit is active, the new estimation-dependent nonlinear guidance law provides superior performance. Also, if the interceptor has a large maneuver capability, the guidance laws have similar homing performance, as here too in the limit case of no saturation, the guidance law degenerates to OGL.

Both SOGL and the nonlinear stochastic guidance law can be solved offline. The resulting gains can then be saved and implemented in real time by gain scheduling.

11.4.2 Fusion of Estimation and Guidance

While in the previous subsection, the measurement noise was assumed to be Gaussian, in a very recent research study [29], this limitation was abandoned, and an entirely new approach, allowing nonlinear dynamics, was developed based on the concept of the *partial separation* property [11]. In this study, the estimator is a *particle filter* [30], based on sequential Monte Carlo methods [31], that can deal with any noise model with arbitrary probability distribution function (PDF). Once the conditional PDF of the estimated state is available, a new guidance law that extends the concept of *reachable sets* [17] is developed. A successful interception (capture) is guaranteed as long as the interceptor's reachable set at the predicted interception time includes the reachable set of the target. As the interception is progressing, both reachable sets contract. The interceptor's reachable set converges to a singleton, while the estimated position of the target is a *set*, expressed by the conditional PDF, due to the uncertainty created by the noisy measurements. As a consequence, capture (in the deterministic sense) cannot be guaranteed. The new guidance law has, therefore, two phases. In the initial phase, where the requirement of *inclusion* is satisfied, the objective of the guidance law is to keep this inclusion property in the future as long as it is possible. This guidance law is of course not unique, and the remaining level of freedom can be used for *trajectory shaping* in order to facilitate and improve the estimation process. From that point of time, when complete inclusion cannot be satisfied any more, the guidance objective becomes to maximize the coverage of the target's estimated position set by the reachable set of the interceptor. An extensive set of Monte Carlo simulations showed that the homing performance of this new approach is superior compared with that of a game optimal deterministic guidance law (such as DGL/1) using classical state estimates (assuming the validity of the *separation* and *certainty equivalence* properties). The only problem of this promising new approach is its heavy computational demand, and therefore, at the present, it cannot be implemented in real time onboard a homing interceptor missile. Nevertheless, it is expected that using advanced parallel computation methods, this difficulty can be overcome in the future.

11.5 Logic-Based Integrated Estimation and Guidance

11.5.1 Engineering Approach

The different mathematical approaches for synthesizing an integrated estimation/guidance law method for the interception of randomly maneuvering targets, presented in the previous sections, were either not fully satisfactory or not yet implementable in real time. In this section, an engineering approach toward an integrated estimation and guidance algorithm is introduced. It is based on the insight gained by understanding the inherent limitations of the classical estimation in a short-duration interception endgame. Since no single estimator can satisfy the requirements of homing accuracy, the different tasks performed by a classical estimator have to be separated and assigned to different elements within a corporate estimation system.

The main task, directly affecting the homing accuracy, is the estimation of the state variables (including the target acceleration) involved in the guidance law. This task can be performed in a satisfactory manner by a narrow-bandwidth filter if (and only if) the correct model of the target maneuver is available. Thus, at the initial part of the endgame, the task of highest priority is model identification using a multiple-model structure [32]. The filters for this task should be of sufficiently large bandwidth in order to complete the model identification as quickly as possible. As the model identification is completed, the role of the state estimator is assigned to an appropriate narrow-bandwidth filter. As long as the target model does not change, this narrow-bandwidth filter provides the estimated state variables to the guidance law. An eventual change in the target maneuver has to be detected by a "fast" detection filter in order to minimize the detection delay.

This engineering synthesis concept was first developed using a planar (horizontal) constant-speed model against a randomly switched "bang-bang" target maneuver [33]. Later, it was extended and validated in a realistic three-dimensional (3D) variable-speed endo-atmospheric ballistic missile defense scenario [34]. In the 3D scenario, two types of stressing random target maneuvers, namely, a randomly switched "bang-bang" target maneuver and a random phase "barrel-roll" type (spiral) maneuver, were considered.

In this section, these integrated estimation/guidance algorithms are explained together with their implementation, and the achieved homing performance is presented.

11.5.2 Planar Algorithm

11.5.2.1 Estimation

In a planar scenario, a "bang-bang" type maneuver is the most effective for evasion; thus, the "model" has to include the direction of the current target acceleration. As long as the model is not identified, a simple Kalman filter and the guidance law DGL/0 [35], both ignoring the target maneuver, are used. After model identification, the state estimation is assigned to the appropriate narrow-bandwidth filter, and the guidance law DGL/1 [7], considering a first-order target maneuver model, is used. If the anticipated direction reversal (switch) occurs sufficiently far away from the end of the interception, there is enough time for the filter to converge to the new value. The DGL/1 guidance law using the correct value of the target acceleration achieves small miss distances, as shown in Figure 11.1. However, if the change takes place after a "critical" time-to-go, there is not enough time available for this process, and large miss distances are created.

Once this "critical" time-to-go, determined by the parameters of the narrow-bandwidth filter (t_{go} = 1.6 s in Figure 11.1), is identified, the estimation/guidance strategy for the remaining part of the endgame is changed. In an earlier paper, also dealing with a planar scenario [16], a multiple-model estimator, where each estimator model assumed a different timing of the direction reversal (switch) of a "bang-bang" type maneuver, is described. Using such an estimator "tuned" to the correct switch eliminates the delay and yields excellent homing performance. Even if the switch occurs shortly after the time anticipated by the estimator, good performance is obtained. Due to this robustness property, a few adequately "tuned" estimators with not-too-narrow bandwidth can cover the range of interest (the remaining part of the endgame). The relevant "tuned" estimator becomes active (replaces the narrow-bandwidth filter for providing state estimates to the guidance law) as soon as the direction change of the target model is observed by the "fast" detection filter.

11.5.2.2 Guidance Law Modifications

In spite of the improved estimation by the "tuned" estimator, if the change of direction occurs close to the end, the interceptor is unable to reach its maximum lateral acceleration and cannot correct the guidance error generated during the detection delay (due to the short time remaining). This deficiency is alleviated by increasing the lateral acceleration command for small values of time-to-go but still respecting the actual acceleration limits.

The increase in the commanded acceleration gain is expressed for $t_{go} \leq (t_{go})_{sw}$ by

$$a_P^c = a_P^c(t_{go}, k) = \frac{a_P^{max} \operatorname{sign} Z}{1 - k \exp\left(-\dfrac{t_{go}}{\tau_P}\right)}. \tag{11.47}$$

The parameter k is selected to satisfy

$$|a_p(t_f, k)| = a_P^{max}. \tag{11.48}$$

It must be less than 1; otherwise, the gain will become infinite. Its value depends on $(t_{go})_{sw}$ and the value of a_P at that very moment. The effect of this modification is illustrated in Figure 11.12.

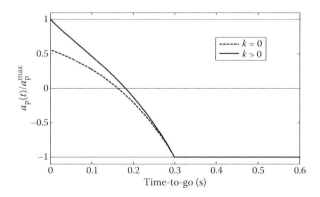

FIGURE 11.12
Effect of command gain enhancement.

Further homing improvement is achieved by introducing a time-varying dead zone version of the *signum* function in the guidance law after the "critical" time-to-go is reached:

$$\text{sign}_{dz}(Z) = \begin{cases} 1.0, & Z > A_{dz}\exp(-b_{dz}(t_f - t_{go})) \\ 0.0, & |Z| \le A_{dz}\exp(-b_{dz}(t_f - t_{go})) \\ -1.0, & Z < -A_{dz}\exp(-b_{dz}(t_f - t_{go})) \end{cases} \qquad (11.49)$$

where A_{dz} is the initial amplitude, and b_{dz} is the exponential decay rate of the dead zone. This modification reduces the error created during the period of detection delay, as illustrated in Figure 11.13. The dead zone is used only

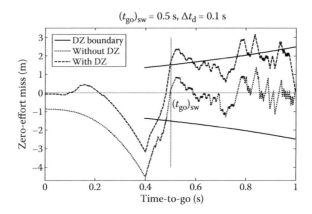

FIGURE 11.13
Effect of time-varying dead zone.

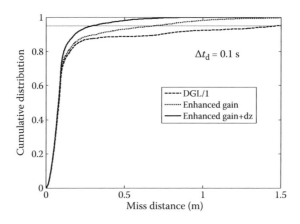

FIGURE 11.14
Planar homing improvement by guidance law modifications.

until the maneuver direction change is detected. In the simulations, the values of $A_{dz} = 50$ m and $b_{dz} = 1/s$ were selected.

This modification, reducing the error created during the detection delay, is used only until the maneuver direction change is detected. The cumulative miss distance distributions demonstrating the improvements achieved by the guidance law modifications for the case of a 0.1 s detection delay are shown in Figure 11.14.

11.5.3 Three-Dimensional Algorithm

The planar algorithm from the previous section has been extended to deal with 3D endo-atmospheric BMD scenarios. A major element of the extension was to include two basic types of target maneuver models. The first one is a slowly varying, piecewise continuous planar "bang-bang" maneuver, assuming a roll-stabilized target. For the sake of comparability with the work of Shinar et al. [33], the maneuvers are oriented in a horizontal plane. The amplitude of the maneuver is monotonically increasing as the target descends to lower altitudes. The second type of maneuver assumes a rolling target with a fixed angle of attack in body coordinates, creating a "barrel-roll" type (spiral) maneuver of time-varying (monotonically increasing) amplitude. A spinning aerodynamically stable reentry vehicle will inherently perform a similar maneuver [36]. (In reality, the maneuver frequency can be slowly time-varying, but in the short-duration endgame, it can be approximated as constant.) Each type of maneuver requires a different type of estimator. The distinction between the two different types of target maneuvers considered in this study is an essential element in this approach.

11.5.3.1 Estimation

The group of estimators for the "bang-bang" type maneuver is similar to the one used in [33]. The only difference is that instead of a constant-acceleration command, a monotonically increasing one is anticipated.

The second type of evasive target maneuver creating a "barrel-roll" type (spiral) trajectory of time-varying radius requires a different estimator. Because this is a typical 3D maneuver, two planar estimators have to be used to estimate the projections of the motion in two perpendicular planes. Within each plane, the motion is periodical with random phase, and therefore, the appropriate shaping filter has to be of the second order [2], assuming a known maneuver frequency. Such a Kalman filter estimates not only the target acceleration but also its time derivative (the "jerk"). If the maneuver frequency is correctly predicted, the output of such an estimator converges well to the actual maneuver even if its amplitude is slowly varying. As a consequence, the homing accuracy is satisfactory. However, if the frequency used in the estimator is incorrect, the estimation is degraded, and the homing performance is poor. The bandwidth of such a periodical estimator can be tuned to allow a reasonable error in the predicted frequency without a great compromise in the homing accuracy.

The first task to be carried out, immediately as the interception endgame starts, is to distinguish between the two different maneuver types using a multiple-model structure [12]. The filters for this task are of large bandwidth, in order to complete the model identification as quickly as possible. If the actual maneuver is either planar or periodical with a frequency of $p_0 \geq 1$ Hz, this first phase of model identification can be completed within the first 0.5 s of the endgame. For periodical maneuvers of lower frequency, the identification may require more time (about half a period).

Once the decision between the two types of maneuvers is made, the second phase of the target model identification for each one becomes different. For the "bang-bang" type target maneuver, the direction of the maneuver as well as the initial amplitude is provided by the first phase. For a "spiral" maneuver, the frequency range of the model has to be identified with a reasonable accuracy. This process requires a large set of different estimator models and needs more time (up to a second), depending on the accuracy level.

After the completion of the model identification, the appropriate narrow-bandwidth state estimator is selected to forward information to the guidance law. Continuous computation of the *a posteriori* probabilities is used to confirm the correctness of the selection. For a "spiral" maneuver, no dramatic changes in the model are expected. For the "bang-bang" type target maneuver, an eventual change of direction is expected and has to be detected by a sufficiently fast detector. Following the detected change, the nearest "tuned" estimator is used, as in [33]. Since the development of such fast detector is yet incomplete [37], in the simulations, the detection delay is parameterized.

11.5.3.2 Guidance Law

In the first phase of the endgame, until the target maneuver is identified, a simple narrow-bandwidth estimator and a differential game-based guidance law, denoted as DGL/0 [35], not requiring the knowledge of the target maneuver, are used.

The guidance law DGL/1 used in [7], where a planar scenario was investigated, was derived based on the assumption of constant speeds. Since in a 3D scenario, the speeds are not constant and the bounds of the lateral accelerations vary with speed and altitude, the model of the game has to be modified. It is assumed that profiles of the time-varying parameters are known along a nominal trajectory. Such a model is suitable for the analysis of a realistic BMD scenario.

The guidance law based on the solution of this game [38], although qualitatively similar to DGL/1, strongly depends on the respective velocity and maneuverability profiles of the *players*. Due to the time-varying profiles, the expressions of the zero-effort miss distance and the other elements of the solution become more complex, as shown in detail in [38]. In spite of this (algebraic) complexity, the implementation of the optimal missile guidance law, denoted as DGL/E, does not present essential difficulties. It requires, of course (in addition to the perfect knowledge of the current lateral acceleration of the target), the speed and maneuverability profiles in the endgame that can be precalculated along a nominal trajectory.

If the target maneuver is planar, the guidance law modifications of Section 11.5.2.2 applied to DGL/E are used. For periodical maneuvers, these modifications are not needed.

11.5.3.3 Simulation Results

For testing the validity of the 3D algorithm, endo-atmospheric interception scenarios terminating between altitudes of 20–30 km with an initial range of 20 km were considered with endgame duration of the order of 4.5 s. The data and the detailed summary of these simulations are presented in the Appendix. Homing accuracy statistics, expressed by the cumulative miss distance distributions, are based on 1000 Monte Carlo simulation runs for each scenario, assuming Gaussian noise and uniform reversal time (switch) or phase distributions.

The main results are summarized in the following.

(1) Against random horizontal bang-bang maneuvers at three interception altitudes (20, 25, and 30 km), assuming a detection delay of 0.1 s, miss distances of 20–30 cm (or less) for 95% of the cases were obtained, similar to the results in [33].

(2) The homing accuracy against spiral target maneuvers of different roll rates using perfectly matched periodical estimators is independent of the roll rate and similar to the accuracy against the "bang-bang" maneuvers.

(3) If the frequency used in the estimator is different from the actual roll rate, the homing accuracy is gradually degraded, depending on the size of the mismatch.

(4) There is a significant performance degradation of the homing performance if the maneuver type (bang-bang or periodical) is not correctly identified.

11.6 Conclusions

This chapter discussed the not-yet-completely solved problem of how to design and implement in a high-performance interceptor missile the estimation process and the guidance law. The deficiency of the common practice, based on assuming the validity of the *certainty equivalence* and *separation* properties, which created reduced homing accuracy and the lack of robustness with respect to random target maneuvers, is illustrated.

Several approaches attempted alleviating these difficulties. Recognizing the time delay associated with the estimation process and compensating for it led to a substantial reduction of the guaranteed miss distance against the optimal (worst-case) target maneuver. This reduction, based on a deterministic estimation model, came at the expense of less efficient homing against other nonoptimal target behavior.

More rigorous mathematical approaches that addressed directly the stochastic optimization problem in different ways were based on the concept of *partial separation* property. These methods showed impressive improvements in homing performance but require a heavy computational load.

The "logic-based" integration of estimation and guidance, based on an engineering approach and presented in the last section, suggested a practically acceptable and implementable solution (which is probably only suboptimal) for realistic interceptions of randomly maneuvering targets in 3D space, such as endo-atmospheric ballistic missile defense.

References

1. Kalman, R. E., A new approach of linear filtering and prediction problems, *Trans. ASME*, 82D, 1960, 35.
2. Zarchan, P., Representation of realistic evasive maneuvers by the use of shaping filters, *J Guidance Control*, 2(1), 1979, 290.
3. Joseph, P. D. and Tau, J. T. On linear control theory, *Trans. AIEE*, Part III, 80(18), 1961.
4. Tse, E. and Bar Shalom, Y., Generalized certainty equivalence and dual effect in stochastic control, *IEEE Trans. Auto Control*, AC-20(6), 1975, 817.
5. Wonham, W. M., On the separation theorem of stochastic control, *SIAM J. Control*, 6(2), 1968, 312.
6. Shinar, J. and Turetsky, V., What happens when certainty equivalence is not valid?—Is there an optimal estimator for terminal guidance?, *Annu. Rev. Control*, 27(2), 2003, 119.
7. Shinar, J., Solution techniques for realistic pursuit-evasion games, in *Advances in Control and Dynamic Systems*, C. T. Leondes, ed., vol. 17, Academic Press, New York, 1981, p. 63.
8. Singer, R. A., Estimating optimal filter tracking performance for manned maneuvering targets, *IEEE Trans. Aerosp. Electron. Syst.*, ES-6(4), 1970, p. 473.
9. Rotstein, H. and Szneier, M., An exact solution to the general 4-blocks discrete-time mixed H_2/H_∞ problems via convex optimization, *IEEE Trans. Auto Control*, AC-43(6), 1998, p. 1475.

10. Lai, T. L. and Shan, Z., Efficient recursive algorithms for detection of abrupt signal and systems, *IEEE Trans. Auto Control*, AC-44(5), 1999, p. 952.
11. Witsenhausen, H. S., Separation of estimation and control for discrete time systems, *Proc. IEEE*, 59(11), 1971, p. 1557.
12. Maybeck, P. S., *Stochastic Models, Estimation and Control*, vol. 2, Academic Press, New York, 1982.
13. Magill D. T., Optimal adaptive estimation of sampled stochastic processes, *IEEE Trans. Auto Control*, AC-10(4), 1965, p. 434.
14. Tam, P. and Moore J. B., Adaptive estimation using parallel processing techniques, *Comput. Elect. Eng.*, 2(2/3), 1975.
15. Rusnak, I., Multiple model-based terminal guidance law, *J. Guidance Control Dynamics*, 23(4), 2000, 742.
16. Shima, T., Oshman, Y., and Shinar, J., Efficient multiple model adaptive estimation in ballistic missile interception scenarios, *J. Guidance Control Dynamics*, 25(4), 2002, 667.
17. Petrosjan, L. A., *Differential Games of Pursuit*, Series on Optimization, vol. 2, World Scientific Publishing, Singapore, 1993, p. 169.
18. Shinar, J. and Glizer, V. Y., Solution of a delayed information linear pursuit–evasion game with bounded controls, *Int. Game Theory Rev.*, 1(3&4), 2000, 197.
19. Shinar, J. and Shima, T., Non-orthodox guidance law development approach for the interception of maneuvering anti-surface missiles, *J. Guidance Control Dynamics*, 25(4), 2002, 658.
20. Shinar J. and Glizer, V. Y., New approach to improve the accuracy in delayed information pursuit-evasion games, in *Annals of Dynamic Games*. A. Hourie et al., ed., vol. 8, Birkhäuser, Boston, 2006, p. 65.
21. Shinar, J. and Glizer, V. Y., A linear pursuit–evasion game with time varying information delay, TAE Report No. 889, Technion, Haifa, 2002.
22. Glizer, V. Y. and Turetsky, V., A linear differential game with bounded controls and two information delays, *Opt. Contr. Appl. Methods*, published online, May 2008.
23. Hexner, G., Shima, T., and Weiss, H., LQG guidance law with bounded acceleration command, *IEEE Trans. Aerosp. Electron. Syst.*, 44(1), 2008, 77.
24. Hexner, G. and Shima, T., Stochastic optimal control guidance law with bounded acceleration, *IEEE Trans. Aerosp. Electron. Syst.*, 43(1), 2007, 71.
25. Fitzgerald, R. J. and Zarchan, P., Shaping filters for randomly initiated target maneuvers, *Proceedings of the AIAA Guidance and Control Conference*, Palo Alto, CA, August 1978, p. 424.
26. Gelb, A. and Velde, W. E. V., *Multiple-Input Describing Functions and Nonlinear System Design*, McGraw-Hill, New York, 1968, p. 365.
27. Cottrell, R. G., Optimal intercept guidance for sort-range tactical missiles, *AIAA J.* 9, 1971, 1414.
28. Kushner, H. J., Numerical methods for stochastic control problems in continuous time, *SIAM J. Control Opt.*, 28(5), 1990, 999.
29. Shaviv, I. G. and Oshman, Y., Estimation-guided guidance, *Proceedings of the AIAA Guidance, Navigation and Control Conference*, Keystone, CO, August 2006.
30. Karlsson, R. and Bergman, N., Auxiliary particle filters for tracking a maneuvering target, *Proceedings of the IEEE Conference on Decision and Control*, Sydney, Australia, December 2000.

31. Doucet, A., DeFreitas, N., and Gordon, N., *Sequential Monte Carlo Methods in Practice*, Statistics in Engineering and Information Science, Springer-Verlag, New York, 2001.
32. Bar-Shalom, Y. and Li, X. R., *Estimation and Tracking: Principles, Techniques, and Software*, Artech House, Boston, 1993, p. 262.
33. Shinar, J., Turetsky, V., and Oshman, Y., Integrated estimation/guidance design approach for improved homing against randomly maneuvering targets, *J. Guidance Control Dynamics*, 30(1), 2007, 154.
34. Shinar, J. and Turetsky, V., Three-dimensional validation of an integrated estimation/guidance algorithm against randomly maneuvering targets, *J. Guidance Control Dynamics*, 32(3), 2009, 1034.
35. Gutman, S., On optimal guidance for homing missiles, *J. Guidance Control*, 3, 1979, 296.
36. Shinar, J. and Ben Asher, J. Z., Interception of naturally maneuvering reentry vehicles, *Proceedings of the 4th AAAF International Conference on Missile Defense "Challenges in Europe"*, Heraklion, Crete, Greece, June 26–29, 2007.
37. Shinar, J., Bokor, J., and Kulcsar, B., Application of unknown input detection theory to interceptor guidance, *Proceedings of the ICNPAA-2006, Mathematical Problems in Engineering and Aerospace Sciences*, Budapest, Hungary, June 21–23, 2006.
38. Shima, T. and Shinar, J., Time varying linear pursuit-evasion game models with bounded controls, *J. Guidance Control Dynamics*, 25(3), 2002, 425.

Appendix: Details of 3D Simulations

11.A.1 Simulation Data

In these simulations, the target is a generic TBM with aerodynamic control, performing either spiral or horizontal bang-bang evasive maneuvers. It is assumed to be launched from a distance of 600 km on a minimum-energy trajectory. It is characterized by a ballistic coefficient $\beta = 5000$ kg/m^2 and a trimmed lift-to-drag ratio $\Lambda = 2.6$. Its velocity at reentry of an altitude of 150 km is $V_{E0} = 1720$ m/s with a flight path angle of $\gamma_{E0} = -18°$ and a horizontal distance of 210 km from its surface target.

The interceptor is a generic roll-stabilized two-stage solid rocket missile that has two identical guidance channels for aerodynamic control (skid to turn). Its seeker provides angular measurements at a sampling rate of 100 Hz. These angular measurements are corrupted by zero-mean white Gaussian angular noise of constant amplitude with a standard deviation of 0.1 mrad.

Both stages of the rocket motor have a specific impulse of $I_{sp} = 250$ s. The propulsion, mass, and aerodynamic data of the stages are summarized in Table 11.A.1.

The second-stage rocket motor is ignited with a delay in order to guarantee that for any interception altitude, the endgame terminates with a positive

TABLE 11.A.1

Interceptor Data

	t_b (s)	T (kN)	m_0 (kg)	SC_D (m²)	SC_{Lmax} (m²)
First stage	6.5	229	1540	0.10	0.24
Second stage	13	103	781	0.05	0.20

longitudinal acceleration and nondecreasing maneuverability. The velocity profiles for different interception altitudes are shown in Figure 11.A.1.

During the endgame, the maneuverability of the target is monotonically increasing, and as a consequence, the value of the interceptor/target maneuverability ratio, denoted by μ, is monotonically decreasing, as seen in Figure 11.A.2. The maneuvering dynamics of the interceptor (*pursuer*) and the target

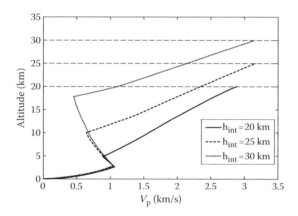

FIGURE 11.A.1
Interceptor velocity profiles.

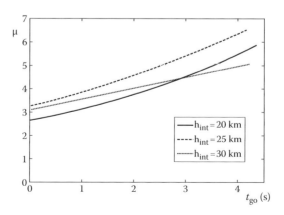

FIGURE 11.A.2
Endgame maneuverability ratio.

(*evader*) are approximated by first-order transfer functions with equal time constants $\tau_P = \tau_E = 0.2$ s.

11.A.2 Simulation Results

The results presented in this Appendix are limited to the following cases: (1) homing accuracy against random horizontal bang-bang maneuvers (using the integrated estimation/guidance scheme in [33]) at different interception altitudes; (2) homing accuracy against spiral target maneuvers of different roll rates (p_0) (using matched and unmatched periodical estimators), at the interception altitude of 25 km; and (3) homing accuracy degradation with wrong estimator type in both cases.

In Figure 11.A.3, the cumulative miss distance distributions against random horizontal bang-bang maneuvers at three interception altitudes (20, 25, and 30 km) assuming a detection delay of 0.1 s are shown. The miss distances are 20–30 cm (or less) for 95% of the cases, indicating the potential for satisfying a hit-to-kill requirement.

The cumulative miss distance distributions against random phase spiral target maneuvers with different roll rates with matched estimators are presented in Figure 11.A.4. It can be seen that if the maneuver frequency determined by the roll rate and the frequency used in the estimator are perfectly matched, the homing accuracy is independent of the roll rate (at least in the tested region) and similar to the accuracy against the "bang-bang" maneuvers.

However, if the frequency used in the estimator is different from the actual roll rate, the homing accuracy is degraded, as shown in Figure 11.A.5.

The importance of correct identification of the maneuver type is illustrated in Figures 11.A.6 and 11.A.7, showing the significant performance degradation if the wrong type of estimator is used.

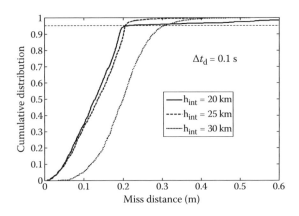

FIGURE 11.A.3
Homing accuracy against random horizontal bang-bang maneuvers.

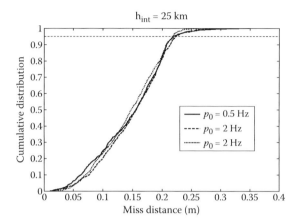

FIGURE 11.A.4
Homing accuracy against random phase spiral maneuvers (matched estimators).

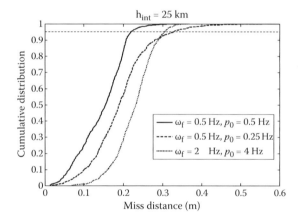

FIGURE 11.A.5
Homing accuracy against random phase spiral maneuvers (unmatched estimators).

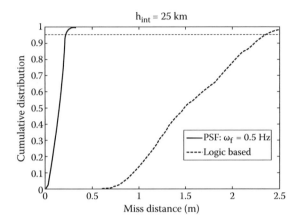

FIGURE 11.A.6
Homing accuracy degradation due to wrong estimator (spiral target maneuver).

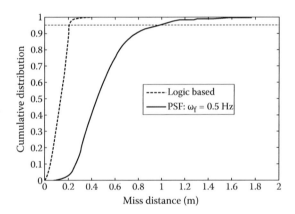

FIGURE 11.A.7
Homing accuracy degradation due to wrong estimator (bang-bang maneuver)

12

Introduction to Particle Filters for Tracking and Guidance

David Salmond

CONTENTS

12.1 Introduction

12.1.1 Aims

The aim of this chapter is to introduce particle filters to those with a background in "classical" recursive estimation based on variants of the Kalman filter. We describe the principles behind the basic particle filter algorithm and present an application to a tracking and guidance example involving multiple objects. A detailed worked example including some simple MATLAB® code is described in the Appendix. We also show that the basic algorithm is a special case of a more general particle filter that greatly extends the filter design options. The chapter concludes with a discussion of computational issues and application areas.

The emphasis of this chapter is on principles and applications at an introductory level. It is not a rigorous treatise on the subject, nor is it by any means an exhaustive survey. For a more detailed introduction (especially from a target-tracking perspective), the reader is referred to the work of Ristic et al. [1] (which uses the same notation as this paper). Other introductory articles include those of Djuric et al. [2], which includes applications to wireless communication problems; Candy [3], with an application to synthetic aperture sonar; and Fox et al. [4], focused on location estimation in a robotics context (which also discusses other Bayesian filters). Also see the extensive review [5]. For material on theoretical foundations and other applications, see [6–9] and a special issue of the IEEE Transactions on Signal Processing (Monte Carlo Methods for Statistical Signal Processing) [10]. For a survey of recent advances in particle filters, the reader is referred to the work of Cappe et al. [11].

12.1.2 Recursive Estimation

There is an enormous range of applications that require online estimates and predictions of an evolving set of parameters given uncertain data and dynamics—examples include object tracking, vehicle guidance, navigation and control, forecasting of financial indices, and environmental prediction. There is, therefore, a huge "market" for effective recursive estimation algorithms. Furthermore, if these problems can be posed in a common framework, it may be possible to apply general techniques over these varied domains. An obvious common framework consists of a dynamics model (describing the evolution of the system) and a measurement model that describes how available data are related to the system. If these models can be expressed in a probabilistic form, a Bayesian approach may be adopted.

12.1.3 Bayesian Estimation

The aim of a Bayesian estimator is to construct the posterior probability density function (pdf) of the required state vector using all available information. The posterior pdf is a complete description of our state of knowledge about (or uncertainty in) the required vector. As such, it is key to optimal estimation—in the sense of minimizing a cost function—and to decision and control problems. The recursive Bayesian filter provides a formal mechanism for propagating and updating the posterior pdf as new information (measurements) is received. If the dynamics and measurement models can be written in a linear form with Gaussian disturbances, the general Bayesian filter reduces to the Kalman filter that has become so widespread over the last 40 years. Mildly nonlinear problems can be linearized for Kalman filtering, but grossly nonlinear or non-Gaussian cases cannot be handled in this way.

A particle filter is an implementation of the formal recursive Bayesian filter using (sequential) Monte Carlo methods. Instead of describing the required pdf as a functional form, in this scheme, it is represented approximately as a set of random samples of the pdf. The approximation may be made as good as necessary by choosing the number of samples to be sufficiently large: as the number of samples tends to infinity, this becomes essentially an exact equivalent of the functional form. For multidimensional pdf's, the samples are random vectors. These random samples are the particles of the filter that are propagated and updated according to the dynamics and measurement models. Unlike the Kalman filter, this approach is not restricted by linear-Gaussian assumptions, so much extending the range of problems that can be tackled. The basic form of the particle filter is also very simple but may be computationally expensive: the advent of cheap, powerful computers over the last 15 years has been key to the introduction of particle filters.

12.1.4 Structure of the Chapter

In Section 12.2, we introduce the most basic version of the particle filter. Extension to a more general form and computational issues are discussed in Sections 12.3 and 12.4. An example of tracking in the presence of intermittent spurious objects is described in Section 12.5. In this section, we also show how the output of a particle filter can be used to generate a guidance demand without resorting to the usual "certainty equivalence" approach. This example is derived from [12–14]. In Sections 12.6 and 12.7, we briefly review other applications of the particle filter and draw some conclusions. The Appendix gives a "worked example" of applying the basic particle filter to a simple, but highly nonlinear, example, and it includes some MATLAB code, which will, hopefully, aid understanding of the algorithm.

12.2 Basic Particle Filter

12.2.1 Problem Definition: Dynamic Estimation

The dynamic estimation problem assumes two fundamental mathematical models: the state dynamics and the measurement equation.

The dynamics model describes how the state vector evolves with time and is assumed to be of the form

$$\mathbf{x}_k = \mathbf{f}_{k-1}(\mathbf{x}_{k-1}, \mathbf{v}_{k-1}), \text{ for } k > 0. \tag{12.1}$$

Here $\mathbf{x}_k$ is the state vector to be estimated, k denotes the time step, and $\mathbf{f}_{k-1}$ is a known possibly nonlinear function. $\mathbf{v}_{k-1}$ is a white noise sequence, usually referred to as the process, system, or driving noise. The pdf of $\mathbf{v}_{k-1}$ is assumed known. Note that Equation 12.1 defines a first-order Markov process, and an equivalent probabilistic description of the state evolution is $p(\mathbf{x}_k|\mathbf{x}_{k-1})$, which is sometimes called the transition density. For the special case when $\mathbf{f}$ is linear and $\mathbf{v}$ is Gaussian, the transition density $p(\mathbf{x}_k|\mathbf{x}_{k-1})$ is also Gaussian. In the context of target tracking, Equation 12.1 is a model of target dynamics, and $\mathbf{x}$ is a target state vector including position, velocity, and possibly other target attributes.

The measurement equation relates the received measurements to the state vector:

$$\mathbf{z}_k = \mathbf{h}_k(\mathbf{x}_k, \mathbf{w}_k), \text{ for } k > 0, \tag{12.2}$$

where $\mathbf{z}_k$ is the vector of received measurements at time step k, $\mathbf{h}_k$ is the known measurement function, and $\mathbf{w}_k$ is a white noise sequence (the measurement noise or error). Again, the pdf of $\mathbf{w}_k$ is assumed known, and $\mathbf{v}_{k-1}$

and $\mathbf{w}_k$ are mutually independent. Thus, an equivalent probabilistic model for Equation 12.2 is the conditional pdf $p(\mathbf{z}_k|\mathbf{x}_k)$. For the special case when $\mathbf{h}_k$ is linear and $\mathbf{w}_k$ is Gaussian, $p(\mathbf{z}_k|\mathbf{x}_k)$ is also Gaussian. For target tracking, Equation 12.2 is the sensor model.

The final piece of information to complete the specification of the estimation problem is the initial conditions. This is the prior pdf $p(\mathbf{x}_0)$ of the state vector at time $k = 0$, before any measurements have been received. In summary, the probabilistic description of the problem is $p(\mathbf{x}_0)$, $p(\mathbf{x}_k|\mathbf{x}_{k-1})$, and $p(\mathbf{z}_k|\mathbf{x}_k)$.

12.2.2 Formal Bayesian Filter

As already indicated, in the Bayesian approach, one attempts to construct the posterior pdf of the state vector $\mathbf{x}_k$ given all the available information. This posterior pdf at time step k is written $p(\mathbf{x}_k|\mathbf{Z}_k)$, where $\mathbf{Z}_k$ denotes the set of all measurements received up to and including $\mathbf{z}_k$: $\mathbf{Z}_k = \{\mathbf{z}_i, i = 1,\ldots k\}$. The formal Bayesian recursive filter consists of a prediction and an update operation. The prediction operation propagates the posterior pdf of the state vector from time step $k - 1$ forward to time step k. Suppose that $p(\mathbf{x}_{k-1}|\mathbf{Z}_{k-1})$ is available; then $p(\mathbf{x}_k|\mathbf{Z}_{k-1})$, the prior pdf of the state vector at time step $k > 0$, may be obtained via the dynamics model (the transition density):

$$\underbrace{p(\mathbf{x}_k|\mathbf{Z}_{k-1})}_{\text{prior at } k} = \int \underbrace{p(\mathbf{x}_k|\mathbf{x}_{k-1})}_{\text{dynamics}} \underbrace{p(\mathbf{x}_{k-1}|\mathbf{Z}_{k-1})}_{\text{posterior from } k-1} d\mathbf{x}_{k-1}. \tag{12.3}$$

This is known as the Chapman–Kolmogorov equation.

The prior pdf may be updated to incorporate the new measurements $\mathbf{z}_k$ to give the required posterior pdf at time step $k > 0$:

$$\underbrace{p(\mathbf{x}_k|\mathbf{Z}_k)}_{\text{posterior}} = \underbrace{p(\mathbf{z}_k|\mathbf{x}_k)}_{\text{likelihood}} \underbrace{p(\mathbf{x}_k|\mathbf{Z}_{k-1})}_{\text{prior}} / \underbrace{p(\mathbf{z}_k|\mathbf{Z}_{k-1})}_{\substack{\text{normalizing} \\ \text{denominator}}}. \tag{12.4}$$

This is Bayes rule, where the normalizing denominator is given by $p(\mathbf{z}_k|\mathbf{Z}_{k-1}) = \int p(\mathbf{z}_k|\mathbf{x}_k)p(\mathbf{x}_k|\mathbf{Z}_{k-1})d\mathbf{x}_k$. The measurement model regarded as a function of $\mathbf{x}_k$ with $\mathbf{z}_k$ given is the measurement likelihood. Equations 12.3 and 12.4 define the formal Bayesian recursive filter with initial condition given by the specified prior pdf $p(\mathbf{x}_0|\mathbf{Z}_0) = p(\mathbf{x}_0)$ (where $\mathbf{Z}_0$ is interpreted as the empty set). If Equation 12.3 is substituted into Equation 12.4, the prediction and update may be written concisely as a single expression.

Equations 12.3 and Equation 12.4 define a very general but formal (or conceptual) solution to the recursive estimation problem. Only in special cases can an exact, closed-form algorithm be obtained from this general result. (In other words, only in special cases can the posterior density be exactly characterized by a sufficient statistic of fixed and finite dimension.) By far, the most important of these special cases is the linear-Gaussian (L-G) model: if $p(x_0)$, $p(x_k|x_{k-1})$, and $p(z_k|x_k)$ are all Gaussian, then the posterior density remains Gaussian [15], and Equations 12.3 and 12.4 reduce to the standard Kalman filter (which recursively specifies the mean and covariance of the posterior Gaussian). Furthermore, for nonlinear/non-Gaussian problems, the first recourse is usually to attempt to force the problem into an L-G framework by linearization. This leads to the extended Kalman filter (EKF) and its many variants. For mildly nonlinear problems, this is often a successful strategy, and many real systems operate entirely satisfactorily using EKFs. However, with increasingly severe departures from the L-G situation, this type of approximation becomes stressed to the point of filter divergence (exhibited by estimation errors substantially larger than indicated by the filter's internal covariance). For such grossly nonlinear problems, the particle filter may be an attractive option.

12.2.3 Algorithm of Basic Particle Filter

The most basic particle filter may be viewed as a direct mechanization of the formal Bayesian filter.

Suppose that a set of N random samples from the posterior pdf $p(x_{k-1}|Z_{k-1})$ ($k > 0$) is available. We denote these samples or particles by $\{x_{k-1}^{i*}\}_{i=1}^{N}$.

The *prediction* phase of the basic algorithm consists of passing each of these samples from time step $k - 1$ through the system model (Equation 12.1) to generate a set of prior samples at time step k. These prior samples are written $\{x_k^i\}_{i=1}^{N}$, where

$$x_k^i = f_{k-1}\left(x_{k-1}^{i*}, v_{k-1}^i\right)$$

and v_{k-1}^i is a (independent) sample drawn from the pdf of the system noise. This straightforward and intuitively reasonable procedure produces a set of samples or particles from the prior pdf $p(x_k|Z_{k-1})$.

To *update* the prior samples in the light of measurement z_k, a weight $\tilde{w}_k^i$ is calculated for each particle. This weight is the measurement likelihood evaluated at the value of the prior sample: $\tilde{w}_k^i = p\left(z_k|x_k^i\right)$. The weights are then normalized so they sum to unity: $w_k^i = \tilde{w}_k^i / \sum_{j=1}^{N} \tilde{w}_k^j$, and the prior particles are resampled (with replacement) according to these normalized weights to produce a new set of particles:

$$\{x_k^{i*}\}_{i=1}^{N} \text{ such that } \Pr\{x_k^{i*} = x_k^j\} = w_k^j \text{ for all } i, j.$$

In other words, a member of the set of prior samples is chosen with a probability equal to its normalized weight, and this procedure is repeated N times to build up the new set $\{\mathbf{x}_k^{i*}\}_{i=1}^N$. We contend that the new particles are samples of the required pdf $p(\mathbf{x}_k|\mathbf{Z}_k)$, and so a cycle of the algorithm is complete.

Note that the measurement likelihood effectively indicates those regions of the state space that are plausible "explanations" of the observed measurement value. Where the value of the likelihood function is high, these state values are well supported by the measurement, and where the likelihood is low, these state values are unlikely. (Where the likelihood is zero, these state values are incompatible with the measurement model—i.e., they cannot exist.) Thus, the update procedure effectively weights each prior sample of the state vector by its plausibility with respect to the latest measurement. The resampling operation is therefore biased toward the more plausible prior samples, and the more heavily weighted samples may well be chosen repeatedly (see discussion of sample impoverishment below). The algorithm is shown schematically in Figure 12.1, and some MATLAB code for an example application is given in the Appendix.

This simple algorithm is often known as the sampling importance resampling filter, and it was introduced in 1993 [16], when it was called the bootstrap filter. It was independently proposed by a number of other research groups, including Kitagawa [17] as a Monte Carlo filter and Isard and Blake [18] as the CONDENSATION algorithm.

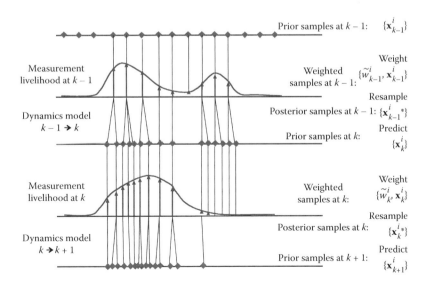

FIGURE 12.1
Schematic of basic particle filter.

12.2.4 Empirical Distributions

The sample sets described above may also be viewed as empirical distributions for the required state pdf's, that is, the prior

$$p(\mathbf{x}_k|\mathbf{Z}_{k-1}) \approx \frac{1}{N}\sum_{i=1}^{N}\delta(\mathbf{x}_k - \mathbf{x}_k^i) \qquad (12.5)$$

and the posterior in weighted or resampled form:

$$p(\mathbf{x}_k|\mathbf{Z}_k) \approx \sum_{i=1}^{N}w_k^i\delta(\mathbf{x}_k - \mathbf{x}_k^i) \approx \frac{1}{N}\sum_{i=1}^{N}\delta(\mathbf{x}_k - \mathbf{x}_k^{i\,*}).$$

This representation also facilitates a simple justification of the update phase of the basic filter using the "plug-in principle" [19]. Substituting the approximate form of the prior (Equation 12.5) into Bayes rule (Equation 12.4), we obtain

$$p(\mathbf{x}_k|\mathbf{Z}_k) = p(\mathbf{z}_k|\mathbf{x}_k)p(\mathbf{x}_k|\mathbf{Z}_{k-1})/p(\mathbf{z}_k|\mathbf{Z}_{k-1})$$

$$\approx p(\mathbf{z}_k|\mathbf{x}_k)\frac{1}{N}\sum_{i=1}^{N}\delta(\mathbf{x}_k - \mathbf{x}_k^i)/p(\mathbf{z}_k|\mathbf{Z}_{k-1})$$

$$= \frac{1}{N}\sum_{i=1}^{N}p(\mathbf{z}_k|\mathbf{x}_k^i)\delta(\mathbf{x}_k - \mathbf{x}_k^i)/p(\mathbf{z}_k|\mathbf{Z}_{k-1})$$

$$= \frac{1}{N}\sum_{i=1}^{N}\tilde{w}_k^i\delta(\mathbf{x}_k - \mathbf{x}_k^i)/p(\mathbf{z}_k|\mathbf{Z}_{k-1})$$

$$= \sum_{i=1}^{N}w_k^i\delta(\mathbf{x}_k - \mathbf{x}_k^i),$$

where, by comparison with Equation 12.4, $p(\mathbf{z}_k|\mathbf{Z}_{k-1}) \approx \frac{1}{N}\sum_{i=1}^{N}\tilde{w}_k^i$. For a more rigorous discussion of the theory behind the particle filter, see [6–8,11,20,21].

12.2.5 Alternative Resampling Scheme

A direct implementation of the resampling step in the update phase of the algorithm would consist of generating N independent uniform samples, sorting them into ascending order, and comparing them with the cumulative sum of the normalized weights. This scheme has a complexity of O(N log N). There are several alternative approaches, including systematic resampling, which has complexity of O(N). In systematic resampling [17], the normalized weights w^i are incrementally summed to form a cumulative sum

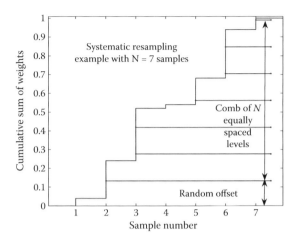

FIGURE 12.2
Systematic resampling scheme.

of $w_C^i = \sum_{j=1}^{i} w^j$. A "comb" of N points spaced at regular intervals of $1/N$ is defined, and the complete comb is translated by an offset chosen randomly from a uniform distribution over $[0, 1/N]$. The comb is then compared with the cumulative sum of weights w_C^i as illustrated in Figure 12.2 for $N = 7$. For this example, the resampled set would consist of labels 2,3,3,5,6,6, and 7 of the original set. This scheme has the advantage of only requiring the generation of a single random sample, irrespective of the number of particles, and it minimizes the Monte Carlo variation—see Section 12.2.7. This method is used in the worked example given in the Appendix (which includes MATLAB code for the procedure).

12.2.6 Impoverishment of Sample Set

As already noted, in the resampling stage, particles with large weights may be selected many times so that the new set of samples may contain multiple copies of just a few distinct values. This impoverishment of the particle set is the result of sampling from a discrete rather than a continuous distribution. If the variance of the system driving noise is sufficiently large, these copies will be redistributed in the prediction phase of the filter, and adequate diversity in the sample set may be maintained. However, if the system noise is small or, in extreme cases, zero (i.e., parameter estimation), the particle set will rapidly collapse, and some artificial means of introducing diversity must be introduced. An obvious way of doing this is to perturb or jitter each of the particles after resampling (termed "roughening" in [16]). This rather ad hoc procedure can be formalized as regularization—where a kernel is placed over each particle to effectively provide a continuous mixture approximation to the discrete (empirical) distribution (akin to kernel density estimation). Optimal kernels for regularization are discussed in [22]. Another scheme for maintaining diversity is to perform a Monte Carlo move (see [23]).

12.2.7 Degeneracy and Effective Sample Size

In the basic version of the filter described in Section 12.2.3, resampling is performed at every measurement update. The function of this resampling process is to avoid wasting the majority of the computational effort in propagating particles with very low weights. Without resampling, as measurement data are integrated, for most interesting problems, the procedure would rapidly collapse to a very small number of highly weighted particles among a multitude of almost useless particles carrying a tiny proportion of the probability mass. This results in failure due to an inadequate representation of the required pdf, that is, degeneracy. Although resampling counters this problem, as noted above, it tends to increase impoverishment, and so there are good arguments for carrying out resampling only if the particle set begins to degenerate [1,21].

A convenient measure of degeneracy is the effective sample size [24] defined by $\hat{N}_{eff} = 1 / \sum_{j=1}^{N} (w_k^j)^2$, which varies between 1 and N. A value close to 1 indicates that almost all the probability mass is assigned to one particle, and there is only one useful sample in the set, that is, severe degeneracy. Conversely, if the weights are uniformly spread among the particles, the effective sample size approaches N. It is often suggested that the resampling process should be performed only if $\hat{N}_{eff}$ falls below some threshold (chosen empirically). If resampling is not carried out, the particle weights from the previous time step are updated via the likelihood $\tilde{w}_k^i = w_{k-1}^i p(\mathbf{z}_k | \mathbf{x}_k^i)$ and then normalized. In this case, the required posterior pdf of the state is given by the random measure $\{\mathbf{x}_k^i, w_k^i\}_{i=1}^{N}$, and these particles are passed through the system model in the prediction phase to generate $\mathbf{x}_{k+1}^i$ for the next measurement update [so the prior distribution at $k + 1$ would be approximated by $p(\mathbf{x}_{k+1} | \mathbf{Z}_k) \approx \sum_{i=1}^{N} w_k^i \delta(\mathbf{x}_{k+1} - \mathbf{x}_{k+1}^i)$].

12.2.8 Sample Representation of Posterior pdf

An important feature of the particle filter is that it provides (an approximation of) the full posterior of the required state. Moreover, the representation of the posterior pdf in the form of a set of samples is very convenient. As well as being straightforward to produce summary statistics, many useful parameters for command, control, and guidance purposes can be easily estimated.

Kalman-like estimators produce estimates of the mean and covariance of the posterior (which completely specify the Gaussian pdf from this type of filter). These statistics are easily estimated from the particle filter sample set (using the plug-in principle) as

$$\hat{\mathbf{x}}_k = E[\mathbf{x}_k | \mathbf{Z}_k] = \int \mathbf{x}_k p(\mathbf{x}_k | \mathbf{Z}_k) d\mathbf{x}_k \approx \sum_{i=1}^{N} w_k^i \mathbf{x}_k^i \text{ or } \frac{1}{N} \sum_{i=1}^{N} \mathbf{x}_k^{i*} \text{ and}$$

$$\text{cov}(\mathbf{x}_k) = E[(\mathbf{x}_k - \hat{\mathbf{x}}_k)(\mathbf{x}_k - \hat{\mathbf{x}}_k)^T | \mathbf{Z}_k] = \int (\mathbf{x}_k - \hat{\mathbf{x}}_k)(\mathbf{x}_k - \hat{\mathbf{x}}_k)^T p(\mathbf{x}_k | \mathbf{Z}_k) d\mathbf{x}_k$$

$$\approx \sum_{i=1}^{N} w_k^i (\mathbf{x}_k^i - \hat{\mathbf{x}}_k)(\mathbf{x}_k^i - \hat{\mathbf{x}}_k)^T \text{ or } \frac{1}{N} \sum_{i=1}^{N} (\mathbf{x}_k^{i*} - \hat{\mathbf{x}}_k)(\mathbf{x}_k^{i*} - \hat{\mathbf{x}}_k)^T.$$

However, the mean and covariance may be a poor summary of the posterior, particularly if it is multimodal or skewed. A scatter plot of the samples, a histogram, or a kernel density estimate [25] is more informative for a 1- or 2-D state vector (or for marginals of the full state vector). Another useful descriptor is the highest probability density (HPD) region. The $(1 - \alpha)$ HPD region is the set of values of the state vector that contains $1 - \alpha$ of the total probability mass, such that the pdf at all points within the region is greater than or equal to the pdf of all those outside the region, that is, if H is the $(1 - \alpha)$ HPD region, then $\int_H p(\mathbf{x})d\mathbf{x} = 1 - \alpha$ and $p(\mathbf{x}') \geq p(\mathbf{x}'')$ for all $\mathbf{x}' \in H$ and $\mathbf{x}'' \notin H$. The HPD region is usually considered only for scalars, and it may be difficult to find for multimodal pdf's. A simpler option is to find the percentile points on scalar marginals of the distribution. For example, the $(1 - \alpha)100$ percentile point is given (roughly) by finding the largest $N\alpha$ samples and choosing the smallest of these.

In many cases, the requirement is to find some particular function of the posterior, and the sample representation is often ideal for this. For example, for threat analysis, one may be interested in the probability that a target is within some particular region—this can be estimated by counting the number of particles falling within that region. Also, for decision and control problems, an estimate of the expected value of any form of cost or utility function $C(\mathbf{x}_k)$ is simply given by

$$E[C(\mathbf{x}_k) | \mathbf{Z}_k] = \int C(\mathbf{x}_k) p(\mathbf{x}_k | \mathbf{Z}_k) d\mathbf{x}_k \approx \sum_{i=1}^{N} w_k^i C(\mathbf{x}_k^i) \text{ or } \frac{1}{N} \sum_{i=1}^{N} C(\mathbf{x}_k^{i*}).$$

This is the starting point for Monte Carlo approaches to the difficult problem of stochastic control—especially with nonquadratic cost functions [13,26].

12.2.9 Discussion

1. *Convenience.* The basic particle filter is a very simple algorithm, and it is quite straightforward to obtain good results for many highly nonlinear recursive estimation problems. Thus, problems that would be difficult to handle using an EKF, state space gridding, or a Gaussian mixture approach are quite accessible to the "novice" via a blind

application of the basic algorithm. However, this is something of a mixed blessing: there is a danger that such challenging cases are not treated with proper respect and that subtleties and implications of the problem are not appreciated [27].

2. *Generality*. The particle approach is very general. It is not restricted to a particular class of distribution or to a form of dynamics model (although the filters discussed in this paper do rely on the Markov property). Thus, for example, the dynamics may include discrete jumps and densities may be multimodal with disconnected regions. Furthermore, the measurement likelihood and transition density do not have to be analytical functions—some form of lookup table is quite acceptable. Also, support regions with hard edges can easily be included (see Section 12.6 below).

12.3 More General Particle Filters

In the basic version of the particle filter, the particles $\{x_k^i\}_{i=1}^N$ used to construct the empirical posterior pdf $\sum_{i=1}^N w_k^i \delta(x_k - x_k^i)$ are assumed to be samples from the prior $p(x_k|Z_{k-1})$. Furthermore, these samples $\{x_k^i\}_{i=1}^N$ are obtained from the posterior samples $\{x_{k-1}^{i*}\}_{i=1}^N$ of the previous time step by passing them through the dynamics model. In other words, each support point x_k^i is a sample of the transition pdf $p(x_k|x_{k-1}^{i*})$ conditional on x_{k-1}^{i*}. However, it is not necessary to generate the $\{x_k^i\}_{i=1}^N$ in this way; they may be obtained from any pdf (known as an importance or proposal density) whose support includes that of the required posterior $p(x_k|Z_k)$. In particular, the importance pdf may depend on z_k, the value of the measurement at time step k. This more general approach considerably broadens the scope for filter design.

The more general formulation is a two-stage process similar to the basic filter of Section 12.2.3, but these stages do not correspond directly to prediction and update phases. As before, we assume that N random samples $\{x_{k-1}^{i*}\}_{i=1}^N$ of the posterior pdf $p(x_{k-1}|Z_{k-1})$ at time step $k-1$ are available.

1. *Sampling*. For each particle x_{k-1}^{i*}, draw a sample x_k^i from an importance density $q(x_k|x_{k-1}^{i*}, z_k)$.

2. *Weight evaluation*. The unnormalized weight corresponding to sample x_k^i is given by

$$\tilde{w}_k^i = \frac{p(z_k|x_k^i)p(x_k^i|x_{k-1}^{i*})}{q(x_k^i|x_{k-1}^{i*}, z_k)}. \tag{12.6}$$

As before, the weights are normalized $w_k^i = \tilde{w}_k^i \Big/ \sum_{j=1}^{N} \tilde{w}_k^j$, and the empirical pdf of the posterior is given by $p(\mathbf{x}_k|\mathbf{Z}_k) \approx \sum_{i=1}^{N} w_k^i \delta(\mathbf{x}_k - \mathbf{x}_k^i)$. Resampling with replacement according to the normalized weights produces a set of samples $\{\mathbf{x}_k^{i*}\}_{i=1}^{N}$ of the posterior pdf $p(\mathbf{x}_k|\mathbf{Z}_k)$. Note that if the importance density is chosen to be the transition pdf, that is, $q(\mathbf{x}_k^i|\mathbf{x}_{k-1}^{i*}, \mathbf{z}_k) = p(\mathbf{x}_k^i|\mathbf{x}_{k-1}^{i*})$, Equation 12.6 reverts to the basic particle filter. The general form of the weight equation (Equation 12.6) is essentially a modification of the basic form to compensate for the different importance density.

The advantage of this formulation is that the filter designer can choose any importance density $q(\mathbf{x}_k|\mathbf{x}_{k-1}, \mathbf{z}_k)$ provided its support includes that of $p(\mathbf{x}_k|\mathbf{Z}_k)$. If this condition is met, as $N \to \infty$, the resulting sample set $\{\mathbf{x}_k^{i*}\}_{i=1}^{N}$ will be distributed as $p(\mathbf{x}_k|\mathbf{Z}_k)$. This flexibility allows one to place samples where they are needed to provide a good representation of the posterior, that is, in areas of high probability density rather than in sparse regions. In particular, since the importance density may depend on the value of the received measurement $\mathbf{z}_k$, if the measurement is very accurate (or if it strongly localizes the state vector in some sense), the importance samples can be placed in the locality defined by $\mathbf{z}_k$ [28]. This is especially important if the "overlap" between the prior and the likelihood is low—adjusting the importance density could avoid wasting a high percentage of the particles (i.e., impoverishment). There is considerable scope for ingenuity in designing the importance density, and a number of particle filter versions have been suggested for particular choices of this density. An optimal importance density may be defined as one that minimizes the variance of the importance weights. For the special case of nonlinear dynamics with additive Gaussian noise, a closed-form expression for the optimal importance density can be obtained [21]. In general, such an analytical solution is not possible, but suboptimal results based on local linearization (via an EKF or unscented Kalman filter) may be employed [1].

As in the basic version of the filter, it is not necessary to carry out resampling at every time step. If resampling is omitted, the particle weights from the previous time step are updated according to

$$\tilde{w}_k^i = w_{k-1}^i \frac{p(\mathbf{z}_k|\mathbf{x}_k^i)p(\mathbf{x}_k^i|\mathbf{x}_{k-1}^i)}{q(\mathbf{x}_k^i|\mathbf{x}_{k-1}^i, \mathbf{z}_k)}. \tag{12.7}$$

This general result is known as sequential importance sampling, and it is most easily derived by (formally) considering the full time history or trajectory of each particle and marginalizing out past time steps [1, 6, 20, 21, 28]. This result is also the starting point for most expositions on particle filter theory (although, unusually, in this paper, the development has been from specific to general).

Rao–Blackwellized or Marginalized Particle Filter. In many cases, it may be possible to divide the problem into linear-Gaussian and nonlinear parts. Suppose that the state vector may be partitioned as $\mathbf{x}_k = \begin{pmatrix} \mathbf{x}_k^L \\ \mathbf{x}_k^N \end{pmatrix}$ so that the required posterior may be factorized into Gaussian and non-Gaussian terms:

$$p(\mathbf{x}_k|\mathbf{Z}_k) = p(\mathbf{x}_k^L, \mathbf{x}_k^N|\mathbf{Z}_k) = p(\mathbf{x}_k^L|\mathbf{x}_k^N, \mathbf{Z}_k)p(\mathbf{x}_k^N|\mathbf{Z}_k),$$

where $p(\mathbf{x}_k^L|\mathbf{x}_k^N, \mathbf{Z}_k)$ is Gaussian (conditional on $\mathbf{x}_k^N$) and $p(\mathbf{x}_k^N|\mathbf{Z}_k)$ is non-Gaussian. In other words, the linear component of the state vector $\mathbf{x}_k^L$ can be "marginalized out." Essentially, the term $p(\mathbf{x}_k^L|\mathbf{x}_k^N, \mathbf{Z}_k)$ may be obtained from a Kalman filter, while the non-Gaussian part $p(\mathbf{x}_k^N|\mathbf{Z}_k)$ is given by a particle filter. The scheme requires that a Kalman filter update be performed for each $\mathbf{x}_k^N$ particle—see the excellent tutorial by Gustafsson [9] for a full specification of the algorithm. This procedure is generally known as Rao–Blackwellization [21,29–31]. The main advantage of this approach is that the dimension of the particle filter state $\mathbf{x}_k^N$ is less than that of the full state vector, so that less particles are required for satisfactory filter performance (see below). This comes at the cost of a more complex algorithm, although the operation count of the marginalized filter for a given number of particles may actually be less than that of the standard algorithm (see [32]).

12.4 Computational Issues

12.4.1 Computational Cost for Basic Filter

The computational cost of the basic particle filter (with systematic resampling) is almost proportional to the number N of particles employed, both in terms of operation count and memory requirements. The computational effort associated with each particle clearly depends directly on the complexity of the system dynamics and the measurement process. For example, problems involving measurement association uncertainty may require a substantial measurement likelihood calculation (i.e., a summation over hypotheses). For such cases, there is a strong motivation to find efficient ways of evaluating the likelihood—including approximate gating and the use of likelihood ratios (see examples in chapters 11 and 12 of [1]).

A notable advantage of the particle filter is that the available computational resources can be fully exploited by simply adjusting the number of particles—so it is easy to take advantage of the ever-increasing capability of cheap computers. Similarly, if the measurement data rate is variable, the filter can match the number of particles to the available time interval to optimize performance. (However, if the number of particles falls below a critical level, the filter performance may degrade to a point from which it cannot recover.) Also note that the filter is amenable to parallelization—until a resampling event occurs, all particle operations are independent [33,34].

12.4.2 How Many Samples?

This is the most common question about particle filters, and there is no simple answer. Classical analysis of Monte Carlo sampling does not apply as the underlying assumption—that the samples are independent—is violated. In the basic particle filter, immediately after the resampling stage, many of the particles are almost certainly identical—definitely not independent. Thus, unfortunately, particle filters are not immune to the curse of dimensionality, although with careful filter design, the curse can be moderated (see the informative and detailed discussion by Daum [27] and Daum and Huang [35]). Generally, based on simple arguments of populating a multidimensional space, one must expect the required number of particles to increase with the dimension of the state vector—hence, the attraction of the Rao–Blackwellized or marginalized form of the filter.

The required sample size depends strongly on the design of the particle filter and the problem being addressed (dimension of state vector, volume of support, etc.). For certain problems, especially high-dimensional ones, an enormous, infeasible, number of samples may be required to obtain satisfactory results with the basic filter. To obtain a practical algorithm in these circumstances, the designer has to be inventive. The theory outlined in Section 12.3 provides a rigorous framework for exploring options, and with a careful choice of proposal distribution and/or exploiting Rao–Blackwellization, it may be possible to design a filter that gives quite satisfactory performance with a modest number of particles (a few hundred or even tens in some cases). However, the basic algorithm has the advantage of simplicity, so that the operation count for each particle may be lower than for a more subtle filter. Practical particle filter design is therefore a compromise between these approaches with the aim of minimizing the overall computational load. Also note that heuristic tricks may well be helpful.

The usual way of determining when enough samples are being deployed is via trial and error: the sample size is increased until the observed error in the parameter of interest (from a set of representative simulation examples) falls to a steady level. If the required sample size is too large for the available processing resources, one may have to settle for suboptimal filter performance or attempt to improve the design of the filter. This empirical approach is not entirely satisfactory, and more work in this area is required to obtain, at least, guidelines that are of use to practicing engineers.

Finally, note that filter initialization is often the most challenging aspect of a recursive estimation problem. In particular, if the prior information (i.e., before measurements are received) is vague, so that the initial uncertainty spans a large volume of state space, the direct (obvious) approach of populating the prior pdf with particles may be very wasteful. Semibatch schemes using the first few measurement frames may be useful.

12.5 Tracking and Guidance Application

12.5.1 Introduction

In this section, we describe an application of the particle filter to a nonlinear tracking and guidance problem [12–14]. The requirement is to track a target (T) in the presence of interfering, intermittent, spurious objects (D), and so guide a pursuer to intercept the target.

12.5.2 Formal Problem Statement

12.5.2.1 Dynamics Models

The dynamics of the primary target (T) are described by the following (known) discrete system model:

$$\mathbf{x}_{Tk+1} = \mathbf{f}_{Tk} \left(\mathbf{x}_{Tk}, \mathbf{w}_{Tk} \right) \tag{12.8}$$

where $\mathbf{x}_{Tk}$ is the target state vector, $\mathbf{w}_{Tk}$ is system driving noise, and $\mathbf{f}_{Tk}$ describes the dynamics of the target.

At some random time step, the target may spawn a secondary object D in the vicinity of T. Thereafter, the secondary object moves independently of T according to the following dynamic model:

$$\mathbf{x}_{Dk+1} = \mathbf{f}_{Dk}(\mathbf{x}_{Dk}, \mathbf{w}_{Dk}) \tag{12.9}$$

where $\mathbf{x}_{Dk}$ is the state vector of D. The initial distribution of $\mathbf{x}_{Dk}$ at birth is a (known) function of $\mathbf{x}_{Tk}$. The secondary object disappears after a random period, and later, (following another random period) another object D may be produced. The birth/death sequence of the object D is described by a Markov process. If $\gamma_k = 0$ indicates that D does not exist at time step k and

$\gamma_k = 1$ indicates that D is in existence, the transitions $0 \rightarrow 1$ and $1 \rightarrow 0$ depend only on the probabilities

$$p_{01} = \Pr\{\gamma_k = 1 | \gamma_{k-1} = 0\} \text{ and } p_{10} = \Pr\{\gamma_k = 0 | \gamma_{k-1} = 1\} \tag{12.10}$$

Clearly, $\Pr\{\gamma_k = 1 | \gamma_{k-1} = 1\} = 1 - p_{10}$ and $\Pr\{\gamma_k = 0 | \gamma_{k-1} = 0\} = 1 - p_{01}$. If the time step Δt is constant, the average period between the death of one secondary object and the birth of another is $\Delta t / p_{01}$, while the average lifetime of D is $\Delta t / p_{10}$. Note that this model implies that only two objects may be present at any instant.

At $k = 0$, it is assumed that only the primary target T is present ($\gamma_0 = 0$). The prior distribution of $\mathbf{x}_{T0}$ is also assumed to be known.

12.5.2.2 State Vector

For this problem, it is convenient to define a system state vector

$$\mathbf{X}_k = (\mathbf{x}_{Tk}^T, \mathbf{x}_{Dk}^T, \gamma_k)^T \tag{12.11}$$

which evolves with time according to Equations 12.8 through 12.10. Note that if $\gamma_k = 0$, $\mathbf{x}_{Dk}$ is redundant. In this case, it is convenient to set $\mathbf{x}_{Dk} = \mathbf{x}_{Tk}$.

12.5.2.3 Sensor Model

At each time step k, N_{Mk} measurements $\mathbf{z}_{ik}$ are received from a sensor carried by the pursuer (whose position is precisely known). If only the primary target T is present (i.e., $\gamma_k = 0$), the probability of detecting the target is P_{TD}. If the secondary, spurious object D is also present (i.e., $\gamma_k = 1$), depending on its proximity to the primary target and the relative geometry, the sensor may be capable of resolving two objects, or it may only be able to resolve a single composite object. If the two objects can be resolved, then the probability of receiving a measurement from T is P_{TD}, and the probability of receiving a measurement from D is P_{DD} (and these are independent events). If the objects cannot be resolved, the probability of receiving a single composite measurement is P_{JD}. These probabilities may be functions of the appropriate T or D states. Additionally, clutter measurements (independent of the two objects) may also be produced by the sensor. We assume that these clutter measurements are uniformly distributed over the measurement space and that they are not subject to resolution limitations (although this would be significant only in exceedingly dense clutter). The number of clutter measurements received at a given time step follows a Poisson distribution with mean m.

Associated with each position measurement $\mathbf{z}_{ik}$ is a classification flag or signature parameter c_{ik}, which may provide an indication of the type of object from which the measurement originated (target, secondary object, composite, or clutter) but gives no direct information on object position. c_{ik} could be a discrete output (e.g., target, secondary object, or clutter) or a continuous parameter such as a measurement intensity. The N_{Mk} measurements and classifications received at time t_k are denoted

$$\mathbf{Z}'_k = \{(\mathbf{z}_{1k}, c_{1k}), (\mathbf{z}_{2k}, c_{2k}), \ldots (\mathbf{z}_{N_{Mk}k}, c_{N_{Mk}k})\}$$

and the set of all data received up to and including time t_k is denoted

$$\mathbf{Z}_k = \{\mathbf{Z}'_1\, \mathbf{Z}'_2, \ldots \mathbf{Z}'_k\}.$$

It is assumed that the association between measurements and the objects is *a priori* unknown. An association hypothesis defines a mapping λ from the subscripts of the measurements to their source [target (T), secondary (D), composite (J), or clutter (C)]:

$$\lambda:\{1,2,\ldots N_{Mk}\} \rightarrow T, D, J, C.$$

Given λ, the conditional pdf's of the measurements $\mathbf{z}_{ik}$ are denoted by

$$
\begin{array}{llll}
p_T(\mathbf{z}_{ik}|\mathbf{x}_{Tk}) & \text{if } \lambda(i) = T, & p_D(\mathbf{z}_{ik}|\mathbf{x}_{Dk}) & \text{if } \lambda(i) = D, \\
p_J(\mathbf{z}_{ik}|\mathbf{x}_{Tk}, \mathbf{x}_{Dk}) & \text{if } \lambda(i) = J, & p_C(\mathbf{z}_{ik}) & \text{if } \lambda(i) = C.
\end{array}
$$

The performance of the classifier is denoted similarly by $p_T(c_{ik}|\mathbf{x}_{Tk})$ for $\lambda(i) = T$, etc. Note that the classifier performance may be state dependent. It is assumed that these conditional distributions are known.

Although the specification of the classifier output is essentially identical in form to that of the measurements, it is convenient to make the distinction between $\mathbf{z}$ and c to emphasize that two quite different types of information are available, one of which is strongly indicative of object position, while the other is primarily dependent on object type. However, if the classifier output is state dependent (albeit only weakly), then this also provides some information on the object state vector. The particle filter is able to exploit this.

12.5.2.4 General Form of Measurement/Classification Likelihood

By careful consideration of the possible measurement-state associations, the likelihood of the state vector $\mathbf{X}$ given the data set $\mathbf{Z}'$ (dropping the time subscript k) may be shown to be (for $m > 0$)

$$p(\mathbf{Z}'|\mathbf{X}) \propto (1-\gamma) g_T(\mathbf{x}_T|\mathbf{Z}') + \gamma[1 - P_{res}(\mathbf{X})] g_J(\mathbf{x}_T, \mathbf{x}_D|\mathbf{Z}')$$

$$+ \gamma \left[g_T(\mathbf{x}_T|\mathbf{Z}') g_D(\mathbf{x}_D|\mathbf{Z}') - P_{TD} P_{DD} \sum_{j=1}^{N_M} \ell_T(\mathbf{x}_T|\mathbf{z}_j, c_j) \ell_D(\mathbf{x}_D|\mathbf{z}_j, c_j) \right].$$

$$(12.12)$$

Here, the likelihood ratio

$$\ell_T(\mathbf{x}_T|\mathbf{z},c) = \frac{1}{m} \frac{p_T(\mathbf{z}|\mathbf{x}_T)}{p_C(\mathbf{z})} \frac{p_T(c|\mathbf{x}_T)}{p_C(c)}.$$

ℓ_D is similar, and

$$\ell_J(\mathbf{x}_T,\mathbf{x}_D|\mathbf{z},c) = \frac{1}{m} \frac{p_J(\mathbf{z}|\mathbf{x}_T,\mathbf{x}_D)}{p_C(\mathbf{z})} \frac{p_J(c|\mathbf{x}_T,\mathbf{x}_D)}{p_C(c)}.$$

The function $g_T(\mathbf{x}_T|\mathbf{Z}')$, which may be interpreted as the likelihood of $\mathbf{x}_T$ for the case of a single target in clutter, is given by

$$g_T(\mathbf{x}_T|\mathbf{Z}') = (1 - P_{TD}) + P_{TD} \sum_{j=1}^{N_M} \ell_T(\mathbf{x}_T|\mathbf{z}_j,c_j).$$

$g_D(\mathbf{x}_D|\mathbf{Z}')$ is similar, and

$$g_J(\mathbf{x}_T,\mathbf{x}_D|\mathbf{Z}') = (1 - P_{JD}) + P_{JD} \sum_{j=1}^{N_M} \ell_J(\mathbf{x}_T,\mathbf{x}_D|\mathbf{z}_j,c_j).$$

Also, $P_{res}(\mathbf{X})$ is the probability that T and D can be resolved (if $\gamma = 1$).

The likelihood (Equation 12.12) has three principal terms. The first of these (for $\gamma = 0$) corresponds to cases when the secondary object is not present so that only measurements from the target or clutter are available. The second term (for $\gamma = 1$ and $P_{res} = 0$) represents the case where the secondary object is present but it is not resolved, so that only measurements from the composite object or clutter are available. The third term (for $\gamma = 1$ and $P_{res} = 1$) corresponds to the case where again the secondary object is present, but it can be resolved from the target. Various special cases follow directly from this expression; for example, if $\gamma = 0$, the likelihood for a single target in clutter is obtained.

Note that if the functions $g_T(.), g_D(.), g_J(.), \ell_T(.), \ell_D(.),$ and $\ell_J(.)$ can be specified, then the likelihood (Equation 12.12) can be directly used in the particle filter update described in Section 12.2.3.

12.5.2.5 Expression for Likelihood with Gaussian Measurements and Uniform Clutter

The above solution is valid for any form of (time-independent) measurement error characteristic (including, e.g., quantization effects and skewed or truncated distributions). Likewise, any form of clutter distribution may be employed provided it is independent of the state **X**. However, consider the common assumptions of Gaussian measurement errors and uniformly distributed clutter:

$$p_T(\mathbf{z}|\mathbf{x}) = p_D(\mathbf{z}|\mathbf{x}) = \mathcal{N}(\mathbf{z};\mathbf{h}(\mathbf{x}), R), \quad p_J(\mathbf{z}|\mathbf{x}_T, \mathbf{x}_D) = \mathcal{N}(\mathbf{z};\mathbf{h}_J(\mathbf{x}_T, \mathbf{x}_D), R_J),$$

$$\text{and } p_c(\mathbf{z}) = 1/V.$$

Here, the measurements **z** are independent, zero mean Gaussian perturbations about a function of the state. In the above expressions, $\mathcal{N}$ indicates a Gaussian pdf. For the resolved objects, the (possibly nonlinear) measurement function is **h**, and the covariance of the Gaussian perturbation or error is R. For the composite return, the measurement function $\mathbf{h}_J$ depends on both $\mathbf{x}_T$ and $\mathbf{x}_D$. This function could depend on some centroid of $\mathbf{x}_T$ and $\mathbf{x}_D$ such as $(\mathbf{x}_T + \mathbf{x}_D)/2$. Also, the covariance of the measurement error has covariance R_J, which may be different from that of the resolved objects. For example, in the case of a radar sensor, the composite measurement might have a larger measurement error due to glint type effects. The parameter V in the clutter distribution is the volume of the sensor field of view (which is assumed to "comfortably" encompass T and D). Note that if the spatial density of the clutter measurements in the observation space is ρ, then $m = \rho V$.

In this case, the likelihood $p(\mathbf{Z}'|\mathbf{X})$ of Equation 12.12 is specified by (for $\rho > O$)

$$\ell_T(\mathbf{x}_T|\mathbf{z},c) = \frac{1}{\sqrt{|2\pi R|}\rho} \frac{p_T(c|\mathbf{x}_T)}{p_C(c)} \exp\left(-\frac{1}{2}(\mathbf{z}-\mathbf{h}(\mathbf{x}_T))^T R^{-1}(\mathbf{z}-\mathbf{h}(\mathbf{x}_T))\right) \quad (12.13)$$

and ℓ_D is similar. Also

$$g_T(\mathbf{x}_T|\mathbf{Z}') = (1-P_{TD}) + \frac{P_{TD}}{\sqrt{|2\pi R|}\rho} \sum_{j=1}^{N_M} \frac{p_T(c_j|\mathbf{x}_T)}{p_C(c_j)}$$

$$\times \exp\left(-\frac{1}{2}(\mathbf{z}_j-\mathbf{h}(\mathbf{x}_T))^T R^{-1}(\mathbf{z}_j-\mathbf{h}(\mathbf{x}_T))\right). \quad (12.14)$$

$g_D(\mathbf{x}_D|\mathbf{Z}')$ is similar, and

$$g_J(\mathbf{x}_T,\mathbf{x}_D|\mathbf{Z}') = (1-P_{JD}) + \frac{P_{JD}}{\sqrt{|2\pi R_J|}\rho} \sum_{j=1}^{N_M} \frac{p_J(c_j|\mathbf{x}_T,\mathbf{x}_D)}{p_C(c_j)}$$

$$\times \exp\left(-\frac{1}{2}(\mathbf{z}_j-\mathbf{h}_J(\mathbf{x}_T,\mathbf{x}_D))^T R_J^{-1}(\mathbf{z}_j-\mathbf{h}_J(\mathbf{x}_T,\mathbf{x}_D))\right). \quad (12.15)$$

12.5.2.6 Cost Functions for Guidance Problem

In the guidance or pursuit problem, the overall goal is to achieve an intercept with a particular target in what may be a complex, multiple-object environment with substantial uncertainty. This is a stochastic control problem: the role of the tracking filter is to construct an estimate of the system state vector that can be used by a control law to generate a guidance demand (to drive interceptor actuators). In general, from a Bayesian/optimal control perspective, the filter produces a posterior pdf of the pursuer–target state (the information state or sufficient statistic) to generate a control that minimizes the expected value of a specified cost function. In the case of intercept guidance, the cost function describes the requirement to minimize the miss distance (closest approach) between the pursuer and the required target, usually within some time or energy constraint. (For the rendezvous problem, the requirement is to minimize the velocity as well as the positional difference.)

The miss distance is the Euclidean distance between the pursuer and the target at the point of closest approach up to some final time step N_F. The aim of the pursuer is essentially to minimize the miss distance via an appropriate choice of future control demands $\mathbf{U}_k^+ = \{\mathbf{u}_k, \mathbf{u}_{k+1}, \ldots, \mathbf{u}_{N_F}\}$ at each time step [given previous measurements and controls $(\mathbf{Z}_k, \mathbf{U}_{k-1})$]. Clearly, the achieved miss for a given control sequence $\mathbf{U}_k^+$ depends on the future trajectory of the target. For most interesting interception problems, the pursuer does not have perfect information on the future target trajectory. This is because future target maneuvers are usually unknown and because the current target state (i.e., $\mathbf{x}_{T\,k}$) is not perfectly known but has to be estimated via the guidance estimator (where we assume that the pursuer state is perfectly known). A full stochastic control law should therefore take account of both these sources of uncertainty (i.e., the expectation of the future cost should be over these sources of uncertainty). For multiple-object scenarios, the latter factor will often be dominant. Thus, for the purposes of this example, we shall assume that if the current pursuer–target state were perfectly known, the future target path would be precisely defined, that is, the achieved miss $m_d = m_d(\mathbf{x}_{Tk}, \mathbf{U}_k^+)$. Under this assumption, if there is no explicit cost on the control action, the optimization problem at time step k is of the form

$$J_k(Z_k, \mathbf{U}_{k-1}) = \min_{\mathbf{U}_k^+} \left\{ E\left[f_C(m_d(\mathbf{x}_{Tk}, \mathbf{U}_k^+)) \mid Z_k, \mathbf{U}_{k-1} \right] \right\} \tag{12.16}$$

for admissible control sequences $\mathbf{U}_k^+$, where $f_c(.)$ is a cost function.

If the posterior distribution of $\mathbf{x}_{Tk}$ is represented by a set of samples $\{\mathbf{x}_{Tk}^{i*}\}_{i=1}^{N}$, a possible approximation for Equation 12.16 is [13]

$$J_k(Z_k, \mathbf{U}_{k-1}) = \min_{\mathbf{U}_k^+} \left\{ \frac{1}{N} \sum_{t=1}^{N} f_C(m_d(\mathbf{x}_{Tk}^{i*}, \mathbf{U}_k^+)) \right\}. \tag{12.17}$$

Note that in this approximation, we have ignored the contribution of future measurements (z_j for $j > k$) to learning about the future state. In the full stochastic optimization problem, the expectation in Equation 12.16 is over the uncertain information from future measurements as well as more obviously over the current state. This may result in control demands that cause the pursuer to deviate from a straightforward intercept course to improve observability of the system. This effect is known as "dual control" [36–38]. To attempt a full optimization over future measurements would massively complicate the solution [26]—especially for guidance against multiple objects. The approximation invoked here of ignoring future information is known as an open-loop optimal feedback approach [39]—and clearly, this precludes the possibility of dual effect. (For LQG assumptions, the certainty equivalence property holds, allowing a simple and most convenient optimal solution [39]—also without dual effect.)

The form of the cost function $f_c(m_d)$ requires careful consideration for multiple-object scenarios. In particular, the familiar quadratic cost may be inappropriate because it always attempts to drive the system toward the expected value of the state. For a multiple-object scenario, the expected target position may lie between two objects, and in the event that this situation persists as range closes, the quadratic cost could give rise to a guidance strategy that would ensure a miss. This is because the cost of missing is unbounded for the quadratic function: probability mass of the target distribution that is quite remote from positions accessible to the pursuer may still significantly affect the guidance demand. This clearly does not reflect the usual requirement in an interception problem: if the miss distance exceeds a certain threshold, the cost is effectively constant. Thus, it is more realistic to employ a bounded cost function, and in this paper, we have chosen to use an inverse Gaussian function of the form

$$f_C(m_d) = 1 - \exp(-m_d^2/\sigma^2),$$

where the parameter σ determines the extent of the "well" in the cost (see Tanner [40]). With this cost function, $0 \le f_c(m_d) < 1$, and the penalty for missing any target probability mass is essentially constant (near unity) if the miss distance exceeds 3σ.

12.5.3 Simulation Example

To illustrate the operation of the particle filter and guidance scheme, a single plane, tail-chase pursuit scenario has been simulated.

12.5.3.1 Object Paths

The target (T) moves at constant speed V_0, initially in a straight line and then, after 25 time steps, at a constant turn rate of ω_0. Three spurious objects ($D1$, $D2$, and $D3$) are generated at time steps 25, 40, and 55. At birth, each spurious object is collocated with the target and has the same velocity as the

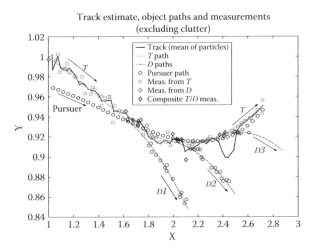

FIGURE 12.3
Object paths and measurements (excluding clutter) in the vicinity of the intercept.

target. Subsequently, the objects maintain their speed (V_0) but immediately turn away from the target (i.e., at a constant turn rate of $-\omega_0$). Each D object survives for 20 time steps and then disappears. The paths of T, $D1$, $D2$, and $D3$ are shown in Figure 12.3. The pursuer also moves at a speed $V_M = 1.6V_0$ and has a maximum turn rate of $\omega_M = 3.75\omega_0$ (and so has a maximum lateral acceleration six times that of the turning target). Its heading is controlled by a guidance demand u_k (turn rate), which is updated at every time step. Thus, the heading ϕ of the pursuer is given by

$$\phi_{k+1} = \phi_k + \omega_M \Delta t u_k \qquad (12.18)$$

where $|u_k| \leq 1$ and Δt is the time step period. (The pursuer is assumed to respond instantly to the demand—it is lag free). The initial position of the pursuer is behind the target and pointing roughly toward it, so the first part of the pursuer's path is not visible in Figure 12.3 (note the unequal axis scales in this figure).

12.5.3.2 Sensor Measurements

The sensor, mounted on the pursuer, takes noise-corrupted measurements of range, range rate, and bearing with respect to nonrotating axes. (It is assumed that the position, velocity, and orientation of the pursuer are perfectly known with respect to a set of fixed axes, that is, it has a perfect inertial measurement unit.) The interval between measurements is Δt. The sensor has a finite field of view, which at each time step is centered on the expected "position" of the target in observation space, as estimated by the filter. Any object falling within the sensor's field of view is detected. The sensor is able to resolve T and D objects only if their angular separation is greater than $\Delta\theta = 0.01$ rad, if their range separation is greater than $\Delta r = 0.02$ units, or if their range rate separation is greater than $\Delta\dot{r} = 0.0005$ units/Δt. Due to the geometry of the problem, the sensor is not able to resolve D objects for the first few time steps following their birth (shown as "composite T/D measurements" in Figure 12.3). For resolved objects, the sensor measurements are a Gaussian perturbation about their actual values, with standard deviations in bearing, range, and range rate of $\Delta\theta$, Δr, and $\Delta\dot{r}$, respectively. For unresolved objects, the perturbation center is the centroid of the pair. The spatial clutter density is $\rho = 5$ per radian per unit per unit/Δt. Associated with each positional measurement is a classification flag that gives information on the origin of the measurement. In this example, the classification flag can take three possible values. The probability of obtaining a particular flag value is indicated in Table 12.1. Note that the classification flag is independent between time steps. For this example, the flag gives a fairly strong indication if the measurement is from a composite, unresolved return (flag = 2 with a probability of 90%).

Measurements received from the sensor are subjected to a coarse acceptance test to reject any clutter returns that are remote from the filter's estimate of T and D. Simulated measurements accepted by the filter are shown in Figure 12.4. The first three traces in this figure show the "errors" in bearing, range, and range rate on the accepted measurements, that is, with respect to the actual position and velocity of the target. The spawning of the spurious objects $D1$, $D2$, and $D3$ from T is evident in the plot of bearing errors. Note that only in the case of $D2$ are all 20 measurements from the object accepted—$D1$ and $D3$ pass out of the sensor's field of view (so effectively dying from the filter's perspective) before they are terminated. Also note that at certain times, measurements from more than one spurious object are accepted (e.g., at time step 58, measurements from T, $D2$, and $D3$ are accepted), thus violating a key assumption of the filter model. The fourth trace in Figure 12.4 shows the classification flag associated with each measurement. As one would expect from Table 12.1, the composite measurements are usually assigned a flag value of 2.

TABLE 12.1

Classification Probabilities

	Classifier Output		
Origin of Actual Measurement	1	2	3
T	0.60	0.30	0.10
D	0.30	0.60	0.10
Composite	0.05	0.9	0.05
Clutter	0.15	0.15	0.70

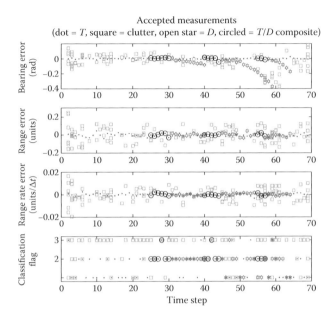

FIGURE 12.4
Measurements.

12.5.3.3 Filter Models

The basic particle filter algorithm of Section 12.2.3 has been employed using $N = 10{,}000$ samples. The state vector $\mathbf{X}_k$ for the problem consists of Cartesian position and velocity coordinates for T and D together with the existence flag γ, so the general form (Equation 12.11) may be written as

$$\mathbf{X}_k^T = (x_T, \dot{x}_T, y_T, \dot{y}_T, x_D, \dot{x}_D, y_D, \dot{y}_D, \gamma)_k.$$

The pursuer's position and velocity are not included in the state vector as they are assumed to be perfectly known. The measurement likelihood is the general expression 12.12 with Gaussian measurement models specified by Equations 12.13 through 12.15 with covariance matrices R and R_J given by the above standard deviations $\Delta\theta$, Δr, and $\Delta \dot{r}$. The measurement functions $\mathbf{h}$ and $\mathbf{h}_J$ are nonlinear functions that transform the Cartesian state vector to the polar measurements. The probability $P_{res}(\mathbf{X})$ of resolving T and D is 0 or 1, depending on whether or not $\mathbf{x}_T$ and $\mathbf{x}_D$ are in the resolution cell specified by $\Delta\theta$, Δr, and $\Delta \dot{r}$. The classifier information is in the form of discrete flags, so the classifier performance specifiers $[p_T(c|\mathbf{x}_T),\ p_D(c|\mathbf{x}_D),\ p_J(c|\mathbf{x}_T,\ \mathbf{x}_D),\ \text{and}\ p_c(c)]$ are the probabilities given by Table 12.1 (and are independent of $\mathbf{X}$ in this example). The assumed detection probabilities are $P_{TD} = P_{DD} = P_{JD} = 0.95$. Thus, except for the detection probabilities and multiple spurious objects, the measurement likelihood employed by the particle filter is exactly matched to the simulated data. So, the particle filter is able to directly exploit this highly nonlinear measurement information without invoking any approximations.

The dynamics models (Equations 12.8 and 12.9) for T and D are both represented by discrete, linear-Gaussian second-order models (i.e., near-constant velocity) with the same level of driving noise [13]. Thus, the filter cannot use dynamics to discriminate between T and D. The assumed Markov transition probabilities for the spurious object birth/death process are $p_{01} = p_{10} = 0.05$. So, the assumed average period between the death of one object and the birth of another is 20 time steps, and the assumed average lifetime of D is also 20 time steps. This assumed average lifetime is well matched to the simulated scenario, but the death-to-birth interval is much too short for the final part of the engagement. Also note that the noise-driven, near-constant velocity, dynamics models are not matched to the constant turn-rate object paths. The initial (at birth) pdf of the state of D relative to T is assumed to be Gaussian centered on T with standard deviations of 0.05 units in x and y and 0.01 units/Δt in $\dot{x}$ and $\dot{y}$.

12.5.3.4 Guidance Demands

To determine a guidance turn-rate demand (Equation 12.18) that minimizes a cost function of the form of Equation 12.17, it is necessary to derive an expression for the miss distance m_d as a function of $\mathbf{X}_k$, the pursuer's state, and the sequence of future controls $\mathbf{U}_k^+$. To simplify the optimization problem and to produce an intuitively satisfactory trajectory, the choice of future controls is restricted to a constant turn rate, that is, $u_j = u_k$ for $j > k$. Thus, the requirement is reduced to selecting a single number u_k from the range [−1, +1]. Using the assumption that future target maneuvers may be neglected (see Section 12.5.2.6), a simple approximate expression for the miss distance may be obtained so that the future cost $f_c(m_d)$ may be easily evaluated for the current state vector and any choice of u_k [13]. Therefore, the expected future cost for the inverse Gaussian may be approximately evaluated using the target samples $\{\mathbf{x}_{Tk}^{i*}\}_{i=1}^N$ from the particle filter. For this simple problem, the value of u_k that minimizes the expected cost and is used for guidance can be found by evaluating the expected cost (Equation 12.17) over a grid of points. The pursuer path generated by this procedure is shown in Figure 12.3—in this example, the achieved miss distance is about 0.004 units.

12.5.3.5 Filter Estimates

The filter's estimate of the target track (the mean of the sample set $\{(x_{Tk}^{i*}, y_{Tk}^{i*}) : i = 1, \ldots, N\}$) is shown in Figure 12.3. It can be seen that the track is noticeably pulled away from the target path by the presence of $D2$. However, as already noted, the mean of the samples is a poor summary of information from the particle filter. Figure 12.5 shows some subsets of the target samples at every third step from time step 30 to 66, that is, as the target is turning and

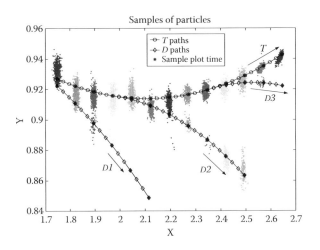

FIGURE 12.5
Subsets of 500 from the 10,000 target particles shown at every third time step.

when *D1* to *D3* are being spawned. The presence of the *D* objects first draws the centroid of the sample clumps away from the target and then splits the clump when the *D* objects are resolved (indicating association uncertainty). In the cases of *D1* and *D3*, the target samples (incorrectly) assigned to the *D* objects soon die out (due to the integration of classification information). However, some target particles cling to *D2* until it dies (causing the deviation of the target track seen in Figure 12.3). This effect is also seen in Figure 12.6, which shows the evolution of the estimate of the marginal pdf's $p(y_{TK}|\mathbf{Z}_k, \mathbf{U}_k)$ as a histogram of the particles, that is, approximately in the cross-range direction. The multiple modes of these marginal pdf's are due to the uncertainty

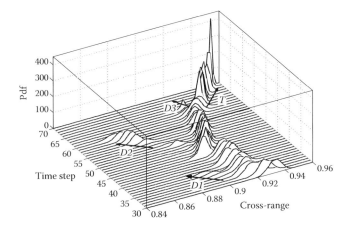

FIGURE 12.6
Bifurcation of the cross-range pdf as *T* and *D* separate.

in associating measurements with the target, and the balance of the probability mass between the modes indicates the filter's assessment of which corresponds to the true target.

12.5.3.6 Guidance Analysis

To analyze the guidance performance of the pursuer, it is useful to consider the estimated distribution of the miss distance for $u_k = 0$, for the latter part of the engagement. This is called the zero-effort miss (ZEM) distribution, and for the particle filter (under the assumption of no future target maneuvers), it is represented by the sample set $\{m_d(x_{Tk}^{i*}, u_k = 0) : i = 1, \ldots, N\}$, where $m_d(.)$ is the approximate predicted miss discussed above. The ZEM distributions for the last 37 time steps prior to fly-past are indicated in Figure 12.7. In this figure, the vertical axis represents the miss distance, and time steps are shown along the horizontal axis. For each time step shown, the thickness of the vertical line represents the probability density of the ZEM (derived from a histogram of the ZEM sample set). Note that toward the end of the engagement, the ZEM distribution splits into two separate clumps corresponding to T and $D2$ (the upper clump being due to T). The mean of the ZEM is also shown: it passes between the clumps through a region of low-probability mass. Note that displacement of the ZEM distribution between time steps is strongly influenced by the chosen guidance demand. The extent of the miss that can be corrected by the pursuer's control authority (i.e., the available divert) is shown as upper and lower bounds in the figure. Clearly, as the time of fly-past is approached, this extent of available divert reduces to zero.

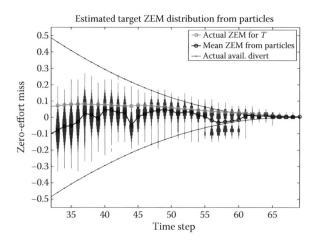

FIGURE 12.7
Estimated distribution of target ZEM.

In an ideal engagement, as the pursuer approaches the target, the ZEM distribution would collapse to a tight concentration, and the guidance demand would force this concentration toward zero miss distance. Specifically, to engineer a perfect intercept, the distribution of ZEM should remain within the available divert—hence the concept of reachable set guidance (see Lawrence [41]). Clearly, this is not possible for the current example due to the two diverging clumps of probability mass. However, the inverse Gaussian cost function produces a reasonable response to this problem. Roughly speaking, a minimum cost control is achieved by maximizing the ZEM probability mass contained within the well of the cost function. The result of this is to force the pursuer toward one of the two clumps as the ZEM distribution separates. Thus, a smooth transition is achieved from a hedging (and learning) guidance strategy to a hard decision prior to intercept.

12.6 Other Applications

Particle filters have been employed in a wide range of domains: essentially, wherever there is a requirement to estimate the state of a stochastic evolving system using uncertain measurement data. Below, we briefly indicate some of the more successful or popular applications (with a bias toward tracking problems).

12.6.1 Tracking and Navigation with Bounded Support

Particle filters are ideal for problems where the state space has a restricted or bounded support. Examples include targets moving on a road network (the Ground Moving Target Indicator problem—see [42,43] and chapter 10 of [1]), inside a building [44], or in restricted waters [45,46]. Hard edges and boundaries, which cannot be easily accommodated by Kalman-type filters, do not pose any difficulty for the particle approach. Essentially, the bounded support is simply flooded with particles.

12.6.2 Tracking with Nonstandard Sensors

The classical nonlinear tracking test case is the bearings-only problem with passive sensors (acoustic, electro-optical, or electronic support measures), and particle filters have certainly been applied to examples of this type (see [16,46] and chapter 6 of [1]). However, particle tracking filters have also been successfully implemented with range-Doppler sensors that provide measurements of only observer–target range and range rate (see chapter 7 of [1]). An interesting application to a network of binary sensors (i.e., each sensor provides one bit of information) is reported in [47]. Also, particle filtering of raw sensor outputs (such as pixel gray levels) has been examined by a number of workers in the context of track-before-detect (see chapter 11 of [1] and [48–51]).

12.6.3 Multiple-Object Tracking and Association Uncertainty

The obvious way of approaching multiple target tracking problems is to concatenate the state vectors of individual targets and attempt to estimate the combined state. This approach is appropriate if the targets' dynamics are interdependent (e.g., formation or group dynamics—see chapter 12 of [1]) or if there is measurement association uncertainty (or unresolved targets) due to object proximity [52]. Particle filters have been successfully applied in these cases for small numbers of objects (as in Section 12.5), although the evaluation of the likelihood function (for every particle) can be expensive as it involves summing over feasible assignment hypotheses. An alternative, more efficient route suggested in [53] is to employ a probabilistic multiple hypothesis tracker likelihood, which effectively imposes independence between object-measurement assignments. This approach may also be viewed as a superposition of Poisson target models (possibly including extended objects) [54]. Particle filtering is also an implementation mechanism for the finite-set statistics probability hypothesis density filter [55–58].

12.6.4 Computer Vision and Robotics

Particle filtering was introduced to the computer vision community as the CONDENSATION algorithm [18]. In this application, the state vector includes shape descriptors as well as dynamics parameters. This has been a successful domain for particle filters, and there is now substantial literature especially in the IEEE Computer Society Conferences on Computer Vision and Pattern Recognition and the IEEE International Conferences on Pattern Recognition. Applications include tracking of facial features (especially using active contours or "snakes"), gait recognition, and people tracking (some recent publications include [59–62]). Particle filters are also well represented in the robotics literature: they have been successfully applied to localization, mapping, and fault diagnosis problems [44,63–65].

12.6.5 Econometrics and Financial Time Series

Progress in this field has tended to parallel, but to remain largely independent of, engineering developments. However, in the case of particle filtering and Monte Carlo methods, there has been perhaps more "cross-over" than usual. Econometric applications include stochastic volatility modeling for stock indices and commodity prices [66–69].

12.6.6 Numerical Weather Prediction

The requirement here is to update model states with observational data from, for example, weather satellites. This is known as data assimilation and can be viewed as a (very large) nonlinear dynamic estimation and prediction problem. A range of techniques is employed, including EKFs and "ensemble Kalman filters," which use samples for nonlinear state propagation but fit a Gaussian for the Kalman update operation (see the special issue of the *IEEE Control Systems Magazine* [70]).

12.7 Conclusions

Over the past few years, particle filters have become a popular topic. There have been a large number of papers demonstrating new applications and algorithm developments. This popularity may be due to the simplicity and generality of the basic algorithm—it is easy to get started. Furthermore, the particle filter is not another variant of the EKF: it does not stem from linear-Gaussian or least-squares theory. It also appeals to both the practical engineer (algorithm tuning) and to the more theoretical community (with substantial challenges to develop performance bounds and guidelines for finite sample sizes). Undoubtedly, a key enabler for this activity has been the massive increase in the capability of cheap computers—as Daum [27] has pointed out, "computers are now eight orders of magnitude faster (per unit cost) compared with 1960, when Kalman published his famous paper."

The basic or naive version of the particle filter may be regarded as a black box algorithm with a single tuning parameter—the number of samples. This filter is very effective for many low-dimensional problems, and perhaps fortuitously, reasonable results were obtained for state vectors with about 10 elements without resorting to an enormous number of particles. For more challenging high-dimensional problems, a more subtle approach (exploiting Rao–Blackwellization and carefully chosen proposal distributions) is generally beneficial—there is a design trade-off between many simple or fewer smart particles. This (problem-dependent) compromise would benefit from further study.

Initially, most particle filter applications were in simulation studies or offline with recorded data. However, particle filters are now appearing as online elements of real systems—mainly in navigation and robotics applications. The technology (and necessary processing capability, e.g., exploiting graphics processing units [33]) is now sufficiently mature to support the leap to such real-time system implementation. We expect to see a significant increase here in coming years.

Acknowledgments

These studies, carried out over the last 15 years at Dstl, QinetiQ, and predecessor organizations, were sponsored by various U.K. MOD programs. The author would like to acknowledge the help and inspiration of many colleagues, and especially Neil Gordon (now at DSTO Australia).

References

1. B. Ristic, S. Arulampalam, and N. Gordon, *Beyond the Kalman Filter: Particle Filters for Tracking Applications*. Boston: Artech House, 2004.
2. P. Djuric, J. Kotecha, J. Zhang, Y. Huang, T. Ghirmai, M. Bugallo, and J. Miguez, Particle filtering, *IEEE Signal Processing Magazine*, 20, 19–38, September 2003.
3. J. Candy, Bootstrap particle filtering, *IEEE Signal Processing Magazine*, 24, 73–85, July 2007.
4. D. Fox, J. Hightower, L. Liao, D. Schulz, and G. Borriello, Bayesian filtering for location estimation, *IEEE Pervasive Computing*, 2(3), 24–33, 2003.
5. Z. Chen, *Bayesian Filtering: from Kalman Filters to Particle Filters, and Beyond*. Hamilton, ON, Canada: Adaptive Syst Lab, McMaster Univ., 2003.
6. A. Doucet, N. de Freitas, and N. J. Gordon, eds., *Sequential Monte Carlo Methods in Practice*. New York: Springer, 2001.
7. O. Cappe, E. Moulines, and T. Ryden, *Inference in Hidden Markov Models*. New York: Springer-Verlag, 2005.
8. A. Bain and D. Crisan, *Fundamentals of Stochastic Filtering*. Springer, 2009.
9. F. Gustafsson, Particle filter theory and practice with positioning applications, *IEEE Aerospace and Electronic Systems Magazine—Part 2: Tutorials V*, 25, 53–82, July 2010.
10. IEEE, Special issue on Monte Carlo methods for statistical signal processing, *IEEE Transactions on Signal Processing*, 50, February 2002.
11. O. Cappe, S. Godsill, and E. Moulines, An overview of existing methods and recent advances in sequential Monte Carlo, *Proceedings of the IEEE*, 95, 899–924, May 2007.
12. D. Salmond, D. Fisher, and N. Gordon, Tracking in the presence of intermittent spurious objects and clutter, *SPIE: Signal and Data Processing of Small Targets*, 3373, 460–474, 1998.
13. D. Salmond, N. Everett, and N. Gordon, Target tracking and guidance using particles, in *American Control Conference*, pp. 4387–4392, Arlington, VA: IEEE, June 2001.
14. D. Salmond and N. Gordon, Particles and mixtures for tracking and guidance, in *Sequential Monte Carlo Methods in Practice* (A. Doucet, N. de Freitas, and N. J. Gordon, eds.), pp. 517–532, New York: Springer, 2001.
15. Y. C. Ho and R. C. K. Lee, A Bayesian approach to problems in stochastic estimation and control, *IEEE Transactions on Automatic Control*, 9, 333–339, 1964.
16. N. Gordon, D. Salmond, and A. Smith, A novel approach to nonlinear/non-Gaussian Bayesian state estimation, *IEE Proceedings on Radar, Sonar and Navigation*, 140(2), 107–113, 1993.

17. G. Kitagawa, Monte Carlo filter and smoother for non-Gaussian non-linear state space models, *Journal of Computational and Graphical Statistics*, 5(1), 1–25, 1996.
18. M. Isard and A. Blake, CONDENSATION—conditional density propagation for visual tracking, *International Journal of Computer Vision*, 29(1), 5–28, 1998.
19. B. Efron and R. Tibshirani, *An Introduction to the Bootstrap*. London: Chapman and Hall, 1998.
20. J. S. Liu and R. Chen, Sequential Monte Carlo methods for dynamical systems, *Journal of the American Statistical Association*, 93, 1032–1044, 1998.
21. A. Doucet, S. Godsill, and C. Andrieu, On sequential Monte Carlo sampling methods for Bayesian filtering, *Statistics and Computing*, 10(3), 197–208, 2000.
22. C. Musso, N. Oudjane, and F. LeGland, Improving regularised particle filters, in *Sequential Monte Carlo Methods in Practice* (A. Doucet, N. de Freitas, and N. J. Gordon, eds.), New York: Springer, 2001.
23. W. Gilks and C. Berzuini, Following a moving target—Monte Carlo inference for dynamic Bayesian models, *Journal of the Royal Statistical Society: Series B*, 63(1), 127–146, 2001.
24. A. Kong, J. S. Liu, and W. H. Wong, Sequential imputations and Bayesian missing data problems, *Journal of the American Statistical Association*, 89(425), 278–288, 1994.
25. B. Silverman, *Density Estimation for Statistics and Applied Data Analysis*. London: Chapman and Hall, 1986.
26. C. Andrieu, A. Doucet, S. Singh, and V. Tadic, Particle methods for change detection, system identification, and control, *Proceedings of the IEEE*, 92, 423–438, March 2004.
27. F. Daum, Nonlinear filters: beyond the Kalman filter, *IEEE Aerospace and Electronic Systems Magazine—Part 2: Tutorials II*, 20, 57–69, August 2005.
28. M. S. Arulampalam, S. Maskell, N. Gordon, and T. Clapp, A tutorial on particle filters for non-linear/non-Gaussian Bayesian tracking, *IEEE Transactions on Signal Processing*, 50, 174–188, February 2002.
29. T. Schon, F. Gustafsson, and P.-J. Nordlund, Marginalized particle filters for mixed linear/nonlinear state-space models, *IEEE Transactions on Signal Processing*, 53, 2279–2289, July 2005.
30. G. Casella and C. Robert, Rao–Blackwellization of sampling schemes, *Biometrika*, 83(1), 81–94, 1996.
31. A. Doucet, N. Gordon, and V. Krishnamurthy, Particle filters for state estimation of jump Markov linear systems, *IEEE Transactions on Signal Processing*, 49, 613–624, March 2001.
32. R. Karlsson, T. Schon, and F. Gustafsson, Complexity analysis of the marginalized particle filter, *IEEE Transactions on Signal Processing*, 53, 4408–4411, 2005.
33. G. Hendeby, R. Karlsson, and F. Gustafsson, Particle filtering: the need for speed, *EURASIP Journal on Advances in Signal Processing*, 2010, 1–9, 2010.
34. S. Sutharsan, A. Kirubarajan, and A. Sinha, An optimization-based parallel particle filter for multitarget tracking, *SPIE: Signal and Data Processing of Small Targets*, 5913, 591309-1–591309-12, August 2005.
35. F. Daum and J. Huang, Curse of dimensionality and particle filters, in *Proceedings of IEEE Aerospace Conference*, Big Sky, MT: IEEE, March 2003.
36. A. A. Fel'dbaum, *Optimal Control Systems*. New York: Academic Press, 1965.
37. D. P. Bertsekas, *Dynamic Programming: Deterministic and Stochastic Models*. Englewood Cliffs, NJ: Prentice-Hall, 1987.

38. D. G. Hull, J. L. Speyer, and D. B. Burris, Linear-quadratic guidance law for dual-control of homing missiles, *Journal of Guidance, Control and Dynamics*, 13(1), 137–144, 1990.

39. Y. Bar-Shalom, Stochastic dynamic programming: caution and probing, *IEEE Transactions on Automatic Control*, AC-26, 1184–1195, October 1981.

40. G. L. Tanner, Missile control against multiple targets using non-quadratic cost functions, in *American Control Conference*, vol. 1, pp. 515–519, Seattle, Washington: IEEE, June 1995.

41. R. Lawrence, Interceptor line-of-sight rate steering: necessary conditions for a direct hit, *Journal of Guidance, Control and Dynamics*, 21, 471–476, May–June 1998.

42. S. Arulampalam, N. Gordon, M. Orten, and B. Ristic, A variable structure multiple model particle filter for GMTI tracking, in *Fusion 2002: Proceedings of the 5th International Conference on Information Fusion*, pp. 927–934, Annapolis, MD: ISIF, 2002.

43. G. Kravaritis and B. Mulgrew, Variable-mass particle filter for road-constrained vehicle tracking, *EURASIP Journal on Advances in Signal Processing*, 2008, 1–13, 2008.

44. D. Fox, S. Thrun, W. Burgard, and F. Dellaert, Particle filters for mobile robot localization, in *Sequential Monte Carlo Methods in Practice* (A. Doucet, N. de Freitas, and N. J. Gordon, eds.), New York: Springer, 2001.

45. M. Mallick, S. Maskell, T. Kirubarajan, and N. Gordon, Littoral tracking using a particle filter, in *Fusion 2002: Proceedings of the 5th International Conference on Information Fusion*, pp. 935–942, Annapolis, MD: ISIF, 2002.

46. R. Karlsson and F. Gustafsson, Recursive Bayesian estimation—bearings-only applications, *IEE Proceedings on Radar, Sonar and Navigation*, 152(5), 305–313, 2005.

47. J. Aslam, Z. Butler, F. Constantin, V. Crespi, G. Cybenko, and D. Rus, Tracking a moving object with a binary sensor network, in *SenSys '03: Proceedings of the 1st International Conference on Embedded Networked Sensor Systems* (New York), pp. 150–161, ACM Press, 2003.

48. D. J. Salmond and H. Birch, A particle filter for track-before-detect, in *Proc. American Control Conf.*, pp. 3755–3760, Arlington, VA: IEEE, June 2001.

49. Y. Boers and J. Driessen, Multitarget particle filter track before detect application, *IEE Proceedings on Radar, Sonar and Navigation*, 151(6), 351–357, 2004.

50. M. Rutten, N. Gordon, and S. Maskell, Efficient particle-based track-before-detect in Rayleigh noise, in *Fusion 2004: Proceedings of the 7th International Conference on Information Fusion*, Stockholm, Sweden: ISIF, 2004.

51. S. J. Davey, M. G. Rutten, and B. Cheung, A comparison of detection performance for several track-before-detect algorithms, *EURASIP Journal on Advances in Signal Processing*, 2008, 1–10, 2008.

52. Z. Khan, T. Balch, and F. Dellaert, "Multitarget tracking with split and merged measurements," in *CVPR 2005: Proceedings of the 2005 Computer Society Conference on Computer Vision and Pattern Recognition*, San Diego, CA: IEEE, 2005.

53. C. Hue, J.-P. Le Cadre, and P. Perez, Tracking multiple objects with particle filtering, *IEEE Transactions on Aerospace and Electronic Systems*, 38, 791–812, July 2002.

54. K. Gilholm, S. Godsill, S. Maskell, and D. Salmond, Poisson models for extended target and group tracking, *SPIE: Signal and Data Processing of Small Targets*, 5913, 59130R-1–59130R-12, August 2005.

55. B.-N. Vo, S. Singh, and A. Doucet, Random finite sets and sequential Monte Carlo methods in multi-target tracking, in *Proceedings of International Radar Conference*, pp. 486–491, Adelaide, Australia: IEEE, September 2003.

56. R. Mahler, Statistics 101 for multisensor, multitarget data fusion, *IEEE Aerospace and Electronic Systems Magazine—Part 2: Tutorials*, 19, 53–64, January 2004.
57. R. Mahler, Random sets: unification and computation for information fusion - a retrospective assessment, in *Fusion 2004: Proceedings of the 7th International Conference on Information Fusion*, pp. 1–20, Stockholm, Sweden: ISIF, 2004.
58. D. Clark, I. Ruiz, Y. Petillot, and J. Bell, Particle PHD filter multiple target tracking in sonar images, *IEEE Transactions on Aerospace and Electronic Systems*, 43, 409–416, January 2007.
59. R. Green and L. Guan, Quantifying and recognizing human movement patterns from monocular images: parts I and II, *IEEE Transactions on Circuits and Systems for Video Technology*, 14(2), 179–198, 2004.
60. L. Wang, H. Ning, T. Tan, and W. Hu, Fusion of static and dynamic body biometrics for gait recognition, *IEEE Transactions on Circuits and Systems for Video Technology*, 14(2), 149–158, 2004.
61. J. Tu, Z. Zhang, Zeng, and T. Huang, Face localization via hierarchical CONDENSATION with Fisher boosting feature selection, in *CVPR 2004: Proceedings of the 2004 Computer Society Conference on Computer Vision and Pattern Recognition*, pp. II–719–II–724, Washington, DC: IEEE, 2004.
62. M. de Bruijne and M. Nielsen, Image segmentation by shape particle filtering, in *ICPR 2004: Proceedings of the 17th International Conference on Pattern Recognition*, Cambridge, UK: IEEE, 2004.
63. S. Thrun, Particle filters in robotics, in *Proceedings of the 18th Annual Conference on Uncertainty in AI (UAI)*, pp. 511–518, Edmonton, Canada: AUAI, 2002.
64. M. Rosencrantz, G. Gordon, and S. Thrun, Locating moving entities in indoor environments with teams of mobile robots, in *AAMAS '03: Proceedings of the Second International Joint Conference on Autonomous Agents and Multiagent Systems*, pp. 233–240, Melbourne, Australia: ACM Press, 2003.
65. S. Sutharsan, A. Kirubarajan, and A. Sinha, Adapting the sample size in particle filters through KLD-sampling, *International Journal of Robotics Research*, 22, 985–1004, October 2003.
66. G. Kitagawa and S. Sato, Monte Carlo smoothing and self-organising state-space model, in *Sequential Monte Carlo Methods in Practice* (A. Doucet, N. de Freitas, and N. J. Gordon, eds.), New York: Springer, 2001.
67. M. Pitt and N. Shephard, Auxiliary variable based particle filters, in *Sequential Monte Carlo Methods in Practice* (A. Doucet, N. de Freitas, and N. J. Gordon, eds.), New York: Springer, 2001.
68. J. Stroud, N. Polson, and P. Mueller, Practical filtering for stochastic volatility models, in *State Space and Unobserved Component Models* (A. Harvey, S. Koopman, and N. Shephard, eds.), Cambridge, UK: Cambridge University Press, 2004.
69. P. Fearnhead, Using random Quasi-Monte-Carlo within particle filters with application to financial time series, *Journal of Computational and Graphical Statistics*, 14, 751–769, 2005.
70. IEEE, Special issue on data assimilation for weather forecasting, *IEEE Control Systems Magazine*, 29, June 2009.

Appendix: Worked Example—Pendulum Estimation

To demonstrate the operation of the particle filter, we present an application to a pendulum estimation problem. A weightless rigid rod of length L is freely pivoted at one end and carries a mass at its other end. The rod makes an angle θ with the horizontal, and its instantaneous angular acceleration is given by

$$\ddot{\theta} = (1/L)(-g + v)\cos\theta$$

where g is the acceleration due to gravity and v is a random disturbance. This differential equation is the motivation for the following simple discrete dynamics model:

$$\left.\begin{aligned}
\theta_k &= \mathrm{mod}[\theta_{k-1} + \Delta t\dot{\theta}_{k-1} + (\Delta t^2/2L)(-g + v_{k-1})\cos\theta_{k-1}, 2\pi] \\
\dot{\theta}_k &= \dot{\theta}_{k-1} + (\Delta t/L)(-g + v_{k-1})\cos\theta_{k-1}
\end{aligned}\right\} \quad (12.19)$$

where θ has been restricted to the range $[0, 2\pi)$, Δt is the fixed time step, and the acceleration disturbance v_k is a zero mean, white, Gaussian random sequence of variance q. Thus, this example has a two-element state vector $\mathbf{x}_k = (\theta_k, \dot{\theta}_k)^T$ for $k > 0$. Measurements are obtained from the length of the rod projected onto a vertical axis, that is, $L|\sin\theta_k|$. These measurements are quantized at intervals of δ but are otherwise error-free, so

$$z_k = Q_\delta(L|\sin\theta_k|),$$

where the quantization operator $Q_\delta(x) = (n - 1)\delta$ for the integer n such that $(n - 1)\delta < x \le n\delta$. Thus, the likelihood of the state vector is

$$p(\mathbf{z}_k|\mathbf{x}_k) = \begin{cases} 1 & \text{if } z_k < L|\sin\theta_k| \le z_k + \delta \\ 0 & \text{otherwise.} \end{cases} \quad (12.20)$$

In other words, given a measurement z, the projected length of the pendulum is equally likely to be anywhere in the interval $(z, z + \delta]$ but cannot be anywhere else. The problem is to construct the posterior pdf of the state vector $(\theta_k, \dot{\theta}_k)^T$ given the set of measurements $\mathbf{Z}_k$ and the initial conditions that θ_0 is uniformly distributed over $[0, 2\pi)$ and $\dot{\theta}_0$ is Gaussian distributed with known mean and variance. The dynamics recursion (Equation 12.19), the likelihood (Equation 12.20), and the above initial conditions completely specify the problem for application of a particle filter. This system is illustrated in Figure 12.8 for $\delta = L/3$.

The basic version of the particle filter has been applied to this example. Here, each particle is a two-element vector $(\theta, \dot{\theta})$. As already indicated, the prediction phase of the filter consists of passing each particle through the dynamics model (Equation 12.19). A MATLAB code for this example is shown below. In this code, the posterior particles $\{x^{i*}\}_{i=1}^{N}$ are contained in the $2 \times N$ array x_post, where the two rows correspond to θ and $\dot{\theta}$, and each column is an individual particle. Similarly, the prior particles $\{x^{i}\}_{i=1}^{N}$ are contained in the $2 \times N$ array x_prior, and nsamples is the number of particles N. The unnormalized weights for each particle are stored in the N element array likelihood, while the normalized weights and their cumulative sum are held in weight. It is easy to recognize the dynamics Equations 12.19 in the prediction phase and the likelihood 13.20 in the update phase.

Hopefully, this code listing will clarify the specification of the filter given in Section 12.3. Note that the complete filter can be expressed in a few lines of MATLAB: the basic algorithm is (embarrassingly) simple. Furthermore, there are no "hidden extras": the code does not call any sophisticated numerical algorithms (numerical integration packages, eigenvector solvers, etc.) or symbolic manipulation packages—except perhaps for the random number generator and the MATLAB array handling routines.

```
%*************************************************************************

%Generate initial samples for k=0:
x_post(1,:)     = 2*pi*rand(1,nsamples);
x_post(2,:)     = theta_dot_init + sig_vel_init*randn(1,nsamples);


for k=1:nsteps

    % PREDICT
    F1          = dt*dt/(2*pend_len);      F2=dt/pend_len;
    drive1      = randn(1,nsamples);    % random samples for system noise
    accn_in     = (-gee+drive1*sig_a).*cos(x_post(1,:));
    x_prior(1,:) = mod( x_post(1,:) + dt*x_post(2,:) + F1*accn_in , 2*pi );
    x_prior(2,:) = x_post(2,:)      + F2*accn_in;


    % UPDATE
    % EVALUATE WEIGHTS resulting form meas(k):
    project     = pend_lcn.*abs(sin(x_prior(1,:)));  % rod projection for each sample
    likelihood  = zeros(nsamples,1);
    likelihood( find( project>=meas(k) & project<meas(k)+delta ) )=1;
    weight      = likelihood/sum(likelihood);    % normalise weights
    weight      = cumsum(weight);                % form cumulative distribution


    % RE-SAMPLING PROCEDURE (SYSTEMATIC)
    addit=1/nsamples; stt=addit*rand(1);
    selection_points=[ stt : addit : stt+(nsamples-1)*addit ];   j=1; %set up comb
    for i=1:nsamples
      while selection_points(i) >= weight(j); j=j+1; end;
      x_post(:,i) = x_prior(:,j);
    end;


    % OUTPUT:  store posterior particles (for analysis only)
    samp_store(:,:,k) = x_post;
end
%*************************************************************************
```

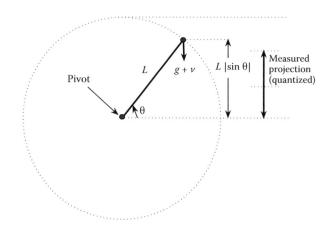

FIGURE 12.8
Pendulum with quantized projection measurement.

This program has been applied to the data set shown in Figure 12.9. Here, the quantization interval is $\delta = L/2$, so the only information available from the measurements is whether the projected rod length is greater or less than $L/2$ (i.e., one bit of information). The other parameters of this simulation are $\theta_0 = 0.3$ rad, $\dot{\theta}_0 = 2$ rad/s, $\Delta t = 0.05$ s, $L = 3$ m, and $g = 10$ m/s², and the standard deviation of the driving noise υ is 7 m/s². In the 10-s period shown, the pendulum changes its direction of rotation twice (after about 1 and 8 s), between which it makes a complete rotation. The initial conditions supplied to the particle filter are that the angular velocity $\dot{\theta}_0$ is from a Gaussian

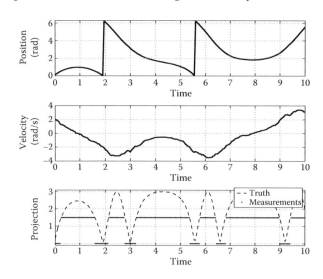

FIGURE 12.9
Truth and measurements.

distribution with a mean of 2.4 rad/s and a standard deviation of 0.4 rad/s. As already stated, the initial angle θ_0 is uniformly distributed over $(0, 2\pi)$ rad. The initial particle set is drawn from these distributions as shown in the above listing.

The result of running the filter with $N = 1000$ particles is shown in Figure 12.10. This figure shows the evolution of the posterior pdf of the angle θ obtained directly from the posterior particles. The pdf for each of the 200 time steps is a simple histogram of the posterior angle particles. The evolving distribution consists of streams or paths of modes that cross and pass through regions of bifurcation. For this case of $\delta = L/2$, the measurements switch between 0 and $L/2$ whenever $|\sin\theta| = 0.5$, that is, when $\theta = \pi/6, 5\pi/6,$ $7\pi/6$, or $11\pi/6$. As is evident from Figure 12.10, at these transition points, the pdf modes sharpen. Occasionally, a path is terminated if it is incompatible with a measurement transition (e.g., for $\theta = 11\pi/6$ at about $k = 15$). The actual angle of the pendulum is shown as a string of dots.

Note that $N = 1000$ is adequate to give a fairly convincing estimate of the posterior density, although it appears a little ragged in the region $\theta = \pi/6$ to $5\pi/6$ about the vertically up position (where the angular velocity tends to be low and the pendulum may swing back or continue over the top). The ragged structure can be smoothed by increasing the number of particles— Figure 12.11 shows the evolving pdf for the extravagantly large value of $N = 50,000$. This produces a pleasingly smooth result but is otherwise very similar to the 1000 particle result. The $N = 1000$ case took about 5 ms per time step to run on a modest PC—a quite acceptable rate for the quality of the result. For $N = 50,000$, the time taken increased almost linearly to 260 ms per time step. Note that apart from the obvious time penalty, it is trivial to improve filter performance to approach the exact posterior pdf.

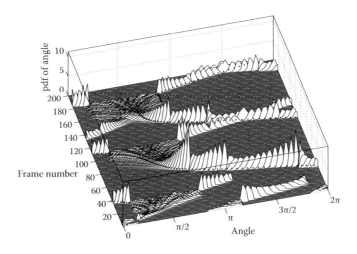

FIGURE 12.10
Posterior pdf from 1000 samples.

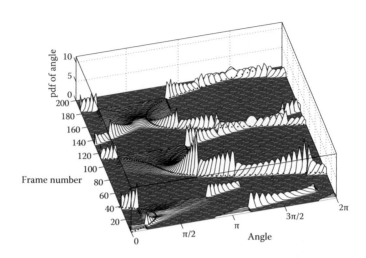

FIGURE 12.11
Posterior pdf from 50,000 samples.

Discussion

This filtering example was chosen to demonstrate the particle filter because it is simple to specify and would be difficult to tackle using an EKF (or an unscented Kalman filter). It is also a low-dimensional case and so is an easy example for a particle filter. With some effort, it would be possible to develop a multiple-hypothesis Kalman filter to capture the multimodal nature of the posterior pdf and to include the 2π wraparound in angle. Also, it might be possible to represent the quantization function as a form of Gaussian mixture. However, this would all be quite awkward and definitely approximate (and would probably be more computationally expensive). The particle approach avoids all such difficulties in this example. Also, the traditional summary descriptors of recursive estimation—the mean and covariance—would be quite inappropriate for this example, where the posterior pdf is often multimodal and sometimes unimodal but highly skewed.

13

Practical Techniques for the Design of Multirate Digital Guidance Laws and Autopilots

C. A. Rabbath, N. Léchevin, and M. Lauzon

CONTENTS

13.1 Introduction

Guidance laws and autopilots are generally developed in the continuous-time (CT) domain. Such systems act upon CT signals and may include exact filtered differentiation and integration. For example, the classical proportional navigation guidance (PNG) law processes filtered line-of-sight (LOS) rates [1]. However, in practice, guidance and control systems are implemented on digital hardware and, consequently, input and output discrete-time (DT) signals. Despite this fact, most homing guidance laws and autopilots proposed recently have, for the most part, ignored the digital implementation issue. Homing guidance laws have been devised to improve performance of weapons in the presence of maneuvering targets, uncertain weapon dynamics and evader information, and limited acceleration capabilities [2–10]. Figure 13.1 shows the generic block diagram of a digital control system with its required subsystems: digital processors, analog-to-digital (A/D) and digital-to-analog (D/A) converters, actuators, and sensors. It is known that a digital implementation may adversely affect the performance of the weapon, when compared with that of a CT control system, due to computational delays, quantization, and finite update rates [11]. For example, ideal sampling, which models the A/D conversion, is a periodic operation that outputs the value of a CT signal at every T instants of time, with a finite number of bits to represent the value of the DT signal. Intuitively, the faster the sampling process, the closer the value of the sampler output to that of the CT input signal. If the LOS rate is used by the guidance law, as is the case with PNG, the sampling rate should be selected fast enough such that sufficient information on the LOS rate is available for the guidance computations. Yet, sampling period T is a parameter that cannot always be selected as small as required by the designer. Indeed, modern weapons are increasingly subject to a greater number of onboard computations to foster an expanded level of autonomy. Although capabilities in computing are constantly improving, designers must exercise care when it comes to choosing the sampling and control update rates as the range of sampling rates available for guidance and control is constrained by the available hardware and software and by the variety of computing tasks performed onboard the vehicles. Ideally, the digital nature of the guidance and the autopilot systems, and the constraints

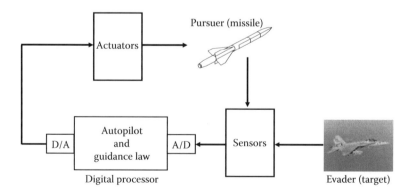

FIGURE 13.1
Missile guidance and autopilot as a digital control system.

on the choice of the sampling and control update rates, should be addressed right at the synthesis phase, the reason being the need to guarantee analytically closed-loop stability and a certain level of closed-loop performance. Ultimately, approaches to the synthesis of autopilots and guidance laws warranting a satisfactory level of performance for the largest range of sampling and control update rates are preferred. For example, it may happen that the available hardware and the computing requirements force the designer to implement the guidance law and autopilot schemes at a relatively slow rate, for which classical techniques of CT to DT guidance and autopilot conversion fail to warrant satisfactory closed-loop performance. Furthermore, plant output, if processed digitally, may only be available to the autopilots and the guidance laws for relatively long update periods. The conversion of systems from CT to DT, known as digital redesign (DR), is favored in practice over the other approaches of direct DT design and sampled-data design [12], as DR takes advantage of the vast body of knowledge available with CT synthesis methods, a variety of software tools for CT synthesis, and the intuitive physical interpretation of CT systems that engineers are most comfortable with. The classical DR approaches are based on the following principle: discretize the individual CT controllers with methods such as hold invariance, matched poles and zeros, and numerical approximation to differentiation and integration [13]. Such DR techniques, however, are known to be highly sensitive to sampling rate selection and may even result in unstable missile control systems when sampling rates are too slow.

With the availability of processors, data acquisition units, and I/O boards operating at various rates and with the presence of multiscale missile dynamics, there arises an additional level of flexibility for the utilization of the available resources, which could enable a reduction in the real-time computational burden. To achieve such an objective, guidance and autopilot must be designed to operate on signals evolving at multiple sampling and control update rates. The idea is as follows. The output of a dynamic system should be sampled at a rate selected with respect to the system dynamics, as is well known [14]. In that sense, employing sampling rates tailored to the time constants of the CT plant output signals upon which the sampling processes act and updating the DT control law at rates fast enough to warrant stability and performance may enable effective handling of the available computing resources. For example, missiles usually exhibit multiple dynamic scales, as the inner-loop angular rate feedback of a missile autopilot typically has a faster response than the outer-loop acceleration feedback [15]. Ideally, one would then sample the angular rate signal with a faster rate than that used for acceleration. If multirate sampling and control update is not implemented, the presence of multiple missile dynamic scales may result in a control system with overly sampled slow signals to accommodate the sampling of the fast signals and may preclude real-time execution. Alternatively, one may view multirate control as a means to compensating for the limited plant output sampling capabilities by means of fast control input updates. As benefits of multirate control, Azad and Hesketh [16] show that using plant output sampling and control input hold at different rates may provide better trade-offs between performance and implementation costs, and Yang and Tomizuka [17] propose a multirate control scheme to reduce the real-time control computational burden for hard disk drive servo systems by updating components of the controller at different rates without performance degradation. For missile control, there have been works on multirate optimal autopilot [15] and dual-rate digital guidance law synthesis [18].

In this chapter, we discuss practical techniques for the design of digital homing guidance laws and autopilots. The methods put forth rely on optimal control theory and handle implementations at multiple and at single sampling and control update rates. The design approaches guarantee performance preservation in the process of going from an idealized CT missile control system to a practical multirate DT implementation. The fundamental concepts and tools needed for design are described in Section 13.2, including the concept of DT lifting [19], which enables single-rate modeling and analysis of multirate weapon control systems. In Section 13.3, models are obtained for the kinematics of a two-dimensional engagement and for the uncertain missile dynamics. A robust homing guidance law is then devised as a three-step synthesis: design of CT robust guidance, modeling of multirate generalized DT plant, and solution to the single rate robust DT control problem. In Section 13.4, dual-rate autopilot synthesis is described. Digital missile autopilots are usually obtained as follows: first, design a linear CT missile autopilot for each operating point in the missile flight envelope, then convert such controllers to single-rate DT systems, and schedule the parameters online. To relax the constraint on the value of the sampling period, at least to some extent, single-rate DR methods have been introduced over the last two decades [20–26]. These techniques take into account the CT control system topology and the plant dynamics in the conversion process. As a result, control systems exhibit satisfactory performances over a wider range of sampling rates when compared with classical (local) DR techniques [23]. Here, two digital autopilot design techniques are presented. Both consider the CT control system topology and the plant dynamics in the conversion process. One autopilot synthesis relies on the solution to a DT H_2 optimal control problem, while the other technique takes advantage of well-known polynomial tools to warrant closed-loop stability while enabling complexity-constrained controllers. In Section 13.5, to test the guidance and autopilots under various implementation conditions, simulations are carried out with high-performance computing and fixed-point, finite word length implementations. Sophisticated guidance and flight control laws require, in general, high-performance microelectronics. Here, however, we demonstrate that satisfactory levels of performance can be achieved through careful synthesis of the guidance laws and autopilots without requiring high-end computational capability.

13.2 Preliminary Concepts

13.2.1 Operators in DT Domain

Consider a DT signal $f(k, T) : N \to R^n$, where $T \in R^+$ is the sampling period and $k \in N$ is the time step. $f(k, T)$ can be written as the following sequence of vectors in R^n:

$$f(k, T) = \{...,f(0, T), f(1, T), f(2, T),...\}. \tag{13.1}$$

The shift operator q acting on $f(k, T)$ is given as

$$qf(k,T)$$
$$= f(k+1,T) \tag{13.2}$$
$$= \{..., \underset{k=0}{f(1,T)}, \underset{k=1}{f(2,T)}, \underset{k=2}{f(3,T)},...\}.$$

For a linear, time-invariant (LTI) DT system G with state $x(k, T) \in R^n$, input $u(k, T) \in R^m$, and output $y(k, T) \in R^p$, the state-space equation in the shift form is given as follows:

$$x(k + 1, T) = Ax(k, T) + Bu(k, T)$$
$$y(k, T) = Cx(k, T) + Du(k, T) \tag{13.3}$$

where $A \in R^{n \times n}$, $B \in R^{n \times m}$, $C \in R^{p \times n}$, and $D \in R^{p \times m}$. Equation 13.3 is the classical expression for DT systems.

It is now a well-established fact that a DT system can also be expressed in the so-called delta form [27]. Using the DT signal $f(k, T)$ the delta operator can be defined as

$$\delta f(k,T) = \frac{f(k+1,T) - f(k,T)}{T} \tag{13.4}$$

which is reminiscent of the numerical approximation to the derivative. It is easily seen that the delta operator can be expressed in terms of the shift operator. To obtain the state-space equations for a given DT system G in the delta form, one can use Equations 13.3 and 13.4, yielding

$$\delta x(k,T) = \underbrace{\frac{(A-I)}{T}}_{=A_\delta} x(k,T) + \underbrace{\frac{B}{T}}_{=B_\delta} u(k,T)$$
$$y(k,T) = C_\delta x(k,T) + D_\delta u(k,T) \tag{13.5}$$

where $C_\delta = C$ and $D_\delta = D$. Inputs, states, and outputs are unaffected by the choice of either the shift or the delta form. Recall that Equations 13.3 and 13.5 are two expressions for system G. The difference between Equations 13.3 and 13.5 is the expression for the time evolution of the state.

For the details on the numerical and analytical properties of the delta form representation, the works of Goodwin and Middleton [27], Hori et al. [28], and Mori et al. [29] should be consulted. Finally, there exist other operators that have been used in the design of DT control systems [30].

13.2.2 Relationships between DT and CT Systems

The concept of single-rate, LTI DT models of CT systems enables designers to establish the fundamental relationships that exist between a class of single-rate DT systems and CT systems, even though the DT and CT domains seem, *a priori*, disjoint [28,29]. DT models arise from the desire to rigorously formalize the intuitive concept that the performance of a DT system obtained through the discretization of a CT system should approach, in one form or another, that of the originating CT system. Even though this type of convergence is expected to be obtained *de facto* with any given discretization method, it is actually not always the case [31,32]. Furthermore, in the design of sampled-data control systems, the choice of hold and sampler must be made with care, especially in the age of generalized sampler and holds [33], where several choices are available beside the classical zero-order hold (ZOH). The definition of a single-rate DT model is given as follows.

Definition 13.1

The single-rate, LTI DT system G expressed in the delta form in Equation 13.5 is said to be a *DT model* of the CT system $\bar{G}$ given as

$$\frac{d\bar{x}(t)}{dt} = \bar{A}\bar{x}(t) + \bar{B}\bar{u}(t)$$
$$\bar{y}(t) = \bar{C}\bar{x}(t) + \bar{D}\bar{u}(t)$$

(13.6)

provided that

$$\lim_{T\to 0} A_\delta = \bar{A}, \; \lim_{T\to 0} B_\delta = \bar{B}, \; \lim_{T\to 0} C_\delta = \bar{C}, \; \lim_{T\to 0} D_\delta = \bar{D}.$$

(13.7)

∎

From now on, a symbol with a bar represents a CT signal or a system under CT control. An explanation on the type of convergence as $T \to 0$ is in order. If the parameters of a DT system satisfy the limits in Equation 13.7, it means that for any fixed time t^*, the output of the DT system, $y(k, T)$ approaches that of the CT system, $\bar{y}(t^*)$, in the following sense:

$$\lim_{T \to 0} \left\| \bar{y}(t^*) - y(k, T) \right\| = 0 \tag{13.8}$$

provided that

$$\lim_{T \to 0} \left\| \bar{u}(t^*) - u(k, T) \right\| = 0 \tag{13.9}$$

where $\bar{u}(t)$ and $\bar{y}(t)$ are, respectively, the input to and the output of the CT system, $u(k, T)$ and $y(k, T)$ are, respectively, those of the DT system, T is the uniform sampling period, and k is an integer such that $kT \leq t^* < (k + 1)T$. Furthermore, for a class of single-input, single-output systems HGS, where H is a single-interval hold satisfying the following:

$$\lim_{T \to 0} \sup_{0 \leq \tau < T} \left| H(\tau) - 1 \right| = 0, \tag{13.10}$$

where $H(\tau)$ is the response function of the hold, and S is the ideal sampler, if G in Equation 13.5 is such that there exists at least one realization that approaches to one of $\bar{G}$ in Equation 13.6 according to Equation 13.7, so that G is a DT model of $\bar{G}$ according to Definition 13.1, then the output of HGS approaches, uniformly in time, that of $\bar{G}$, as $T \to 0$, whenever the input to HGS approaches uniformly in time that of $\bar{G}$, with both input signals being continuous. The proof can be found in [32]. This statement forms the basis for the sampled-data model concept introduced in [32]. In other words, if we exclude quantization effects, an input–output convergence of a digital controller preceded by an A/D converter and followed by a D/A converter is assured, in the CT domain, provided that the DT controller is a DT model of a CT controller and the hold satisfies Equation 13.10. The natural consequence is that the overall closed-loop system, with DT control, will approach the closed-loop CT system, as $T \to 0$. The importance of the uniform-in-time convergence lies in the avoidance of undesired ripples or hidden oscillation effects, at least in the limit of $T \to 0$.

13.2.3 Principal Discretization Methods

The main approaches to converting a CT system to a single-rate DT system can be classified into three classes: (1) numerical integration/differentiation, (2) hold equivalence, and (3) matched pole–zero [13,34]. Methods falling in the first class readily apply to both linear and nonlinear CT systems. The traditional approach to converting CT controllers to the DT domain is to employ a discretization method and to implement the resulting controller onboard the missile.

13.2.3.1 Numerical Integration/Differentiation

As a CT system transfer function represents a differential equation, it is natural to try approximating the derivatives for numerical implementation of the ordinary differential equations. Assuming continuous derivatives for the states of a CT system, an approximation at any given sampling instant can be performed. One method widely used in aerospace is Tustin's method, which can be viewed as a trapezoidal approximation of integration. In practice, Tustin's method is applied as follows [13]. In the transfer function of a CT system, a DT system approximation is obtained by replacing s with

$$\frac{2}{T}\frac{z-1}{z+1} \tag{13.11}$$

where z is the complex variable traditionally used in DT system transfer functions. An alternative to z is the γ variable given as $\gamma = (z - 1)/T$ [27]. Thus, a DT system can be represented by transfer functions in either the z or the γ variable. It is important to point out that the numerical integration/differentiation method readily applies to the discretization of nonlinear CT systems. However, for relatively large values of T, the output samples of the CT and the DT systems may significantly differ.

13.2.3.2 Hold Equivalence

The hold-equivalent DT model of a CT system $\bar{G}$ is defined as $G = S\bar{G}H$. To obtain G, proceed as follows. Assume that a known input is applied to the CT system. Typical signals include the step and the ramp inputs. Select a hold H that outputs such CT signals when subject to an appropriate DT input, and place this hold at the input channel of the CT system $\bar{G}$. At the output of $\bar{G}$ place an ideal sampler S. Then, the outputs of $G = S\bar{G}H$ and $\bar{G}$ are equivalent at the sampling instants. Such an approach warrants exact output response matching for a single input type through careful hold selection or design. The simplest hold-equivalence method is the so-called step-input invariance [13], where H is the ZOH. Step-input invariance guarantees exact matching of the output signals of LTI systems G and $\bar{G}$, at the sampling instants, provided that $\bar{G}$ is subject to a unit-step input, whereas G is faced with a unit DT step input.

13.2.3.3 Pole–Zero Matching

An LTI CT system $\bar{G}$ with transfer function

$$\bar{G}(s) = \bar{K}\frac{\displaystyle\prod_{i=1}^{m}(s-z_i)}{\displaystyle\prod_{i=1}^{n}(s-p_i)} \tag{13.12}$$

can be converted to a DT system G, with transfer function in the z variable and with sampling period T by mapping the finite poles p_i and zeros z_i through the exponential given as

$$p_i^z = e^{p_i T}, z_i^z = e^{z_i T}.$$ (13.13)

The DT transfer function is then

$$G(z,T) = K \frac{\prod\limits_{i=1}^{m}(z - z_i^z)}{\prod\limits_{i=1}^{n}(z - p_i^z)}.$$ (13.14)

The gain K is generally calculated with the objective of satisfying a gain requirement at a particular frequency. For matrix transfer functions, the process can be performed for each matrix entry.

13.2.4 Multirate Systems

A multirate DT system comprises a minimum of two signals defined for different update rates. The rates are integer related. The most widely studied multirate system is the dual-rate control system. Figure 13.2 shows two classes of such systems. Feedback control systems with fast plant output sampling and slow control input update rate have the block diagram in Figure 13.2a. The case of fast control update rate, slow plant output sampling is shown in Figure 13.2b. The sampling periods are T and h. Period h is an integer

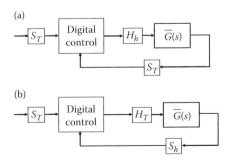

FIGURE 13.2
Dual-rate systems.

multiple of period T. Since holds and samplers can operate at either period T or h, from now on, a subscript is used to indicate either period. In the figure, the digital control block may correspond to guidance and autopilot subsystems, for example. Although not shown in the figure, each output of the multivariable plant $\bar{G}(s)$ can be connected to either S_T or S_h.

The so-called lifting method [19] enables the design and the analysis of dual-rate systems. Lifting is a mathematical tool that allows a designer to reformulate a dual-rate DT system into a slow, single-rate DT system. Then, all of the single-rate DT tools become available for the study of dual-rate systems. The DT lifting operation L takes a fast DT vector signal $f(k, T)$ of size M and converts it to a slow MN-vector DT signal with period $h = N \cdot T$, expressed as $f^L(l, h)$; that is, $L : l_{\mathcal{R}M}^{\infty} \rightarrow l_{\mathcal{R}MN}^{\infty}$. For example, with $N = 3$, lifting the DT scalar signal $f(k, T)$ given as

$$f(k, T) = \{f(0, T), f(1, T), f(2, T), \ldots\} \tag{13.15}$$

results in the DT vector signal $f^L(l, h)$, where $l \in Z^+$, given by

$$f^L(l,h)$$

$$= \left\{ \begin{bmatrix} f(0,T) \\ f(1,T) \\ f(2,T) \end{bmatrix}, \begin{bmatrix} f(3,T) \\ f(4,T) \\ f(5,T) \end{bmatrix}, \cdots \right\}. \tag{13.16}$$

DT lifting is norm preserving [19]. L^{-1} is the inverse DT lifting and $L^{-1}L = I$. A lifted DT system can be simply understood as a single-rate system that operates on at least one lifted signal.

The concept of single-rate DT models of CT systems discussed in Section 13.2.2 can be extended to the case of multirate systems. A closed-loop multirate system can then be made to approach a known reference system as the sampling rates are increased. Consider the CT missile control system in Figure 13.3a; the multirate digital control system in Figure 13.3b, where the update rates are $1/hg$, $1/h$, and $1/T$ Hz; and the digital missile autopilot in Figure 13.4, where ε represents either z or γ. The magnitude of the rates is as follows: $1/T > 1/h > 1/hg$. For the purpose of illustration, dynamics in only one plane are considered. In Figure 13.3b, a_g is the control input to the autopilot, δ_p is the fin deflection command, a_m is the actual missile acceleration, q is the missile body rate, and a_t is the target acceleration. For pitch plane dynamics, q corresponds to pitch rate, whereas in lateral dynamics, q is the yaw rate. For the CT control system of Figure 13.3a, similar definitions

(a)

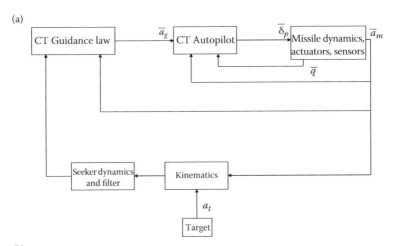

(b)

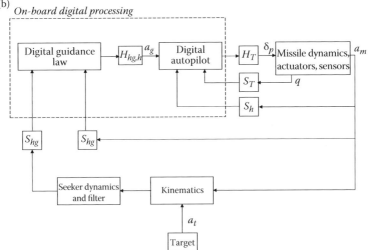

FIGURE 13.3
(a) CT and (b) multirate missile control systems.

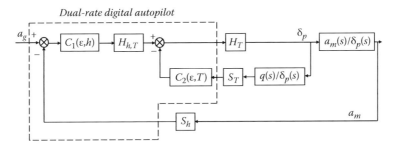

FIGURE 13.4
Dual-rate digital acceleration autopilot.

apply, albeit with a bar to denote the CT control action. The block ensuring rate transition in Figure 13.3b, at least conceptually, from update rate $1/hg$ to update rate $1/h$ is labeled $H_{hg,h}$, the dual-rate hold [35], which is formally defined as follows.

Definition 13.2

A *dual-rate hold* $H_{hg,h}$, is a system that receives a bounded DT input signal with period hg and outputs a bounded DT signal with period h. Consider $u(k, hg)$, a DT scalar input to $H_{hg,h}$. With lifting done at period $hg = M \cdot h$, the lifted output of $H_{hg,h}$, labeled as $y^L(k, hg) \in \mathcal{R}^{M \times 1}$, is given by Equation 13.17, where $H_j^{i-1}(h, hg) \in R$, $i = 1,...,M$ and $j = -l,...,m - 1$. $H_{hg,h}$ has a nonzero DT impulse response at $t = -l \cdot MT, -l \cdot MT + T,...,(m - 1)MT + (M - 1)T$. Here, $m \geq 1$ and $l \geq 0$. In case of vector input signals in R^N, $H_{hg,h} = diag\{H_{hg,h}^1,...,H_{hg,h}^N\}$, where $diag\{\cdot\}$ is the diagonal matrix and $H_{hg,h}^i$ is the hold affecting the ith entry of the input vector signal as given by Equation 13.17: ■

$$y^L(k, hg) = \sum_{j=-l}^{m-1} \begin{bmatrix} H_j^0(h, hg) \\ H_j^1(h, hg) \\ \vdots \\ H_j^{M-1}(h, hg) \end{bmatrix} u(k - j, hg). \qquad (13.17)$$

The simplest dual-rate hold is the DT ZOH [36], which has the following lifted output to a DT scalar input $u(k, hg)$:

$$y^L(k, hg) = \underbrace{\begin{bmatrix} 1 \\ 1 \\ \vdots \\ 1 \end{bmatrix}}_{= R} u(k, hg) \qquad (13.18)$$

with $y^L(k, hg) \in \mathcal{R}^{M \times 1}$. Control inputs to the actuators, given as δ_p in Figure 13.3b, are issued at the fastest rate to control the relatively fast dynamics. The computations associated with the control of the system from δ_p to a_m do not have to be completed within a cycle of T seconds; in reality, the computing requirement is relaxed through the use of a second rate of $1/h$ Hz, as opposed to processing the control computations at a single rate of $1/T$ Hz. This is one advantage of dual-rate control.

It may also occur that signals be required to transition from a fast rate to a slower rate. In such cases, a dual-rate sampler is required.

Definition 13.3

A *dual-rate sampler* $S_{h,hg}$ is a system that receives a bounded DT input signal with period h and outputs a bounded DT signal at period $hg = M \cdot h$. The dual-rate sampler considered here outputs every Mth input. ∎

For a DT scalar input $u(k, h)$ lifted as

$$u^L(l, hg)$$

$$= \left\{ \begin{bmatrix} u(0,h) \\ u(1,h) \\ \vdots \\ u(M-1,h) \end{bmatrix}, \begin{bmatrix} u(M,h) \\ u(M+1,h) \\ \vdots \\ u(2M-1,h) \end{bmatrix}, \cdots \right.$$

$$\left. \begin{bmatrix} u(l \cdot M, h) \\ u(l \cdot M + 1, h) \\ \vdots \\ u(l \cdot M + M - 1, h) \end{bmatrix}, \cdots \right\} \tag{13.19}$$

the output of the dual-rate sampler is given as

$$y(l, hg) = \underbrace{[1, 0, \ldots, 0]}_{=D} u^L(l, hg) \tag{13.20}$$

with $D \in \mathcal{R}^{1 \times M}$.

Remarks (1) The dual-rate hold can be implemented as an up-sampler followed by an finite impulse response filter [14]. (2) The dual-rate sampler corresponds to a down-sampler. Decimation can be accomplished by preceding the dual-rate sampler with an appropriate low-pass filter [14]. (3) Decimator D and repeater R are given by Equations 13.20 and 13.18, respectively. Such blocks, which have been discussed in the literature, for instance in [26], can be readily implemented in software code.

13.3 Guidance Synthesis via Successive Optimizations

This section presents the synthesis of the robust digital guidance laws whose objective is to achieve stable and accurate missile guidance in spite of uncertainties in missile flight control system dynamics and constraints on missile acceleration. Before presenting the guidance synthesis, we formulate the models of the engagement.

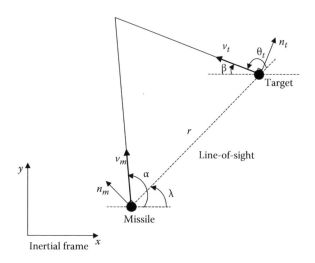

FIGURE 13.5
Two-dimensional engagement.

13.3.1 System Kinematics and Dynamics

Consider a two-dimensional engagement with geometry shown in Figure 13.5. In the figure, v_m is missile speed, and v_t is target speed. The range r between the missile and the target is related to the closing velocity v_{cl} as $v_{cl} = -dr/dt$. The LOS angle $\lambda(t)$ is the angle between the LOS and the fixed reference x-axis, and the relative separation between the missile and the target along the fixed reference y-axis is denoted as $y(t)$. The equation expressing the relationship between the target and the missile accelerations is [8]

$$\frac{d^2 y(t)}{dt^2} = a_t(t) - a_m(t) \tag{13.21}$$

$$= n_t \cos(\beta) - n_m \cos(\lambda)$$

where n_m is perpendicular to LOS and corresponds to acceleration commands obtained with true PNG. Note that Equation 13.21 applies to missiles under digital control. For the case of CT control, a_m is replaced with $\bar{a}_m$. One can then obtain the following state-space equations representing missile-target kinematics:

$$\frac{d}{dt} \begin{bmatrix} x_1(t) \\ x_2(t) \end{bmatrix} = \begin{bmatrix} 0 & 1 \\ 0 & 0 \end{bmatrix} \begin{bmatrix} x_1(t) \\ x_2(t) \end{bmatrix} + \begin{bmatrix} 0 \\ 1 \end{bmatrix} (a_t(t) - a_m(t))$$

$$y(t) = \begin{bmatrix} 1 & 0 \end{bmatrix} \begin{bmatrix} x_1(t) \\ x_2(t) \end{bmatrix}. \tag{13.22}$$

Suppose the CT missile autopilot in closed loop with actuators, missile dynamics, and sensors, as shown in Figure 13.3a, is modeled as an uncertain flight control system. Some uncertainties arise from the variations in the missile flight control system performances over the entire operating envelope, while others are a consequence of the low-order approximation. With the knowledge that the flight control system actually implemented is obtained through a DR of the original CT flight control system, the discrepancies between CT and DT systems can be interpreted and modeled as uncertainties in the original CT flight control system. To model the uncertain CT missile control system, we express the relationship between the actual missile acceleration, $\bar{a}_m$, and the commanded lateral acceleration, $\bar{a}_g$, by the following interval transfer function [8]:

$$\frac{\bar{a}_m(s)}{\bar{a}_g(s)} = \frac{\tau_2 s + 1}{\frac{s^2}{\omega^2} + \frac{2\xi s}{\omega} + 1} \tag{13.23}$$

where parameters τ_2, ω, and ξ are assumed to lie in the following, known compact sets:

$$0 < \underline{\tau_2} \le \tau_2 \le \overline{\tau_2}, \underline{\tau_2}, \overline{\tau_2} \in \mathcal{R}^+$$

$$0 < \underline{\omega} \le \omega \le \overline{\omega}, \underline{\omega}, \overline{\omega} \in \mathcal{R}^+ \tag{13.24}$$

$$0 < \underline{\xi} \le \xi \le \overline{\xi}, \underline{\xi}, \overline{\xi} \in \mathcal{R}^+.$$

Equations 13.23 and 13.24 model the uncertain flight control system over the range of admissible operating points. In state-space form, we can write

$$\frac{d}{dt}\begin{bmatrix} \bar{\varsigma}_1(t) \\ \bar{\varsigma}_2(t) \end{bmatrix} = \begin{bmatrix} 0 & 1 \\ -\omega^2 & -2\xi\omega \end{bmatrix}\begin{bmatrix} \bar{\varsigma}_1(t) \\ \bar{\varsigma}_2(t) \end{bmatrix} + \begin{bmatrix} 0 \\ 1 \end{bmatrix}\bar{a}_g(t)$$

$$\bar{a}_m(t) = \begin{bmatrix} \omega^2 & \tau_2\omega^2 \end{bmatrix}\begin{bmatrix} \bar{\varsigma}_1(t) \\ \bar{\varsigma}_2(t) \end{bmatrix}. \tag{13.25}$$

The target maneuver is modeled as a first-order, zero-mean Markov process [37] given by

$$\frac{da_t(t)}{dt} = -\frac{a_t(t)}{\tau_t} + w_t(t) \tag{13.26}$$

where τ_t is the acceleration time constant, and w_t is a zero-mean white Gaussian stochastic input with covariance σ_t^2. From Equations 13.22, 13.25, and 13.26, the CT missile-target model $\bar{P}$ with inputs $\bar{a}_g$ and w_t and output $[\bar{y}, \bar{a}_m]^T$ can be obtained as

$$\frac{d\bar{X}(t)}{dt} = \bar{A}_o X(t) + \bar{B}_1 \bar{a}_g(t) + \bar{B}_2 \bar{w}_m(t) + \bar{B}_3 \bar{w}_t(t)$$

$$\bar{P}: \begin{bmatrix} \bar{y}(t) \\ \bar{a}_m(t) \end{bmatrix} = \underbrace{\begin{bmatrix} 1 & 0 & 0 & 0 & 0 \\ 0 & 0 & -\alpha_o & -\beta_o & 0 \end{bmatrix}}_{=\bar{C}_2} X(t) - |\alpha_o| \underbrace{\begin{bmatrix} 0 & 0 & 0 & 0 & 0 \\ 0 & 1 & 0 & 0 & 0 \end{bmatrix}}_{=\bar{D}_{22}} \bar{w}_m(t) a$$

$$\bar{z}(t) = \underbrace{I}_{=\bar{C}_1} \cdot \bar{X}(t) \tag{13.27}$$

where $\bar{X} = [\bar{x}_1, \bar{x}_2, \bar{\zeta}_1, \bar{\zeta}_2, a_t]^T$, $\bar{B}_2 = |\alpha_o| \cdot I$, I is the identity matrix,

$$\bar{A}_o = \begin{bmatrix} 0 & 1 & 0 & 0 & 0 \\ 0 & 0 & -\omega^2 & -\tau_2\omega^2 & 1 \\ 0 & 0 & 0 & 1 & 0 \\ 0 & 0 & -\omega^2 & -2\xi\omega & 0 \\ 0 & 0 & 0 & 0 & -\dfrac{1}{\tau_t} \end{bmatrix}, \bar{B}_1 = \begin{bmatrix} 0 \\ 0 \\ 0 \\ 1 \\ 0 \end{bmatrix}, \bar{B}_3 = \begin{bmatrix} 0 \\ 0 \\ 0 \\ 0 \\ 1 \end{bmatrix}$$

$$\begin{aligned} \alpha &= -\omega^2, & \beta &= -\tau_2\omega^2, & \gamma &= -2\xi\omega \\ &= \alpha_o + \tilde{\alpha} & &= \beta_o + \tilde{\beta} & &= \gamma_o + \tilde{\gamma} \end{aligned} \tag{13.28}$$

and

$$\bar{w}_m(t) = \frac{1}{|\alpha_o|} \underbrace{\begin{bmatrix} 0 & 0 & 0 & 0 & 0 \\ 0 & 0 & \tilde{\alpha} & \tilde{\beta} & 0 \\ 0 & 0 & 0 & 0 & 0 \\ 0 & 0 & \tilde{\alpha} & \tilde{\gamma} & 0 \\ 0 & 0 & 0 & 0 & 0 \end{bmatrix}}_{=\tilde{A}} \bar{C}_1 X(t). \tag{13.29}$$

It should be noted that x_1 and x_2 are defined similarly to $\bar{x}_1$ and $\bar{x}_2$, from Equation 13.22, although the missile is under CT control. Matrix $\tilde{A}$ corresponds to the normalized bounded plant uncertainty matrix.

13.3.2 Guidance Law Synthesis

With model $\bar{P}$ available, the guidance scheme can be designed. The objective of the digital guidance law is to minimize missile–target relative separation y despite (1) missile acceleration magnitude limits, (2) onboard implementation at possibly slow sampling rates of $1/h_g$ Hz, and (3) uncertain missile flight control system dynamics. To satisfy this objective, let us formulate the digital guidance law C_{hg} as the following state-space equation:

$$C_{hg}: \quad \begin{aligned} x(k+1,hg) &= A_d x(k,hg) + B_d y(k,hg) \\ a_g(k,hg) &= C_d x(k,hg) + D_d y(k,hg) \end{aligned} \tag{13.30}$$

where $k \in Z^+ = \{0, 1, 2,...\}$ is the time step, $y(k, hg)$ is the missile–target relative separation available at every $k \cdot hg$ time instant, and $a_g(k, hg)$ is the DT acceleration command. The guidance law in Equation 13.30 is obtained by means of a two-step procedure.

Step 1: Robust guidance law synthesis in the CT domain.
Let a CT guidance law $\bar{C}$ be expressed as

$$\bar{C}: \quad \begin{aligned} \frac{d\bar{x}_c(t)}{dt} &= \bar{A}_c \bar{x}_c(t) + \bar{B}_c \bar{y}(t) \\ \bar{a}_g(t) &= \bar{C}_c \bar{x}_c(t) + \bar{D}_c \bar{y}(t) \end{aligned}. \tag{13.31}$$

Calculate $\bar{C}$ by solving a mixed H_2/H_∞ minimization of the gain of the missile–target plant model $\bar{P}$ given in Equation 13.27. The CT guidance law aims at minimizing the H_∞ gain from the finite energy input signal $\bar{w}_m$ to the output $\bar{z}$ in Equation 13.27, and the H_2 gain from the finite variance input signal w_t to the output vector $[\bar{y}, \bar{a}_m]^T$, with the knowledge of the normalized norm-bounded uncertainties (Equation 13.29). Such a problem is standard in robust control theory [38]. Linear matrix inequalities (LMIs) can be used to obtain $\bar{C}$ [18]. Indeed, the LMI formalism is an appropriate tool to combining H_2/H_∞ minimization with pole constraint requirements defined by means of convex regions [39]. In doing so, it is assumed that the missile–target plant model satisfies the requirements of (A_o, B_1) being stabilizable and (A_o, C_2) being detectable. It is worth mentioning that the majority of known optimal guidance laws (OGLs) are calculated over a finite temporal horizon corresponding to the time-to-go, or an estimate thereof, in the final phase of the engagement. Here, however, the minimization is carried out over an infinite temporal horizon during which the missile–target relative range asymptotically converges to zero. In such context, designers must carefully select the required closed-loop pole locations to ensure sufficiently fast missile response, to minimize miss, while complying to constraints in the available missile acceleration.

Step 2: Multirate plant modeling.

Step 2.1: Fast discretization of the CT system formed by missile–target plant model in closed loop with robust guidance law $\bar{C}$.

The closed-loop system is made of the CT robust guidance law given by Equation 13.31 in a closed loop with the uncertain missile–target plant model (Equations 13.27 and 13.29). The CT closed-loop system is converted to a fast DT system running at the short sampling period $hf = hg/N$ where N is a positive integer. N can be selected arbitrarily, although it should be large enough so that the discretized system behaves as closely as desired to the CT closed-loop system. For the CT to DT conversion, one may use any of the principal discretization methods discussed in Section 13.2.3. These methods are known to yield DT models of CT systems, according to Definition 13.1, and thus have known time-domain characteristics, as $hf \rightarrow 0$. With the fast closed-loop DT system known, one can then formulate the state-space equations in the delta form, with $\delta = (q-1)/hf$. The use of the delta operator is key at this stage of the design as it allows us to write the state-space form of the fast DT plant as one comprising the CT system matrices with some error terms that vanish as the sampling rate is increased, therefore, bridging the gap between the DT and CT systems as the sampling period hf approaches zero [27]. As an additional benefit of the delta operator, a system expressed in the delta form is known to possess superior numerical properties than the same system expressed in the classical shift operator, especially for implementations at relatively fast sampling rates, as explained in [27]. The uncertain DT system is represented with the linear fractional transform formalism [40] in Equations 13.32 and 13.33. Equation 13.32 corresponds to the nominal portion of the closed-loop DT system, whereas the parametric uncertainties and the extra terms arising from the discretization are given by Equation 13.33. The discretization errors are known to approach zero with an increase in sampling rate, as the fast DT plant is a DT model of the original CT system:

$$\delta X_{cl}(k, hf) = FX_{cl}(k, hf) + GW_{cl}(k, hf)$$

$$\begin{bmatrix} y_{cl}(k,hf) \\ a_{m,cl}(k,hf) \end{bmatrix} = \begin{bmatrix} C_2 & 0 \end{bmatrix} X_{cl}(k,hf) + \begin{bmatrix} D_{22} & 0 & 0 & 0 \end{bmatrix} W_{cl}(k,hf)$$

$P_{cl}:$ \hfill (13.32)

$$z_{cl}^2(k,hf) = \begin{bmatrix} y_{cl}(k,hf) \\ a_{m,cl}(k,hf) \end{bmatrix}$$

$$z_{cl}^\infty(k,hf) = \begin{bmatrix} z_{cl}^x(k,hf) \\ z_{cl}^t(k,hf) \end{bmatrix} = G_1^\infty X_{cl}(k,hf) + H_1^\infty W_{cl}(k,hf).$$

In Equation 13.32, $X_{cl}^T(k, hf) = [X^T(k, hf), x_c^T(k, hf)]$, $\delta X_{cl}(k, hf) = (X_{cl}(k + 1, hf) - X_{cl}(k, hf))/hf$, $y_{cl}(k, hf)$ is the missile–target relative separation, and $a_{m,cl}(k, hf)$ is the acceleration. Vector $W_{cl}(k, hf)$, which includes uncertainties, discretization error, and noise, is given by

$$W_{cl}(k, hf) = \begin{bmatrix} w_{m,cl}(k, hf) \\ w_{m,cl}^d(k, hf) \\ w_{t,cl}^d(k, hf) \\ w_t(k, hf) \end{bmatrix}$$

$$w_{m,cl}(k, hf) = \tilde{A} \begin{bmatrix} C_1 & 0 \end{bmatrix} z_{cl}^x(k, hf) \tag{13.33}$$

$$w_{m,cl}^d(k, hf) = \begin{bmatrix} M_2^{-1} \tilde{G}_2 \tilde{A} \begin{bmatrix} C_1 & 0 \end{bmatrix} \\ M^{-1} \tilde{F} \end{bmatrix} z_{cl}^x(k, hf)$$

$$w_{t,cl}^d(k, hf) = M_3^{-1} \tilde{G}_3 z_{cl}^t(k, hf)$$

where $w_{m,cl}(k, hf)$ is the perturbation due to parametric uncertainties, $w_{m,cl}^d(k, hf)$ and $w_{t,cl}^d(k, hf)$ represent discretization perturbations of the DT system when compared with the CT system, $w_t(k, hf)$ is the sampled input to the maneuvering target, $z_{cl}^x(k, hf) = X_{cl}(k, hf)$, and $z_{cl}^t(k, hf) = w_{t,cl}(k, hf)$. The closed-loop system matrices found in Equations 13.32 and 13.33 are defined as

$$F = \begin{bmatrix} A_o + B_1 D_c C_2 & B_1 C_c \\ B_c C_2 & A_c \end{bmatrix}, G_2 = \begin{bmatrix} B_2 + B_1 D_c D_{22} \\ B_c D_{22} \end{bmatrix}$$

$$\tag{13.34}$$

$$G_3 = \begin{bmatrix} B_3 \\ 0 \end{bmatrix}, G_1^\infty = \begin{bmatrix} I & 0 \\ 0 & 0 \end{bmatrix}, H_1^\infty = \begin{bmatrix} 0 & 0 & 0 & H_{12}^\infty \end{bmatrix}, H_{12}^\infty = \begin{bmatrix} 0 & 0 \\ 0 & I \end{bmatrix}$$

$$\tilde{F} = \frac{F_d - I}{hf} - F, \tilde{G}_2 = \frac{F^{-1}}{hf}(F_d - I)G_2 - G_2$$

$$\tilde{G}_3 = \frac{F^{-1}}{hf}(F_d - I)G_3 - G_3, F_d = e^{Fhf} \tag{13.35}$$

$$M - \|\tilde{F}\|, M_2 = \|\tilde{G}_2\|, M_3 = \|\tilde{G}_3\|$$

$$
G = \begin{bmatrix}
G_2 & 0 & 0 & 0 \\
0 & \begin{matrix} M_2 \\ \ddots \\ M_2 \\ M \\ \ddots \\ M \end{matrix} \begin{matrix} \\ 0 \\ \\ 0 \end{matrix} & 0 & 0 \\
0 & 0 & \begin{matrix} M_3 & 0 \\ & \ddots \\ 0 & M_3 \end{matrix} & 0 \\
0 & 0 & 0 & G_3
\end{bmatrix}
$$

where

$$
\lim_{hf \to 0} \left\| \tilde{F} \right\| = 0, \ \lim_{hf \to 0} \left\| \tilde{G}_2 \right\| = 0, \ \lim_{hf \to 0} \left\| \tilde{G}_3 \right\| = 0 \tag{13.36}
$$

and $\|\cdot\|$ is either matrix or vector norm, or absolute value. It is clear from Equation 13.36 that making the sampling period hf short enough results in

the discretization errors, as given by vectors $\begin{bmatrix} M_2 I \\ MI \end{bmatrix} w^d_{m,cl}$ and $M_3 w^d_{t,cl}$ in

the term $GW_{cl}(k, hf)$ of Equation 13.32, approaching zero. The fast DT missile control system behaves as closely as desired to the CT robust control system simply by making hf sufficiently short and thus can be considered to be a DT model of the CT system, according to Definition 13.1.

Remark Matrix inversion, as is done in Equation 13.35, can be avoided by using series methods [41] to perform the numerical integration.

Step 2.2: Fast discretization of CT missile–target plant model.

CT plant $\bar{P}$ given by Equations 13.27 through 13.29 is discretized at the fast rate hf using step-input invariance discretization. The resulting DT system P_{ol} is given by

$$
\delta X(k, hf) = A_o X(k, hf) + B_1 a_g(k, hf) + EW(k, hf)
$$

$$
\begin{bmatrix} y(k, hf) \\ a_m(k, hf) \end{bmatrix} = C_2 X(k, hf) + \begin{bmatrix} D_{22} & 0 & 0 & 0 & 0 \end{bmatrix} W(k, hf)
$$

$$
P_{ol}: \quad z^2(k, hf) = \begin{bmatrix} y(k, hf) \\ a_m(k, hf) \end{bmatrix} \tag{13.37}
$$

$$
z^\infty(k, hf) = \begin{bmatrix} z^x(k, hf) \\ z^a(k, hf) \\ z^l(k, hf) \end{bmatrix} = C_1^\infty X(k, hf) + D_{11}^\infty a_g(k, hf) + D_{12}^\infty W(k, hf).
$$

where $z^2(k, hf)$ represents the missile–target relative separation to be minimized and the missile acceleration to be bounded and $z^\infty(k, hf)$ is the plant output connected to the uncertainties. Vector $W(k, hf)$ in Equation 13.37 arising from parameter uncertainties, discretization, and noise can be described as follows:

$$W(k, hf) = \begin{bmatrix} w_m(k, hf) \\ w_m^d(k, hf) \\ w_a^d(k, hf) \\ w_t^d(k, hf) \\ w_t(k, hf) \end{bmatrix}$$

$$w_m(k, hf) = \tilde{A}z^x(k, hf)$$

(13.38)

$$w_m^d(k, hf) = \begin{bmatrix} N_2^{-1}\tilde{B}_2\tilde{A} \\ N_o^{-1}\tilde{A}_o \end{bmatrix} z^x(k, hf)$$

$$w_a^d(k, hf) = N_1^{-1}\tilde{B}_1 z^a(k, hf)$$

$$w_t^d(k, hf) = N_3^{-1}\tilde{B}_3 z^t(k, hf)$$

where $w_a^d(k, hf)$ is a perturbation vector expressed as a function of the guidance signal a_g. Vector signal w_a^d results from the discretization error terms, which approach zero as $hf \to 0$. Matrices found in Equations 13.37 and 13.38 are defined as

$$C_1^\infty = \begin{bmatrix} I & 0 & 0 \\ 0 & 0 & 0 \\ 0 & 0 & 0 \end{bmatrix}, D_{11}^\infty = \begin{bmatrix} 0 & 0 & 0 \\ 0 & I & 0 \\ 0 & 0 & 0 \end{bmatrix}, D_{12}^\infty = \begin{bmatrix} 0 & 0 & 0 \\ 0 & 0 & 0 \\ 0 & 0 & I \end{bmatrix}$$

(13.39)

$$\tilde{A}_o = \frac{A_{do} - I}{hf} - A_o, A_{do} = e^{A_o hf}, \tilde{B}_1 = \frac{A_o^{-1}}{hf}(A_{do} - I)B_1 - B_1$$

$$\tilde{B}_2 = \frac{A_o^{-1}}{hf}(A_{do} - I)B_2 - B_2, \tilde{B}_3 = \frac{A_o^{-1}}{hf}(A_{do} - I)B_3 - B_3$$

(13.40)

$$N_0 - \|\tilde{A}_o\|, N_1 = \|\tilde{B}_1\|, N_2 = \|\tilde{B}_2\|, N_3 = \|\tilde{B}_3\|$$

$$E = \begin{bmatrix} B_2 & 0 & 0 & 0 & 0 \\ 0 & \begin{smallmatrix} N_2 & & 0 \\ & \ddots N_2 & \\ & & N_0 \\ 0 & & \ddots N_0 \end{smallmatrix} & 0 & 0 & 0 \\ 0 & 0 & \begin{smallmatrix} N_1 & 0 \\ & \ddots \\ 0 & N_1 \end{smallmatrix} & 0 & 0 \\ 0 & 0 & 0 & \begin{smallmatrix} N_3 & \\ & \ddots N_3 \end{smallmatrix} & 0 \\ 0 & 0 & 0 & 0 & B_3 \end{bmatrix}$$

where

$$\lim_{hf \to 0} \|\tilde{A}_o\| = 0, \; \lim_{hf \to 0} \|\tilde{B}_1\| = 0, \; \lim_{hf \to 0} \|\tilde{B}_2\| = 0, \; \lim_{hf \to 0} \|\tilde{B}_3\| = 0. \tag{13.41}$$

Step 2.3: Multirate generalized plant modeling for robust DT control synthesis.

The fast closed-loop missile control system given by Equations 13.32 and 13.33 is taken as the reference system, exhibiting satisfactory performances in terms of acceleration and miss distance. To effectively use this knowledge for the synthesis of the slow-rate guidance law C_{hg}, a multirate generalized plant is modeled with a structure shown in Figure 13.6a. The plant has lifted input $w_t^l(k, hg)$. Plant output $e(k, hf) = z_{cl}^2(k, hf) - z^2(k, hf)$ is lifted to $e^l(k, hg)$. In Figure 13.6a, Δ_{hf}^{cl} is the closed-loop uncertainty operator. It represents parameter mismodeling and perturbation terms that arise from the use of delta representation (Equations 13.32 and 13.33). Δ_{hf}^{ol} is the open-loop uncertainty operator. The guidance law C_{hg} which is in closed loop with the generalized plant model, should be such that the gain from $w_t(k, hf)$ to $e(k, hf)$ is minimized and the closed-loop system is stable. Such slow-rate digital guidance law would then warrant an acceleration and a miss distance as close as possible to those of the reference system in spite of the uncertainties.

Step 3: Solution to single-rate DT robust control problem.

Referring to Figure 13.6b, let T_{we} be the system from w_t^L to e^L. Furthermore, suppose that for a given hf

$$\left\| \begin{bmatrix} \Delta_{hf}^{cl} & 0 \\ 0 & \Delta_{hf}^{ol} \end{bmatrix} \right\| < 1/\gamma \tag{13.42}$$

where $\gamma \in \mathcal{R}^+$. The bound in the right-hand side of inequality (Equation 13.42) can always be made relatively small by reducing hf to yield a value of the bound at least as small as that for CT systems P_{cl} and P_{ol}. Such convergence

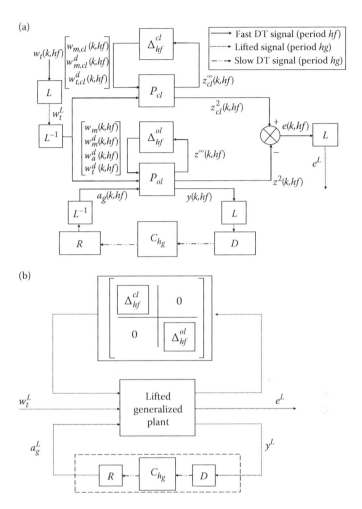

FIGURE 13.6
Generalized plant structure.

property can be readily appreciated from Equations 13.36 and 13.41. Define $\left\| T_{we} \right\|_\infty$ to be the $\mathcal{H}_\infty$ norm of T_{we} [38]. Then, the robust guidance law C_{hg} is obtained by finding the smallest achievable $\left\| T_{we} \right\|_\infty < \gamma$ by means of LMI optimization. The problem includes pole placement constraints defined by means of convex regions in the γ plane [27]. Such regions are defined as the intersection of a vertical strip and a sector centered at the origin. The former determines the fastest pole of the closed-loop dynamics and accounts for hardware limitations, such as those imposed by actuator and sensor bandwidth. The latter is selected so as to limit system oscillations.

13.4 Dual-Rate Autopilot Synthesis

Two methods are presented for the synthesis of dual-rate autopilots. First, an optimal DR of a CT autopilot is presented. The technique relies on the solution of a DT H_2 optimal control problem [42] allowing the conversion to either a slow-rate or a dual-rate control system while guaranteeing stability and performance, to some extent. Second, a polynomial approach to the DR of a CT autopilot [25] is described. The polynomial method warrants closed-loop stability, even for implementations at relatively large sampling intervals, taking advantage of well-known linear algebraic tools while enabling complexity-constrained controllers.

13.4.1 Optimal DR

Consider a CT acceleration autopilot with the structure shown in Figure 13.7. Such feedback loop is typical of missile autopilots.

Controllers $\bar{C}_1$ and $\bar{C}_2$ are LTI systems. Over the entire flight envelope, such LTI controllers are scheduled accordingly. It is assumed that the CT autopilot satisfies the design specifications.

Step 1: Fast discretization of closed-loop system.

Precede each of the dynamic systems $\bar{C}_1$, $\bar{C}_2$, $\bar{q}/\bar{\delta}_p$, and $\bar{a}_m/\bar{\delta}_p$ by a ZOH H_T and place the ideal sampler S_T at each output. Then, there results the fast, single-rate DT closed-loop system shown in Figure 13.8, which can be simplified. The selection of T depends on the design specifications and on the dynamics of the system under control. Period T should be selected short enough to accommodate the fast dynamics of $\bar{q}/\bar{\delta}_p$.

Step 2: Dual-rate generalized plant modeling for slow DT controller synthesis.

From Figure 13.8, let $\bar{G}_1 = \bar{a}_m/\bar{\delta}_p$, $\bar{G}_2 = \bar{q}/\bar{\delta}_p$, $C_{1,T} = S_T\bar{C}_1H_T$, $C_{2,T} = S_T\bar{C}_2H_T$, $G_{1,T} = S_T\bar{G}_1H_T$, and $G_{2,T} = S_T\bar{G}_2H_T$. Formulate a dual-rate generalized plant model as shown in Figure 13.9. Periods are related as $h = N \cdot T$, where $N \in \mathbb{N}^+$. The generalized plant is connected to a slow controller to be synthesized, labeled $C_{1,h}$, having output $u(k, h)$ and input $y(k, h)$. The dual-rate samplers

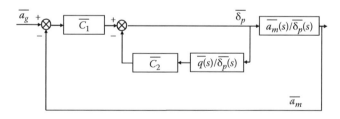

FIGURE 13.7
CT acceleration autopilot structure.

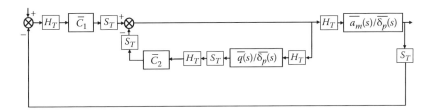

FIGURE 13.8
Fast DT autopilot.

$S_{T,h}$ and gain $\rho \in \mathcal{R}$ are design parameters. Controller $C_{1,h}$ can be obtained such that the induced norm from reference input $a_g(k, h)$ to measured outputs $z_1(k, h)$ and $z_2(k, h)$ is minimized. The dual-rate DT control system then provides an optimal approximation of the fast DT control system. It should be noted that the block diagram of Figure 13.9 is intended for reference input tracking.

Step 3: Solution to the optimal control problem.

The dual-rate DR problem can be formulated as follows. Given a fast DT control system comprising $C_{1,T}$ and $C_{2,T}$, design a dual-rate DT control system, with dynamic controllers $C_{1,h}$ and $C_{2,T}$, such that its closed-loop step

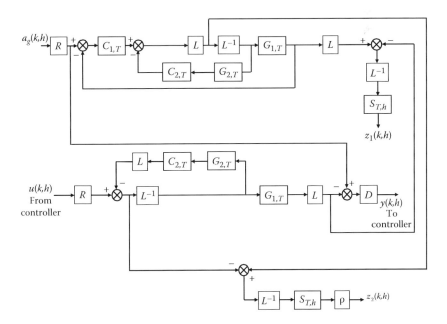

FIGURE 13.9
Dual-rate generalized plant for controller synthesis.

responses optimally match those of the fast DT control system in the sense that J given in Equation 13.43 is minimized:

$$J = \sum_{k=0}^{\infty} |z_1(k,h)|^2 + |z_2(k,h)|^2. \tag{13.43}$$

The minimization of J is a DT H_2 problem. With a unit DT impulse input $a_g(k, h)$ and an appropriate filter placed at the reference input channel to convert the impulse signal to a step, the DT H_2 problem consists of obtaining controller $C_{1,h}$ such that the DT H_2 norm of the closed-loop system relating $a_g(k, h)$ to $[z_1(k, h), z_2(k, h)]^T$ is minimized.

13.4.2 Polynomial Approach to Dual-Rate DR

Consider again the closed-loop system shown in Figure 13.7. Such a system provides a model for the linear behavior of the missile under control around equilibrium points. A polynomial approach to DR is applied to each CT closed-loop system warranting stability, in the DT sense, for any nonpathological value of T, as the roots of the characteristic equation of the CT closed-loop system are mapped inside the DT stability domain, either γ or z. This dual-rate DR takes its origin from the single-rate DR known as the plant input mapping method (see [25] and references therein). The five-step approach is described next.

Step 1: Calculate a CT closed-loop transfer function.

With reference to Figure 13.7, calculate closed-loop transfer function $\bar{M}(s)$, from $\bar{a}_g$ to $\bar{u}$, with $\bar{u}$ being the output of $\bar{C}_1$:

$$\bar{M}(s) = \frac{\bar{C}_1(s)[1 + \bar{C}_2(s)\bar{G}_2(s)]}{1 + \bar{C}_1(s)\bar{G}_1(s) + \bar{C}_2(s)\bar{G}_2(s)}$$

$$= \frac{\bar{u}}{\bar{a}_g}. \tag{13.44}$$

Writing the transfer function from $\bar{u}$ to $\bar{a}_m$ as

$$\bar{G}(s) = \frac{\bar{G}_1(s)}{1 + \bar{C}_2(s)\bar{G}_2(s)}$$

$$= \frac{n_{\bar{G}}(s)}{d_{\bar{G}}(s)}, \tag{13.45}$$

one can rewrite $\bar{M}(s)$ as

$$\bar{M}(s) = \frac{\bar{m}(s)d_{\bar{G}}(s)}{d_{\bar{M}}(s)}. \tag{13.46}$$

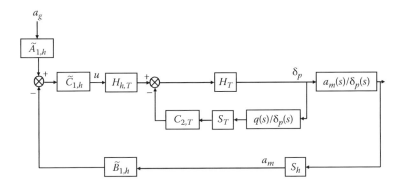

FIGURE 13.10
Dual-rate digital missile autopilot.

Step 2: Fix the structure of the DT autopilot and obtain the DT plant model.
Set the structure of the dual-rate digital missile autopilot to be that shown
in Figure 13.10, which is more general than that of Figure 13.4. The control-
lers to be designed are $\tilde{A}_{1,h}$, $\tilde{B}_{1,h}$, and $\tilde{C}_{1,h}$. Plant $\tilde{G}$ corresponds to the system
from $u(k, h)$ to $a_m(k, h)$. When the dual-rate hold is given as $H_{h,T} = L^{-1}R$, with
$h = N \cdot T$, the transfer function in z for the slow DT plant $\tilde{G}$ is

$$\tilde{G}(z,h) = \frac{C \cdot \text{Adj}(zI - A^N)\left[A^{N-1}B \ \ A^{N-2}B \ \cdots \ B \right]\begin{bmatrix} 1 \\ 1 \\ \vdots \\ 1 \end{bmatrix} + \left|zI - A^N\right| \cdot D}{\left|zI - A^N\right|} \qquad (13.47)$$

$$= \frac{n_{\tilde{G}}(z,h)}{d_{\tilde{G}}(z,h)}$$

where A, B, C, and D are matrices for the shift-form realization of the fast DT
plant model shown in Figure 13.11. The shift-form realization is

$$x(k + 1, T) = Ax(k, T) + Bu(k, T)$$

$$a_m(k, T) = Cx(k, T) + Du(k, T). \qquad (13.48)$$

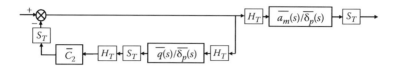

FIGURE 13.11
Fast DT plant model.

Step 3: Obtain $\tilde{M}$ from the knowledge of $\bar{M}$ and $\tilde{G}$.
Let the desired closed-loop transfer function from $a_g(k, h)$ to $u(k, h)$ be given as

$$\tilde{M}(z,h) = \tilde{K}\,\frac{\tilde{m}(z,h)d_{\tilde{G}}(z,h)}{d_{\tilde{M}}(z,h)} \tag{13.49}$$

where $d_{\tilde{G}}$ is the denominator polynomial of Equation 13.47, fixed by the implementation structure of the digital autopilot. $\tilde{K}$, $d_{\tilde{M}}$, and $\tilde{m}$ are determined as follows.

- Map roots of $\bar{m}(s)$ to $\tilde{m}(z,h)$ through the exponential $z_i = \exp(s_i \cdot h)$, where s_i are the roots of $\bar{m}(s)$ and z_i are the roots of $\tilde{m}(z,h)$, $i = 1,\dots,\eta$.
- Map roots of $d_{\bar{M}}(s)$ to $d_{\tilde{M}}(z,h)$ through the exponential, as done in the previous step.
- Determine $\tilde{K}$ with objective of making $\tilde{M}(z,h)$ have the same gain as that of $\bar{M}(s)$ at a frequency of interest.

Step 4: Obtain DT controllers $\tilde{A}_{1,h}$, $\tilde{B}_{1,h}$, and $\tilde{C}_{1,h}$.
Calculate $\tilde{A}(z,h)$, $\tilde{B}(z,h)$, and $\tilde{C}(z,h)$ such that the transfer function from $a_g(k, h)$ to $u(k, h)$ in Figure 13.10 is the desired closed-loop transfer function (Equation 13.49).
First, solve the following Diophantine equation for polynomials u and v:

$$u(z,h)d_{\tilde{G}}(z,h) + v(z,h)n_{\tilde{G}}(z,h) = d_{\tilde{M}}(z,h). \tag{13.50}$$

$u(z, h)$ is of degree l, and $v(z, h)$ has degree r. One approach to solving for u and v is to employ linear algebra. Write

$$u(z,h) = u_l z^l + u_{l-1}z^{l-1} + \dots + u_1 z + u_0$$
$$v(z,h) = v_r z^r + v_{r-1}z^{r-1} + \dots + v_1 z + v_0$$
$$n_{\tilde{G}}(z,h) = n_{n-1}z^{n-1} + n_{n\,2}z^{n-2} + \dots + n_1 z + n_0 \tag{13.51}$$
$$d_{\tilde{G}}(z,h) = d_n z^n + d_{n-1}z^{n-1} + \dots + d_1 z + d_0$$
$$d_{\tilde{M}}(z,h) = d_M z^M + d_{M-1}z^{M-1} + \dots + d_1 z + d_0.$$

By equating coefficients of like powers of z on both sides of Equation 13.51, the following matrix equation can be obtained:

$$
\underbrace{\left[\begin{array}{ccccccc}
d_n & & & n_{n-1} & & & \\
d_{n-1} & \ddots & & n_{n-2} & \ddots & & \\
\vdots & \ddots & d_n & \vdots & \ddots & n_{n-1} & \\
d_1 & & d_{n-1} & & \ddots & \vdots & \\
d_0 & \ddots & \vdots & n_0 & & n_{n-2} & \\
& \ddots & d_1 & & \ddots & \vdots & \\
& & d_0 & & & n_0 &
\end{array}\right]}_{= A_d}
\underbrace{\left[\begin{array}{c}
u_l \\
\vdots \\
u_0 \\
v_r \\
\vdots \\
v_0
\end{array}\right]}_{=X_d}
=
\underbrace{\left[\begin{array}{c}
d_M \\
d_{M-1} \\
\vdots \\
d_1 \\
d_0
\end{array}\right]}_{=Y_d}.
$$

where $\mathrm{dim} = l+1$ and $\mathrm{dim} = r+1$. (13.52)

In the system of equations, $A_d X_d = Y_d$, $A_d \in \mathcal{R}^{(M+1)\times(l+r+2)}$, $X_d \in \mathcal{R}^{(l+r+2)\times1}$, and $Y_d \in \mathcal{R}^{(M+1)\times1}$. Interestingly, matrix A_d is full rank whenever the DT plant is irreducible [27]. One can then solve uniquely for u and v provided $M + 1 = l + r + 2$. To meet such a dimension condition, polynomials u and v must have degrees fixed to $l = M - n$ and $r = n - 1$, respectively, whenever $n \geq 1$. If this condition is not satisfied, a least-squares approximation can be obtained as $\hat{X}_d = (A_d^T A_d)^{-1} A_d^T Y_d$ when Equation 13.52 is overdetermined. With such $\hat{X}_d$, controller orders can be constrained, although closed-loop stability is not guaranteed.

Second, with the knowledge of u and v, controllers can be obtained in three ways.

(1) *Dynamic blocks* $\tilde{A}_{1,h}$, $\tilde{B}_{1,h}$, *and* $\tilde{C}_{1,h}$

Let $w(z, h)$ be an arbitrary, stable polynomial of degree l. If $l \geq \eta$ and $l \geq r$:

$$
\tilde{A}_{1,h} = \tilde{K}\frac{\tilde{m}(z,h)}{w(z,h)}, \tilde{B}_{1,h} = \frac{v(z,h)}{w(z,h)}, \tilde{C}_{1,h} = \frac{w(z,h)}{u(z,h)}. \tag{13.53}
$$

The conditions on l are not always met in practice. In fact, $l \geq r$ if and only if $M \geq 2n - 1$. If the latter condition is not met, this means $M + \phi = 2n - 1$, where $\phi \in \mathbb{N}^+$. To obtain the controllers, do as follows. Modify Equation 13.49 to the following desired closed-loop transfer function:

$$
\tilde{M}(z,h) = \tilde{K}\frac{\tilde{m}(z,h)d_{\tilde{G}}(z,h)}{d_M(z,h)}\frac{p(z,h)}{p(z,h)} \tag{13.54}
$$

where $p(z, h)$ is an arbitrary, stable polynomial of degree ϕ. Then, solve the following Diophantine equation for u and v, with $l = n - 1$ and $r = n - 1$:

$$u(z,h)d_{\tilde{G}}(z,h) + v(z,h)n_{\tilde{G}}(z,h) = d_{\tilde{M}}(z,h)p(z,h). \tag{13.55}$$

When $\eta \le M - n$, the controllers are

$$\tilde{A}_{1,h} = \tilde{K}\frac{\tilde{m}(z,h)p(z,h)}{w(z,h)}, \tilde{B}_{1,h} = \frac{v(z,h)}{w(z,h)}, \tilde{C}_{1,h} = \frac{w(z,h)}{u(z,h)}. \tag{13.56}$$

(2) *Dynamic blocks $\tilde{B}_{1,h}$ and $\tilde{C}_{1,h}$*

A structure with static $\tilde{A}_{1,h}$ and dynamic blocks $\tilde{B}_{1,h}$ and $\tilde{C}_{1,h}$ is possible when $r \le \eta \le l$. The controllers are

$$\tilde{A}_{1,h} = \tilde{K}, \tilde{B}_{1,h} = \frac{v(z,h)}{\tilde{m}(z,h)}, \tilde{C}_{1,h} = \frac{\tilde{m}(z,h)}{u(z,h)}. \tag{13.57}$$

(3) *Dynamic blocks $\tilde{A}_{1,h}$ and $\tilde{C}_{1,h}$*

A structure with static $\tilde{B}_{1,h}$ and dynamic blocks $\tilde{A}_{1,h}$ and $\tilde{C}_{1,h}$ is possible when $\eta \le r \le l$ and $v(z, h)$ is stable. The controllers are

$$\tilde{A}_{1,h} = \tilde{K}\frac{\tilde{m}(z,h)}{v(z,h)}, \tilde{B}_{1,h} = 1, \tilde{C}_{1,h} = \frac{v(z,h)}{u(z,h)}. \tag{13.58}$$

Remark The linear system of equations can be formulated using polynomials expressed either in the z or in the γ operator. The γ operator is known to provide superior numerical results if the period h is relatively short; otherwise, the use of the z operator is recommended.

Step 5: Reduce control system complexity (if required).

When the number of operations associated with the computations of the DT control law is unacceptable for the computing power available, individual controller orders must be constrained. Whereas the order of the inner-loop DT controller $C_{2,T}$ depends strictly on the local hold-equivalent discretization of $\bar{C}_2(s)$, orders of the outer-loop DT controllers $\tilde{A}_{1,h}$, $\tilde{B}_{1,h}$, and $\tilde{C}_{1,h}$ can be reduced, if required. Consider the desired closed-loop transfer function $\tilde{M}$ in Equation 13.49. Convert $\tilde{M}(z,h)$ to $\tilde{M}(w,h)$, where

$$w = \frac{2}{T}\frac{z-1}{z+1}. \tag{13.59}$$

Poles in the unit circle of the z-plane are known to be mapped to the left-hand side of the w-plane. Furthermore, conventional frequency domain synthesis and analysis methods for CT systems can be applied to systems expressed in the w operator even when h is relatively long [30]. With degrees M and n known, it is required to implement a DT autopilot such that a reduced-order version of $\tilde{M}$, denoted as $\tilde{M}_{RO}$, is actually implemented. Let the order of $\tilde{M}_{RO}$ be M_{RO}, the desired degree for the denominator polynomial of the closed-loop system from $a_g(k, h)$ to $u(k, h)$. Apply a CT model reduction method on $\tilde{M}(w, h)$, for instance, balanced truncation [43] or Routh–Pade [45] model reduction. This process results in $\tilde{M}_{RO}(w, h)$. With either method, closed-loop stability is preserved. Convert system $\tilde{M}(w, h)$ to the z-plane using inverse transformation (Equation 13.59). Finally, solve a Diophantine equation of the form given by Equation 13.50 and calculate controllers according to Equation 13.53 through 13.58. Clearly, the designer must make sure the controllers are causal and the closed-loop system is stable.

13.5 Numerical Simulations

13.5.1 Digital Guidance Laws

Assessment of the performance of the digital implementation of guidance laws is a key step in the design process, particularly when small, low-cost, digital boards are used to comply with volume, weight, and cost constraints. By digital implementation, we mean the coding of the guidance laws on finite-word-length, fixed-point microprocessors and accounting for the quantization effects of the A/D converters. In the sequel, the performance of the proposed DT guidance law is first analyzed in its original form. Then, to diminish the impact of truncation errors on closed-loop performance, a reduced-order, DT guidance law is derived and implemented on a 16-bit, fixed-point processor, emulated by means of the Simulink® fixed-point blockset.

Missile flight control (Equation 13.23) and target (Equation 13.26) models are used with parameters assumed to be normally distributed with the mean values corresponding to the nominal plant model. The nominal values are as follows:

$$\tau_t = 0.1 \text{ s}, \omega = 6.71 \text{ rad/s}, \xi = 1.88, \tau_2 = -2.48 \cdot 10^{-2} \text{ s}. \tag{13.60}$$

The covariance and standard deviations are

$$\sigma_t^2 = 3700 \text{ m}^2 \cdot \text{s}^{-1}, \sigma_\omega = 2 \text{ rad/s}, \sigma_\xi - 0.1. \tag{13.61}$$

In Equation 13.61, the subscript indicates to which parameter the covariance and standard deviations refer. Missile acceleration is saturated at ± 200 m/s². Initial conditions are defined as $y(0) = 100$ m and $\dfrac{dy}{dt}(0) = 10$ m/s. The time of flight is set to $t_f = 5$ s.

Guidance law C_{hg} is applied to the missile–target plant model with a ratio of 5 between hg and hf. Increasing this ratio may cause numerical problems during the synthesis phase, as the dimensions of the state-space matrices are proportional to the value of this ratio. Pole placement with $p_s = -1200$, $p_f = -100$, and $\xi_r = 0.84$ is used in the synthesis along with balanced-truncation model reduction [43]. The values for p_s and p_f are selected such that the solution to the LMIs does not result in large gain matrices. The choice for p_s and p_f constrains the transients of $y(t)$ to smaller values, which in turn may prevent actuator saturation. ξ_r allows limiting oscillations in $y(t)$. The resulting DT guidance law is calculated to be

$$\frac{a_g(z)}{y(z)} = \frac{10.9z^8 + 6.9z^7 - 27.4z^6 - 15.3z^5 + 24.4z^4 + 11.1z^3 - 8.8z^2 - 2.6z + 1.0}{1.0z^8 + 1.5z^7 - 1.3z^6 - 2.7z^5 + 0.1z^4 + 1.4z^3 + 0.3z^2 - 1.4 \cdot 10^{-1}z - 2.3 \cdot 10^{-2}}$$

$$\frac{a_g(z)}{a_m(z)} = \frac{-4.1z^8 - 0.5z^7 + 8.3z^6 + 2 \cdot 10^{-4}z^5 - 5.0z^4 + 1.1z^3 + 4.6 \cdot 10^{-1}z^2 - 6.1 \cdot 10^{-1}z + 2.6 \cdot 10^{-1}}{1.0z^8 + 1.5z^7 - 1.3z^6 - 2.7z^5 + 0.1z^4 + 1.4z^3 + 0.3z^2 - 1.4 \cdot 10^{-2}z - 2.3 \cdot 10^{-2}}$$

$$(13.62)$$

for $h_g = 0.05$ and

$$\frac{a_g(z)}{y(z)} = \frac{40.7z^7 - 34.8z^6 - 62.8z^5 + 68.3z^4 + 9.8z^3 - 31.6z^2 + 12.6z - 1.6}{z^7 + 3.0 \cdot 10^{-1}z^6 - 1.4z^5 - 4.4 \cdot 10^{-2}z^4 + 6.0 \cdot 10^{-1}z^3 - 1.3 \cdot 10^{-1}z^2 - 3.1 \cdot 10^{-2}z + 9.9 \cdot 10^{-3}}$$

$$\frac{a_g(z)}{a_m(z)} = \frac{-2.7z^7 + z^6 + 3.8z^5 - 2.4z^4 - 8.7 \cdot 10^{-1}z^3 + 1.1z^2 - 3.8 \cdot 10^{-1}z + 5.2 \cdot 10^{-2}}{z^7 + 3.0 \cdot 10^{-1}z^6 - 1.4z^5 - 4.4 \cdot 10^{-2}z^4 + 6.0 \cdot 10^{-1}z^3 - 1.3 \cdot 10^{-1}z^2 - 3.1 \cdot 10^{-2}z + 9.9 \cdot 10^{-3}}$$

$$(13.63)$$

for $h_g = 0.1$, which are both numerically implemented with the shift operator $q = z^{-1}$ and the autoregressive and moving average (ARMA) representation shown in Figure 13.12.

The reduced-order guidance law, which is obtained by eliminating states associated with small Hankel singular values [44], is expressed as

$$\frac{a_g(z)}{y(z)} = \frac{10.9z^3 + 8.4z^2 - 7.6z - 5.2}{z^3 + 1.6z^2 - 6.4 \cdot 10^{-1}z - 1.5 \cdot 10^{-2}}$$

$$\frac{a_g(z)}{a_m(z)} = \frac{-4.1z^3 - 9.4 \cdot 10^{-1}z^2 + 1.3z - 8.7 \cdot 10^{-1}}{z^3 + 1.6z^2 - 6.4 \cdot 10^{-1}z - 1.5 \cdot 10^{-2}}$$

$$(13.64)$$

(a)

Emulation of CT system by means of
variable step-size integration

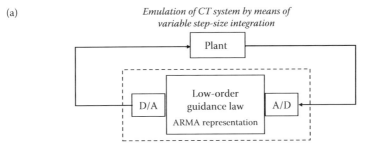

Emulation of 16-bit, fixed-point arithmetic digital board

(b)

ARMA representation

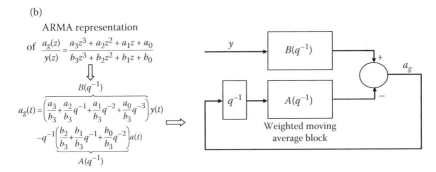

FIGURE 13.12
(a) Block diagram of digital guidance law and (b) ARMA representation.

for $h_g = 0.05$ and as

$$\frac{a_g(z)}{y(z)} = \frac{40.7z^3 + 41.8z^2 - 35.9z - 37.1}{z^3 + 2.2z^2 + 1.4z + 2.4 \cdot 10^{-1}}$$

$$\frac{a_g(z)}{a_m(z)} = \frac{-2.7z^3 - 4.2z^2 - 6.0 \cdot 10^{-1}z + 8.6 \cdot 10^{-1}}{z^3 + 2.2z^2 + 1.4z + 2.4 \cdot 10^{-1}}$$

(13.65)

for $h_g = 0.1$. From now on, the full-order guidance law (Equations 13.62 and 13.63) is denoted as FODTGL, the reduced-order guidance law (Equation 13.64) is labeled RODTGL, and the digital implementation on an emulated 16-bit, fixed-point processor of the reduced-order guidance law is DigDTGL. Figure 13.13 depicts Bode diagrams of transfer functions expressed in the Laplace operator and obtained from Equations 13.62 through 13.65 to which the local DR with Tustin's method is applied. One can notice that the order reduction does not entail significant deterioration for frequencies in $[0, 10^3]$ Hz.

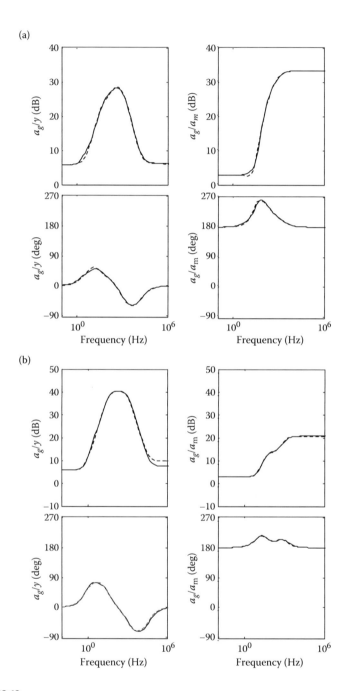

FIGURE 13.13
Bode diagrams for full-order (solid line) and reduced-order (dashed line) DT guidance laws. (a) $h_g = 0.05$ s; (b) $h_g = 0.1$ s.

Through simulations, the proposed approach is compared with the OGL [2,46] and the zero miss distance PNG (ZMD-PNG) [8]. The CT OGL is given by

$$a_g(t) = \frac{N(t_{go})}{t_{go}^2} \left(y(t) + t_{go} \cdot \frac{dy(t)}{dt} + 0.5\hat{a}_t(t) \cdot t_{go}^2 - a_m(t) \cdot \left(e^{-t_{go}} + t_{go} - 1 \right) \right) \quad (13.66)$$

$$N(t_{go}) = \frac{6t_{go}^2 \left(e^{-t_{go}} + t_{go} - 1 \right)}{2t_{go}^3 - 6t_{go}^2 + 6t_{go} + 3 - 12t_{go}e^{-t_{go}} - 3e^{-2t_{go}}}, \hat{a}_t(t) = a_t(t - 0.2), t_{go} = t_f - t$$

$$(13.67)$$

where $\hat{a}_t(t)$ is a delayed estimate of the target acceleration. The CT ZMD-PNG is

$$\frac{a_g(s)}{\left[\frac{d\lambda(t)}{dt} \right](s)} = 5V_c \frac{(0.2304s + 1)^2}{(0.01s + 1)^2}, V_c = 1000 \text{ m/s}. \quad (13.68)$$

Measurements of LOS rate, as required in Equation 13.68, can be obtained by means of a seeker. Seeker dynamics are modeled as [46]

$$\frac{\left[\frac{d\lambda(t)}{dt} \right](s)}{\lambda(s)} = \frac{s}{0.1s + 1} \quad (13.69)$$

where $\lambda(t) = y(t)/(V_c \cdot t_{go})$. ZMD-PNG is discretized with Tustin's method [13] at periods $h_g = 0.05$ and $h_g = 0.1$. The DT OGL is simply a sampled version of Equations 13.66 and 13.67. Numerically, the CT missile–target plant model is simulated with a variable step-size method, available with the MATLAB®/Simulink software, in closed loop with the digital controllers. Zero-order holds and ideal samplers ensure the transition between CT and DT domains. A one-simulation time step delay is placed at the output of the controllers to model the computational delay associated with a digital implementation. Finally, first-order transfer functions given as $1/(0.1s + 1)$ are placed at the y and a_m channels to model sensor dynamics (range measurement, gyros).

Simulation results for the nominal missile–target plant model are shown in Figures 13.14 through 13.16. It can be seen in Figure 13.14 that FODTGL provides a missile–target relative separation converging asymptotically to zero and a miss distance relatively close to that obtained with DT ZMD-PNG, as opposed to DT OGL, which results in the largest absolute miss distance, for both sampling periods tested. DigDTGL yields trajectories that are

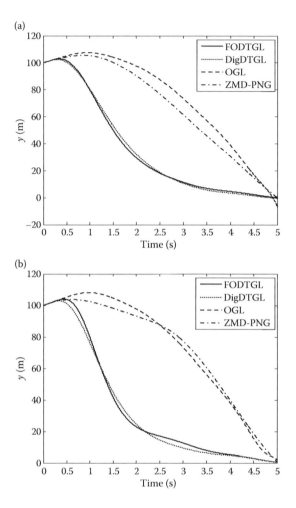

FIGURE 13.14
Missile–target relative separations obtained with FODTGL, DigDTGL, DT OGL, and DT ZMD-PNG for (a) $h_g = 0.05$ s and (b) $h_g = 0.1$ s.

relatively close to those obtained with FODTGL. From Figures 13.14 through 13.16, FODTGL is shown to provide an output signal that does not saturate, is relatively smooth, and settles relatively rapidly, for both sampling periods. FODTGL presents satisfactory noise rejection property due to the global DR strategy adopted by the authors and the $\mathcal{H}_2$ norm minimization carried out during the first phase of the synthesis.

To test the robustness of the proposed guidance law, 500 simulation runs were carried out. In each simulation run, the parameters ω and ξ of the missile model were selected according to normal distributions $N(\omega, \sigma_\omega)$ and $N(\xi, \sigma_\xi)$, respectively, as described in Equations 13.60 and 13.61. The mean values for

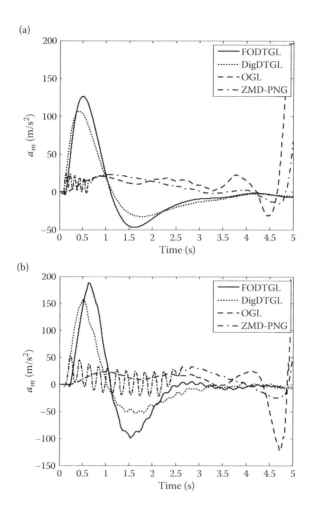

FIGURE 13.15
Missile accelerations obtained with FODTGL, DigDTGL, DT OGL, and DT ZMD-PNG for (a) $h_g = 0.05$ s and (b) $h_g = 0.1$ s.

the absolute miss distances, denoted as $\bar{y}(t_f)$, and the standard deviations for the miss distances, given as σ_y, are shown in Table 13.1. It is clear from the table that FODTGL results in a value for $\bar{y}(t_f)$ relatively close to that obtained with DT ZMD-PNG at the shortest sampling period, whereas DT OGL offers the worst performance for both sampling periods tested. For the relatively large sampling time of $h_g = 0.1$ s, FODTGL results in the smallest $\bar{y}(t_f)$ and σ_y, although $\bar{y}(t_f)$ is larger than that obtained for $h_g = 0.05$ s. Clearly, at the largest sampling period $h_g = 0.1$ s, using FODTGL results in miss distance, which is more robust to parameter variations than that obtained with systems equipped with DT ZMD-PNG and DT OGL.

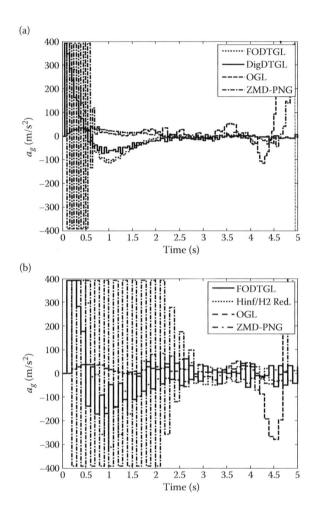

FIGURE 13.16
Commanded missile accelerations obtained with FODTGL, DigDTGL, DT OGL, and DT ZMD-PNG for (a) $h_g = 0.05$ s and (b) $h_g = 0.1$ s.

TABLE 13.1

Mean Values and Standard Deviations for Miss Distances (No Measurement Noise)

	$\bar{y}(t_f)(h_g = 0.05$ s$)$	$\sigma_y(h_g = 0.05$ s$)$	$\bar{y}(t_f)(h_g = 0.1$ s$)$	$\sigma_y(h_g = 0.1$ s$)$
FODTGL	0.246	0.453	0.892	0.166
DT ZMD-PNG	0.156	0.161	0.928	0.830
DT OGL	5.735	1.488	1.674	1.755

TABLE 13.2

Mean Values for Absolute Miss Distances and Standard Deviations Obtained with Different Implementations of Proposed Guidance Law

	$\bar{y}(t_f)(h_g = 0.05\ \text{s})$	$\sigma_y(h_g = 0.05\ \text{s})$	$\bar{y}(t_f)(h_g = 0.1\ \text{s})$	$\sigma_y(h_g = 0.1\ \text{s})$
FODTGL	0.246	0.453	0.892	0.166
RODTGL	0.494	0.761	0.616	1.396
DigDTGL	0.548	0.869	0.620	0.733

While the controller is not designed to take into account the measurement noise in $y(t)$, numerical simulations that include measurement noise and parameter uncertainties in ω and ξ are carried out. The nonstationary additive noise to the measurement of $y(t)$ has normal distribution $N(0, \sigma_y(k, h_g))$ where $\sigma_y(k, h_g) = r(k, h_g)\ \sigma_{LOS}$. The standard deviation in the LOS angle measurement is set to $\sigma_{LOS} = 0.2$ mrad. For $h_g = 0.05$ s and 0.1 s, the simulations result in $\bar{y}(t_f) = 0.253$, $\sigma_y = 0.532$ and $\bar{y}(t_f) = 0.751$, $\sigma_y = 1.789$, respectively. When the guidance law is implemented with a relatively long sampling period, the worst-case miss distance obtained with the noisy system is larger than that obtained in the absence of noise, the latter case being shown in Table 13.1. This could be the result of large magnitudes of p_f leading to a closed-loop system sensitive to noise. However, a smaller magnitude of p_f results in an infeasible LMI problem. It is clear, then, that a trade-off has to be made.

Mean values and standard deviations of miss distances obtained with guidance laws FODTGL, RODTGL, and DigDTGL are shown in Table 13.2. From the table, one can notice that $\bar{y}(t_f)$ is smallest for FODTGL and largest for DigDTGL when $h_g = 0.05$ s. This behavior is expected since DigDTGL is the result of an approximation of RODTGL, which is itself a simplified version of FODTGL. This indicates that the analysis of the effect of the digital implementation constitutes an important design step for finite-precision implementation. Yet, values for $\bar{y}(t_f)$ obtained with RODTGL and DigDTGL are smaller than miss distances obtained with FODTGL for the sample period of 0.1s. For such a sampling period, standard deviations show that the upper limit of the 95% confidence interval of RODTGL and DigDTGL is greater than that of FODTGL. Those results have to be interpreted carefully as the technique of controller order reduction applied to FODTGL is achieved in an open-loop fashion and, thus, does not guarantee high performance of the closed-loop system.

13.5.2 Digital Autopilots

The effectiveness of the proposed autopilots is demonstrated for two linearized missile dynamics, exhibiting different time constants, and for classical and optimal CT autopilots.

13.5.2.1 Optimal Redesign

Consider a symmetrical, tail-controlled missile with linearized pitch plane dynamics given as [47]

$$\bar{G}_1(s) = \frac{\bar{a}_m(s)}{\bar{\delta}_p(s)}$$

$$= \frac{1 \times 10^6(-0.001s^2 - 0.0038s + 5.23)}{s^2 + 6.907s + 726.43}$$

$$\bar{G}_2(s) = \frac{\bar{q}(s)}{\bar{\delta}_p(s)} \qquad (13.70)$$

$$= \frac{-1354.99s - 3865.92}{s^2 + 6.907s + 726.43}$$

and with idealized actuator dynamics. To control plant (Equation 13.70), a classical two-loop proportional and integral missile acceleration autopilot, shown in Figure 13.7, is designed as

$$\bar{C}_1(s) = \frac{3.5 \times 10^{-6}s + 1.28 \times 10^{-3}}{s}, \bar{C}_2(s) = -0.0283. \qquad (13.71)$$

Various DR techniques are employed on this CT autopilot. Inner-loop gain control is applied at the fast rate and is the same for all the digitally redesigned autopilots, whereas the outer-loop DT controller is different from one technique of redesign to another. A set of outer-loop DT controllers obtained with the optimal DR is given as follows:

$$C_{1,T} = (0.2 \times 10^{-3}z^5 + 0.005 \times 10^{-3}z^4 + 0.00003 \times 10^{-3}z^3 - 0.00000001 \times 10^{-3}z^2)/$$
$$(z^6 + 0.489z^5 - 1.45z^4 - 0.037z^3 - 0.0002z^2 + 0.0000001z)$$
(implemented at 3.33 Hz)

$$C_{1,T} = (0.206 \times 10^{-3}z^5 + 0.016 \times 10^{-3}z^4 + 0.006 \times 10^{-3}z^3 - 0.0001 \times 10^{-3}z^2$$
$$+ 0.000017 \times 10^{-3}z - 0.0000008 \times 10^{-3})/(z^6 + 0.718z^5 - 1.505z^4 - 0.166z^3$$
$$- 0.044z^2 - 0.0018z + 0.00014)(\text{implemented at 10 Hz})$$

$$C_{1,T} = (0.117 \times 10^{-3}z^5 - 0.09 \times 10^{-3}z^4 + 0.044 \times 10^{-3}z^3 - 0.0119 \times 10^{-3}z^2$$
$$+ 0.0019 \times 10^{-3}z - 0.00013 \times 10^{-3})/(z^6 - 0.743z^5 - 0.409z^4 + 0.216z^3$$
$$- 0.059z^2 - 0.006z + 0.00189)(\text{implemented at 20 Hz}).$$

$$(13.72)$$

Selected simulation results are shown in Figures 13.17 and 13.18. In Figure 13.17, all three autopilots approach to the same time-domain behavior, as the outer-loop rate is increased, as expected from the concept of DT models. However, classical Tustin redesign cannot warrant closed-loop stability in the case of the slowest outer-loop rate, as this method does not take into account the closed-loop model in the discretization. With the proposed optimal DR, however, reference input tracking is *always* obtained. For slower rates, it is important to point out that overshoot and response times are increased. A disturbance input signal given as

$$d(t) = 1 \text{ m/s}^2, 1 \leq t < 2; 0 \text{ otherwise} \qquad (13.73)$$

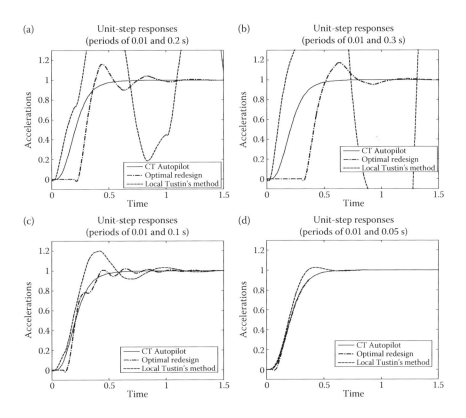

FIGURE 13.17
Accelerations (in meters per square second) for fixed value of inner-loop rate (100 Hz) and various outer-loop rates.

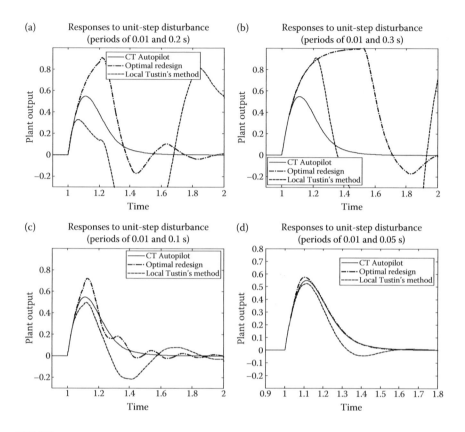

FIGURE 13.18
Disturbance responses (in meters per square second) for fixed value of inner-loop rate (100 Hz) and various outer-loop rates.

is applied to the autopilots, and the responses are shown in Figure 13.18. The disturbance responses of CT autopilot and that of the DT autopilot obtained with the optimal redesign asymptotically reject the disturbances in all cases, whereas it does not with the classical, local DR approach.

13.5.2.2 Redesign with Polynomial Method

Consider, again, pitch plane dynamics for a symmetrical, tail-controlled missile, although the equations of motion come from Nichols et al. [48]. The dynamics of this missile are faster than those of the previous system; thus, they require faster sampling and control update rates. For an angle of attack of 10°, a speed of Mach 3, and an altitude of 20,000 ft., the linearized CT dynamics are

$$\bar{G}_1(s) = \frac{0.2038(s^2 - 34.3^2)}{s^2 - 1.12s + 87.18}$$

$$\bar{G}_2(s) = \frac{-131(s+1)}{s^2 - 1.12s + 87.18}$$

$$\bar{A}(s) = \frac{\bar{\delta}_p(s)}{\bar{\delta}_c(s)} \qquad (13.74)$$

$$= \frac{1}{4.444 \times 10^{-5} s^2 + 0.009333s + 1}$$

where $\bar{A}$ represents actuator dynamics, from commanded to actual fin deflections. With reference to Figure 13.7, a reduced-order H_∞ CT acceleration autopilot is given as

$$\bar{C}_1(s) = \frac{-0.01594s^4 - 152.6s^3 - 1.048 \times 10^5 s^2 - 8.641 \times 10^7 s - 8.638 \times 10^7}{s^4 + 499.8s^3 + 6.276 \times 10^5 s^2 + 6.506 \times 10^5 s + 2.21 \times 10^4}$$

$$\bar{C}_2(s) = \frac{1.117s^4 + 4663s^3 + 5.839 \times 10^5 s^2 + 1.306 \times 10^7 s + 6.022 \times 10^5}{s^4 + 499.8s^3 + 6.276 \times 10^5 s^2 + 6.506 \times 10^5 s + 2.21 \times 10^4}.$$

$$(13.75)$$

The original controllers have order 7, and the magnitude of their coefficients is relatively large, making them unsuitable for practical implementation. In this case, order reduction allows obtaining low-order CT transfer functions and controller coefficients of reasonable magnitude. To prevent having to implement a relatively high-order DT autopilot, or to use successive order reduction techniques to warrant tractable DT controller orders at the cost of a loss in performance, optimal redesign is not carried out. However, polynomial redesign is employed. The inner-loop DT controller is a ZOH equivalent of $\bar{C}_2$ running at the rate of 400 Hz. With the proposed polynomial method, the outer loop is composed of three fourth-order DT blocks, whereas local DR with Tustin's method and with ZOH equivalence yields a single fourth-order outer-loop DT controller. We constrain polynomials u and v in Equation 13.50 to be of degree 4, which requires solving Equation 13.52 with the least-squares method. With reference to Figure 13.10, for an outer-loop rate of 21 Hz, the DT controllers obtained with the polynomial redesign are

$$\tilde{A}_{1,h} = \frac{-9.18z^4 + 8.757z^3 + 0.000001z^2}{z^4 - 2.287z^3 + 1.917z^2 - 0.697z + 0.093}$$

$$\tilde{B}_{1,h} = \frac{-1.425z^4 + 2.982z^3 - 2.232z^2 + 0.223z - 0.026}{z^4 - 2.287z^3 + 1.917z^2 - 0.697z + 0.093} \qquad (13.76)$$

$$\tilde{C}_{1,h} = \frac{z^4 - 2.287z^3 + 1.917z^2 - 0.697z + 0.093}{z^4 - 1.319z^3 + 0.555z^2 - 0.2847z + 0.0446}.$$

Figures 13.19 and 13.20 present simulation results obtained for various implementation rates. Unit-step responses (Figure 13.19) and responses to the disturbance input given in Equation 13.73 (Figure 13.20) clearly demonstrate that superior closed-loop performance is obtained with the proposed polynomial redesign when compared with classical approaches, although at the cost of an increased number of outer-loop controllers.

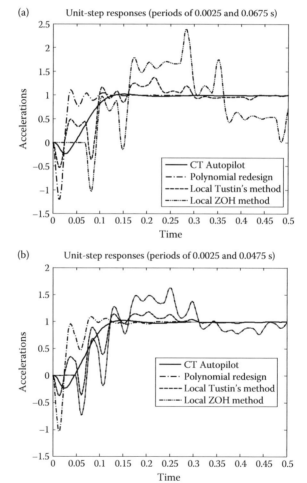

FIGURE 13.19
Accelerations in meters per square second.

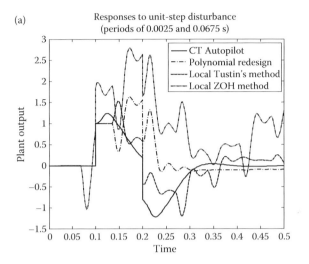

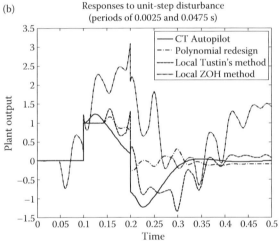

FIGURE 13.20
Disturbance responses in meters per square second.

13.6 Conclusions and Future Directions

This chapter presented practical techniques for the design of digital homing guidance laws and autopilots. The methods put forth enabled implementations at multiple and single sampling and control update rates. Fundamental concepts and tools needed for the design and the analysis of missile systems were discussed. These include lifting, DT modeling, discretization, dual-rate holds and sampler, and optimal DT control. The proposed digital guidance laws and autopilots allow designers to take advantage of relaxed constraints on the values of the sampling and control update rates, at least to some extent, while warranting satisfactory performances despite adverse implementation conditions. Indeed, extensive numerical simulations of pursuer–evader engagements, carried out in part with digital-autopilot implementations with fixed-point, finite word length arithmetic, demonstrated that satisfactory levels of precision to target can be achieved with the proposed synthesis of the guidance laws and the autopilots without requiring high-end computational capability. Future work should extend the proposed techniques to the case of nonlinear dynamic systems and study the integrated design of multirate guidance laws with autopilots. Therefore, a challenging research direction is the development of an integrated nonlinear guidance-autopilot system minimizing the effects caused by quantization and finite word length implementations, among other design objectives.

Acknowledgments

The authors would like to thank the Department of National Defence of Canada and the Natural Sciences and Engineering Research Council of Canada for supporting this work.

References

1. P. Zarchan, *Tactical and Strategic Missile Guidance*, 4th Edition, vol. 199, Progress in Astronautics and Aeronautics, Reston, VA, 2002.
2. R.G. Cottrel, Optimal intercept guidance for short-range tactical missiles, *AIAA Journal*, 9, 1414–1415, July 1971.
3. I. Rusnak, Optimal guidance laws with uncertain time-of-flight, *IEEE Transactions on Aerospace and Electronic Systems*, 36(2), 721–725, 2000.
4. S. Gutman, Superiority of canards in homing missiles, *IEEE Transactions on Aerospace and Electronic Systems*, 39(3), 740–746, July 2003.
5. J.Z. Ben-Asher and I. Yaesh, *Advances in Missile Guidance Theory*, vol. 180, Progress in Astronautics and Aeronautics, AIAA, Reston, VA, 1998.
6. A.V. Savkin, P.N. Pathirana, and F.A. Faruqi, Problem of precision missile guidance: LQR and H_∞ control frameworks, *IEEE Transactions on Aerospace and Electronic Systems*, 39(3), 901–910, July 2003.
7. P. Gurfil, M. Jodorkovsky, and M. Guelman, Neoclassical guidance for homing missiles, *Journal of Guidance, Control, and Dynamics*, 24(3), 452–459, May–June 2001.

8. P. Gurfil, Robust guidance for electro-optical missiles, *IEEE Transactions on Aerospace and Electronic Systems*, 39(2), 450–461, 2003.

9. N. Lechevin, C.A. Rabbath, and P. Sicard, A passivity perspective for the synthesis of robust terminal guidance, *IEEE Transactions on Control Systems Technology*, 13(5), 760–765, September 2005.

10. D. Dionne and C.A. Rabbath, Predictive Guidance for Pursuit–Evasion Engagements Involving Decoys, *Proceedings of AIAA Conference on Guidance, Navigation and Control 2006*, Paper AIAA 2006-6214, Keystone, CO, 21–24 August 2006. American Institute of Aeronautics and Aeronautics Inc., Reston, VA.

11. E.J. Holder and V.B. Sylvester, An analysis of modern versus classical homing guidance, *IEEE Transactions on Aerospace and Electronic Systems*, 26(4), 599–606, 1990.

12. T. Chen and B. Francis, *Optimal Sampled-Data Control Systems*, Springer-Verlag, London, U.K., 1995.

13. K.J. Astrom and B. Wittenmark, *Computer-Controlled Systems: Theory and Design*, Prentice Hall, NJ, 1990.

14. J.G. Proakis and D.G. Manolakis, *Digital Signal Processing—Principles, Algorithms, and Applications*, 2nd Edition, Macmillan Publishing Co., New York, 1992.

15. D. Farret, G. Duc, and J.P. Harcaut, Multirate LPV Synthesis: A Loop-Shaping Approach for Missile Control, *Proceedings of the American Control Conference*, Anchorage, AK, pp. 4092–4097, 2002. Institute of Electrical and Electronics Engineers, Piscataway, NJ.

16. A.M. Azad and T. Hesketh, H∞ Optimal Control of Multi-Rate Sampled-Data Systems, *Proceedings of the American Control Conference*, Anchorage, AK, pp. 459–464, 2002. Institute of Electrical and Electronics Engineers, Piscataway, NJ.

17. L. Yang and M. Tomizuka, Short seeking by multirate digital controllers for computation saving with initial value adjustment, *IEEE/ASME Transactions on Mechatronics*, 11(1), 9–16, February 2006.

18. N. Lechevin and C.A. Rabbath, A multi-objective control approach for the synthesis of robust digital guidance laws, *Proceedings of the Institution of Mechanical Engineers Journal of Aerospace Engineering—Part G*, 219(2), 89–102, April 2005.

19. P. Khargonekar, K. Poolla, and A. Tannenbaum, Robust control of linear time-invariant plants using periodic compensation, *IEEE Transactions on Automatic Control*, 30(11), 1088–1096, 1985.

20. J.P. Keller and B.D.O. Anderson, A new approach to the discretization of continuous time controllers, *IEEE Transactions on Automatic Control*, AC-37, 214–223, 1992.

21. N. Rafee, T. Chen, and O.P. Malik, A technique for optimal digital redesign of analog controllers, *IEEE Transactions on Control Systems Technology*, 5(1), 89–99, 1997.

22. R. Kennedy and R. Evans, Digital Redesign of a Continuous Controller Based on Closed Loop Performance, *Proceedings of the 29th Conference on Decision and Control*, pp. 1898–1901, 1990. Institute of Electrical and Electronics Engineers, Piscataway, NJ.

23. A.H.D. Markazi and N. Hori, A new method with guaranteed stability for discretization of continuous-time control systems, *Proceedings of the American Control Conference*, 2, 1397–1402, 1992.

24. L.-S. Shieh, J. Zheng, and W. Wang, Digital modeling and digital redesign of analog uncertain systems using genetic algorithms, *Journal of Guidance, Control and Dynamics*, 20(4), 721–728, 1997.

25. C.A. Rabbath and N. Hori, Reduced-order PIM methods for digital redesign, *IEE Proceedings—Control Theory and Applications*, Institute of Electrical and Electronics Engineers, Piscataway, NJ, 150(4), 335–346, 2003.
26. B.D.O. Anderson, Controller design: moving from theory to practice, *IEEE Control Systems Magazine*, Institute of Electrical and Electronics Engineers, Piscataway, NJ, 16–25, 1993.
27. G.C. Goodwin and R.H. Middleton, *Digital Control and Estimation—A Unified Approach*, Prentice Hall, Englewood Cliffs, NJ, 1990.
28. N. Hori, T. Mori, and P.N. Nikiforuk, A new perspective for discrete-time models of a continuous-time system, *IEEE Transactions on Automatic Control*, AC-37(7), 1013–1017, 1992.
29. T. Mori, P.N. Nikiforuk, M.M. Gupta, and N. Hori, A class of discrete-time models for a continuous-time system, *IEE Proceedings-D*, 136(2), 79–83, March 1989.
30. R.F. Whitbeck and L.G. Hofmann, Digital control law synthesis in the w' domain, *AIAA Journal of Guidance and Control*, 1(5), 319–326, 1978.
31. N. Hori, T. Mori, and P.N. Nikiforuk, Discrete-time models of continuous-time systems, *Control and Dynamic Systems*, 66, 1–45, 1994. Academic Press, Amsterdam, The Netherlands.
32. C.A. Rabbath, N. Hori, and N. Lechevin, Convergence of sampled-data models in digital redesign, *IEEE Transactions on Automatic Control*, 49(5), 850–855, May 2004.
33. P.T. Kabamba, Control of linear systems using generalized sampled-data hold functions, *IEEE Transactions on Automatic Control*, AC-32(9), 772–783, 1987.
34. G.F. Franklin, J.D. Powell, and M.L. Workman, *Digital Control of Dynamic Systems*, 2nd Edition, Addison-Wesley, Reading, MA, 1990.
35. C.A. Rabbath, N. Lechevin, and N. Hori, Practical Techniques for Optimal Dual-Rate Digital Redesign, *Proceedings of American Control Conference*, Boston, pp. 3496–3501, 2004. Institute of Electrical and Electronics Engineers, Piscataway, NJ.
36. Y. Gu, M. Tomizuka, and J. Tornero, Digital Redesign of Continuous Time Controller by Multirate Sampling and High Order Holds, *Proceedings of the 38th IEEE Conference on Decision and Control*, Phoenix, AZ, pp. 3422–3427, 1999.
37. R.A. Singer, Estimating Optimal Tracking Filter Performance for Manned Maneuvering Targets, *IEEE Transactions on Aerospace and Electronic Systems*, AES-6(4), 473–483, 1970.
38. K. Zhou and J.C. Doyle, *Essentials of Robust Control*, Prentice Hall, Upper Saddle River, NJ, 1998.
39. M. Chilali and P. Gahinet, H_∞ design with pole placement constraints: an LMI approach, *IEEE Transactions on Automatic Control*, 41(3), 358–367, 1996.
40. J.C. Doyle and G. Stein, Multivariable feedback design: concepts for a classical/modern synthesis, *IEEE Transactions on Automatic Control*, 26(1), 4–16, 1981.
41. S. Bingulac and H.F. VanLandingham, Discretization and Continualization of MIMO Systems, In *Proceedings of the 12th World Congress of the International Federation of Automatic Control*, Sydney, Australia, July, 1993. Elsevier Ltd, Kidlington, UK.
42. C.A. Rabbath, N. Lechevin, and N. Hori, Optimal dual-rate digital redesign with application to missile control, *AIAA Journal of Guidance, Control and Dynamics*, 27(6), 1083–1087, November–December 2004.
43. M. G. Safonov and R. Y. Chiang, A Schur method for balanced model reduction, *IEEE Transactions on Automatic Control*, AC-34(7), 729–733, 1989.

44. B. Moore, Principal component analysis in linear systems: controllability, observability, and model reduction, *IEEE Transactions on Automatic Control*, 26(1), 17–31, 1981.

45. V. Krishnamurthy and V. Seshadri, Model reduction using the Routh stability criterion, *IEEE Transactions on Automatic Control*, AC-23(4), 729–731, August 1978.

46. P. Gurfil, Zero-miss distance guidance law based on line-of-sight rate measurement only, *Control Engineering Practice*, 11(7), 819–832, 2003.

47. R. Lestage, M. Lauzon, and A. Jeffrey, Integrated Autopilot Tuning Methodology for Airframe Parametric Simulations, *AIAA Paper 2002-4661*, August 2002. American Institute of Aeronautics and Astronautics Inc., Reston, VA.

48. R.A. Nichols, R.J. Reichert, and W.J. Rugh, Gain Scheduling for H-Infinity Controllers: A Flight Control Example, *IEEE Transactions on Control Systems Technology*, 1(2), 69–79. Institute of Electricaland Electronics Engineers, Piscataway, NJ.

14

Design of CLOS Guidance System

G. Hexner and H. Weiss

CONTENTS

14.1 Overview

A command to line-of-sight (CLOS) guidance system employs a single sensor to measure the position of both the target and the interceptor missile. Usually, the single sensor is stationary and requires that the intercepting missile stay close to the line connecting the sensor to the target. The system operates by sending acceleration commands to the missile based on the sensor measurement of the target and missile positions.

CLOS systems have an advantage in close-in point defense applications. The use of CLOS guidance makes possible a simpler and cheaper missile and at the same time employs a better-quality (more expensive) stationary sensor. Some of the disadvantages of CLOS systems are the inherent limitation of being able to engage only one target at a time*; the requirement to maneuver in order to engage a crossing target; and the limitation of the effective

---

* In a more advanced system, the sensor may be a phased array radar system. Such a sensor is capable of simultaneously tracking several objects, freeing the missile trajectory from the constraint of staying on the line connecting the sensor to the target and enables the engagement of multiple targets.

range of such systems, because both the target and the missile measurement errors are proportional to the range.

The approach presented here is based on the linear quadratic Gaussian (LQG) formalism. Probably the first paper to suggest that the CLOS guidance problem may be formulated as a stochastic optimal control problem is [15]. Other possible approaches are [5, 9], and [12–14].

The principal components of this system are the sensor, the sensor platform, and the missile. The sensor has a small field of view, and hence, the sensor platform must rotate the sensor, so that the target always stays within the sensor's field of view. It is the guidance system's job to ensure that the acceleration commands sent to the missile are such that the missile at all times remains within the sensor's field of view. The guidance system consists of the target estimator, the missile estimator, and the guidance law. The functions of the target estimator and the missile estimator are self-explanatory. The target estimator has a single design parameter, which is its bandwidth. The missile estimator and the guidance law must be designed as a single unit. The missile estimator contains both a model of the missile acceleration transfer function and an associated uncertainty model. The guidance law must be designed to operate with this uncertainty model and to maintain the stability of the guidance loop in the face of a certain amount of uncertainty in the missile transfer function.

When the missile is launched, a means must be provided to ensure that the missile enters the observation system's field of view. This problem is addressed here in only its simplest solution: it is assumed that in addition to the main antenna used for tracking, there is also a wider antenna that shares the boresight with the main antenna and is used for tracking the missile in the period immediately following its launch.

The focus of this chapter is on the *design and implementation* of a CLOS guidance system. The kinematics of the motion of the missile pursuing a target and the calculation of the missile path are excellently presented in [11, 23, 27].

14.2 System Operation

In Figure 14.1, a schematic block diagram of the guidance system is shown. The sensor measures the position of the target and the missile in its field of view. These measurements are passed to the target state estimator (TSE) and to the missile state estimator (MSE). The sensor platform receives the output of the TSE and uses the measurements of its built-in inertial measurement unit (IMU) to rotate the sensor so that the target remains in its field of view. In addition, it outputs the readings of its IMU to both the TSE and the MSE. Both the TSE and the MSE require the IMU readings to generate their

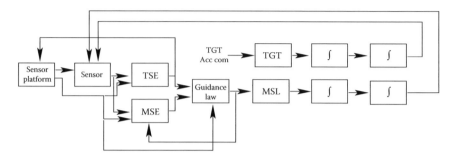

FIGURE 14.1
Schematic block diagram of a CLOS guidance system.

respective estimates. The guidance law (GL) outputs missile acceleration commands to be transmitted to the missile and to be passed to the MSE. The inputs to the GL are the sensor platform IMU measurements and the state estimates of the TSE and the MSE. The missile carries out the acceleration commands, and after double integration, the missile position is obtained, which in turn is measured by the sensor.

14.3 System Component Requirements

14.3.1 Missile

The missile receives the acceleration commands from the guidance law, and its autopilot must generate fin deflections to cause the missile to carry out the commanded accelerations. The function of the missile is to carry its warhead to the close proximity of the target. The missile warhead size is determined by the expected miss distance, which in turn is determined by the expected target maneuver level, the missile's own agility, the angular sensor observation errors, and the intercept range. As can be seen, these factors are all interrelated, and the choice of a good operating point is a compromise among all these factors.

14.3.2 Sensor

The sensor measures both the target and the missile positions within its field of view. The sensor must be able to distinguish between the target and the missile. In addition, as will be shown, the availability of the time to intercept for the guidance law greatly reduces the miss distance. It is then advantageous to have a sensor capable of measuring the range to both the target and the missile.

14.3.3 Sensor Platform

The sensor is mounted on the sensor platform. The sensor platform must rotate the sensor so that the target remains within the sensor's field of view at all times. The sensor platform must contend with disturbances resulting from both the target's maneuvers and external loads, such as the wind acting on the sensor structure. Optionally the position of the missile within the sensor field of view may be taken into account when designing the sensor platform's control system.

14.3.4 Target State Estimator

The target is modeled as a first-order Gauss Markov process, which is the shaping filter associated with Singer model [24,27]. The target state is estimated in the sensor coordinate system.

14.3.5 Missile State Estimator

The missile model has a deterministic part and a stochastic part used to account for the missile uncertainty. The missile state is also estimated in the sensor coordinate system.

14.3.5.1 Missile Acceleration Model

The missile acceleration model is derived from the known missile a_M/a_{cB} transfer function. The design model is a first-order approximation to the full order missile a_M/a_{cB} transfer function, obtained by the use of the balanced order reduction on the full order missile transfer function.

14.3.5.2 Missile Uncertainty Model

The missile uncertainty model serves to model acceleration deviations of the missile from the reduced first-order model. The source of the deviation may be the simplification of the missile transfer function to a first-order model, or it may be due to some imperfection within the missile itself. One such imperfection is the drift of the roll reference, which would cause the missile to carry out the acceleration commands in a plane rotated with respect to that intended by the guidance law.

14.3.6 Guidance Law

The TSE and the MSE are an integral part of the GL, which is based on an LQG design. The missile estimator and the GL must be designed as a single unit. The missile estimator contains both a model of the missile transfer function and a suitable uncertainty model. The GL must be designed to operate in the face of variations in the missile transfer function.

14.3.6.1 Ensuring Robustness of the Guidance Law

While the missile is far from the target, the GL acts as a regulator. The acceleration commands are designed to keep the missile within the sensor's field of view, in the face of all possible target maneuvers and all possible disturbances acting on the interceptor missile. In order to accomplish this, it is assumed that the sensor platform tracks the target with small errors. Then the problem is equivalent to ensuring the robustness of any regulator, so that the usual tools, gain margin and phase margin, apply.

14.3.6.2 Terminal Guidance Law

The ultimate aim of the CLOS guidance law is to hit the target with a small miss distance, without violating the constraint of being within the sensor's field of view at all times. Similar to other types of guidance, the use of terminal guidance leads to considerable improvement in the miss distance.

14.4 System Models

System models are required both in the synthesis of the MSE and GL and in the subsequent analysis of the robustness of the system. Generally, the design models are simpler (lower order) than the models used in the robustness check. There is a one-to-one correspondence between the complexity of the missile model used in the synthesis phase and the complexity of the GL; hence, the simplest possible model is used in this phase. The robustness model, on the other hand, should include the variability of the missile transfer function.

14.4.1 Kinematic Model

In this section, the kinematics of a body moving in a rotating coordinate system is developed. The natural coordinate system for setting up the guidance problem is the coordinate system rigidly attached to the sensor. The *x*-axis coincides with the boresight of the sensor, the *y*-axis is perpendicular to the *x*-axis and is horizontal when the roll angle of the sensor platform is zero, and finally, the *z* complements the *x* and *y* coordinates to form a right-handed coordinate system.

Let p and v be the position and velocity, respectively, of a body expressed in the coordinate system just defined undergoing acceleration a. Then these quantities are related by

$$\begin{bmatrix} \dot{p} \\ \dot{v} \end{bmatrix} = \begin{bmatrix} -\Omega_r & I \\ 0 & -\Omega_r \end{bmatrix} \begin{bmatrix} p \\ v \end{bmatrix} + \begin{bmatrix} 0 \\ I \end{bmatrix} a, \tag{14.1}$$

where

$$
\Omega_r = \begin{bmatrix} 0 & -\omega_r^z & \omega_r^y \\ \omega_r^z & 0 & -\omega_r^x \\ -\omega_r^y & \omega_r^x & 0 \end{bmatrix}
\tag{14.2}
$$

and

$$
\omega_r = \begin{bmatrix} \omega_r^x \\ \omega_r^y \\ \omega_r^z \end{bmatrix}
\tag{14.3}
$$

is the angular velocity of the sensor platform.

When the sensor platform rotates on two gimbals mounted on a stationary base *and* the elevation angle is zero, then $\omega_r^x = 0$. In the remainder, it is assumed that the elevation angle is sufficiently small to allow neglecting ω_r^x. In this case, $\omega_r^x = 0$, and Equation 14.1 can be decomposed into a pair of uncoupled equations. Here the equations for the motion in the x–y plane are presented. Similar equations (with the exception of some minus signs, and ω_r^z replaced by ω_r^y) exist for the x–z plane. These equations describe equally the motions of the two objects of interest: the missile and the target. The subscript i takes on value of M for the missile and T for the target. The kinematics in the x–y plane follow the equations

$$
\dot{p}_i^y = -\omega_r^z p_i^x + v_i^y
\tag{14.4}
$$

$$
\dot{v}_i^y = -\omega_r^z v_i^x + a_i^y.
\tag{14.5}
$$

In order for the CLOS system to operate properly, the angle between the sensor platform boresight and each object—the missile and the target   must be kept small. The dynamics of the angle between the boresight and object i is derived next. Define

$$
\varepsilon_i^z = p_i^y / p_i^x
\tag{14.6}
$$

where ε_i^z is approximately equal to the angle between the position of object i projected onto the x–y plane and the boresight. Differentiating Equation 14.6,

$$
\dot{\varepsilon}_i^z = -\omega_i^z + \omega_i^z - \varepsilon_i^z \frac{\dot{r}_i}{r_i},
\tag{14.7}
$$

where

$$\omega_i^z = \frac{v_i^y}{p_i^x},$$ (14.8)

r_i is the range to object i, and the approximation

$$r_i \approx p_i^x$$ (14.9)

was used. Using similar manipulations, the variable ω_i^z satisfies

$$\dot{\omega}_i^z = -\omega_r^z \frac{\dot{r}_i}{r_i} - \omega_i^z \frac{\dot{r}_i}{r_i} + \frac{a_i^y}{r_i}.$$ (14.10)

Note that the only approximation used in obtaining Equations 14.7 and 14.10 was Equation 14.9, although the identification of ε_i with the angle between the boresight and object i, and ω_i with the angular velocity of object i, depends on ε_i being small.

14.4.2 Missile Model

In the present chapter, only a simplified missile model is used since the focus of the chapter is the design of a CLOS guidance system. Only the elements of the missile model that impact on the guidance system design are modeled in detail here. The following components *are* included:

1. Missile time constant
2. Limit on missile acceleration
3. Missile drag
4. Missile engine thrust
5. Varying missile velocity

Here a "three-loop autopilot" is assumed. (The design of such an autopilot is covered in several places; see for example [19] or [20]). In this case, the missile transfer function for aerodynamic acceleration commands perpendicular to the missile body in either the y or z direction is

$$\frac{a_M}{a_{cB}} = \frac{(s/z_A + 1)(-s/z_A + 1)}{(s/p_A + 1)(s^2/\omega_A^2 + 2\xi_A/\omega_A + 1)}.$$ (14.11)

The values of the poles and zeroes of the transfer function vary according to the missile flight conditions. One set of possible values is shown in Table 14.1.

TABLE 14.1

Missile Transfer Function Parameters

Symbol	Value	Units	Description
z_A	54.4	rad/s	Zero location
p_A	12.7	rad/s	Real pole location
ω_A	20.7	rad/s	Frequency of complex poles
ξ_A	0.4		Damping of complex poles

A linear approximation is assumed for the missile aerodynamic coefficients. Their values for the vertical plane are shown in Table 14.2. The aerodynamic coefficients for the horizontal plane have the same numerical values but differ in their signs.

The dependence of this transfer function on the flight conditions is fully accounted for in the performance evaluation presented later. For the purpose of developing the design model in Section 14.5, the transfer function is assumed to vary sufficiently slowly to allow neglecting the time-varying nature of the missile transfer function in the design process.

The missile acceleration is limited by the missile's ability to reach and to maintain a large angle of attack. Here it is assumed that the missile is capable of sustained flight up to angle of attack of 25°. The acceleration limit, $a_{M\,lim}$, is given by the formula

$$a_{M\,lim} = \frac{Q_{pres}S_{ref}}{m} \frac{C_{m\delta}C_{z\alpha} - C_{m\alpha}C_{z\delta}}{C_{m\delta}} \alpha_{max} \qquad (14.12)$$

where Q_{pres} is the dynamic pressure, S_{ref} is the reference area of the missile, and m is the missile mass. This equation summarizes the limitations on the missile acceleration due to aerodynamics. There may be an additional limit on the missile acceleration due to the finite strength of its structure.

The missile axial acceleration in the missile x body direction is given by

$$a_M^x = \frac{T + D_S}{m}, \qquad (14.13)$$

TABLE 14.2

Missile Aerodynamic Coefficients for Vertical Plane

Symbol	Value
$C_{m\alpha}$	−20
$C_{m\delta}$	−18.75
$C_{z\alpha}$	−25
$C_{z\delta}$	−4
α_{max}	25°

where T is the missile engine thrust, and the skin drag D_s is given by

$$D_s = Q_{pres} S_{ref} C_{x0} \tag{14.14}$$

Note that the total force acting parallel to the missile's velocity vector and causing the loss of the missile's velocity may be obtained by expressing the aerodynamic forces acting in the "wind" coordinates. The dependence of C_{x0} on the Mach number is shown in Figure 14.2.

As the missile accelerates, its mass decreases according to

$$\dot{m} = K_T T \tag{14.15}$$

where $K_T = -4.3 \times 10^{-4}$ kg/N s is a proportionality constant, and T is the thrust exerted by the missile engine.

The variation of the missile mass and the thrust are shown in Figures 14.3 and 14.4.

In addition to the full order transfer function in Equation 14.11, a simpler first-order model is required. This model is obtained from Equation 14.11 by the balanced order reduction:

$$\dot{x}_M^j = -\frac{1}{\tau_M} x_M + \frac{1}{\tau_M} a_{cB}^j$$

$$a_M^j = \left(1 - \frac{\tau_{zc}}{\tau_M}\right) x_M^j + \frac{\tau_{zc}}{\tau_M} a_{cB}^j \tag{14.16}$$

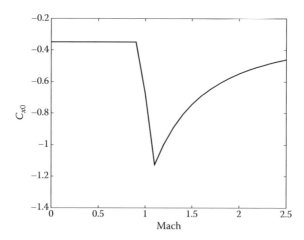

FIGURE 14.2
Variation of the aerodynamic coefficient C_{x0} with Mach.

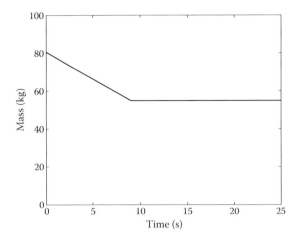

FIGURE 14.3
Missile mass profile.

where j denotes either the y or z direction perpendicular to the missile body axis. Equivalently the transfer function associated with Equation 14.16 may be expressed as

$$\frac{a_M^j}{a_{cy}^j} = \frac{1 + s\tau_{zc}}{1 + s\tau_M}. \tag{14.17}$$

Strictly speaking, this model is valid in the lateral body coordinates. Here the liberty is taken of using the same transfer function to describe the

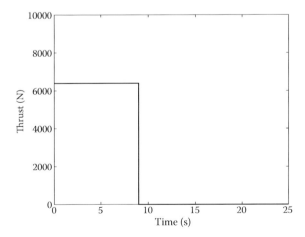

FIGURE 14.4
Missile thrust profile.

dynamics of the missile acceleration in the body, wind, as well as the sensor coordinate systems. This first-order model is a good description of the low-frequency acceleration response of the missile. This model is a part of the MSE to be described in Section 14.6.3. The same model is also used in the system model for calculating the guidance gains in Section 14.5.3. There, a further approximation is made: whereas Equation 14.16 describes the missile's response in its body coordinates, in Section 14.5.3, the same model is applied to the missile acceleration perpendicular to the sensor boresight. If the scenario is the interception of a target approaching the sensor close to its boresight, then this is very good approximation. On the other hand, the use of the same model for other scenarios is an approximation, which must be evaluated on the basis of the performance of the guidance system.

14.4.3 Missile Uncertainty Model

In order to ensure the robustness of the design, the deviation of the missile model from the simple model Equation 14.16 must be built into the design model. The model for a_d^j, the deviation of the actual missile acceleration in the j direction from its assumed model, consists of a shaping filter with a white noise signal input:

$$\dot{x}_d^j = -\frac{1}{\tau_d} x_d^j + w_M^j$$

$$a_d^j = x_d^j + \frac{\tau_d \tau_{zd}}{\tau_d - \tau_{zd}} w_M^j. \tag{14.18}$$

When Equation 14.18 is expressed as a transfer function driven by the white noise process w_M^j, then τ_{zd} is the zero of the transfer function. The important feature of the model is that the spectral content of a_d^j remains nonzero at high frequencies. This desensitizes the guidance loop to high-frequency uncertainty in the missile transfer function. The spectral density of the white noise process w_M^j is the same for all axes, σ_M^2.

The uncertainty acceleration a_d^j is a stochastic representation of the uncertainty in the missile transfer function. There is no physical signal in the system, which it directly represents. It is used in the missile estimator to estimate the missile acceleration $\hat{a}_M^j$ as

$$\hat{a}_M^j = a_{Mdet}^j + \hat{a}_d^j, \tag{14.19}$$

that is, $\hat{a}_M^j$, the estimated missile acceleration in the j direction, is represented as the sum of the deterministic acceleration a_{Mdet}^j and the estimate of the uncertainty in the acceleration $\hat{a}_d^j$. The deterministic acceleration a_{Mdet}^j is calculated by feeding the acceleration commands a_{cB}^j into Equation 14.16 to obtain a_{Mdet}^j. Note that

$$E\{a_d^j \mid y_M^j\} = \hat{x}_d^j, \tag{14.20}$$

where y_M^j represents the missile measurements obtained from the sensor. Hence, for the purposes of calculating the missile acceleration, it would have been sufficient to use the first of the two equations in Equation 14.18. The addition of the white noise in the second equation of Equation 14.18 *does* influence the missile estimator gain; see Equation 14.114, which in turn influences the stability margins of the guidance control loop. The disturbance acceleration $\hat{a}_d^j$ is one of the outputs of the missile estimator.

14.4.4 Target Model

The target acceleration model is the Singer model [24]. As discussed in Section 14.6.2, the same model is used in all the axes. The stochastic model for a_T^j, the target acceleration in each of the coordinate axes, is

$$\dot{a}_T^j = -\frac{1}{\tau_T} a_T^j + w_T^j \tag{14.21}$$

where τ_T is the target acceleration decorrelation time, and w_T^j are continuous time white noise processes. The model includes two parameters: the decorrelation time τ_T and the standard deviation of the target acceleration $\sigma_{a_T}^j$. The spectral density of the white noise w_T^j is for a given standard deviation of a_T^j:

$$(\sigma_T^j)^2 = \frac{2}{\tau_T} (\sigma_{a_T}^j)^2. \tag{14.22}$$

14.4.5 Sensor Platform Model

The sensor platform consists of an outer (azimuth) and inner (elevation) gimbal. It is assumed that the two gimbal actuator systems are uncoupled, and in fact, their dynamics are neglected here. It is further assumed that the sensor platform tracking loop has a bandwidth of about 3–4 Hz and that the tracking loop is sufficiently stiff to be able to withstand the wind and other external disturbances.

The sensor platform's angular velocity is measured by an IMU. The errors in the measurement of the sensor platform's angular velocity are of some importance since it is the angular rate measurements that enable the estimators to relate measurements at different time instances. If the angular rate measurement is in error, then a stationary object tracked by the sensor will have an apparent acceleration. It is assumed here that this apparent acceleration is negligible with respect to the assumed acceleration capability of the tracked object. If this is not true, then the angular rate measurement error in the sensor platform must be accounted for in the estimators' process noise (see [26]).

The equations in Section 14.4.1 are valid for any sensor angular velocity ω_r. For the purposes of calculating the guidance law, it is better to also include a model that predicts the average future values of the sensor platform angular velocity. To develop such a model, it is assumed, for the sake of developing the prediction, that the sensor platform tracks the target with zero angular error. Then the platform angular velocity is

$$\omega_r = \frac{p_T \times v_T}{\left\|p_T\right\|^2}.$$

(14.23)

Differentiating Equation 14.23 yields

$$\dot{\omega}_r = \frac{\dot{p}_T \times v_T + p_T \times \dot{v}_T}{\left\|p_T\right\|^2} - 2\frac{p_T \times v_T}{\left\|p_T\right\|^2}\frac{p_T \cdot v_T}{\left\|p_T\right\|^2}.$$

(14.24)

Upon using the approximation that the sensor platform tracks the target with zero angular error, so that $p_T^y = p_T^z = 0$, the identity $\dot{p}_T \times v_T = v_T \times v_T = 0$, and Equation 14.23, then

$$\dot{\omega}_r = -2\frac{\dot{r}_T}{r_T}\omega_r + \begin{bmatrix} 0 \\ -a_T^z/r_T \\ a_T^y/r_T \end{bmatrix}.$$

(14.25)

14.4.6 Sensor Model

The basic sensor considered in this chapter is a fire control radar, operating in the K_u band, with a dish diameter of 1.5 m. In addition, there is also a smaller antenna, whose diameter is 0.3 m, which shares the same boresight as the main 1.5 m diameter antenna and is used to track the missile in the initial phase of its flight. Here only a minimal description of the parameters of the radar is presented. For a more complete description, see [8] and [25].

The S/N ratio of a radar signal is defined by the relation

$$\left(\frac{S}{N}\right)_{skin} = \left(\frac{r_T^o}{r_T}\right)^4 \sigma_T^{cs}$$

(14.26)

where σ_T^{cs} is the radar cross section of the target, and r_T^o is the range where the radar signal to noise is 1 for a target with radar cross section of 1 m^2.

For example, to enable the tracking of a target with a radar cross section (RCS) of 2 m² with a signal-to-noise ratio of 13 dB at a range of 20 km would require $r_T^\circ = 35$ km. If the missile is also tracked using the skin signal, then the formula in Equation 14.26 (with the subscript $_T$ replaced by $_M$) also applies. On the other hand, if the missile carries a beacon transmitter, then the formula

$$\left(\frac{S}{N}\right)_{\text{beacon}} = \left(\frac{r_M^\circ}{r_M}\right)^2 \tag{14.27}$$

applies, where r_M° is the range where the beacon signal-to-noise ratio is equal to 1. For example, a signal-to-noise ratio of 20 dB at a range of 10 km would require an r_M° value of 100 km.

14.4.6.1 Measurement Error Approximations

There are three main sources for the radar measurement errors

1. Thermal noise
2. Glint noise
3. Radar imperfections

Under the radar imperfection category, one may find radar antenna mapping errors, cross-polarization errors, receiver gain imbalance, etc. Here all of these errors are accounted for by setting a lower limit on the angular radar measurement errors of 1/50 of the radar 3 dB beam width.

The source of the thermal noise error is the finite value of the signal-to-noise ratio. The radar angle measurement errors due to thermal noise are well approximated by a Gaussian random noise with a standard deviation of σ_ε:

$$\sigma_\varepsilon = \frac{\theta_{3\text{dB}}}{k_m \sqrt{2(S/N)}}, \tag{14.28}$$

where $\theta_{3\text{dB}}$ is the antenna beam width at the 3 dB points, and k_m is the monopulse slope, with a typical value of 1.6. The antenna beam width is approximately given by

$$\theta_{3\text{dB}} = k_A \frac{\lambda}{D} \tag{14.29}$$

where λ is the radar signal wavelength, D is the antenna diameter, and k_A is a constant dependent on the illumination of the radar aperture; typical value may be 1.2. In the case at hand, $\lambda - 17.6$ mm, so that the beam width is about

14 mR $\approx$ 0.8° for a 1.5 m diameter antenna and 70 mR $\approx$ 4° for the wider 0.3 m diameter antenna.

The glint noise is due to the target scatterers separated in the cross-range direction producing fluctuations in the phase of the signal return [4, 8]. If the missile carries a beacon transmitter, then this noise, for the case of the missile, is negligible. The linear glint error measured at the target is usually taken to be about 1/6 of the linear extent of the target perpendicular to the line of sight from the radar position. It is generally a non-Gaussian correlated noise. For the purpose of this chapter, it is assumed that suitable measures are taken to minimize the correlation of the glint error from pulse to pulse. Once this is done, the effect of the glint error is approximated by white Gaussian noise for design purposes. To evaluate the performance of any particular system, a detailed simulation is generally required. In such a simulation, the glint noise can be modeled to any desired accuracy. The glint noise is converted to angular noise by dividing its value by the target range:

$$\sigma_{Gl} = \sigma_g / r, \tag{14.30}$$

where σ_g is approximately 1/6 the extent of the target perpendicular to the line of sight to the radar.

The range measurement errors are generally small enough that the performance of the system is little affected by these measurement errors. The range measurement error is an additive noise, whose standard deviation σ_r is

$$\sigma_r = \frac{\Delta r}{\sqrt{2(S/N)}}, \tag{14.31}$$

where Δr is the radar range resolution.

14.4.6.2 Measurement Model

The radar provides three measurements: range, an elevation angle, and azimuth angle, which are denoted y^x, y^y, and y^z, respectively. The range measurement is modeled as

$$y^x = r + \underline{n}^x \tag{14.32}$$

where $\underline{n}^x$ is a zero mean white Gaussian random sequence with standard deviation as defined in Equation 14.31.

The two angle measurements in the elevation and azimuth directions (y and z direction) are modeled as

$$y^j = \varepsilon^j + \underline{n}^j \, ; \, j = y, z \tag{14.33}$$

where the zero mean white Gaussian noise sequence n^j is the sum of the noises due to the thermal and glint noise contributions. The variance of n^j is

$$(\sigma_{n^j})^2 = (\sigma_{GI})^2 + (\sigma_\varepsilon)^2 \; ; \; j = y, z \tag{14.34}$$

where σ_{GI} and σ_ε are defined in Equations 14.30 and 14.28, respectively, and (S/N) is calculated by Equations 14.26 and 14.27 for the target and the missile as appropriate.

14.5 Detailed Design of Missile State Estimator and Guidace Law

In this section, the models making up the CLOS system, as presented in Section 14.4, are integrated, and the detailed design of the CLOS guidance system is carried out.

The design method used here is based on the use of LQG optimal control. In this formalism, the design parameters are

1. The dynamic system model parameters
2. The weights in the quadratic criterion function
3. The intensities of the process noise levels

In particular, the guidance gains are not adjusted individually, but rather, the model parameters are adjusted until the desired system behavior is obtained. It is therefore of paramount importance to choose the system model well.

14.5.1 Design Configuration

The coordinate system for the guidance design is rigidly attached to the sensor platform. Here it is assumed that the sensor platform is stationary. In this case, its roll rate is small, and it is here assumed to be negligible. Then the kinematic equations (Equation 14.1) neatly decouple into a horizontal y direction and a vertical z direction. The kinematics in the horizontal plane are described by Equations 14.7 and 14.10.

If the sensor platform is not stationary and there is appreciable roll rate about its boresight (its x-axis), then there are a number of possible approaches to the treatment of the problem. The simplest one is to define a virtual coordinate system, which is attached to the sensor platform but is free to rotate about the sensor boresight and rotates with respect to the sensor platform so that its z-axis is always vertical and its y-axis is always horizontal. Then in this virtual coordinate system, the kinematics are again neatly decoupled into horizontal and vertical components.

A second approach leaves the coordinate system attached to the sensor platform and, in the 3D realization of the target and missile estimators, transforms the states after the prediction step of the Kalman filter to the new coordinate system before processing each observation, to account for the current roll angle of the sensor platform. The development of the 3D implementation of the target and missile estimators in Section 14.6 is compatible with this possibility.

There is a fundamental difference in the treatment of the change in the guidance system's coordinates between the estimators and the guidance algorithm. The estimators need the history of the orientation of the coordinate system, while the guidance law consists of multiplying the estimated states by a suitable gain vector. In the guidance law, then, only an instantaneous transformation is required to transform the missile acceleration command to a coordinate system in which the missile is able to interpret.

14.5.2 Continuous versus Discrete Time Models

All the design models including the process and observation noise models are expressed in continuous time. In the implementation, the observations arrive at discrete time instants and are processed by a discrete time Kalman filter. Throughout the remainder of the chapter, the same letter denotes both the continuous time and discrete time noise; the discrete time version is underlined, and the continuous time one is not. When the discrete time models are introduced, the conversion between the discrete time and continuous time noise intensities is presented.

14.5.3 The System Model

The guidance system works in the coordinate system attached to the sensor. If the roll rate of the sensor is sufficiently small, then the guidance problem separates into two uncoupled problems, one for the horizontal plane and one for the vertical plane. This is assumed. In this section, the attention is restricted to the horizontal plane. Hence, all angles and angular rates are around the z-axis, and the missile and target accelerations are in the y direction. In order to simplify the notation, these directions are assumed to be understood and are not explicitly included in the symbols in the equations. A parallel development is possible for the vertical plane. Only the signs in a few of the equations for the vertical and horizontal planes differ. In particular, the signs of the acceleration terms in the vertical plane are opposite to those in the horizontal plane.

CLOS guidance systems generally cannot be designed with sufficiently high gain crossover frequency to make the system stiff enough to derive the missile acceleration commands only from the error signal, which in this case is the angular deviation of the interceptor missile from the sensor boresight.

The guidance system must rely on feed-forward from the target observations in order to close the guidance control loop. The way this is done in the LQG formalism is to include the target model and other disturbance models in the system dynamics. The state variables associated with the target and disturbance models are uncontrollable, but their presence is essential for a successful LQG-based design. The issues arising from the lack of controllability are addressed subsequently. In Section 14.4.1, the kinematics of two bodies moving in a rotating coordinate frame were derived. These equations are repeated here for convenience. The deviation from the boresight of either body, ε_i, follows the differential equation

$$\dot{\varepsilon}_i = -\omega_r + \omega_i - \varepsilon_i \frac{\dot{r}_i}{r_i}. \tag{14.35}$$

Each of the bodies' angular velocities are governed by the differential equation

$$\dot{\omega}_i = -\omega_r \frac{\dot{r}_i}{r_i} - \omega_i \frac{\dot{r}_i}{r_i} + \frac{a_i}{r_i}. \tag{14.36}$$

To describe the kinematics of both the target and the missile, two sets of equations (Equations 14.35 and 14.36) are required, one set for the target, where the subscript i is replaced by T, and a second set for the missile, where the subscript i is replaced by M. It is more convenient to replace the missile states ε_M and ω_M by the difference states, as defined in the following two equations:

$$\varepsilon_d = \varepsilon_M - \varepsilon_T \tag{14.37}$$

and

$$\omega_d = \omega_M - \omega_T \tag{14.38}$$

The difference angle ε_d and the difference angular rate ω_d follow the differential equations

$$\dot{\varepsilon}_d = \omega_d - \varepsilon_d \frac{\dot{r}_M}{r_M} + \varepsilon_T \left(\frac{\dot{r}_T}{r_T} - \frac{\dot{r}_M}{r_M} \right) \tag{14.39}$$

and

$$\dot{\omega}_d = \omega_r \left(\frac{\dot{r}_T}{r_T} - \frac{\dot{r}_M}{r_M} \right) - \omega_d \frac{\dot{r}_M}{r_M} + \omega_T \left(\frac{\dot{r}_T}{r_T} - \frac{\dot{r}_M}{r_M} \right) + \frac{a_M}{r_M} - \frac{a_T}{r_T}. \tag{14.40}$$

After a little algebra, and substituting the definitions of Section 14.4, the following system equation is obtained:

$$
\begin{bmatrix} \dot{x}_c \\ \dot{x}_0 \end{bmatrix} = \begin{bmatrix} A_c & E \\ 0 & A_0 \end{bmatrix} \begin{bmatrix} x_c \\ x_0 \end{bmatrix} + \begin{bmatrix} B_c \\ 0 \end{bmatrix} a_c + \begin{bmatrix} G_c \\ G_0 \end{bmatrix} w
\tag{14.41}
$$

where

$$
w = \begin{bmatrix} w_M \\ w_T \end{bmatrix} .
\tag{14.42}
$$

The components of x_c and x_0 are

$$
x_c = \begin{bmatrix} \varepsilon_d \\ \omega_d \\ x_M \end{bmatrix} ,
\tag{14.43}
$$

$$
x_0 = \begin{bmatrix} x_d \\ \varepsilon_T \\ \omega_T \\ a_T \\ \omega_r \end{bmatrix} .
\tag{14.44}
$$

The entries in the submatrices are

$$
A_c = \begin{bmatrix} -\dfrac{\dot{r}_M}{r_M} & 1 & 0 \\[2mm] 0 & -\dfrac{\dot{r}_M}{r_M} & \dfrac{\tau_M - \tau_{zc}}{\tau_M r_M} \\[2mm] 0 & 0 & -\dfrac{1}{\tau_M} \end{bmatrix}
\tag{14.45}
$$

$$
E = \begin{bmatrix} 0 & \dfrac{\dot{r}_T}{r_T} - \dfrac{\dot{r}_M}{r_M} & 0 & 0 & 0 \\[2mm] \dfrac{1}{r_M} & 0 & \dfrac{\dot{r}_T}{r_T} - \dfrac{\dot{r}_M}{r_M} & -\dfrac{1}{r_T} & \dfrac{\dot{r}_T}{r_T} - \dfrac{\dot{r}_M}{r_M} \\[2mm] 0 & 0 & 0 & 0 & 0 \end{bmatrix}
\tag{14.46}
$$

$$A_0 = \begin{bmatrix} -\dfrac{1}{\tau_d} & 0 & 0 & 0 & 0 \\[2mm] 0 & -\dfrac{\dot{r}_T}{r_T} & 1 & 0 & -1 \\[2mm] 0 & 0 & -\dfrac{\dot{r}_T}{r_T} & \dfrac{1}{r_T} & -\dfrac{\dot{r}_T}{r_T} \\[2mm] 0 & 0 & 0 & -\dfrac{1}{\tau_T} & 0 \\[2mm] 0 & 0 & 0 & \dfrac{1}{r_T} & -2\dfrac{\dot{r}_T}{r_T} \end{bmatrix} \tag{14.47}$$

$$B_c = \begin{bmatrix} 0 \\[2mm] \dfrac{\tau_{zc}}{r_m \tau_M} \\[2mm] \dfrac{1}{\tau_M} \end{bmatrix} \tag{14.48}$$

$$G_c = \begin{bmatrix} 0 & 0 \\[2mm] \dfrac{\tau_d \tau_{zd}}{r_M(\tau_d - \tau_{zd})} & 0 \\[2mm] 0 & 0 \end{bmatrix} \tag{14.49}$$

$$G_0 = \begin{bmatrix} 1 & 0 \\ 0 & 0 \\ 0 & 0 \\ 0 & 1 \\ 0 & 0 \end{bmatrix}. \tag{14.50}$$

It may be observed that

1. The states x_c are controllable, while the states x_0 are uncontrollable.
2. The dynamics of x_0 contain unstable dynamics for an approaching target.
3. The white noise w excites the uncontrollable states.

In order to implement the GL, estimates are required of all eight state variables. As described in Section 14.4, the observation noise for the missile and the target are independent of each other. Further, the process noise exciting the missile is by physical considerations independent of the process noise exciting the missile disturbance states. It is therefore both feasible and advantageous to estimate the missile and the target states in separate estimators. For the purpose of designing the CLOS guidance system, the target states (ε_T, ω_T, and a_T) and the corresponding missile states (ε_M, ω_M, and a_M) are estimated directly, while in the implementation, an estimator expressed in a Cartesian coordinate system is employed. The actual states used in the implementation of the estimators are listed in Tables 14.3 and 14.4.

The variables used in Equation 14.41 are straightforwardly obtained from the estimated states:

$$\hat{\varepsilon}_i = \hat{p}_i / \hat{r}_i, \tag{14.51}$$

$$\hat{\omega}_i = \hat{v}_i / \hat{r}_i, \tag{14.52}$$

where the subscript i may take on the value of M for missile or T for target. Recalling the assumption that the ranges to both the target and the missile are known with negligible error, then the transformations in Equations 14.51 and 14.52 are linear, so that the Cartesian and polar implementations may be considered equivalent.

The missile estimator observes y_M, where

$$y_M = \varepsilon_M + n_M \tag{14.53}$$

TABLE 14.3

Target Estimator State Variables

Variable	Description
p_T	Target displacement from sensor boresight
v_T	Target velocity perpendicular to the sensor boresight
a_T	Target acceleration perpendicular to the sensor boresight

TABLE 14.4

Missile Estimator State Variables

Variable	Description
p_M	Missile displacement from sensor boresight
v_M	Missile velocity perpendicular to the sensor boresight
x_M	Missile first-order approximate transfer function state variable

and the target estimator observes y_T,

$$y_T = \varepsilon_T + n_T \tag{14.54}$$

where n_M and n_T are independent white Gaussian random processes, with autocorrelation function with

$$E\{n_M(t)n_M(t+\tau)\} = \delta(\tau)\sigma_{n_M}^2 \tag{14.55}$$

$$E\{n_T(t)n_T(t+\tau)\} = \delta(\tau)\sigma_{n_T}^2. \tag{14.56}$$

14.5.4 Criterion Function

For the system to operate properly, the missile must be at all times in the sensor's field of view. The sensor's field of view is generally small so that the angular deviation of the missile from the sensor's boresight must remain small at all times if the system is to operate as intended. In addition, at the final time, it is desirable that the miss distance to the target be small. The following criterion function expresses these aims:

$$J = x_c'(T_f)Q_{cf}x_c(T_f) + \int_0^{T_f} \left(x_c'(s)Q_c x_c(s) + a_c^2(s) \right) ds \tag{14.57}$$

where

$$Q_{cf} = \begin{bmatrix} q_{\varepsilon_f} & 0 & 0 \\ 0 & q_{\omega_f} & 0 \\ 0 & 0 & 0 \end{bmatrix} \tag{14.58}$$

$$Q_c = \begin{bmatrix} q_\varepsilon & 0 & 0 \\ 0 & q_\omega & 0 \\ 0 & 0 & 0 \end{bmatrix}. \tag{14.59}$$

Note that only the controllable states are weighted. A GL is sought to minimize the expected value of the criterion $E\{J\}$ in Equation 14.57.

14.5.5 Optimal Control Solution

The optimal solution for minimizing the expected value of J is well known [7]. It is based on the use of the certainty equivalence and separation principles.

The control part of the solution is obtained by solving the Riccati differential equation:

$$\dot{P} = -PA - A'P + PBB'P - Q \tag{14.60}$$

$$P(T_f) = Q_f \tag{14.61}$$

and then the optimal controller is

$$a_c = -B'P\hat{x} \tag{14.62}$$

where $\hat{x}$ is the estimate of the system state vector

$$x = \begin{bmatrix} x_c \\ x_0 \end{bmatrix}. \tag{14.63}$$

The matrices A, B, Q, and Q_f are defined as

$$A = \begin{bmatrix} A_c & E \\ 0 & A_0 \end{bmatrix} \tag{14.64}$$

$$B = \begin{bmatrix} B_c \\ 0 \end{bmatrix} \tag{14.65}$$

$$Q = \begin{bmatrix} Q_c & 0 \\ 0 & 0 \end{bmatrix} \tag{14.66}$$

$$Q_f = \begin{bmatrix} Q_{cf} & 0 \\ 0 & 0 \end{bmatrix}. \tag{14.67}$$

14.5.6 Design Parameters

The design parameters may be divided into two groups: the first group are parameters whose values are determined externally, and the second group are those whose values the designer uses to achieve the required robustness and performance goals. Among the parameters in the first group are the missile and target observation noise and the missile and target ranges. The parameters under the designer's discretion are shown in Table 14.5.

TABLE 14.5

Values of Design Parameters

Symbol	Value	Units	Description
τ_d	2	s	Missile disturbance model: time constant
τ_{zd}	0.1	s	Missile disturbance model: inverse of zero
$\sigma_{x_d^j}$	50	m/s^2	Missile disturbance model: acceleration standard deviation
τ_T	3	s	Target acceleration decorrelation time
σ_{a_T}	50	m/s^2	Target maneuver standard deviation
q_ε	10^6–10^7		Criterion weight on angular deviation ε_d
q_ω	$q_\varepsilon/5$		Criterion weight on angular velocity difference ω_d
$q_{\varepsilon T_f}$	10^8, $r_M \le 2000$ m		Terminal criterion weight on angular deviation $\varepsilon_d(T_f)$
	$10^8\,(r_M/2000)^2$, $r_M > 2000$ m		
$q_{\omega T_f}$	$q_{\varepsilon T_f}/5$		Terminal weight on angular velocity difference $\omega_d(T_f)$

14.5.7 Design Method

While the missile is far from the target, the purpose of the guidance system is to generate missile acceleration commands such that the missile remains within the sensor's field of view in the face of maneuvers by the target. Although the parameters $\dot{r}_M/r_M$ and $\dot{r}_T/r_T$ slowly vary in Equations 14.45 through 14.47, this is essentially a stationary control problem. As such, the usual requirements for linear control design apply: the control loop should be robust to any expected plant variations while minimizing the influence of the disturbance inputs to the guidance loop. There are two main sources of disturbance inputs to the guidance loop. The first source is target maneuvers, and the second one is deviations of missile behavior from its design model. In addition, only noisy observations of the missile and the target are available. The noise must be filtered sufficiently well so that the noise content of the acceleration command sent to the missile does not exceed 10% of missile acceleration command dynamic range.

The simplest and probably the most widely used robustness test is to demand good margins in the Nyquist or Nichols plots. The simplest way to express the guidance loop's ability to reject the disturbance inputs is to calculate its response to step inputs at the target command input and at the missile acceleration command input. Recall that the aim of the guidance loop is to keep the missile inside the sensor's field of view at all times. This can be verified by inspection of the step responses of the angle between the sensor boresight and the missile position.

To examine these properties of the guidance loop, the steady-state solution of Equation 14.60 is required. One way of obtaining the steady-state solution is to integrate Equation 14.60 numerically until the steady-state solution is reached. A simpler but more elegant method is to decompose the solution into a steady-state part and a terminal part and to use the steady-state part in the calculations.

14.5.8 Solution of Riccati Equation

It is convenient to be able to obtain the steady-state solution of the Riccati equation (Equation 14.60) without having to solve it numerically as a differential equation. Fortunately, there is a lemma in Appendix A of [22] that makes possible such a calculation. In [22], the equivalence of the following two problems is shown.

PROBLEM 1
Minimize

$$J_1 = \frac{1}{2}x_f'P_fx_f + \frac{1}{2}\int_0^{T_f}[x'Qx + u'Ru]dt \qquad (14.68)$$

subject to

$$\dot{x} = Ax + Bu. \qquad (14.69)$$

PROBLEM 2
Minimize

$$J_2 = \frac{1}{2}x_f'W_fx_f + \frac{1}{2}\int_0^{T_f}u'Rudt \qquad (14.70)$$

subject to

$$\dot{x} = A_dx + Bu. \qquad (14.71)$$

Then optimization problems 1 and 2 are equivalent if

$$A_d = A - B_cR^{-1}B_c'\bar{P} \qquad (14.72)$$

$$W_f = P_f - \bar{P} \qquad (14.73)$$

and $\bar{P}$ is the solution of the algebraic Riccati equation:

$$A'\bar{P} + \bar{P}A - \bar{P}BR^{-1}B'\bar{P} + Q = 0. \qquad (14.74)$$

In [22], the proof of the lemma shows that

$$P = \bar{P} + W, \tag{14.75}$$

where W is the solution of the Riccati equation associated with Problem 2:

$$\dot{W} + A'_{cl}W + WA_{cl} - WBR^{-1}B'W = 0. \tag{14.76}$$

Thus, the optimal control for Problem 1 is decomposed as

$$
\begin{aligned}
u &= u_{ss} &+& u_{tc} \\
&= -R^{-1}B'\bar{P}x &-& R^{-1}B'Wx.
\end{aligned}
\tag{14.77}
$$

In other words, the optimal control is the sum of two control signals, u_{ss}, a steady-state term, and u_{tc}, a terminal control term. Hence, the use of this lemma makes it possible to write the solution of the optimization in Equation 14.57 subject to the system dynamics Equation 14.41 as the sum of two optimization problems: the first one over an infinite interval and the second one over a finite interval. Note that the second problem has a closed-form solution since there is no weighting of the states in the running cost. Note also that the solution of the terminal cost problem is solved for the closed-loop system obtained from the solution of the steady-state problem.

When the solution of Equation 14.74 is attempted, using one of the standard Riccati equation solvers, one is immediately confronted by a slew of error messages. The difficulty is that the system equations 14.41 contain uncontrollable unstable states. An approach to overcoming such a difficulty is presented in Section 9.6 of [10]. Note that the solution of Equation 14.60 is a well-defined problem with a solution. The difficulty arises from the attempt to convert the problem to the sum of an infinite time and a finite time problem.

To examine the problem more closely, write Equations 14.60 and 14.62 in terms of the submatrices. Then Equation 14.60 becomes

$$\dot{P}_{cc} = -P_{cc}A_c - A'_c P_{cc} + P_{cc}B_c B'_c P_{cc} - Q_c \tag{14.78}$$

$$\dot{P}_{c0} = -P_{cc}E - P_{c0}A_0 - (A'_c - P_{cc}B_c B'_c)P_{c0} \tag{14.79}$$

$$\dot{P}_{00} = -P_{00}A_0 - A'_0 P_{00} - P'_{c0}E - E'P_{c0} + P'_{c0}B_c B'_c P_{c0}. \tag{14.80}$$

The differential equation 14.80 does not have a steady-state solution. The commanded acceleration in terms of the partitioned matrices is obtained by substituting into Equation 14.62:

$$a_c = -B'_c P_{cc}\hat{x}_c - B'_c P_{c0}\hat{x}_0 \tag{14.81}$$

so that P_{00} is not necessary for calculating a_c. Equation 14.78 may be solved using the decomposition lemma, without encountering any difficulties. The differential equation satisfied by W_{cc} is

$$\dot{W}_{cc} + (A'_{cl})_{cc} W_{cc} + W_{cc} (A_{cl})_{cc} - W_{cc} B_c B'_c W_{cc} = 0. \tag{14.82}$$

In particular, the steady-state $(a_c)_{ss}$ part of the acceleration command may be expressed as

$$(a_c)_{ss} = -B'_c \bar{P}_{cc} \hat{x}_c - B'_c \bar{P}_{c0} \hat{x}_0 \tag{14.83}$$

where $\bar{P}_{cc}$ is the positive solution of the 3×3 algebraic Riccati equation:

$$\bar{P}_{cc} A_c + A'_c \bar{P}_{cc} - \bar{P}_{cc} B_c B'_c \bar{P}_{cc} + Q_c = 0 \tag{14.84}$$

and $\bar{P}_{c0}$ is the solution of the linear Sylvester equation:

$$\bar{P}_{cc} E + \bar{P}_{c0} A_0 + (A'_c - \bar{P}_{cc} B_c B'_c) \bar{P}_{c0} = 0. \tag{14.85}$$

This equation can also be written as

$$\bar{P}_{cc} E + \bar{P}_{c0} A_0 + (A'_{cl})_{cc} \bar{P}_{c0} = 0 \tag{14.86}$$

where $(A_{cl})_{cc} = A_c - B_c B'_c \bar{P}_{cc}$ is the closed-loop steady-state system matrix. The decomposition lemma may not be used to decompose Equation 14.79 since the lack of existence of $\bar{P}_{00}$ calls into question the validity of the decomposition lemma when applied to the system equations 14.78 through 14.80. However, substituting $P_{c0} = \bar{P}_{c0} + W_{c0}$, where $\bar{P}_{c0}$ is the solution of Equation 14.86, into Equation 14.79 results in the differential equation for W_{c0}:

$$\dot{W}_{c0} + (A'_{cl})_{cc} W_{c0} + W_{cc} (A_{cl})_{c0} + W_{c0} A_0 - W_{cc} B_c B'_c W_{c0} = 0 \tag{14.87}$$

which happens to be the same equation as would have been obtained from partitioning Equation 14.76.

14.5.9 Summary of Design

The design as described to this point was carried out and is summarized here. A schematic block diagram of the GL in one coordinate is shown in Figure 14.5.

The target and missile measurements are inputs to the target and missile estimators. The rotation rate ω_r of the sensor platform is obtained using

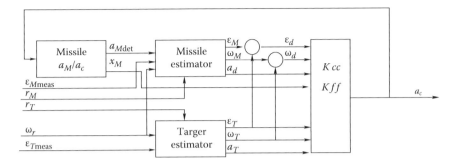

FIGURE 14.5
Schematic block diagram of the CLOS guidance system.

inertial instrumentation attached to the sensor platform and is a further input to the two estimators. The block labeled "Missile a_M/a_c" is an open-loop simulation of the missile using a first-order approximation to the missile transfer function (see Equation 14.16). This block has two outputs. The first block output is the missile acceleration, which is connected to the missile estimator. The block's other output is the missile model internal state, x_M. The variables ε_d and ω_d are formed by the sum blocks (circles) as $\varepsilon_d = \varepsilon_M - \varepsilon_T$ and $\omega_d = \omega_M - \omega_T$, from the outputs of the two estimators. The final block is the one labeled Kcc Kff, where the estimated variables are multiplied by the gains as in Equation 14.62 or Equation 14.81 to obtain the commanded acceleration a_c in the sensor reference frame. The total number of states in the GL is 8, with the missile and target estimators each containing three states, a single state in the missile block, and ω_r, the angular velocity of the sensor platform. Note that the first term on the right-hand side of Equation 14.81 constitutes the feedback term and the feedback of the controllable states, and the second term involving the uncontrollable states constitutes the feed-forward term of the control signal.

In Figure 14.6, the Bode plot of the guidance loop opened at the input of the block labeled MSL in Figure 14.1 is shown.

In Figure 14.7, the step response of the difference between the missile angular position and the target angular position is shown, when the target undergoes a 1 m/s² step acceleration. Assuming that the sensor platform tracks the target with small angular error, this response demonstrates the guidance system's ability to maintain the missile inside the guidance beam when the target maneuvers. Note that the system contains an uncontrollable and unstable pole (from the target kinematics), so that for large values of time, the response becomes unbounded.

Both Figures 14.6 and 14.7 were plotted with all the random inputs in the system model set to zero, but their covariances maintained at their nominal values. The other system parameters may be found in Tables 14.1 and 14.5. The figures were plotted when the missile was at a range of 2500 m and the target at a range of 4500 m from the sensor platform; the missile velocity was 450 m/s, and the target was approaching at a velocity of 300 m/s.

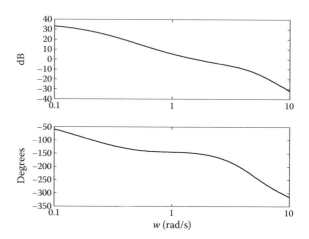

FIGURE 14.6
Bode plot of the open-loop transfer function.

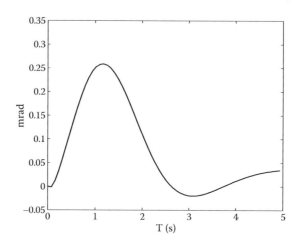

FIGURE 14.7
Step response of the guidance system to 1 m/s² step in target acceleration.

14.6 3D Implementation of Target and Missile Estimators

14.6.1 Generic 3D Estimator

The generic state estimator (GSE) is a 3D tracking filter whose measurements include range, azimuth, and elevation. The filter follows the TSE of [26] and consists of three independent single-axis filters whose time update stage involves rotation of the estimated target state.

The discrete target state equations associated with a single axis in the sensor frame are

$$z^i(k+1) = Fz^i(k) + Gu^i(k) + \Gamma \underline{w}^i(k) \qquad (14.88)$$

where the elements of the state vector z consist of the relative position p^i, the relative velocity v^i, and the inertial acceleration a^i in the ith coordinate of the sensor coordinate system:

$$z^i(k) = [p^i(k) \quad v^i(k) \quad a^i(k)]'; i = x, y, z \qquad (14.89)$$

and $\underline{w}^i(k)$ is a white Gaussian sequence, which is the process noise in the ith coordinate.* The system matrix F and the input matrices G and Γ are specified in Sections 14.6.2 and 14.6.3 for the target and missile estimators, respectively. These matrices differ slightly for the target and missile estimators and are defined subsequently.

The sensor measures the range to the target y^x, the elevation angle, which is a rotation about the sensor y-axis, y^y, and the azimuth angle y^z. Both ε^y and ε^z are measured with respect to the sensor boresight. Assuming small angular deviations during tracking, the measurements taken in the sensor frame are

$$y^x = p^x + \underline{n}^x \qquad (14.90)$$

$$y^y = -p^z/p^x + \underline{n}^y \qquad (14.91)$$

$$y^z = p^y/p^x + \underline{n}^z \qquad (14.92)$$

where $\underline{n}^x$, $\underline{n}^z$, and $\underline{n}^y$ are additive observation noises in the three observations. They are zero mean white Gaussian sequences with variances

$$E\{(\underline{n}^x)^2\} = (\sigma_{n^x})^2$$
$$E\{(\underline{n}^y)^2\} = (\sigma_{n^y})^2 \qquad (14.93)$$
$$E\{(\underline{n}^z)^2\} = (\sigma_{n^z})^2$$

and the relationship to the continuous time versions in Equations 14.55 and 14.56 is

$$\sigma_{n^i}^2/T_s = \sigma_{n^i}^2; i = x, y, z \qquad (14.94)$$

---

* Recall that the same symbol, w, is used to represent both the continuous time process noised as used in Section 14.5 and the discrete time sequence used here. Whenever the discrete sequence is meant, the symbol is underlined.

where T_s is the sampling time, and n^i represents either n_M^i or n_T^i. Note that for a reasonably operating CLOS guidance system, $|p^y| \ll p^x$ and $|p^z| \ll p^x$, so that y^x is a very good approximation of the range to the observed object. Alternatively, the observations may be expressed as

$$y^i = h^i z^i + \underline{n}^i \tag{14.95}$$

where

$$h^x = [1 \quad 0 \quad 0] \tag{14.96}$$

$$h^y = [0 \quad 0 \quad -1/p^x] \tag{14.97}$$

and

$$h^z = [0 \quad 1/p^x \quad 0]. \tag{14.98}$$

The notation is simplified if the estimator state is arranged in a matrix Z

$$Z = \begin{bmatrix} p & v & a \end{bmatrix} = \begin{bmatrix} p^x & v^x & a^x \\ p^y & v^y & a^y \\ p^z & v^z & a^z \end{bmatrix}. \tag{14.99}$$

That is, the state of each elemental coordinate filter is a row of the matrix Z. The time update is easily expressed in this compact notation:

$$\hat{Z}_{k|k-1} = C_{k,k-1} \hat{Z}_{k-1|k-1} F' + U_{k-1} G' \tag{14.100}$$

where $C_{k,k-1}$ is the rotation matrix from the previous position (instantaneously frozen) sensor platform coordinates to their present position:

$$C_{k,k-1} \approx I - \Omega_r T_s = \begin{bmatrix} 1 & T_s \omega_r^z & -T_s \omega_r^y \\ -T_s \omega_r^z & 1 & T_s \omega_r^x \\ T_s \omega_r^y & -T_s \omega_r^x & 1 \end{bmatrix} \tag{14.101}$$

where Ω_r is defined in Equation 14.2, and U_k is the matrix of deterministic inputs to the estimator:

$$U_k = \begin{bmatrix} (u^x) \\ (u^y) \\ (u^z) \end{bmatrix}. \tag{14.102}$$

In the approximation used in the estimator, the filters in the three coordinates are decoupled (see [2] and [26]). Only the 3×3 covariance matrices in each of the three coordinate directions comprising p^i, v^i a^i are calculated. This approximation would be exact if the sensor coordinate did not rotate, since both the process and measurement noises are independent from one coordinate direction to a different coordinate direction. Here it is assumed that the sensor rotation rate is sufficiently small to justify the approximation (see [2] and [26]). Each of the 3×3 covariance blocks of the ith coordinate is propagated according to

$$P^i_{k|k-1} = FP^i_{k-1|k-1}F' + Q^i. \tag{14.103}$$

The equations used to update the estimates are presented next. First the covariance of the innovation S^i is calculated as

$$S^i = h^i P^i (h^i)' + (\sigma_{n^i})^2 \tag{14.104}$$

then the Kalman gain L^i

$$L^i = P^i (h^i)'/S^i \tag{14.105}$$

and finally, the state is updated:

$$z^i_{k|k} = z^i_{k|k-1} + L^i (y^i - h^i z^i_{k|k-1}). \tag{14.106}$$

When $i = y$ or $i = z$ in Equation 14.106, then the estimated value of p^i and $\hat{p}^x_{k|k-1}$ is substituted in h^i. Finally the covariance is updated to reflect the observation

$$P^i_{k|k} = (I - L^i_k h^i_k) P^i_{k|k-1} (I - L^i_k h^i_k)' + L^i_k (\sigma_{n^i})^2 (L^i_k)'. \tag{14.107}$$

14.6.2 Target Estimator

The target acceleration model is the shaping filter associated with the Singer model [24], as defined in Section 14.4.4. Here the same model is assumed in all three coordinates. This is equivalent to assuming that the target is able to accelerate in any direction equally well. If it is known with certainty that this is not the case, then it is possible to include the target's preferential accelera- tion direction (see [6]). On the other hand, if the target does accelerate in a direction where the target acceleration is assumed to be zero, then there is a danger that the filter will diverge. It is then better to use an isotropic model unless there is 100% certainty that the actual target conforms to the assumed model.

Equation 14.100 neatly takes care of the effects due to the rotation of the coordinate system attached to the sensor platform. For the purpose of developing the Kalman filter equations, it is sufficient to consider the prob- lem in stationary coordinates but to use Equation 14.100 for the time update step.

Using the isotropic model in continuous time for the ith coordinate, the target estimator dynamics are

$$
\begin{bmatrix} \dot{p}_T^i \\ \dot{v}_T^i \\ \dot{a}_T^i \end{bmatrix} = \begin{bmatrix} 0 & 1 & 0 \\ 0 & 0 & 1 \\ 0 & 0 & -1/\tau_T \end{bmatrix} \begin{bmatrix} p_T^i \\ v_T^i \\ a_T^i \end{bmatrix} + \begin{bmatrix} 0 \\ 0 \\ 1 \end{bmatrix} w_T^i. \tag{14.108}
$$

There are no deterministic inputs for the TSE. The observation equa- tion conforms to the observation model described in the generic estimator (Equation 14.95).

An exact discretization of Equation 14.108 results in

$$
z_T^i(k+1) = F_T z_T^i(k) + \tilde{w}_T^i \tag{14.109}
$$

where $\tilde{w}_T^i$ is a discrete time white noise sequence, and

$$
F_T = \begin{bmatrix} 1 & T_s & (T_s/\tau_T - 1 + e^{-T_s/\tau_T})\tau_T^2 \\ 0 & 1 & (1 - e^{-T_s/\tau_T})\tau_T \\ 0 & 0 & e^{-T_s/\tau_T} \end{bmatrix} \approx \begin{bmatrix} 1 & T_s & T_s^2/2 \\ 0 & 1 & T_s \\ 0 & 0 & e^{-T_s/\tau_T} \end{bmatrix}. \tag{14.110}
$$

In the rest of this chapter, the approximate form of F_T is used. The exact expression for the covariance of $\tilde{w}_T^i$ is a bit complicated and is not repro- duced here. It may be found in [3, p. 323]. There are a number of direct dis- crete time models possible, which avoid the cumbersome discretization of Equation 14.108. Both of these models are detailed in [3, near p. 272]. The first model replaces the continuous noise $w_T^i(t)$ by a piecewise constant signal, with the signal being constant between sampling times. The constant values of the signal are given by a random sequence $\underline{w}_T^i(k)$. The statistics of $\underline{w}_T^i(k)$ are chosen so as to retain the statistical properties of $a_T^i(k)$:

$$z_T^i(k+1) = F_T z_T^i(k) + \Gamma \underline{w}_T^i(k) \tag{14.111}$$

where

$$\Gamma = \begin{bmatrix} \tau_T^3 - \tau_T^2 T_s + \dfrac{1}{2}\tau_T T_s^2 - \tau_T^3 e^{-\frac{T_s}{\tau_T}} \\ \tau_T^2 e^{-\frac{T_s}{\tau_T}} + \tau_T T_s - \tau_T^2 \\ -\tau_T e^{-\frac{T_s}{\tau_T}} + \tau_T \end{bmatrix} \approx \begin{bmatrix} T_s^3/6 \\ T_s^2/2 \\ T_s \end{bmatrix}. \tag{14.112}$$

For a given target acceleration level $\sigma_{a_T}^2$ assumed to be identical in all sensor directions, the spectral density of $\underline{w}_T^i(k)$ is σ_T^2, where

$$
\begin{aligned}
\sigma_T^2 &= \frac{1-e^{-\frac{2T_s}{\tau_T}}}{\tau_T^2\left(1-e^{-\frac{T_s}{\tau_T}}\right)^2}\sigma_{a_T}^2 \\[2ex]
&\approx \frac{2}{\tau_T T_s}\sigma_{a_T}^2 \qquad \text{if } \frac{T_s}{\tau_T} \ll 1
\end{aligned}
\tag{14.113}
$$

A second type of direct discrete time model is also possible. In the second type of model, the white noise sequence represents the increment in the target acceleration at each sampling interval. The increments are assumed to occur at each sampling time. This type of model is not pursued further. All the stochastic models here are based on the piecewise constant model.

The implementation of the TSE is shown in Figure 14.8. Note that the sensor platform acceleration was assumed to be zero.

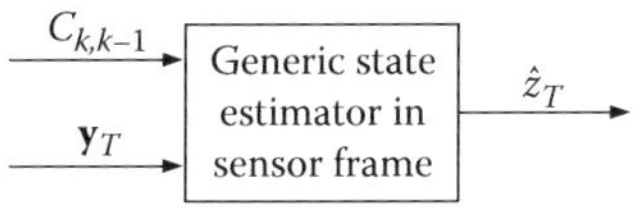

FIGURE 14.8
Implementation of TSE using GSF

14.6.3 Missile State Estimator

In contrast to the target acceleration, the missile acceleration is approximately known since the acceleration commands are generated within the guidance algorithm. On the other hand, it is not possible to entirely rely on the missile acceleration command to calculate the acceleration since this would ignore any deviations between the model used to calculate the missile acceleration and the missile's actual possibly nonlinear characteristics, as well as external influences acting on the missile. The model chosen for the MSE combines an approximate model of the missile transfer function to calculate an approximation to the missile acceleration and a stochastic model for the difference in the missile acceleration calculated by the approximate model and the actual acceleration. The modeling of the deviations of the missile acceleration from the model is crucial for satisfactory performance of the system. This model also affects the robustness of the guidance loop; hence, the parameters of the model must be chosen with a view to satisfying the robustness requirements.

As in the TSE, the rotation of the coordinate axes is conveniently taken care of in the time update equation (Equation 14.100). Then for the purposes of developing the MSE, it is sufficient to consider the problem in inertial coordinates but to rely on Equation 14.100 for the time update step.

As in the TSE, it is assumed that the disturbance acceleration model is isotropic. In continuous time, the missile estimator model in each coordinate is

$$
\begin{bmatrix} \dot{p}_M^i \\ \dot{v}_M^i \\ \dot{x}_d^i \end{bmatrix} = \begin{bmatrix} 0 & 1 & 0 \\ 0 & 0 & 1 \\ 0 & 0 & -1/\tau_d \end{bmatrix} \begin{bmatrix} p_M^i \\ v_M^i \\ x_d^i \end{bmatrix} + \begin{bmatrix} 0 \\ 1 \\ 0 \end{bmatrix} a_{Mdet}^i + \begin{bmatrix} 0 \\ \tau_d \tau_{zd} \\ \tau_d - \tau_{zd} \\ 1 \end{bmatrix} w_M^i. \quad (14.114)
$$

Here the state x_d^i represents the low-frequency deviation of the missile acceleration from the acceleration a_{Mdet} calculated by the missile deterministic acceleration model in the sensor coordinates. The continuous time white noise w_M^i plays a role similar to the continuous time noise w_T^i in the target estimator. The deterministic missile acceleration is calculated according to the model in Section 14.4.2, while taking note of the required coordinate transformation between the sensor coordinates and the missile body coordinates, where Equation 14.16 is valid. The calculation of a_{Mdet}^i is presented toward the end of Section 14.7.

The discretization procedure follows the piecewise constant type discretization of the TSE in Section 14.6.2. The discrete time estimator design model is

$$z_M^i(k+1) = F_M z_M^i(k) + G_M a_{Mdet}^i + \Gamma_M \underline{w}_M^i(k) \tag{14.115}$$

where

$$F_M = \begin{bmatrix} 1 & T_s & (T_s/\tau_M - 1 + e^{-T_s/\tau_M})\tau_M^2 \\ 0 & 1 & (1 - e^{-T_s/\tau_M})\tau_M \\ 0 & 0 & e^{-T_s/\tau_M} \end{bmatrix} \approx \begin{bmatrix} 1 & T_s & T_s^2/2 \\ 0 & 1 & T_s \\ 0 & 0 & e^{-T_s/\tau_M} \end{bmatrix}, \tag{14.116}$$

$$\Gamma_M = \begin{bmatrix} \tau_M^3 - \tau_M^2 T_s + \dfrac{1}{2}\tau_M T_s^2 - \tau_M^3 e^{-\frac{T_s}{\tau_M}} \\ \tau_M^2 e^{-\frac{T_s}{\tau_M}} + \tau_M T_s - \tau_M^2 \\ -\tau_M e^{-\frac{T_s}{\tau_M}} + \tau_M \end{bmatrix} \approx \begin{bmatrix} T_s^3/6 \\ T_s^2/2 \\ T_s \end{bmatrix} \tag{14.117}$$

and

$$G_M = \begin{bmatrix} T_s^2/2 \\ T_s \\ 0 \end{bmatrix}. \tag{14.118}$$

As was the case for the TSE, the approximate forms for the matrices F_M, Γ_M, and G_M are used elsewhere within the chapter. The missile acceleration due to the stochastic input contains a white noise term; however, it is possible to calculate the noise intensity required for a given variance $(\sigma_{x_d^i})^2$ of x_d^i:

$$\sigma_M^2 = \frac{1 - e^{-\frac{2T_s}{\tau_M}}}{\tau_M^2 \left(1 - e^{-\frac{T_s}{\tau_M}}\right)^2} \sigma_{x_d^i}^2 \qquad \forall i$$

$$\approx \frac{2}{\tau_M T_s} \sigma_{x_d^i}^2 \qquad \text{if } \frac{T_s}{\tau_M} \ll 1 \tag{14.119}$$

where σ_M^2 is the spectral density of the stochastic excitation $w_M^i(k)$, for all i.

The implementation of the MSE is shown in Figure 14.9, where it is also assumed that the sensor acceleration is zero. The calculation of the deterministic acceleration a_{Mdet} is described in Section 14.7.

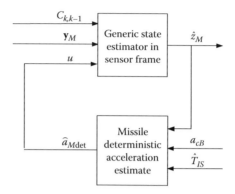

FIGURE 14.9
Implementation of MSE using GSE.

14.7 Real-Time Implementation of the Guidance Law

In this section, several issues that are required to implement the GL in its full 3D version are described.

14.7.1 Guidance Gain Calculation

For the purposes of this calculation, it is assumed that the parameters $\dot{r}_M/r_M$ and $\dot{r}_T/r_T$ may be treated as sufficiently slowly varying as to be constant.

The most straightforward way of calculating the guidance gains required in Equation 14.62 is to integrate Equation 14.60 backward from the terminal time to the present time. This requires the solution of the matrix differential equation 14.60, which contains 32 unknowns. Hence, the computational load is equivalent to solving a 32nd-order nonlinear differential equation at each step. In this section, an alternative approach is explored.

In Section 14.5.8, it was shown that the acceleration command can be decomposed into the sum of a steady state and a time-varying terminal solution. The steady-state part $(a_c)_{ss}$ was shown in Equations 14.83 through 14.85. Hence, the steady-state part of the acceleration command requires the solution of a 3×3 algebraic Riccati equation (Equation 14.84) and a 3×5 linear Sylvester equation (Equation 14.85). The simplest method for solving the steady-state algebraic Riccati equation is described in [16], and a more modern one may be found in [1]. Since the solution of the Riccati equation is required at each time step, the iterative algorithm of [16] has the advantage of being able to use the computation of the solution of the Riccati equation at a previous time step.

The solution of the terminal control part is

$$(a_c)_{tc} = -B'W\hat{x} \tag{14.120}$$

where W is the solution of the matrix Riccati differential equation (Equation 14.76). This matrix differential equation has an analytically expressible solution [21]:

$$W = e^{A_d'(T_f-t)}W_f(I+GrW_f)^{-1}e^{A_d(T_f-t)} \tag{14.121}$$

where

$$Gr = \int_t^{T_f} e^{A_d(T_f-\lambda)}BB'e^{A_d'(T_f-\lambda)}d\lambda \tag{14.122}$$

is called the Grammian. At this point, Equation 14.121 must be considered as a solution in the formal sense only because P_{00} and hence W_{f00} do not exist. If the formal manipulations are carried out, and all the matrices in Equations 14.121 and 14.122 are partitioned as in Equations 14.64 through 14.67, then

$$Gr = \begin{bmatrix} Gr_{cc} & 0 \\ 0 & 0 \end{bmatrix} \tag{14.123}$$

$$Gr_{cc} = \int_t^{T_f} (e^{A_d(T_f-\lambda)})_{cc} B_c B_c' (e^{A_d'(T_f-\lambda)})_{cc} d\lambda. \tag{14.124}$$

Expanding Equation 14.120, it is observed that a_c is expressed as

$$(a_c)_{tc} = -B_c(e^{A_d'(T_f-t)})_{cc} \begin{bmatrix} H_c & H_0 \end{bmatrix} \begin{bmatrix} (e^{A_d(T_f-t)})_{cc} & (e^{A_d(T_f-t)})_{c0} \\ 0 & (e^{A_d(T_f-t)})_{00} \end{bmatrix} \begin{bmatrix} \hat{x}_c \\ \hat{x}_0 \end{bmatrix} \tag{14.125}$$

where

$$H_c = W_{fcc}(I + Gr_{cc}W_{fcc})^{-1} \tag{14.126}$$

$$H_0 = -W_{fcc}(I + Gr_{cc}W_{fcc})^{-1}Gr_{cc}W_{fc0} + W_{fc0}. \tag{14.127}$$

It is observed that the solution 14.125 does not require W_{f00}, and in fact, only the submatrices W_{cc} and W_{c0} of the solution to Equation 14.76 are required.

To show that Equation 14.125 is actually a solution of the terminal guidance problem, it is necessary to show that

$$W_{cc} = (e^{A_{cl}(T_f-t)})_{cc} H_c (e^{A_{cl}(T_f-t)})_{cc} \tag{14.128}$$

and

$$W_{c0} = (e^{A'_{cl}(T_f-t)})_{cc} H_c (e^{A_{cl}(T_f-t)})_{c0} + (e^{A'_{cl}(T_f-t)})_{cc} H_0 (e^{A_{cl}(T_f-t)})_{00} \tag{14.129}$$

solve the differential equations 14.82 and 14.87. The simplest method is to substitute back into the differential equations and verify the solutions. This is straightforward but tedious. Two key intermediate results are

$$\dot{H}_c = H_c (e^{A_{cl}(T_f-t)})_{cc} B_c B'_c (e^{A'_{cl}(T_f-t)})_{cc} H_c \tag{14.130}$$

and

$$\dot{H}_0 = H_c (e^{A_{cl}(T_f-t)})_{cc} B_c B'_c (e^{A'_{cl}(T_f-t)})_{cc} H_0. \tag{14.131}$$

If W_{fcc} is invertible, then Equation 14.125 can be rewritten as the product of a time-varying matrix gain times a zero-effort miss like term

$$(a_c)_{tc} = -B'_c (e^{A'_{cl}(T_f-t)})_{cc} H_c \widehat{Zem} \tag{14.132}$$

where $\widehat{Zem}$ is a three-component vector:

$$\widehat{Zem} = \begin{bmatrix} I & W_{fcc}^{-1} W_{c0} \end{bmatrix} \begin{bmatrix} (e^{A_{cl}(T_f-t)})_{cc} & (e^{A_{cl}(T_f-t)})_{c0} \\ 0 & (e^{A_{cl}(T_f-t)})_{00} \end{bmatrix} \begin{bmatrix} \hat{x}_c \\ \hat{x}_0 \end{bmatrix}. \tag{14.133}$$

The major task in evaluating $(a_c)_{tc}$ is the calculation of Gr_{cc}, $(e^{A_{cl}(T_f-t)})_{cc}$ $(e^{A_{cl}(T_f-t)})_{c0}$, and $(e^{A_{cl}(T_f-t)})_{00}$. The easiest way to calculate the latter three is by the use of a good symbolic algebra program, using the Laplace transform:

$$(e^{A_{cl}t})_{cc} = \mathcal{L}^{-1}\left[(sI - (A_{cl})_{cc})^{-1}\right] \tag{14.134}$$

and

$$(e^{A_{cl}t})_{c0} = \mathcal{L}^{-1}\left[(sI - (A_{cl})_{cc})^{-1}(A_{cl})_{c0}(sI - A_0)^{-1}\right] \tag{14.135}$$

$$(e^{A_{cl}t})_{00} = \mathcal{L}^{-1}\left((sI - A_0)^{-1}\right) \tag{14.136}$$

where A_{cl} is the closed-loop system matrix. It is partitioned as

$$A_{cl} = \begin{bmatrix} (A_{cl})_{cc} & (A_{cl})_{c0} \\ 0 & (A_{cl})_{00} \end{bmatrix} \tag{14.137}$$

where

$$(A_{cl})_{cc} = A_c - B_c B_c' \bar{P}_{cc} \tag{14.138}$$

$$(A_{cl})_{c0} = E - B_c B_c' \bar{P}_{c0} \tag{14.139}$$

$$(A_{cl})_{00} = A_0. \tag{14.140}$$

Some of the better symbolic algebra programs are even capable of outputting source code in a high-level language such as C. For the inversion of the first two Laplace transforms, the eigenvalues of $(A_{cl})_{cc}$ are required. This involves the solution of a cubic equation, or alternatively, some of the methods of calculating the solution of the 3×3 algebraic Riccati equation calculate the eigenvalues of the closed-loop system as a byproduct (see for example [1]). The eigenvalues of A_0 may be obtained by inspection. Other methods for evaluating the matrix exponential may be found in [17] and [18].

The evaluation of $(Gr)_{cc}$ may be carried out either by numerically evaluating the integral in Equation 14.124 or by converting the problem to the solution of a differential equation:

$$\dot{Gr}_{cc} = (A_{cl})_{cc} Gr_{cc} + Gr(A_{cl})'_{cc} + B_c B_c', \quad Gr_{cc}(T_f) = 0. \tag{14.141}$$

Arranging the entries of Gr_{cc} in a column vector then converts the problem to the calculation of an exponential matrix (see [17] and [18] for the possible algorithms).

Another possibility for calculating Gr_{cc} is to use the Laplace transform plus symbolic algebra program to solve the differential equation resulting from arranging the elements of Gr_{cc} in a column vector.

14.7.2 Transformation from Estimator States to System Model States

The system model used to derive the guidance commands was defined in Equations 14.41 through 14.44 and has a mixture of polar and Cartesian variables. On the other hand, the estimator states are exclusively Cartesian, as in Equations 14.88 and 14.89.

The transformation from the Cartesian estimated states to the angular deviations from the boresight follow their definitions in Equation 14.6, with the required adjustment of the signs in the two axes:

$$\hat{\varepsilon}_i^y = -\hat{p}_i^z / \hat{p}_i^x \tag{14.142}$$

and

$$\hat{\varepsilon}_i^z = \hat{p}_i^y / \hat{p}_i^x, \tag{14.143}$$

where i may be M or T for missile and target. Similarly the angular rate states are derived from Equation 14.8:

$$\hat{\omega}_i^y = -\hat{v}_i^z / \hat{p}_i^x \tag{14.144}$$

and

$$\hat{\omega}_i^z = -\hat{v}_i^y / \hat{p}_i^x. \tag{14.145}$$

Also $\hat{\varepsilon}_d^j$ and $\hat{\omega}_d^j$ are defined by

$$\hat{\varepsilon}_d^j = \hat{\varepsilon}_M^j - \hat{\varepsilon}_T^j; \, j = y, z \tag{14.146}$$

and

$$\hat{\omega}_d^j = \hat{\omega}_M^j - \hat{\omega}_T^j; \, j = y, z. \tag{14.147}$$

The missile acceleration state, $\hat{x}_M^j, j = y, z$, is obtained by inputting into the missile dynamic model (Equation 14.16) the acceleration commands sent to the missile and propagating the differential equation.

The measured value of the platform's angular velocity, ω_r, is used directly in the guidance algorithm.

14.7.3 Transformation of Missile Acceleration Commands to Missile Body Coordinates

The acceleration commands calculated by the guidance system have so far been in the sensor coordinate system. Here, a transformation of the acceleration command is calculated such that when the acceleration commands are carried out by the missile autopilot perpendicular to the missile body, the resulting missile acceleration perpendicular to the sensor boresight is as intended by the guidance system.

Recall that T_{WS} is the coordinate transformation from the sensor platform to the missile wind axes and that T_{WB} is the transformation from the missile body axes to the missile wind axes. Then

$$
\begin{bmatrix} a_M^x \\ a_{cB}^y \\ a_{cB}^z \end{bmatrix} = T_{BW} T_{WS} \begin{bmatrix} a_c^x \\ a_c^y \\ a_c^z \end{bmatrix}
\tag{14.148}
$$

and hence

$$
T_{WS} \begin{bmatrix} a_c^x \\ a_c^y \\ a_c^z \end{bmatrix} = T_{WB} \begin{bmatrix} a_M^x \\ a_{cB}^y \\ a_{cB}^z \end{bmatrix}.
\tag{14.149}
$$

The transformation T_{WB} is calculated under the assumption that the missile acceleration commands are actually performed. If this is not possible due to acceleration constraints, then the command is scaled in both axes so that the new command is within the capabilities of the missile:

$$
T_{WB} = \begin{bmatrix} 1 & -\beta_c & \alpha_c \\ \beta_c & 1 & 0 \\ -\alpha_c & 0 & 1 \end{bmatrix}.
\tag{14.150}
$$

The angles of attack α_c and β_c are calculated using the relations

$$
\alpha_c = K_a a_{cB}^z
\tag{14.151}
$$

and

$$
\beta_c = -K_a a_{cB}^y
\tag{14.152}
$$

where

$$
K_a = \frac{m}{Q_{pres} S_{ref}} \frac{C_{m\delta}}{C_{z\alpha} C_{m\delta} - C_{z\delta} C_{m\alpha}}
\tag{14.153}
$$

where Q_{proc} is the dynamic pressure, and S_{ref} is the missile cross-section reference area. The coefficient K_a was obtained by assuming trim conditions at the commanded acceleration (note: $K_a < 0$). Using Equations 14.150 through 14.152, Equation 14.149 can be rewritten as

$$T_{WS} \begin{bmatrix} a_c^x \\ a_c^y \\ a_c^z \end{bmatrix} = \begin{bmatrix} 1 & K_a a_{cB}^y & K_a a_{cB}^z \\ -K_a a_{cB}^y & 1 & 0 \\ -K_a a_{cB}^z & 0 & 1 \end{bmatrix} \begin{bmatrix} a_M^x \\ a_{cB}^y \\ a_{cB}^z \end{bmatrix}. \tag{14.154}$$

The subscript c in α_c and β_c indicates that these angles of attack were calculated from the commanded accelerations. In Equation 14.154, the unknowns are a_{cB}^y, a_{cB}^z, and a_c^x, while the known variables are a_M^x, a_c^y, and a_c^z. Actually, a_c^x, the missile acceleration in the boresight direction, is not used but is part of the solution. The accelerations a_{cB}^y and a_{cB}^z of Equation 14.154 may be solved to obtain

$$a_{cB}^y = \frac{T_{WS}^{21} a_c^x + T_{WS}^{22} a_c^y + T_{WS}^{23} a_c^z}{1 - K_a a_M^x} \tag{14.155}$$

and

$$a_{cB}^z = \frac{T_{WS}^{31} a_c^x + T_{WS}^{32} a_c^y + T_{WS}^{33} a_c^z}{1 - K_a a_M^x}. \tag{14.156}$$

Substituting Equations 14.155 and 14.156 into the first equation of Equation 14.154, a single quadratic equation in a_c^x is obtained:

$$c_0 (a_c^x)^2 + c_1 a_c^x + c_2 = 0. \tag{14.157}$$

The coefficients c_i depend on the entries in the transformation T_{WS} and known values of the variables a_c^y, a_c^z, and a_M^x The following is noted:

1. Equation 14.157 is a quadratic equation. It may have two real roots, no real roots, or a double real root.
2. If there are two real roots, then the one with the most positive value is the desired one since a_c^x is the missile acceleration along the boresight toward the target, and it is desirable to have the missile velocity as large as possible.
3. If there are no real roots, this indicates that there is no value of the lateral missile acceleration that when combined with the missile axial acceleration yields the required acceleration perpendicular to the boresight. In this case, the best approximation to the required acceleration is obtained by discarding the imaginary part of the solution and using the remaining real part.

4. Equations 14.154 and 14.157 take into account the missile axial acceleration, given by Equation 14.13. Observe that a_M^x is independent of the missile lateral acceleration.

5. The actual missile acceleration command is obtained by substituting the solution, a_c^x, into Equations 14.155 and 14.156.

6. Note that in Equation 14.153, the dynamic pressure appears in the denominator, so that there is an inverse relationship between the dynamic pressure and K_a. Also, in Equations 14.155 and 14.156, the coefficient K_a multiplies a_M^x, so that the missile axial acceleration has the most influence when the dynamic pressure is low, for example, immediately after the missile launch.

14.7.4 Gravity Compensation

Gravity, which has so far been neglected, in fact acts on the missile. It is better to anticipate its effects rather than allowing it to cause the missile to deviate from its flight along the boresight and rely on the resulting error signal to generate the required acceleration command.

There are two places where gravity needs to be considered in the guidance scheme: First, a gravity term, $-T_{SI}g$, is added to the acceleration command calculated in the sensor frame before being transmitted to the missile; and second, a gravity term, $T_{SI}g$, is added to the calculated deterministic missile acceleration before inputting it to the missile estimator (see Figure 14.11). The symbol g denotes the gravity vector in inertial space, $g = [0\ 0\ g]'$. Note that the input to the calculation of the missile deterministic acceleration *includes* gravity.

14.7.5 Sensor Frame versus Missile Body Frame Calculations

The missile acceleration commands calculated so far are in the sensor platform coordinate system; however, in reality, the acceleration commands sent to the missile are carried out in the missile lateral (y and z) directions. The model used to describe the missile lateral acceleration transfer function (Equation 14.16) may be considered in the missile body coordinates or in the missile wind coordinates. Here it is convenient to regard the equation as being in the wind coordinates. On the other hand, in the kinematic differential equations 14.39 and 14.40, the a_M term is expressed in the sensor (S) coordinates. Hence, a more exact depiction of the dynamics would add the transformation T_{WS}' between the missile dynamics (Equation 14.16) and the kinematics (Equations 14.39 and 14.40). Since the transformation T_{WS}' varies much more slowly than the missile acceleration transfer function, Equation 14.16, the transformation T_{WS}', with a small approximation, may be applied to the missile acceleration command, rather than at the input to the kinematics. This interchange of the transformation and the missile dynamics is shown in Figure 14.10.

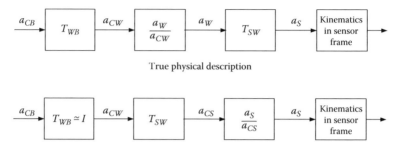

True physical description

Approximation of the true physical description assuming that $T_{WB} \approx I$ and that T_{SW} varies much more slowly than the missile acceleration transfer function

Model used in the derivation of the guidance commands

FIGURE 14.10
Interchange of missile transfer function and T_{SW}.

Once this interchange is carried out, the inputs to the missile transfer function are $T'_{WS}a^i_{cW}$, $i = y, z$. The guidance algorithm in fact calculates a_c. Since a_c is expressed in the sensor coordinates, an algorithm is required that transforms the missile acceleration commands to acceleration commands in the missile's lateral body coordinates. This algorithm was presented in Section 14.3.

14.7.6 Guidance System Block Diagram

The complete guidance system block diagram implemented in three dimensions, including gravity compensation, is shown in Figure 14.11.

In Table 14.6, the major blocks of the guidance diagram and the sections where the details of the algorithm were presented are shown.

In addition, several transformation matrices are required. The calculation of these and selected blocks are detailed in the following.

1. Transformation T_{WS}

   The transformation T_{WS} is the transformation from the sensor frame to the missile wind frame. It is calculated from the estimated missile velocity and the assumption that the missile roll is at its nominal value, namely, that the y-axis is horizontal (assuming that the missile velocity is well away from the vertical).

2. Block a_{Mdet}/a_c

   The block is a very simplified representation of the missile lateral transfer function (Equation 14.16) implemented twice, once for the y and once for the z-axis.

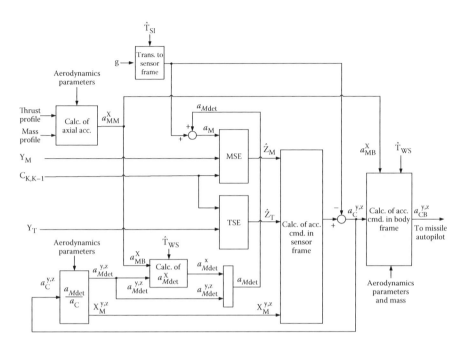

FIGURE 14.11
Basic 3D implementation of GL.

3. Calculation of Missile Axial Acceleration

The missile axial acceleration is calculated according to Equation 14.13.

4. Calculation of a^x_{MSdet}

The calculation is the same as the one presented in Section 14.3 but with the acceleration commands a^i_c replaced by the missile deterministic accelerations a^i_{Mdet}. The acceleration in the sensor x direction is the root of the quadratic equation 14.157.

TABLE 14.6

Guidance Block Diagram Details

Block	Equation	Section
$\dfrac{a^i_{Mdet}}{a^i_c}$	14.16	14.4.2
Calc. of a^x_{Mdet}		14.7.3
Calc. of acc. cmd. in sensor frame	14.81	14.5.5
MSE		14.6.3
TSE		14.6.2
Calc. of acc. cmd. in body frame		14.7.3
Calc. of axial acc.	14.13	14.4.2

14.8 Performance

14.8.1 Background

The CLOS system performance was evaluated using a six degrees of freedom (6DOF) simulation. The simulation assumes that the CLOS system is composed of a missile launcher and fire control radar. The missile launcher and the fire control radar are both mounted on a single gimbaled platform that rotates in both elevation and azimuth to track the desired target. The main radar antenna is 1.5 m in diameter and has a beam width of about 0.8°. There is also a smaller radar antenna of 0.3 m diameter with a beam width of about 4°. The two antennas share a common boresight. The smaller antenna is used to track the missile during the first 1–2 seconds of its flight. The missile's launch velocity is generally too small for sufficient dynamic pressure for effective flight control. To overcome this initial low dynamic pressure, the missile is launched with an elevation angle of 5°–7° above the boresight of the radar antennas. It is assumed that the missile carries a beacon, which is tracked by the radar system.

14.8.2 Simulation Description

The central element of the simulation is the 6DOF missile simulation. The aerodynamics and mass properties used were described in Section 14.2. The missile simulation consists of the complete 6DOF dynamics and a simplified three-loop autopilot with an ideal servo (one with a unity transfer function). The missile maneuverability was bounded by limiting both the angle of attack and the deflection angle of the fins at 25°. The missile time constant varied according to the dynamic pressure from about 500 ms at low dynamic pressure to about 125 ms at high dynamic pressure. The missile velocity varied in accordance with the maneuvers carried out by the missile along its flight. An extremely simplified roll control dynamics was assumed in the simulation:

$$\omega_B^x = -k_r \phi \tag{14.158}$$

where K_r is a constant and ϕ is the missile roll angle. A typical missile velocity profile is shown in Figure 14.12, when the missile is launched at $t = 3$ seconds with an initial velocity of about 30 m/s. Here the missile tip-off as the missile leaves the launcher was ignored, although the tip-off may have a substantial effect on the missile flight path during the first 1–2 seconds of flight.

The radar noise is modeled by white Gaussian noise according to the formulas in Section 14.6. The sampling interval used was 25 ms. The radar observation noise was the sum of the thermal (Equation 14.28) and glint noise (Equation 14.30). In addition, the standard deviation of the noise was lower bounded at 1/50 of the beam width (Equation 14.29) of each antenna.

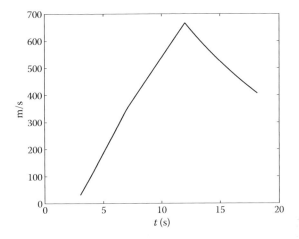

FIGURE 14.12
Missile velocity.

The radar antenna was assumed to be mounted on an outer azimuth and an inner elevation gimbal. The base of the gimbals was assumed to be stationary. Each gimbal angle was assumed driven by a simplified second-order control loop, and the control loops were driven by the radar observations. To achieve good tracking performance, the gimbal control loop included both feed-forward and feedback terms.

14.8.3 Scenario Description

The guidance system was tested in three scenarios, and with four variations to each scenario, where the parameters (generally the intercept range) of the scenario are varied. Table 14.7 summarizes the main parameters of the tested scenarios. The 3D views for all the possible variations of the missile and target paths are shown in Figures 14.13 through 14.15. Each scenario was designed to demonstrate one aspect of the performance of the CLOS guidance system. The first scenario is the most elementary one; here the target follows the boresight, so the missile acceleration needs only to compensate for the gravity effects and the missile accelerations due to external disturbances, such as wind, and small asymmetries in the missile shape. The second

TABLE 14.7

Scenario Description

Target	Scenario 1	Scenario 2	Scenario 3
Trajectory	Along boresight	Step maneuver	Crossing
Velocity	300 m/s	300 m/s	300 m/s
Acceleration	0	Step ±50 m/s^2	0

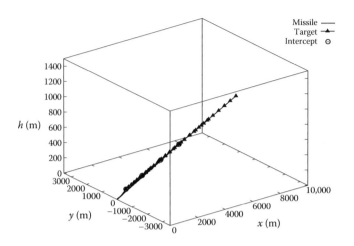

FIGURE 14.13
3D view of missile and target paths, scenario 1.

scenario demonstrates the guidance scheme's ability to intercept a missile, which performs a sudden maneuver near the intercept point. Here the target is assumed to perform a 50 m/s² maneuver either in the positive or negative y direction, each with equal probability. Note that in this scenario, the target flies at a constant height of 1000 m, and not along the boresight. This presents an additional challenge to the GL especially for the shorter-range intercepts. The initiation of the target maneuver is uniformly distributed between 0 and 2 s before the intercept. The target autopilot time constant was set at 0.5 s. The third and last scenario demonstrates the guidance scheme's ability to intercept a crossing target. A crossing target poses a special challenge for

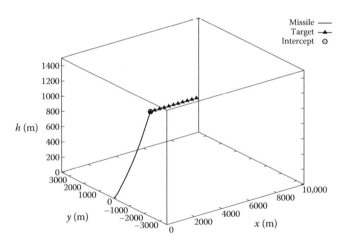

FIGURE 14.14
3D view of missile and target paths, scenario 2.

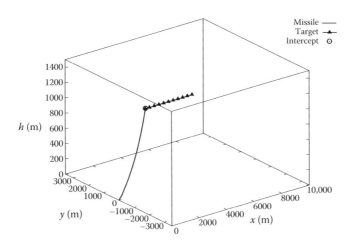

FIGURE 14.15
3D view of missile and target paths, scenario 3.

all CLOS guidance systems because the missile must expend an increasing lateral acceleration as it flies toward the target. Because of the small bandwidth of the guidance loop, this is basically a test of the calculation of the feed-forward term in the acceleration command. In this scenario, both the intercept range and the cross range of the target were varied.

The missile and the target paths are shown in 3D plots in Figures 14.13 through 14.15.

The projections of the missile paths on the *X–Y* and *X–H* planes are shown in Figures 14.16 through 14.18.

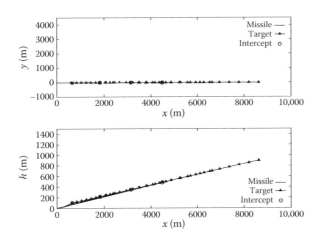

FIGURE 14.16
Projections of missile and target paths, scenario 1.

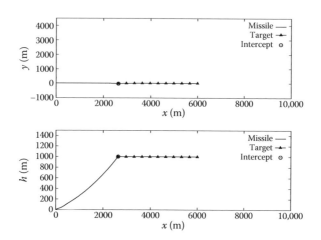

FIGURE 14.17
Projections of missile and target paths, scenario 2.

Finally the root mean square (RMS) miss distances for the three scenarios and the four variations for each of the three scenarios are shown in Figure 14.19. The RMS miss distances are based on 100 Monte Carlo runs. The values of the main parameters used in the Monte Carlo simulations are shown in Table 14.8.

To get an idea of the source of the miss distance, consider the size of the assumed observation noise. A graph of the missile and target observation noise is shown in Figure 14.20. During the initial portion of the missile's flight, the observations are from the wider secondary antenna, so that the standard deviation of the observation noise is much larger than for the measurements in the main antenna. It may also be observed that for most of the engagement, both the target and the missile observations are limited by the

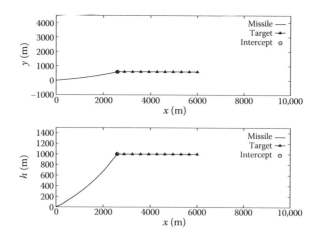

FIGURE 14.18
Projections of missile and target paths, scenario 3.

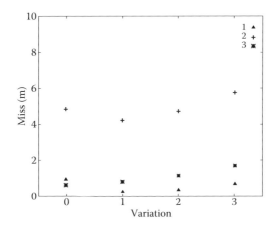

FIGURE 14.19

RMS miss distances for the four variations of the three scenarios.

TABLE 14.8

Parameters Used in Monte Carlo Simulations

Parameter	Value	Reference
Missile Parameters		
$C_{z\alpha}$	−25	Table 14.2
$C_{m\alpha}$	−20	Table 14.2
$C_{m\delta}$	−18.75	Table 14.2
d	0.16	Table 14.2
m_0	55 kg	Figure 14.3
T	6400 N	Figure 14.4
α_{max}	25°	Table 14.2
Radar Parameters		
σ_T^{CS}	2 m²	Equation 14.26
r_M°	100,000	Equation 14.27
r_T°	35,000 km	Equation 14.26
k_A	1.2	Equation 14.29
k_m	1.6	Equation 14.28
λ	0.0176 m	Equation 14.29
D	1.5 m	Equation 14.29
	0.3 m	Radar secondary antenna
σ_g	1 m	Equation 14.30
Guidance Parameters		
τ_d	2 s	Equation 14.18
τ_{zd}	0.1 s	Equation 14.18
σ_{aT}	50 m/s²	Equation 14.113
$\sigma_{x_d^i}$	50 m/s²	Equation 14.119

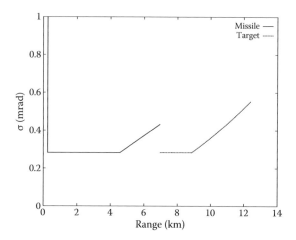

FIGURE 14.20
Missile and target observation angular measurement standard deviation.

assumed minimum value of 1/50 of the antenna beam width for the observation noise. Translated into linear distance, the angular observation noise near the intercept is about 2.4 m for the missile and about 1.8 m for the target. The miss distances for the nonmaneuvering targets are smaller than the standard deviation of the individual observations, indicating that there is considerable filtering taking place in the missile and target Kalman filters. The small miss distances for crossing target indicate that the guidance *does* correctly calculate the feed-forward component of the missile acceleration command necessary for intercepting a crossing target in a CLOS guidance system. The considerably larger miss distances for the maneuvering target indicate the inherent difficulty of intercepting a target performing a maneuver close to the intercept for *any* guidance system.

14.9 Summary

In this chapter, the CLOS guidance problem was formulated in the LQG framework. The coordinate system used was the coordinate system rigidly attached to the sensor platform. The formulation suited the guidance problem well in that the resulting optimization problem turned out to be linear, although time varying. The inclusion of target-related, uncontrollable states resulted in the generation of both the feed-forward and feedback terms of the GL in the solution of the LQG problem.

It was shown that the solution of the linear quadratic optimal control problem is equivalent to a steady-state regulator problem and terminal control problem. The terminal control problem has an analytic solution, while the regulator problem requires the numerical solution of a 3×3 algebraic Riccati equation. The separation of the solution into a regulator and a terminal control problem made possible the real-time numerical calculation of the guidance gains.

The chapter covered a design and implementation method for a CLOS guidance system based on the LQG framework. The discussion included the conceptual phase of the design, which used simplified planar models; the numerical calculation in real time of the required gains; the issues arising in the 3D implementation of the blocks making up the GL; and finally, an evaluation of the performance of the GL using 6DOF Monte Carlo simulation.

Acknowledgments

The authors would like to thank Dr. Margalit Ronen. The work presented here would not have been brought to fruition without her insight and indefatigable capacity to debug and implement these new algorithms. In addition, her ideas on the choice of missile and target kinematic variables together with many helpful discussions had a significant influence on the development of the concepts for the new CLOS guidance scheme presented here.

List of Symbols

Missile disturbance model parameters

$(\sigma_M)^2$	Spectral density of w_M^j. The same value is used $\forall_j$.
$1/\tau_d$	The pole in the missile disturbance model
$1/\tau_{zd}$	The zero in the missile disturbance model
$\sigma_{x_d^j}$	The standard deviation of the missile disturbance acceleration model internal state x_d^i
a_d^j	Missile disturbance acceleration
w_M^j	Continuous time white noise input to the missile disturbance model in the j direction
x_d^j	Missile disturbance model internal state variable

Missile deterministic model parameters

$1/\tau_{zc}$	The zero in the first order approximation to the missile transfer function
ω_A, ξ_A	Frequency and damping of complex pole in missile transfer function

τ_M	Time constant in the first order approximation to the missile transfer function
$a^j_{M\text{det}}$	The deterministic part of the missile acceleration in the j direction
d	Missile reference length
K_T	Missile engine constant
M	Missile Mach number
m	Missile mass
p_A	Real pole of the missile transfer function
T	Missile engine thrust
x_m	The internal state of the first order approximation to the missile transfer function
z_A	Zero of the missile transfer function

Generic state estimator

σ_{n^i}	The standard deviation of the discrete time observation noise
$\underline{n^i}$	Discrete time observation noise for coordinate i, $i = x, y, z$
T_s	Filter sampling time
y^j	The estimator observations for the jth axis coordinate
Q^i	The process noise matrix in the generic estimator for the ith coordinate
Z	Matrix of all 9 states of the generic estimator in sensor platform coordinates
Z^i	The estimator state for the ith coordinate

System kinematic model

ω_r	Sensor platform angular velocity vector
$\in^j_i$	The angle between object i and the boresight around the j axis
ω^j_i	The angular rate of object i with respect to inertial space around the j axis
r_i	The range to object i

Missile related parameters and transformations

α	Missile angle of attack
$\alpha_{\max}$	maximum sustainable missile angle of attack
T_{BS}	Transformation from sensor to missile body coordinates
T_{BW}	Transformation from missile wind to missile body coordinates
T_{WS}	Transformation from sensor to missile wind coordinates
a^j_c	The missile acceleration command perpendicular to the boresight in the j direction, for $j = y, z$
a^x_c	Missile acceleration in the boresight direction
a^j_{cB}	Commanded missile aerodynamic acceleration perpendicular to missile body in the j direction
$a_{M\lim}$	Missile acceleration limit

$C_{m\alpha}$	Missile aerodynamic moment coefficient
$C_{m\delta}$	Missile aerodynamic moment coefficient
C_{x0}	Missile aerodynamic force coefficient in the x direction
$C_{z\alpha}$	Missile aerodynamic force coefficient
$C_{z\delta}$	Missile aerodynamic force coefficient
Q_{pres}	Dynamic pressure
S_{ref}	Missile reference cross section area

Sensor model

λ	Radar wave length
$\left(\dfrac{S}{N}\right)_{beacon}$	Radar signal to noise ratio for observing missile beacon
$\left(\dfrac{S}{N}\right)_{skin}$	The target radar signal to noise ratio
σ_ϵ	Radar angular observation noise standard deviation due to thermal noise
σ_g	Target linear glint standard deviation
σ_T^{CS}	Target radar cross section area
σ_{Gl}	Target angular glint noise
σ_{n^i}	Additive observation noise standard deviation for the ith coordinate
θ_{3dB}	Radar antenna beam width at the 3 dB points
D	Radar antenna diameter
k_A	Radar antenna illumination constant with typical value of 1.2
k_m	Monopulse slope with typical value of 1.6
r_M°	Missile range where the missile beacon signal to noise ratio is unity
r_T°	Range for 1 m^2 target radar return signal for signal to noise ratio of 1

Target parameters

σ_{aT}^j	The standard deviation of the target acceleration in the j direction
τ_T	Decorrelation time constant in the Singer model for the target acceleration
a_T^j	Target acceleration perpendicular to the boresight in the j direction
w_T^j	White noise exciting the Singer target acceleration model in the j direction
$(\sigma_T^j)^2$	Spectral density of white noise exciting the Singer target acceleration model in the j direction

References

1. W. F. Arnold and A. J. Laub. Generalized eigenproblem algorithms and software for algebraic Riccati equations. *Proceedings of the IEEE*, 72(12):1746–1754, Dec 1984.
2. R. S. Baheti. Efficient approximation of Kalman filter for target tracking. *IEEE Transactions on Aerospace and Electronic Systems*, AES-22:8–14, Jan 1986.
3. Y. Bar-Shalom, X. Li, and T. Kirubarajan. *Estimation with Application to Tracking and Navigation: Theory, Algorithms and Software*. John Wiley and Sons, New York, 2001.
4. D. K. Barton. *Modern Radar System Analysis*. Artech House, Norwood, MA, 1988.
5. D. G. BenShabat and A. Bar-Gill. Robust command-to-line-of-sight guidance via variable-structure control. *IEEE Transactions on Control Systems and Technology*, AES-3(3):356–361, Sept 1995.
6. R. F. Berg. Estimation and prediction for maneuvering target trajectories. *IEEE Transactions on Automatic Control*, AC-28(3):294–304, 1983.
7. A. E. Bryson and Y. C. Ho. *Applied Optimal Control*. Baisdel Publishing Company, Waltham, MA, 1969.
8. G. R. Curry. *Radar System Performance Modeling*. Artech House, Boston, MA, 2005.
9. B. T. Fang. Improved command line-of-sight for homing guidance. *IEEE Transactions on Aerospace and Electronic Systems*, AES-31(1):506–510, Jan 1995.
10. B. Friedland. *Control System Design: An Introduction to State–Space Methods*. McGraw-Hill, New York, 1986.
11. P. Garnell. *Guided Weapon Control Systems*. Pergamon Press, Oxford, UK, 1980.
12. I. J. Ha and S. Chong. Design of a CLOS guidance law via feedback linearization. *IEEE Transactions on Aerospace and Electronic Systems*, AES-28(1):51–63, Jan 1992.
13. J. Huang. A modified CLOS guidance law via right inversion. *IEEE Transactions on Aerospace and Electronic Systems*, AES-31(1):491–495, Jan 1995.
14. G. W. Irwin and R. J. Fleming. Analysis of coloured filter controllers for bank-to-turn CLOS guidance. *IEE Proceedings*, 135, Pt. D(6):486–492, Nov 1988.
15. J. E. Kain and D. J. Yost. Command to line-of-sight guidance: A stochastic optimal control problem. *Journal of Spacecraft and Rockets*, 14(7):438–444, 1977.
16. D. L. Kleinman. On an iterative technique for Riccati equation computations. *IEEE Transactions on Automatic Control*, AC-13:114–115, Feb 1968.
17. C. Moler and C. Van Loan. Nineteen dubious ways to compute the exponential of a matrix. *SIAM Review*, 20(4):801–836, Oct 1978.
18. C. Moler and C. Van Loan. Nineteen dubious ways to compute the exponential of a matrix, twenty-five years later. *SIAM Review*, 45(1):3–49, Mar 2003.
19. C. P. Mracek and D. B. Ridgely. Missile longitudinal autopilots: Connections between optimal control and classical topologies. In *AIAA Conference on Guidance Navigation and Control Conference*. Tucson, AZ, 2005, AIAA.
20. F. W. Nesline and P. Zarchan. Robust instrumentation configurations for homing missile flight control. In *AIAA Guidance, Navigation and Control Conference*. Danvers, MA, 1980. Paper 80-1749.
21. I. Rusnak. Almost analytic representation for the solution of the differential matrix Riccati equation. *IEEE Transactions on Automatic Control*, AC-33(2):191–193, Feb. 1988.

22. I. Rusnak and L. Meir. Optimal guidance for high-order and acceleration constrained missile. *Journal of Guidance*, 14(3):589–596, May–Jun 1991.
23. N. A. Shneydor. *Missile Guidance and Pursuit Kinematics, Dynamics and Control*. Horwood Publishing, Chichester, UK, 1998.
24. R. A. Singer. Estimating optimal tracking filter performance for manned maneuvering targets. *IEEE Transactions on Aerospace and Electronic Systems*, AES-6:473–483, July 1970.
25. M. I. Skolnik. *Introduction to Radar Systems*. McGraw-Hill, New York, 1980.
26. H. Weiss and G. Hexner. Simple structure for a high performance three dimensional tracking filter. *AIAA Journal of Guidance, Control and Dynamics*, 27(3):491–493, May–Jun 2004.
27. P. Zarchan. *Tactical and Strategic Guidance*, volume 157 of *Progress in Astronautics and Aeronautics*. American Institute of Aeronautics and Astronautics, Washington DC, Inc., 2nd ed., 1994.

15

Practical Considerations in Robust Control of Missiles

Kevin A. Wise

CONTENTS

15.1 Introduction

The robust control of missiles is a multidisciplinary challenge in dynamics, aerodynamics, propulsion, and controls. Figure 15.1 illustrates an artist's conception of a high-performance agile missile. The understanding and modeling of these disciplines are critical in the development of a missile's flight control system. The control software (guidance and flight control) that controls the weapon system during flight is the linkage that connects these disciplines and makes the system perform as required. This chapter will discuss these disciplines and practical considerations in the control of missiles and guided munitions from a control engineer's perspective.

FIGURE 15.1
High-performance agile missile.

If accurate system models are available, then the problem of control system design is less difficult. However, in today's ever-increasing competitive market, programs can seldom afford to accurately measure the dynamics to the level that mitigates all uncertainties and risks. This leads the engineer to the question—How accurately must we know these dynamics, and how can the robustness of the controller be measured, gauged, and improved? Most control system design methods used in practice are model-based and yield the specified performance only when the model is adequately known. Linear robust control methods developed in the late 1980s and early 1990s, such as μ-synthesis [1–4], have proven to provide robust stability, but system performance degrades for off-nominal design conditions. This chapter will address the robust control-of-missiles challenge by using optimal control theory [5–7], which has excellent performance *and* robustness properties, augmented with adaptive control [8,9] to further extend performance and robustness guarantees.

During the 1980s and 1990s, the control community produced new tools and design and analysis processes for linear robust control synthesis and analysis. Increased insight from these linear model-based methods led to further development of adaptive and nonlinear methods, followed by next-generation system and parameter identification methods. These methods, as did the previous classical methods, attempt to provide industry's engineers with the tools needed for robust control synthesis and analysis. If one is to synthesize a high-performance robust control, one must know how to analyze and test the robustness.

Historically, the development and transition of control technology evolve on a time scale that spans years, if not a decade. In almost a cyclic fashion, the control community develops new paradigms for control, new tools, techniques, and processes, with these all building upon the methods of previous cycles. The industry then slowly filters these methods and, through a trial-and-error process, implements them to develop new and evolving aerospace systems. New methods are typically blended with existing well-proven techniques (such as classical and optimal) to address the industry's design challenge. Methods that prove beneficial in terms of improving performance and robustness that are able to address key technical challenges, or that are able to reduce control system development costs, survive and get used. Those that prove to be just a stepping stone fall away and fail to pass the test of time.

Recent successes in developing weapon system control systems have combined linear optimal control, used as a gain scheduled baseline control, with nonlinear adaptive methods to address uncertainties and guarantee command tracking in the presence of these uncertainties. This combination of linear and nonlinear control has proven to provide a method that yields robust high-performance control in the presence of significant uncertainty and that at The Boeing Company has been transitioned into production [10]. This chapter will focus on the practical considerations regarding these methods.

This chapter will begin by presenting the dynamic models typically used for bank-to-turn (BTT) missiles and weapon systems, highlighting key aspects that can be exploited to improve overall understanding and lead to improved robust control synthesis. Analysis methods used to evaluate the control designs will then be discussed. Methods for directly measuring the robustness will be presented. Following this, case studies will be presented where optimal and adaptive flight control systems are designed, combined, and analyzed, showing the reader how to apply the material presented in the first two sections.

15.2 Flight Control Design Models

This section presents the models used to design and analyze a missile autopilot. Figure 15.2 shows a block diagram illustrating the component models included in a typical linear frequency domain analysis. State space models for each of these components are populated with data that represent the flight condition and understanding of the physical hardware. Time and frequency domain analysis is performed on the model to evaluate performance, stability, and robustness properties. Rise time, settling time, and actuator usage in response to a step command input are collected along with stability margins to show that the controller design meets mission requirements.

15.2.1 Dynamic Models

Typically, the body axis equations of motion (EOM) are used to design a missile flight control system. If the missile is a BTT or preferred orientation control weapon, then the autopilot is designed to command body-normal acceleration A_Z and to roll the airframe about the velocity vector. If implemented correctly, this approach coordinates turns and minimizes unwanted out-of-plane accelerations. If the weapon is a skid-to-turn system, then accelerations are commanded in the vertical (normal) and directional (yaw) axes, and the body roll rate is regulated.

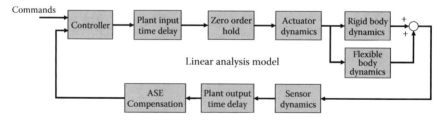

FIGURE 15.2
Missile autopilot linear analysis model.

Assuming a rigid body, the linear and angular momentum vectors $P = mV$ and $H = I\omega$, respectively, are differentiated with respect to time and are equated to the forces and moments acting on the body. Performing the differentiation with respect to a nonrotating coordinate system yields

$$\frac{dP}{dt} = \dot{m}V + m\dot{V} + \omega \times mV$$

$$\frac{dH}{dt} = \dot{I}\omega + I\dot{\omega} + \omega \times (I\omega). \tag{15.1}$$

Assuming the mass and inertia time variations are negligible, Equation 15.1 is equated to the forces F and moments M acting on the body. This results in

$$m\dot{V} = -\omega \times mV + F$$

$$I\dot{\omega} = -\omega \times (I\omega) + M. \tag{15.2}$$

The body forces F are modeled as gravitational forces G and aerodynamic forces A using $F = mA + mG$. Dividing Equation 15.2 by mass, multiplying by the inverse inertia matrix, and introducing the gravitational forces and aerodynamic forces yields

$$\dot{V} = -\omega \times V + G + A$$

$$\dot{\omega} = -I^{-1}\omega \times (I\omega) + I^{-1}M. \tag{15.3}$$

Next, assume that the inertia cross products I_{xy} and I_{zy} are zero, and expand the linear and angular velocity vectors into their component directions using $V = [u\ v\ w]^T$ for ($x\ y\ z$ body axes) and $\omega = [p\ q\ r]^T$ for (roll pitch yaw about the body axes). This results in the standard six-degrees-of-freedom body EOM written as

$$\dot{u} = rv - qw + X + G_x + T_x$$

$$\dot{v} = pw - ru + Y + G_y + T_y$$

$$\dot{w} = qu - pv + Z + G_z + T_z$$

$$\dot{p} = -L_{pq}pq - L_{qr}qr + L + L_T \tag{15.4}$$

$$\dot{q} = -M_{pr}pr - M_{r^2p^2}\left(r^2 - p^2\right) + M + M_T$$

$$\dot{r} = -N_{pq}pq - N_{qr}qr + N + N_T$$

where G_i models gravity, $(X\,Y\,Z)$ models the linear accelerations produced by the aerodynamic forces, $(L\,M\,N)$ models the angular accelerations produced by the aerodynamic moments, (T_x, T_y, T_z) models propulsion system forces, and (L_T, M_T, N_T) models the moments produced by the propulsion system. Note that these variables have units of acceleration. The aerodynamic forces are modeled as nondimensional quantities and are scaled to units of force. This scaling is described by

$$
\begin{bmatrix} X \\ Y \\ Z \end{bmatrix} = \frac{\bar{q}S}{m} \begin{bmatrix} C_x \\ C_y \\ C_z \end{bmatrix}
\tag{15.5}
$$

where $\bar{q}$ (in pounds per square feet) is the dynamic pressure, S (in square feet) is a reference area, m is the mass in slugs, and (C_x, C_y, C_z) are nondimensional aerodynamic forces. The aerodynamic moments acting on the body are similarly modeled as

$$
\begin{bmatrix} L \\ M \\ N \end{bmatrix} = \begin{bmatrix} \dfrac{\bar{q}sl}{I_{xx}I_{zz} - I_{xz}^2}(C_l I_{zz} + C_n I_{xz}) \\[3mm] \dfrac{\bar{q}sl}{I_{yy}}C_m \\[3mm] \dfrac{\bar{q}sl}{I_{xx}I_{zz} - I_{xz}^2}(C_n I_{xx} + C_l I_{zz}) \end{bmatrix}
\tag{15.6}
$$

where $(C_l\,C_m\,C_n)$ model nondimensional moments, and l is a reference length. Note that the cross-axis inertia term I_{xz} couples the roll–yaw moment equations.

The gravitational forces are modeled as

$$
\begin{bmatrix} G_x \\ G_y \\ G_z \end{bmatrix} = g \begin{bmatrix} -\sin(\theta) \\ \cos(\theta)\sin(\phi) \\ \cos(\theta)\cos(\phi) \end{bmatrix}.
\tag{15.7}
$$

The pitch-plane angle of attack α and yaw-plane sideslip angle β are defined in Figure 15.3, along with the total angle of attack, α_T. The stability axis coordinates are a transformation of the body axes using α. The wind axis coordinates are a transformation from stability axes using β.

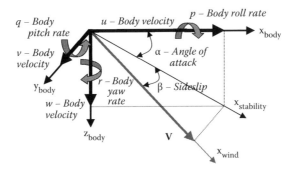

FIGURE 15.3
Angle of attack and sideslip angle definitions.

15.2.1.1 Aerodynamics

The missile's aerodynamic forces (C_x, C_y, C_z) and moments (C_l C_m C_n) are typically modeled as functions of α, β, Mach, body rates (p, q, and r), $\dot{\alpha}$, $\dot{\beta}$, the aerodynamic control surface deflections (δ_e, δ_a, and δ_r), center-of-gravity changes, and whether the main propulsion system is on or off (plume effects). Also, the aerodynamic forces may depend upon whether reaction jets are on or off (jet interaction effects). These complicated and highly nonlinear functions are used in the EOM to model the airframe's aerodynamics.

Asymmetric vortex shedding is a nonlinear phenomenon that must be addressed when considering high angle-of-attack (AOA) flight. Figure 15.4 illustrates the effect of asymmetric nose vortices at high AOAs with zero

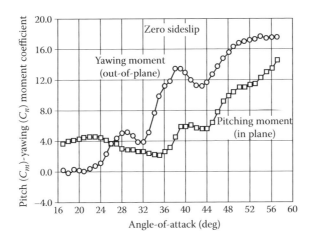

FIGURE 15.4
Out-of-plane moment caused by asymmetric vortex shedding.

sideslip. Note that due to the asymmetric vortices, the out-of-plane moment (yaw moment) is larger than the in-plane moment (pitch moment). These asymmetric vortices can cause the nose to slice right or left and may require large control inputs to counter the effect. This phenomenon is often referred to as phantom yaw and can be mitigated by the addition of small nose strakes and/or nose bluntness. Figure 15.5 illustrates the possible reduction in out-of-plane moment by modifying the missile's configuration.

In addition to the challenge of overcoming phantom yaw, the missile's static stability significantly changes with AOA. Figure 15.6 illustrates the changing pitch-plane stability with AOA. A positive slope is unstable, and a negative slope is stable. For the missile under investigation, aerodynamic control authority ends at or near 30° AOA, and some form of alternate control is needed to fly at high AOAs.

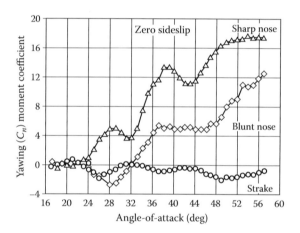

FIGURE 15.5
Out-of-plane yawing moment sensitivity to nose shape.

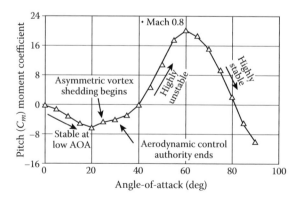

FIGURE 15.6
Missile pitch-plane stability characteristics with angle-of-attack.

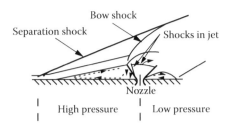

FIGURE 15.7
Interaction of jet plume with free-stream aerodynamics.

When reaction control system (RCS) thrusters are used for control, there is an interaction of the jet plume with the free stream aerodynamic flow. Figure 15.7 illustrates this jet interaction, which is very nonlinear with AOA. Note the high/low pressure areas in the front/back of the nozzle. This occurs when the jet plume is entrapped by the free stream flow. The entrapped high-pressure region (times its area and moment arm) produces a moment on the missile that is larger than the jet thrust force times its moment arm [distance to the center of gravity (CG)]. By exploiting this phenomenon, a smaller reaction jet can be used, thus reducing propellant requirements. This amplification, if not accounted for, would be an increase in the loop gain in the flight control system, which can significantly impact stability.

If the jet plume penetrates the free stream flow, the high-pressure field does not form, with the net thrust force produced by the jet reduced by the low-pressure field. Thus, a 200 lb. thruster may produce only 80 lb. of thrust. This attenuation, if not accounted for, would be a decrease in the loop gain in the flight control system, which can also impact stability for an open-loop unstable missile.

A key parameter in modeling the jet interaction phenomena is the jet penetration height. This parameter indicates if there is amplification or an attenuation of the jet thrust (force). Unfortunately, this parameter varies significantly with flight condition and is difficult to predict. These nonlinear jet interaction effects also cause the moments produced by the thruster not to be proportional to the thruster force.

15.2.1.2 Propulsion System Forces and Moments

For thrust vector control (TVC) actuators, the autopilot is designed to command the TVC actuator angle δ_T (in radians). For the RCS thrusters, the autopilot is designed to command the thrust level T_{RCS} (in pounds).

The TVC forces and moments are modeled using a constant thrust that is deflected by the actuator. It is assumed that the actuator can deflect the thrust vector only in the pitch (δ_{T_e}) and yaw (δ_{T_r}) planes, using two actuators devoted to this task. (No roll control from the TVC actuation system is assumed.) The resulting thrusts (used in Equation 15.1) along the body axes are

$$
\begin{bmatrix} T_x \\ T_y \\ T_z \end{bmatrix} = \frac{T}{m} \begin{bmatrix} \cos\left(\delta_{T_e}\right)\cos\left(\delta_{T_r}\right) \\ -\cos\left(\delta_{T_r}\right) \\ -\sin\left(\delta_{T_e}\right)\cos\left(\delta_{T_r}\right) \end{bmatrix} \tag{15.8}
$$

where T is the axial thrust of the main propulsion system. The roll, pitch, and yaw moments (L_T, M_T, N_T) produced by the TVC will be the moment arm $l_T = x_{cg} - x_{TVC}$ times the above pitch and yaw forces, respectively, and are described as

$$
\begin{bmatrix} L_T \\ M_T \\ N_T \end{bmatrix} = \begin{bmatrix} \dfrac{-l_T I_{xz} T \sin(\delta_{T_r})}{I_{xx}I_{zz} - I_{xz}^2} \\ \dfrac{l_T T \sin(\delta_{T_e})\cos(\delta_{T_r})}{I_{yy}} \\ \dfrac{-l_T I_{xx} T \sin(\delta_{T_r})}{I_{xx}I_{zz} - I_{xz}^2} \end{bmatrix}. \tag{15.9}
$$

An RCS jet also produces propulsion forces and moments. The reaction jets are assumed to be positioned such that no axial force is generated. The RCS is designed to provide roll, pitch, and yaw moment control. The forces produced by the pitch and yaw jets are modeled as

$$
\begin{bmatrix} 0 \\ T_y \\ T_z \end{bmatrix} = \frac{1}{m} \begin{bmatrix} 0 \\ \bar{T}_y \\ \bar{T}_z \end{bmatrix} \tag{15.10}
$$

where $\bar{T}_y$ and $\bar{T}_z$ are the RCS thrust forces in the y-body (yaw) and z-body (pitch) directions, respectively. These moments produced by the thrusters are modeled by the x_{RCS} thruster forces multiplied by the moment arm $l_T = x_{cg} - x_{RCS}$. It is assumed here that the pitch and yaw jets are located at the same missile x-station x_{RCS}.

Roll jets may also be used to control missile roll. These jets are symmetrically placed so that only a rolling moment is produced, modeled as

$$
L_T = \frac{l_{Roll} \bar{T}_{Roll}}{I_{xx}}. \tag{15.11}
$$

Adding this to the moments produced from the pitch and yaw jets results in

$$
\begin{bmatrix} L_T \\ M_T \\ N_T \end{bmatrix} = \begin{bmatrix} \dfrac{l_{Roll}\overline{T}_{Roll}}{I_{xx}} + \dfrac{-l_T I_{xz} \overline{T}_y}{I_{xx} I_{zz} - I_{xz}^2} \\ -\dfrac{l_T \overline{T}_z}{I_{yy}} \\ \dfrac{l_T I_{xx} \overline{T}_y}{I_{xx} I_{zz} - I_{xz}^2} \end{bmatrix}.
\tag{15.12}
$$

15.2.1.3 Angle-of-Attack and Sideslip Dynamics

The following derivation will form a set of differential equations describing the dynamics for $\dot{V}$, $\dot{\alpha}$, and β valid for large α's and $\beta < 90°$. Consider the following definition of the body velocities from Figure 15.3:

$$
u = V \cos(\alpha) \cos(\beta)
$$
$$
v = V \sin(\beta)
\tag{15.13}
$$
$$
w = V \sin(\alpha) \cos(\beta)
$$

where V is the magnitude of the missile velocity vector. This can be represented as a transformation of the wind-axis velocity vector to the body axes as follows:

$$
\begin{bmatrix} u \\ v \\ w \end{bmatrix}_{Body} = S_\alpha S_\beta \begin{bmatrix} V \\ 0 \\ 0 \end{bmatrix}_{Wind}
\tag{15.14}
$$

where the transformations S_α and S_β are

$$
S_\alpha = \begin{bmatrix} c\alpha & 0 & -s\alpha \\ 0 & 1 & 0 \\ s\alpha & 0 & c\alpha \end{bmatrix} \quad S_\beta = \begin{bmatrix} c\beta & -s\beta & 0 \\ s\beta & c\beta & 0 \\ 0 & 0 & 1 \end{bmatrix}
\tag{15.15}
$$

where $c(\bullet)$ and $s(\bullet)$ denote $\cos(\bullet)$ and $\sin(\bullet)$, respectively. Differentiating Equation 15.14 yields

$$
\begin{bmatrix} \dot{u} \\ \dot{v} \\ \dot{w} \end{bmatrix} = \begin{bmatrix} c\alpha c\beta & -s\alpha c\beta & -c\alpha s\beta \\ s\beta & 0 & c\beta \\ s\alpha c\beta & c\alpha c\beta & -s\alpha s\beta \end{bmatrix} \begin{bmatrix} \dot{V} \\ V\dot{\alpha} \\ V\dot{\beta} \end{bmatrix}.
\tag{15.16}
$$

Inverting the coefficient matrix in the above equation yields

$$
\begin{bmatrix} \dot{V} \\ V\dot{\alpha} \\ V\dot{\beta} \end{bmatrix} = \frac{-1}{c\beta} \underbrace{\begin{bmatrix} -c\alpha c^2\beta & -s\beta c\beta & -s\alpha c^2\beta \\ s\alpha & 0 & -c\alpha \\ s\beta c\beta c\alpha & -c^2\beta & s\alpha s\beta c\beta \end{bmatrix}}_{W(\alpha,\beta)} \begin{bmatrix} \dot{u} \\ \dot{v} \\ \dot{w} \end{bmatrix}. \tag{15.17}
$$

Substituting from Equation 15.4 yields

$$
\begin{bmatrix} \dot{V} \\ V\dot{\alpha} \\ V\dot{\beta} \end{bmatrix} = W(\alpha,\beta)\left(-\begin{bmatrix} p \\ q \\ r \end{bmatrix} \times \begin{bmatrix} u \\ v \\ w \end{bmatrix} + \begin{bmatrix} X \\ Y \\ Z \end{bmatrix} + \begin{bmatrix} G_x \\ G_y \\ G_z \end{bmatrix} + \begin{bmatrix} T_x \\ T_y \\ T_z \end{bmatrix} \right). \tag{15.18}
$$

Expanding Equation 15.15 results in

$$
\dot{V} = c\alpha c\beta(X + G_x + T_x) + s\beta(Y + G_y + T_y) + s\alpha c\beta(Z + G_z + T_z)
$$
$$
\dot{\alpha} = (1/Vc\beta)\left[-s\alpha(X + G_x + T_x) + c\alpha(Z + G_z + T_z) \right] + q - [pc\alpha + rs\alpha]\tan(\beta)
$$
$$
\dot{\beta} = (1/V)\left[-c\alpha s\beta(X + G_x + T_x) + c\beta(Y + G_y + T_y) - s\alpha s\beta(Z + G_z + T_z) \right] + ps\alpha - rc\alpha. \tag{15.19}
$$

15.2.1.4 Acceleration Dynamics

This section derives rigid body differential equations for the body axis accelerations at the CG. The body axis acceleration at the CG is given by

$$
\begin{bmatrix} A_x \\ A_y \\ A_z \end{bmatrix} = \begin{bmatrix} X + G_x + T_x \\ Y + G_y + T_y \\ Z + G_z + T_z \end{bmatrix}. \tag{15.20}
$$

Expanding these terms gives

$$
\begin{bmatrix} A_x \\ A_y \\ A_z \end{bmatrix} = \begin{bmatrix} \dfrac{\bar{q}S}{m}C_x(\alpha,\beta,\delta_e,\delta_a,\delta_r) + G_x + T_x \\[2mm] \dfrac{\bar{q}S}{m}C_y(\alpha,\beta,\delta_e,\delta_a,\delta_r) + G_y + T_y \\[2mm] \dfrac{\bar{q}S}{m}C_z(\alpha,\beta,\delta_e,\delta_a,\delta_r) + G_z + T_z \end{bmatrix} \tag{15.21}
$$

where the functional dependence of the aerodynamics on α, β, and the aerodynamic control surfaces δ_i is shown to highlight what terms will be differentiated. The aerodynamic effects due to body rates (from p, q, r) and the plunge effects (from $\dot{\alpha}$ and $\dot{\beta}$) are assumed zero. If these effects are known, then they should be included. Differentiating this expression yields

$$
\begin{bmatrix} \dot{A}_x \\ \dot{A}_y \\ \dot{A}_z \end{bmatrix} = \frac{2\bar{q}S}{mV} \begin{bmatrix} C_x \\ C_y \\ C_z \end{bmatrix} \dot{V} + \frac{\bar{q}S}{m} \begin{bmatrix} C_{x_\alpha} \\ C_{y_\alpha} \\ C_{z_\alpha} \end{bmatrix} \dot{\alpha} + \frac{\bar{q}S}{m} \begin{bmatrix} C_{x_\beta} \\ C_{y_\beta} \\ C_{z_\beta} \end{bmatrix} \dot{\beta} + \frac{\bar{q}S}{m} \begin{bmatrix} C_{x_{\delta e}} & C_{x_{\delta a}} & C_{x_{\delta r}} \\ C_{y_{\delta e}} & C_{y_{\delta a}} & C_{y_{\delta r}} \\ C_{z_{\delta e}} & C_{z_{\delta a}} & C_{z_{\delta r}} \end{bmatrix} \begin{bmatrix} \dot{\delta}_e \\ \dot{\delta}_a \\ \dot{\delta}_r \end{bmatrix}
$$

$$
+ \begin{bmatrix} \dot{G}_x + \dot{T}_x \\ \dot{G}_y + \dot{T}_y \\ \dot{G}_z + \dot{T}_z \end{bmatrix}
\tag{15.22}
$$

where the subscripts denote partial derivatives. Grouping terms results in

$$
\begin{bmatrix} \dot{A}_x \\ \dot{A}_y \\ \dot{A}_z \end{bmatrix} = \frac{\bar{q}S}{mV} \begin{bmatrix} 2C_x & C_{x_\alpha} & C_{x_\beta} \\ 2C_y & C_{y_\alpha} & C_{y_\beta} \\ 2C_z & C_{z_\alpha} & C_{z_\beta} \end{bmatrix} \begin{bmatrix} \dot{V} \\ V\dot{\alpha} \\ V\dot{\beta} \end{bmatrix}
$$

$$
+ \frac{\bar{q}S}{m} \begin{bmatrix} C_{x_{\delta e}} & C_{x_{\delta a}} & C_{x_{\delta r}} \\ C_{y_{\delta e}} & C_{y_{\delta a}} & C_{y_{\delta r}} \\ C_{z_{\delta e}} & C_{z_{\delta a}} & C_{z_{\delta r}} \end{bmatrix} \begin{bmatrix} \dot{\delta}_e \\ \dot{\delta}_a \\ \dot{\delta}_r \end{bmatrix} + \begin{bmatrix} \dot{G}_x + \dot{T}_x \\ \dot{G}_y + \dot{T}_y \\ \dot{G}_z + \dot{T}_z \end{bmatrix}.
\tag{15.23}
$$

Substituting from Equation 15.17 yields

$$
\begin{bmatrix} \dot{A}_x \\ \dot{A}_y \\ \dot{A}_z \end{bmatrix} = \begin{bmatrix} \dot{G}_x + \dot{T}_x \\ \dot{G}_y + \dot{T}_y \\ \dot{G}_z + \dot{T}_z \end{bmatrix} + \frac{1}{V} \begin{bmatrix} 2X & X_\alpha & X_\beta \\ 2Y & Y_\alpha & Y_\beta \\ 2Z & Z_\alpha & Z_\beta \end{bmatrix} \begin{bmatrix} c\alpha c\beta & s\beta & s\alpha c\beta \\ -s\alpha/c\beta & 0 & c\alpha/c\beta \\ s\beta c\alpha & c\beta & -s\alpha s\beta \end{bmatrix} \begin{bmatrix} \dot{u} \\ \dot{v} \\ \dot{w} \end{bmatrix}
$$

$$
+ \begin{bmatrix} X_{\delta e} & X_{\delta a} & X_{\delta r} \\ Y_{\delta e} & Y_{\delta a} & Y_{\delta r} \\ Z_{\delta e} & Z_{\delta a} & Z_{\delta r} \end{bmatrix} \begin{bmatrix} \dot{\delta}_e \\ \dot{\delta}_a \\ \dot{\delta}_r \end{bmatrix}.
\tag{15.24}
$$

Substituting for $[\dot{u} \ \dot{v} \ \dot{w}]^T$ yields

$$
\begin{bmatrix} \dot{A}_x \\ \dot{A}_y \\ \dot{A}_z \end{bmatrix} = \begin{bmatrix} \dot{G}_x + \dot{T}_x \\ \dot{G}_y + \dot{T}_y \\ \dot{G}_z + \dot{T}_z \end{bmatrix} + \begin{bmatrix} X_{\delta_e} & X_{\delta_a} & X_{\delta_r} \\ Y_{\delta_e} & Y_{\delta_a} & Y_{\delta_r} \\ Z_{\delta_e} & Z_{\delta_a} & Z_{\delta_r} \end{bmatrix} \begin{bmatrix} \dot{\delta}_e \\ \dot{\delta}_a \\ \dot{\delta}_r \end{bmatrix}
$$

$$
+ \frac{1}{V} \begin{bmatrix} 2X & X_\alpha & X_\beta \\ 2Y & Y_\alpha & Y_\beta \\ 2Z & Z_\alpha & Z_\beta \end{bmatrix} W(\alpha,\beta) \left\{ \begin{bmatrix} A_x \\ A_y \\ A_z \end{bmatrix} + V \begin{bmatrix} s\beta r - s\alpha c\beta q \\ s\alpha c\beta p - c\alpha c\beta r \\ c\alpha c\beta q - s\beta p \end{bmatrix} \right\}.
$$

$$(15.25)$$

Note that the time rate of change of the accelerations is modeled proportionally to actuator rates. Thus, an actuator dynamics model is required to model the accelerations as states in a state space model.

15.2.2 Autopilot Design Models

Maximizing overall missile performance requires choosing the appropriate autopilot command structure for each mission phase. This may include designing a different autopilot for separation (launch), an agile turn (high AOA turn), midcourse (long flyout), and endgame (terminal homing) maneuvers. The autopilot can command body rates, wind angles, attitudes, or accelerations.

During launch, a body rate command system is typically used. Rate command autopilots are very robust to the uncertain proximity aerodynamics. During an agile turn, directional control of the missile's velocity vector relative to the missile body is desired. This equates to commanding AOA or sideslip and regulating roll to zero. During midcourse and in the terminal phase, an acceleration command autopilot is typically used. At the end of terminal homing, during a guidance integrated fuse maneuver, the missile attitude may be commanded to improve the lethality of the warhead.

Separation, midcourse, and endgame autopilots have been designed and implemented in production missiles and are, in general, well understood. Autopilot designs for agile turns (high AOA flight) are significantly less understood. Missile performance during the agile turn can be maximized by maximizing the missile's turn rate (higher turn rates lead to faster target intercepts). The missile's turn rate (for a pitch-plane maneuver) is given by

$$
\dot{\gamma} = \frac{A_z \cos(\alpha) - A_x \sin(\alpha)}{V}. \tag{15.26}
$$

High turn rates can be achieved by commanding a constant high AOA or by commanding large values of normal acceleration $[A_z \cos(\alpha) - A_x \sin(\alpha)]$.

Simulation studies have shown that due to the large changes in the missile's velocity (V) at high AOAs (due to the high drag), commanding body accelerations during an agile turn may not be desirable.

The nonlinear missile dynamics can be written as

$$\dot{x} = f\left(x, u\right). \tag{15.27}$$

To form a linear model, partial derivatives of the f_i are needed with respect to each state variable and each control input. These partial derivatives are evaluated at a specific design point (flight condition). This would typically be at a trimmed equilibrium condition; however, at high AOAs, the missile is generally not in what is considered an equilibrium condition.

For the sake of brevity, BTT autopilots will be discussed and used in the remaining sections of the chapter.

15.2.2.1 Pitch Autopilot Design Model

The pitch-plane nonlinear AOA and pitch rate dynamics are described in Equations 15.4 and 15.19. Neglecting the roll–yaw dynamics and linearizing about α_0 results in

$$\dot{\alpha} = \frac{1}{V}\left(Z_\alpha \alpha + q + Z_{\delta_e}\delta_e - \sin\left(\alpha_0\right)T_x + \cos\left(\alpha_0\right)T_z\right)$$

$$\dot{q} = M_\alpha \alpha + M_q q + M_{\delta_e}\delta_e + M_T \tag{15.28}$$

where

$$Z_\alpha = \left.\frac{\partial\dot{\alpha}}{\partial\alpha}\right|_{\alpha=\alpha_0} = \left[\cos(\alpha)\left(\frac{\partial Z}{\partial\alpha} - G_x - T_x - X\right) - \sin(\alpha)\left(\frac{\partial X}{\partial\alpha} + G_z + T_z + Z\right)\right]_{\alpha=\alpha_0}$$

$$Z_{\delta_e} = \left.\frac{\partial\dot{\alpha}}{\partial\delta_e}\right|_{\alpha=\alpha_0} = \left[\frac{\partial Z}{\partial\delta_e}\cos(\alpha) - \frac{\partial X}{\partial\delta_e}\sin(\alpha)\right]_{\alpha=\alpha_0}$$

$$M_\alpha = \left.\frac{\partial M}{\partial\alpha}\right|_{\alpha=\alpha_0} \quad M_q = \left.\frac{\partial M}{\partial q}\right|_{\alpha=\alpha_0} \quad M_{\delta_e} = \left.\frac{\partial M}{\partial\delta_e}\right|_{\alpha=\alpha_0}.$$

Since most TVC actuators are limited to small deflection angles, $\sin(\delta_{T_e}) \approx \delta_{T_e}$ and $\cos(\delta_{T_e}) \approx 1$, resulting in

$$T_x = T/m \quad T_z = -(T/m)\delta_{T_e} \quad M_T = -\left(l_T T/I_{yy}\right)\delta_{T_e}. \tag{15.29}$$

To model RCS thruster forces (axial thrust T is due to main engine; see Equations 15.7 and 15.9)

$$T_x = T/m \quad T_z = T_{RCS}/m \quad M_T = -\left(l_T/I_{yy}\right)T_{RCS}. \tag{15.30}$$

Neglecting the influence of gravity on the AOA dynamics (since it is divided by V) and the $T \sin(\alpha_0)$ term (since it represents a constant) and combining these into a linear matrix model results in

$$
\begin{bmatrix} \dot{\alpha} \\ \dot{q} \end{bmatrix} =
\begin{bmatrix} \dfrac{Z_\alpha}{V} & 1 \\ M_\alpha & M_q \end{bmatrix}
\begin{bmatrix} \alpha \\ q \end{bmatrix} +
\begin{bmatrix} \dfrac{Z_\delta}{V} \\ M_\delta \end{bmatrix} \delta_e +
\begin{bmatrix} \dfrac{c\alpha_0}{mV} \\ \dfrac{l_T}{I_{yy}} \end{bmatrix} T_{RCS}
$$

$$
+ \begin{bmatrix} \dfrac{T(s\alpha_0 \delta_{T_0} - c\alpha_0 \delta_{T_0})}{mV} \\ \dfrac{-Tl_T}{I_{yy}} \end{bmatrix} \delta_{T_e}. \tag{15.31}
$$

This state space model can be used to design pitch autopilots at a specific flight condition (α_0, Mach, altitude, CG). If A_z rather than α is preferred as a state variable, then replace the $\dot{\alpha}$ equation with A_z from Equation 15.22 (this also requires removing α from the pitch rate dynamics and adding an actuator model to include the terms proportional to the actuator deflection rates).

15.2.2.2 Roll–Yaw Autopilot Design Model

The lateral directional nonlinear dynamics are described in Equations 15.1 and 15.16. Zeroing the pitch dynamics and linearizing about α_0 (with $\beta = 0$) results in

$$
\begin{bmatrix} \dot{\beta} \\ \dot{p} \\ \dot{r} \end{bmatrix} =
\begin{bmatrix} \dfrac{Y_\beta}{V} & s\alpha_0 + \dfrac{Y_p}{V} & c\alpha_0 + \dfrac{Y_r}{V} \\ L_\beta & L_p & L_r \\ N_\beta & N_p & N_r \end{bmatrix}
\begin{bmatrix} \beta \\ p \\ r \end{bmatrix} +
\begin{bmatrix} \dfrac{Y_{\delta_a}}{V} & \dfrac{Y_{\delta_r}}{V} \\ L_{\delta_a} & N_{\delta_a} \\ L_{\delta_r} & N_{\delta_r} \end{bmatrix}
\begin{bmatrix} \delta_a \\ \delta_r \end{bmatrix}
$$

$$
+ \begin{bmatrix} \dfrac{1}{V} & 0 & 0 \\ 0 & 1 & 0 \\ 0 & 0 & 1 \end{bmatrix}
\begin{bmatrix} T_y \\ L_T \\ N_T \end{bmatrix} \tag{15.32}
$$

where the elements of the matrices were obtained in a similar manner to Equation 15.25. For TVC (assuming a small TVC angle δ_{T_r}), this results in

$$T_y = \frac{-T}{m}\delta_{T_r} \qquad L_T = \frac{-l_T I_{xz} T}{I_{xx}I_{zz} - I_{xz}^2}\delta_{T_r} \qquad N_T = \frac{-l_T I_{xx} T}{I_{xx}I_{zz} - I_{xz}^2}\delta_{T_r}.$$

Modeling an RCS yields

$$T_y = \frac{\overline{T}}{m} \qquad L_T = \frac{l_{Roll}\overline{T}_{Roll}}{I_{xx}} + \frac{l_T I_{xz}\overline{T}_y}{I_{xx}I_{zz} - I_{xz}^2} \qquad N_T = \frac{l_T I_{xx}\overline{T}_y}{I_{xx}I_{zz} - I_{xz}^2}.$$

Neglecting gravity results in the following linear autopilot design model:

$$
\begin{bmatrix} \dot{\beta} \\ \dot{p} \\ \dot{r} \end{bmatrix} = \begin{bmatrix} \dfrac{Y_\beta}{V} & s\alpha_0 + \dfrac{Y_p}{V} & c\alpha_0 + \dfrac{Y_r}{V} \\ L_\beta & L_p & L_r \\ N_\beta & N_p & N_r \end{bmatrix} \begin{bmatrix} \beta \\ p \\ r \end{bmatrix} + \begin{bmatrix} \dfrac{-T}{m} \\ \dfrac{-l_T I_{xz} T}{I_{xx}I_{zz} - I_{xz}^2} \\ \dfrac{-l_T I_{xx} T}{I_{xx}I_{zz} - I_{xz}^2} \end{bmatrix} \delta_{T_r}
$$

$$
+ \begin{bmatrix} \dfrac{Y_{\delta_a}}{V} & \dfrac{Y_{\delta_r}}{V} \\ L_{\delta_a} & N_{\delta_a} \\ L_{\delta_r} & N_{\delta_r} \end{bmatrix} \begin{bmatrix} \delta_a \\ \delta_r \end{bmatrix} + \begin{bmatrix} 0 & \dfrac{1}{m} \\ \dfrac{l_{Roll}}{I_{xx}} & \dfrac{l_T I_{xz}}{I_{xx}I_{zz} - I_{xz}^2} \\ 0 & \dfrac{l_T I_{xx}}{I_{xx}I_{zz} - I_{xz}^2} \end{bmatrix} \begin{bmatrix} \overline{T}_{Roll} \\ \overline{T}_y \end{bmatrix}.
$$

(15.33)

This state space model can be used to design roll–yaw autopilots at a specific flight condition (α_0, Mach, altitude, CG). Note that β was assumed to be zero as for a BTT missile. This modeling assumption is used because at high AOAs, β must be kept very small to keep the roll channel controls from saturating.

15.2.3 Sensor Measurements

Most tactical missiles use strapdown inertial measurement units (IMUs) for navigation, which have three accelerometers and three gyros. The accelerometers and gyros are arranged into a triad to measure accelerations along and rotational rates about the x-, y-, and z-body axes, respectively. Due to packaging considerations, the IMU is usually not located at the missile's CG.

The location of the accelerometers relative to the CG greatly effects the measured accelerations and must be accounted for in the design of the flight control system. Ideally, if the accelerometers are located at the CG, then they measure just the translational accelerations. If the accelerometers are located off the CG, then they measure a combination of translational and rotational accelerations. This can be expressed as

$$a_{IMU} = a_{CG} + \dot{\omega} \times r_{IMU} + \omega \times \omega \times r_{IMU} \qquad (15.34)$$

where r_{IMU} is a vector from the CG to the IMU and $\omega = [p \quad q \quad r]^{T}$. Note that the sensed accelerations are a nonlinear function of the body rates.

Linear sensor models are required for linear autopilot design. On symmetric airframes, the y- and z-axis CG offsets are usually small and can be neglected. The z-axis accelerometer is compensated for the x-axis CG offset as follows:

$$A_{z_{IMU}} = A_{z_{CG}} + (x_{CG} - x_{IMU})\dot{q}. \qquad (15.35)$$

This effect can have a dramatic impact on the flight control system design. Equation 15.35 shows that the rotational dynamics are blended with the translational dynamics. This changes the zeros of the transfer function from the control input to the sensor output.

15.2.3.1 Shaping Zero Dynamics

Consider the transfer function from elevator δ_e to acceleration $A_{z_{CG}}$ (from [5, Equation 4]) given by

$$\frac{A_{z_{CG}}}{\delta_e} = \frac{\omega_n^2 \left(Z_{\delta_e} s^2 + Z_\alpha M_{\delta_e} - Z_{\delta_e} M_\alpha \right)}{\left(s^2 - \dfrac{Z_\alpha}{V} s - M_\alpha \right)\left(s^2 + 2\zeta\omega_n s + \omega_n^2 \right)}. \qquad (15.36)$$

For tail-controlled missiles, this transfer function is nonminimum phase [has a right half plane (RHP) zero]. As the elevator δ_e deflects, the fin force $Z_{\delta_e}\delta_e$ accelerates the missile in the wrong direction. However, this fin force creates a pitching moment that rotates the missile. As the missile rotates, the body force builds $Z_\alpha\alpha$, accelerating the missile in the correct (commanded) direction. Aerodynamically unstable $M_\alpha > 0$ tail-controlled missiles pose a considerable control challenge in that they have both RHP poles and zeros.

The transfer function from δ_e to $A_{z_{IMU}}$ does not have the same zeros as using $A_{z_{CG}}$. Figure 15.8 illustrates the location of the acceleration zeros as the sensor is moved along the body of the missile. When the IMU is aft of the CG, the

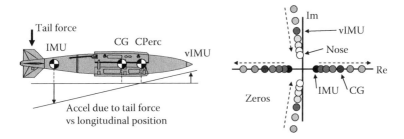

FIGURE 15.8
Acceleration zeros.

two zeros are real with one in the RHP. As the IMU moves forward of the CG to the center of percussion, the zeros bifurcate and become complex, moving in along the $j\omega$-axis.

The autopilot designer can shape the zeros in the acceleration transfer function by placing the sensor at a different location. Depending upon how the feedback gains are designed, this can be exploited to improve stability and transient performance. See Wise [7] for more discussion on zeros.

15.2.4 Actuator Models

15.2.4.1 Fin Actuator Model

There are four tail fins each driven by an electromechanical (EM) actuator. The fin actuator dynamics can be modeled with a second-order transfer function. The significant nonlinearities typically modeled include position and rate limits, as well as mechanical backlash.

The fin mixing logic that relates δ_e, δ_a, and δ_r commands to individual fin deflections is configuration specific and depends upon whether the missile is flown with an "x" or "+" tail. The logic for mixing these is not unique. One example, for an x tail, is given as

$$\begin{bmatrix} \delta_1 \\ \delta_2 \\ \delta_3 \\ \delta_4 \end{bmatrix} = \begin{bmatrix} 1 & -1 & -1 \\ 1 & -1 & 1 \\ 1 & 1 & -1 \\ 1 & 1 & 1 \end{bmatrix} \begin{bmatrix} \delta_e \\ \delta_a \\ \delta_r \end{bmatrix} \tag{15.37}$$

where δ_e, δ_a, and δ_r are the autopilot pitch, roll, and yaw fin commands, respectively, distributed to the four fins, and δ_i, $i = 1, \ldots, 4$, are the actual fin deflections. Note that it is the δ_i that exhibit the nonlinearities (fin and rate limits, backlash).

15.2.4.2 RCS Thruster Actuator Model

RCS actuators can be built with EM valves, hot gas poppet valves, hydraulic valves, and solenoid valves. They can be stand-alone systems (requiring their own propellant) or can be integrated with the main engine (bleed off the main engine chamber pressure). They can also be on–off or continuous (throttling).

Both the continuous and on–off RCS actuators can be modeled with a first-order transfer function. The 63% rise time of the thrust is used to specify the time constant in the transfer function. The RCS thrust is magnitude and rate limited. In addition, some on–off designs cannot change states (on or off) until they are fully opened or closed.

15.2.4.3 TVC Actuator Model

TVC actuators operate in a similar manner to the fin actuators and are usually modeled with second-order transfer functions. The nonlinearities typically modeled include position and rate limits, as well as mechanical backlash.

Figure 15.9 shows a nonlinear TVC actuator response (no backlash modeled) to a 10° square wave input. This response illustrates both the position and rate limiting. Also shown in the figure is a plot of thrust produced normal to the missile's x-body axis. For a 5000 lbf main engine motor and a deflection limit of 10°, the maximum normal force is 868.24 lbf.

Using the response shown in Figure 15.9, the maximum thrust rate for the TVC actuator can be computed as a function of the TVC rate limit. A fast thrust rate is required to maintain stability and capture the high AOA command during the agile turn.

Figure 15.10 shows a comparison of the TVC thrust rate capability with that of a 500 lbf RCS jet, in which the RCS jet is parameterized by the time it takes the jet to reach full thrust.

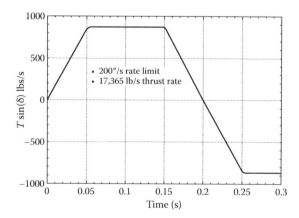

FIGURE 15.9
Thrust rate for TVC effector.

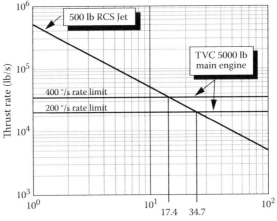

FIGURE 15.10
TVC versus RCS power.

For a 200°/s TVC actuator rate limit, the thrust rate is slightly less than 20,000 lbf/s. This equates to a 500 lbf RCS jet that reaches its full thrust in 34.7 ms. At 400°/s TVC rate limit, the equivalent RCS jet time to full thrust is 17.4 ms. This figure indicates that the TVC actuator system is somewhat slower than the RCS in developing normal thrust. High TVC rates will be required in order for the TVC actuator to slew the thrust vector to its maximum position. These high TVC rates will in turn drive the TVC power consumption and battery sizing requirements upward (increasing the cost and weight of the missile).

15.2.5 Flexible Body Dynamics

In deriving the autopilot design models, it was assumed that the missile was a rigid body. In fact, it is a flexible body, and these dynamics have a significant impact on the sensed accelerations and body rates. The discussion here is limited to the airframe's pitch plane. Also discussed is the tail-wags-the-dog effect due to fin mass unbalance and inertias and TVC nozzle inertias (see [2] for more details on modeling these dynamics).

Consider the following flexible model:

$$
\begin{bmatrix} \dot{\alpha} \\ \dot{q} \\ \dot{b}_1 \\ \ddot{b}_1 \end{bmatrix} = \begin{bmatrix} \dfrac{Z_\alpha}{V} & 1 & Z_{b_1} & Z_{\dot{b}_1} \\ M_\alpha & M_q & M_{b_1} & M_{\dot{b}_1} \\ 0 & 0 & 0 & 1 \\ b_{\alpha 1} & b_{q1} & b_{b_1} & b_{\dot{b}_1} \end{bmatrix} \begin{bmatrix} \alpha \\ q \end{bmatrix} + \begin{bmatrix} \dfrac{Z_\delta}{V} & \dfrac{Z_{\ddot{\delta}}}{V} \\ M_\delta & M_{\ddot{\delta}} \\ 0 & 0 \\ b_{\delta 1} & b_{\ddot{\delta}_1} \end{bmatrix} \begin{bmatrix} \delta_e \\ \ddot{\delta}_e \end{bmatrix}. \qquad (15.38)
$$

This linear analysis model describes the pitch-plane rigid body dynamics (α, q) combined with the first bending mode (b_1), including the tail-wags-the-dog effects proportional to $\ddot{\delta}_e$.

The pitch rate gyro and z-axis accelerometer measurements are

$$q_{FLEX} = q_{IMU} + F'_{A_1}\dot{b}_1$$

$$A_{zFLEX} = A_{zIMU} + F_{A_1}\ddot{b}_1/g. \tag{15.39}$$

Partitioning the A-matrix in Equation 15.38 into 2×2 blocks, the (1,1) block is the same as in Equation 15.28 and describes the rigid body dynamics. The (1,2) block describes the changes in the aerodynamic forces and moments due to the body flexure. The (2,1) block describes how the rigid body states (α, q) excite the bending mode. The (2,2) block describes the first bending mode's second-order dynamics.

In addition to the rigid body states (α, q), the fin deflection δ_e also excites the bending dynamics. When the fin rotates, a bending torque is applied to the missile body that is proportional to both the fin's inertia and any mass unbalance (if the fin CG is located off the fin's rotational axis). This effect is called the tail-wags-the-dog effect (see Reichert [2] for more discussion) and can be significant. (When TVC is used, this effect is large because the nozzle is heavy and its CG is not located about its rotational axis.)

The IMU sensor measurements are corrupted by the flexible dynamics. Filters are designed to remove these signals from the sensed rates and accelerations. Unfortunately, these filters add gain attenuation and phase lag at the loop gain crossover frequency, thus impacting stability margins.

15.2.6 Control Power Analysis

The autopilot design models can be used to assess the control effector's (aero, RCS, or TVC) capability to control the missile's dynamics as the flight envelope changes. This section presents a static control power analysis of the missile's pitch dynamics, examining in which part of the flight envelope the aero control, RCS, and TVC effectors are the most useful.

Aero control effectors depend upon dynamic pressure to generate control power. At low velocities, aero control effectors have low control power. They also depend upon the AOA and lose effectiveness at high AOAs.

RCS thrusters provide a fixed level of thrust normal to the x-body axis (independent of flight condition, excluding jet interaction effects). The same level of control power is obtained at low and high velocities (neglecting jet interaction effects).

TVC exhibits the same characteristics as RCS (independent of flight condition). TVC actuators have a limited deflection that then limits the normal thrust. Control power data for the TVC designs are not presented, but the same trends apply to both the TVC and RCS effectors.

A state space model for the missile's dynamics can be written as $\dot{x} = Ax + Bu$, where the aero control and RCS inputs enter into the dynamics through the B matrix. The B matrix can be partitioned as $B = [B_\delta \quad B_{RCS}]$. The control effectiveness, or control power, can be analyzed by computing the "size" of the B matrix as a function of AOA, CG location, altitude, and Mach number. The singular values of the B matrix are computed [22], and the maximum singular value is examined as the flight envelope parameters vary.

In order to minimize the amount of propellant used to perform maneuvers, it is important to know at what velocities and AOAs the aero controls are effective. Similarly, for RCS and TVC, it is important to know at what flight conditions the main engine must be providing thrust.

Study results show that the tail fins are very effective near zero AOA but lose their pitch moment capability as the AOA increases or as the velocity decreases (dynamic pressure decreases). The RCS jet's pitch moment capability is constant with AOA and Mach number (neglecting jet interaction effects). At low Mach numbers, the RCS is more effective than aero control. At high Mach numbers, the aero controls are much more effective than the RCS jets. As the altitude increases, the aero becomes less effective and requires an increase in velocity to maintain its effectiveness (due to decrease in dynamic pressure).

Figure 15.11 summarizes data comparing missile fin and RCS effectiveness for an empty weight configuration. The RCS has a fixed magnitude versus Mach number. The three curves that change magnitude with Mach number represent the amount of aero control power for 0, 10, and 35 kft. altitudes. For small AOAs and above Mach 0.8, the fins are as effective as the RCS (data is for 0° AOA). As expected, at higher Mach numbers, the aero control is significantly more effective than the RCS.

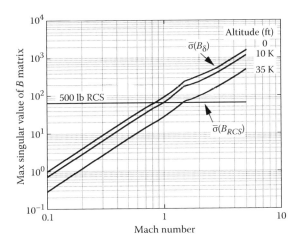

FIGURE 15.11
Comparison of RCS and aero control power.

15.2.7 Time Delays

Time delays, e^{-sT}, occur due to the digital implementation of the flight control system and must be accounted for in stability and performance analyses. These delays are usually small in practice but add to the phase delay and can affect the phase margins. A second-order Pade approximation is typically used to model the time delay, creating a state space model to be included in Figure 15.2.

15.2.8 Zero-Order Hold Effects

The implementation of the autopilot in a computer introduces digital sampling effects that impact the phase margin of the system. The autopilot command that drives the actuators changes at the digital sample rate at which the autopilot is implemented. The command is held constant until the next frame. This holding effect is modeled as $1 - e^{-sT_s}$, where T_s is the sample period. For an autopilot implemented at a 100 Hz digital rate, $T_s = 10$ ms. A Pade approximation is again used to create the component model for Figure 15.2.

15.3 Frequency Domain Analysis and Robustness Measures

The stability and robustness analysis of a missile's flight control system requires the use of both classical, single-input single-output (SISO) methods and modern, multi-input multi-output (MIMO) analysis methods. Stability requirements must be validated throughout the flight envelope, which requires all loops in the missile's guidance and flight control system to be opened and analyzed for gain and phase margins. Many missile configurations today have highly coupled cross-axis dynamics and aerodynamics, which requires multiple loops to be opened for stability analysis.

This section reviews MIMO analysis methods and relates them to their classical counterparts. These methods play a critical role in establishing both robust stability and robust performance.

15.3.1 Transfer Functions and Transfer Function Matrices

Many of the frequency domain analysis models for MIMO systems are natural extensions of transfer functions used to analyze SISO systems. However, unlike transfer functions for SISO systems, MIMO analysis models have different sizes depending upon where the loop is broken for analysis. When analyzing a missile's flight control system for stability, it is typical to break all loops in a classical manner.

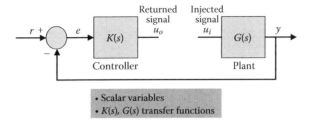

FIGURE 15.12
SISO system with loop opened at plant input.

Consider the SISO system shown in the block diagram of Figure 15.12. The loop gain for this system can be calculated by breaking the loop at the control generation point (plant input) and injecting a signal u_i. The returned signal is

$$u_o = -\underbrace{K(s)G(s)}_{L(s)} u_i \qquad (15.40)$$

in which $L(s)$ is the *loop gain transfer function*. Differencing the injected signal u_i and the returned signal u_o results in

$$u_i - u_o = u_i + K(s)G(s)u_i = (I + K(s)G(s))u_i$$
$$= \left(I + L(s)\right)u_i \qquad (15.41)$$

which is the *return difference* for the loop. We will find later in this chapter that the *return difference matrix* (RDM) plays a very important role in the development of stability robustness analysis tests for MIMO systems. The error transfer function for this system is

$$\frac{e(s)}{r(s)} = \frac{1}{1 + K(s)G(s)} = S(s) \qquad (15.42)$$

where $S(s)$ is the *sensitivity* function, which describes the error dynamics. Note that the sensitivity is the inverse of the return difference. The closed-loop response to a command input is

$$\frac{y(s)}{r(s)} = \frac{K(s)G(s)}{1 + K(s)G(s)} = T(s) \qquad (15.43)$$

where $T(s)$ is the *closed-loop transfer function*. The transfer function $T(s)$ is also called the *complementary sensitivity* since $S(s)$ and $T(s)$ satisfy the identity

$$S(s) + T(s) = 1. \qquad (15.44)$$

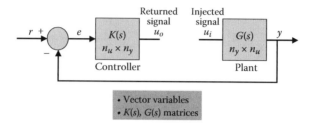

FIGURE 15.13
MIMO system with loop opened at plant input.

Now, consider the multivariable equivalent of Figure 15.11 as shown in Figure 15.13. In Figure 15.13, the variables r, e, u_i, u_o, and y are vectors, with the controller $K(s)$ an $n_u \times n_y$ matrix and the plant $G(s)$ an $n_y \times n_u$ matrix. The figure shows the loop broken at the plant input. The loop gain $L(s)$ is formed using the same procedure as in Equation 15.40, where $L(s) = K(s)G(s)$ is an $n_u \times n_u$ matrix. Forming the RDM yields

$$u_i - u_o = \left(I_{n_u} + K(s)G(s) \right) u_i \tag{15.45}$$

where $I_{n_u} + L(s)$ is also an $n_u \times n_u$ matrix. If this same procedure for calculating the loop gain is applied at the output of the plant, as shown in Figure 15.13, the return difference dynamics are

$$u_i' - u_o' = \left(I_{n_y} + G(s)K(s) \right) u_i' \tag{15.46}$$

which produces a loop gain and RDM that are $n_y \times n_y$ in dimension.

It is very important to learn that for MIMO systems, the loop gain is different at the plant input and plant output loop break points, which is unlike SISO systems. This dissimilarity is caused by the fact that matrices do not commute, but scalars do. Figure 15.14 summarizes the loop gain, return difference, sensitivity, and complementary sensitivity transfer functions and matrices for the SISO and MIMO systems shown in Figures 15.12 and 15.13.

Figure 15.14 lists the various matrices used to analyze MIMO control systems. In the remainder of this chapter, the subscript on the identity matrix indicating the dimension will be dropped for notational convenience.

15.3.1.1 Example

Consider the linear time-invariant (LTI) pitch-plane dynamics of the missile flight control system. The plant dynamics (A, B, C, D) can be written as

Function	SISO System (Figure 5.1)	MIMO System (Figure 5.2) at plant input	MIMO System (Figure 5.2) at plant output
Loop Gain	$L(s) = K(s)G(s)$ $= G(s)K(s)$	$L(s) = K(s)G(s)$	$L(s) = G(s)K(s)$
Return difference	$1 + L(s)$	$I_{n_u} + L(s)$	$I_{n_y} + L(s)$
Sensitivity $S(s)$	$\dfrac{1}{1+L(s)}$	$\left(I_{n_u} + L(s)\right)^{-1}$	$\left(I_{n_y} + L(s)\right)^{-1}$
Complementary sensitivity $T(s)$	$\dfrac{L(s)}{1+L(s)}$	$\left(I_{n_u} + L(s)\right)^{-1} L(s)$	$\left(I_{n_y} + L(s)\right)^{-1} L(s)$

FIGURE 15.14
SISO and MIMO matrices.

$$
\begin{bmatrix} \dot{\alpha} \\ \dot{q} \end{bmatrix} = \underbrace{\begin{bmatrix} Z_\alpha & 1 \\ M_\alpha & 0 \end{bmatrix}}_{A} \begin{bmatrix} \alpha \\ q \end{bmatrix} + \underbrace{\begin{bmatrix} Z_\delta \\ M_\delta \end{bmatrix}}_{B} \delta_e
$$

$$
\begin{bmatrix} A_z \\ q \end{bmatrix} = \underbrace{\begin{bmatrix} VZ_\alpha & 0 \\ 0 & 1 \end{bmatrix}}_{C} \begin{bmatrix} \alpha \\ q \end{bmatrix} + \underbrace{\begin{bmatrix} VZ_\delta \\ 0 \end{bmatrix}}_{D} \delta_e.
$$

(15.47)

These dynamics form a single-input multi-output system. The transfer function matrix for the plant dynamics is

$$
G(s) = C(sI - A)^{-1} B + D = \begin{bmatrix} \dfrac{A_z}{\delta_e} \\ \dfrac{q}{\delta_e} \end{bmatrix}
$$

(15.48)

which is a 2×1 matrix. The autopilot (controller) for this plant contains proportional plus integral control elements in the inner rate loop closure and outer acceleration loop closure, given by

$$
K_{A_z}(s) = \frac{K_{A_z}(s + a_z)}{s}
$$

(15.49)

and

$$K_q(s) = \frac{K_q\left(s + a_q\right)}{s} \tag{15.50}$$

with the controller transfer function matrix given by

$$K(s) = \left[K_{A_z}(s)K_q(s) \quad K_q(s)\right] \tag{15.51}$$

which is a 1×2 matrix. The loop gain at the input to the plant is

$$L(s) = K(s)G(s) = \frac{A_z}{\delta_e}K_{A_z}(s)K_q(s) + \frac{q}{\delta_e}K_q(s) \tag{15.52}$$

which is a scalar transfer function. To analyze stability for this system, any SISO analysis technique can be applied. If we examine the loop gain at the plant output, then

$$L(s) = G(s)K(s) = \begin{bmatrix} \dfrac{A_z}{\delta_e}K_{A_z}(s)K_q(s) & \dfrac{A_z}{\delta_e}K_q(s) \\[2ex] \dfrac{q}{\delta_e}K_{A_z}(s)K_q(s) & \dfrac{q}{\delta_e}K_q(s) \end{bmatrix} \tag{15.53}$$

which is a 2×2 matrix and is singular.

Figure 15.15 illustrates an LTI MIMO system with command $r(t) \in \mathbb{R}^{n_y}$, plant disturbance $w(t) \in \mathbb{R}^{n_y}$ and measurement noise $v(t) \in \mathbb{R}^{n_y}$.

The output response from the system shown in Figure 15.15 is

$$Y(s) = T(s)R(s) + S(s)W(s) + T(s)V(s). \tag{15.54}$$

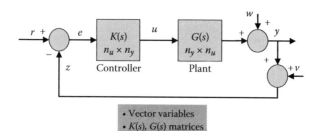

FIGURE 15.15
MIMO control system.

This equation shows how the output response depends upon the sensitivity and complementary sensitivity functions. At frequencies $s = j\omega$ where commands are to be followed, we want $T(s) \to 1$, which shows that sensor noise is passed through the system into the output. It is not possible to reject sensor noise and track commands at the same frequencies. At frequencies where plant disturbances are to be rejected, we want $S(s) \to 0$.

The error response $E(s)$ can be formed by writing the following loop equations:

$$u = Ke$$

$$y = GKe + w$$

$$z = GKe + w + v \qquad (15.55)$$

$$e = r + z = r + GKe + w + v$$

$$E(s) = S(s)(R(s) + W(s) + V(s))$$

which show that to make errors in tracking commands small, we want $S(s) \to 0$. Equations 15.54 and 15.55 illustrate the control design dilemma faced by engineers, that is, to make $S(s) \to 0$ at low frequencies for command tracking and disturbance rejection and $T(s) \to 0$ at high frequencies for sensor noise rejection and robustness to high-frequency unmodeled dynamics. The dilemma is that $S(s) + T(s) = 1$ at all frequencies, and as the sensitivity is made small, the complementary sensitivity is made unity, and vice versa.

15.3.2 Multivariable Stability Margins

Classical stability margin analyses use frequency response methods (Bode and Nyquist) in determining the relative stability of SISO systems. These methods manipulate the loop transfer function of the system to derive gain and phase margins, typical measures of relative stability. In multivariable systems (MIMO systems), the loop transfer function of the system is a complex valued matrix, making it difficult to apply the same SISO methods to determine relative stability.

In SISO systems, the gain of the loop transfer function is easily determined by computing the magnitude of the complex-valued transfer function versus frequency. For MIMO systems, the notion of gain or magnitude for the loop transfer function matrix becomes a question of determining the magnitude of a matrix versus frequency. To accomplish this task, the singular values of the matrix can be computed versus frequency and used as a measure of its magnitude.

In this section, we are concerned with deriving stability margins for multivariable systems. The robust stability analysis tests and stability margin formulas developed here are derived from application of the multivariable

Nyquist theorem. These tests and formulas are natural extensions of the SISO tests reviewed in the previous section.

15.3.2.1 Singular Values

The singular value decomposition (SVD) of a matrix A of dimension $n \times m$ is $A = U\Sigma V^*$, where U and V are unitary matrices, whose columns denote left and right singular vectors of the matrix A, respectively. (Note the similarity to an eigenvalue decomposition.) Assuming that the matrix is of rank k, the nonzero portion of the singular value matrix is

$$\Sigma = \begin{bmatrix} \Sigma_1 & 0 \\ 0 & 0 \end{bmatrix}; \quad \Sigma_1 = \mathrm{diag}[\sigma_1 \quad \cdots \quad \sigma_k] \tag{15.56}$$

with the singular values ordered in size with $\bar{\sigma} = \sigma_1$ the largest and $\underline{\sigma} = \sigma_k$ the smallest. The use of singular values plays an important role in analyzing the near singularity of matrices. If A is a square singular matrix, then $\underline{\sigma} = 0$, and it is not invertible. The maximum and minimum singular values of the matrix A can be defined as

$$\bar{\sigma}(A) = \max_{x \neq 0} \frac{\|Ax\|_2}{\|x\|_2} = \|A\|_2 \tag{15.57}$$

$$\underline{\sigma}(A) = \min_{x \neq 0} \frac{\|Ax\|_2}{\|x\|_2}.$$

The max and min optimization implied by Equation 15.57 can be eliminated by use of a property known as Rayleigh's quotient.

15.3.2.1.1 Rayleigh's Quotient

If A is a Hermitian matrix, then

$$\min_{x \neq 0} \frac{x^H A x}{x^H x} = \lambda_{\min}(A) \tag{15.58}$$

where $\lambda_{\min}(A)$ is the minimum eigenvalue of A, and $(\bullet)^H$ denotes complex conjugate transpose. (A matrix W is Hermitian if $W = W^H$.) The minimum is attained when x is the eigenvector of A corresponding to $\lambda_{\min}(A)$. Also,

$$\max_{x \neq 0} \frac{x^H A x}{x^H x} = \lambda_{\max}(A) \tag{15.59}$$

where $\lambda_{max}(A)$ is the maximum eigenvalue of A. The maximum is attained when x is the eigenvector of A corresponding to $\lambda_{max}(A)$.

The maximum singular value of A (assume A to be complex valued) can be expressed as

$$\bar{\sigma}(A(j\omega)) = \max_{x \neq 0} \frac{\|Ax\|_2}{\|x\|_2}$$

$$= \max_{x \neq 0} \sqrt{\frac{x^H A^H A x}{x^H x}}. \tag{15.60}$$

Note that the product $A^H A$ is a matrix that is real, symmetric, and positive semidefinite, thus Hermitian. Applying Rayleigh's quotient to Equation 15.60 yields

$$\bar{\sigma}(A(j\omega)) = \max_{x \neq 0} \sqrt{\frac{x^H A^H A x}{x^H x}}$$

$$= \sqrt{\lambda_{max}(A^H A)} = \sqrt{\lambda_{max}(AA^H)}. \tag{15.61}$$

Similarly,

$$\underline{\sigma}(A(j\omega)) = \min_{x \neq 0} \sqrt{\frac{x^H A^H A x}{x^H x}}$$

$$= \sqrt{\lambda_{min}(A^H A)} = \sqrt{\lambda_{min}(AA^H)}. \tag{15.62}$$

The maximum singular value of the matrix A is the 2-norm of the matrix and in some sense represents how "big" the matrix is or how large the "gain" of the matrix is. The minimum singular value represents how nearly singular the matrix is. The condition number for a matrix, $\kappa(A)$, is the ratio of the maximum and minimum singular values, given by

$$\kappa(A) = \frac{\bar{\sigma}(A)}{\underline{\sigma}(A)}, \tag{15.63}$$

and is used by numerical analysts to gain insight into how invertible a matrix is.

Associated with each singular value are singular vectors that describe the "direction" of the singular value. Consider the matrix $A \in C^{n \times m}$ with rank $k = \min(n, m)$. The k nonzero singular values of A, denoted as $\sigma_i(A)$, are the strictly positive square roots of the k nonzero eigenvalues of $A^H A$ (or equivalently AA^H). This is expressed as

$$\sigma_i(A) = \sqrt{\lambda_i(A^H A)} = \sqrt{\lambda_i(AA^H)} > 0. \tag{15.64}$$

Each singular value has an input and output direction, which can be determined by examining the singular vectors associated with the SVD of the matrix. The SVD of a complex matrix $A \in \mathbb{C}^{n \times m}$ is

$$A = U\Sigma V^* \tag{15.65}$$

where U is an $n \times n$ unitary matrix (i.e., $U^H = U^{-1}$) consisting of orthonormal column vectors u_i:

$$U = [u_1 \quad \cdots \quad u_n], \tag{15.66}$$

which are referred to as the left singular vectors of the matrix, V is a unitary matrix consisting of orthonormal column vectors v_i

$$V = [v_1 \quad \cdots \quad v_m], \tag{15.67}$$

which are referred to as the right singular vectors of the matrix, and Σ is a real $n \times m$ matrix given by

$$\Sigma = \begin{bmatrix} \sigma_1 & & & & & 0 & & \\ & \sigma_2 & & & & & 0 & \\ & & \ddots & & & & & \\ 0 & & & \sigma_k & & & & \\ \hline & & 0 & & & & 0 & \end{bmatrix}. \tag{15.68}$$

The σ_i in Equation 15.68 is the ith singular value of the matrix A, with a corresponding left singular vector u_i (Equation 15.66) and right singular vector v_i (Equation 15.67). It is easy to show that

$$\begin{aligned} Av_i &= \sigma_i u_i \\ A^H u_i &= \sigma_i v_i. \end{aligned} \tag{15.69}$$

The above equations can also be written as

$$\begin{aligned} A^H A v_i &= \sigma_i^2 v_i \\ AA^H u_i &= \sigma_i^2 u_i \end{aligned} \tag{15.70}$$

which shows that σ_i^2 is an eigenvalue of AA^H or A^HA, u_i is an eigenvector of AA^H, and v_i is an eigenvector of A^HA.

Consider a square matrix $A \in \mathbb{C}^{n \times n}$ having rank k. Using an SVD, the matrix A can be represented using a dyadic expansion as

$$A = \sigma_1 u_1 v_1^H + \sigma_2 u_2 v_2^H + \cdots + \sigma_k u_k v_k^H = \sum_{i=1}^{k} \sigma_i u_i v_i^H. \qquad (15.71)$$

The SVD of a matrix describes the gain through the matrix, with the maximum gain equal to the 2-norm of the matrix $[\|A\|_2 = \sigma_1(A) = \bar{\sigma}(A)]$. In addition to the gain, the SVD describes the direction associated with the gain. The dyadic expansion in Equation 15.68 indicates that the left and right singular vectors describe the direction of the gain. The maximum gain through the matrix occurs with the input direction from v_1 and output direction u_1.

Further insight into the relationship between the gain of a matrix and the input/output vector directions can be gained through a simple geometric visualization. Consider the equation

$$y = Au, \qquad (15.72)$$

which has two inputs and two outputs. If the input vector u has unit magnitude, the possible input vector directions correspond to points on the unit circle shown in Figure 15.16. When the input vector has the direction OA corresponding to the direction of vector v_1 in the SVD, the output y has the direction OA′ corresponding to the direction of vector u_1 in the SVD. The magnitude (length) of vector OA′ is $\sigma_1(A) = \bar{\sigma}(A)$. Similarly, when the input vector has the direction OB, the output y lies along OB′ and has a magnitude $\sigma_2(A) = \underline{\sigma}(A)$. The gains corresponding to all other input directions fall between the maximum and minimum singular values as indicated by Figure 15.16.

Figure 15.17 illustrates the input-to-output mapping for a general transfer function matrix $G(j\omega) \in \mathbb{C}^{n \times n}$. Here the singular value expansion provides

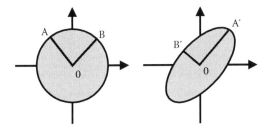

FIGURE 15.16
Principal gain direction.

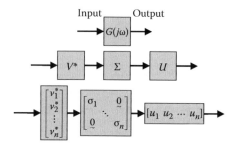

FIGURE 15.17
Singular value decomposition.

insight into the relative gain between input-to-output channels for a transfer function matrix.

15.3.2.1.2 Singular Value Facts
If the matrix A is invertible, that is, A^{-1} exists, then

$$\bar{\sigma}(A^{-1}) = \frac{1}{\underline{\sigma}(A)} \text{ and } \underline{\sigma}(A^{-1}) = \frac{1}{\bar{\sigma}(A)}$$

$$\|A\|_2 = \bar{\sigma}(A)$$

$$\|A\|_F^2 = \sum_{i=1}^{n} \sigma_i^2(A)$$

where $\|\bullet\|_F$ denotes the Frobenius norm. If the matrices U and V are unitary, then

$$\sigma_i(UA) = \sigma_i(A)$$

$$\sigma_i(AV) = \sigma_i(A)$$

which says that unitary matrices preserve the singular values and $\|\bullet\|_2$ of a matrix.

15.3.2.2 Multivariable Nyquist Theory

The multivariable Nyquist criterion gives a yes or no answer to the stability question. Other methods such as computing the eigenvalues of the system A matrix or solving a Lyapunov equation can also be used to answer the stability question. However, understanding the multivariable Nyquist criterion leads to important understanding of robustness analysis tests used to analyze model uncertainties. In addition, time delays e^{sT} are easily incorporated into the analysis in order to analyze MIMO systems with time delays.

The multivariable Nyquist criterion is derived from an application of the principle of the argument from complex variable theory.

Principle of the Argument

Let Γ be a closed clockwise contour in the s-plane. Let $f(s)$ be a complex-valued function. Suppose that

1. $f(s)$ is analytic on Γ.
2. $f(s)$ has Z zeros inside Γ.
3. $f(s)$ has P poles inside Γ.

■

Then $f(s)$ will encircle the origin, 0, $Z - P$ in a clockwise sense as s transverses Γ.

Let $N(p, f(s), \Gamma)$ denote the number of encirclements of the point p made by the function $f(s)$ as s transverses the closed clockwise contour Γ. If Γ equals the standard Nyquist D-contour (D_R) encircling the RHP, and $f(s)$ is a rational function in s, then $N(0, f(s), D_R) = Z - P$.

If $f(s)$ is factored where $f(s) = f_1(s)f_2(s)$, then

$$N\left(0, f_1(s)f_2(s), D_R\right) = N\left(0, f_1(s), D_R\right) + N\left(0, f_2(s), D_R\right)$$
$$= \left(Z_1 - P_1\right) + \left(Z_2 - P_2\right) = Z - P. \tag{15.73}$$

Consider the feedback system shown in Figure 15.18. The state equations for this system are

$$\dot{x} = Ax + Bu$$

$$u = Kx.$$

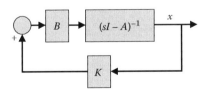

FIGURE 15.18
State feedback LTI system.

The closed-loop system is

$$\dot{x} = (A + BK)x.$$

Let $L(s)$ denote the loop transfer function matrix for this system, written as

$$L(s) = K(sI - A)^{-1} B.$$

The determinant of the RDM, $\det[I + L(s)]$, is equal to the closed-loop characteristic polynomial divided by the open-loop characteristic polynomial, that is,

$$\det[I + L(s)] = \frac{\phi_{cl}(s)}{\phi_{ol}(s)}.$$

This can be shown as

$$\phi_{cl}(s) = \det[sI - A + BK]$$

$$= \underbrace{\det[sI - A]}_{\phi_{ol}(s)} \det\left[I + (sI - A)^{-1} BK\right]. \qquad (15.74)$$

Now, using the identity

$$\det\left[I_n + \underset{n\times m}{E}\ \underset{m\times n}{G}\right] = \det\left[I_m + \underset{m\times n}{G}\ \underset{n\times m}{E}\right]$$

and using this in Equation 15.74 yields

$$\phi_{cl}(s) = \phi_{ol}(s) \det\left[I + (sI - A)^{-1} BK\right]$$

$$= \phi_{ol}(s) \det\left[I + \underbrace{K(sI - A)^{-1} B}_{L(s)}\right] = \phi_{ol}(s) \det[I + L(s)] \qquad (15.75)$$

where $\phi_{ol}(s)$ is the open-loop system's characteristic polynomial, and $\phi_{cl}(s)$ is the closed-loop system's characteristic polynomial. If $\phi_{cl}(s)$ is stable (the closed-loop system is stable), then $N(0, \phi_{cl}(s), D_R) = 0$. From Equation 15.75, stability of $\phi_{cl}(s)$ requires that

$$N\left(0, \phi_{ol}(s), D_R\right) + N\left(0, \det\left[I + L(s)\right], D_R\right) = 0. \qquad (15.76)$$

With this understanding, we can state the multivariable Nyquist theorem.

Theorem 15.1: Multivariable Nyquist Theorem

The feedback control system shown in Figure 15.18 will be closed-loop stable in the sense that $\phi_{cl}(s)$ has no closed RHP zeros if and only if for all R sufficiently large (radius of the D-contour)

$$N(0, \det[I + L(s)], D_R) = -P_{ol} \tag{15.77}$$

or equivalently

$$N(-1, -1 + \det[I + L(s)], D_R) = -P_{ol}$$

where $P_{ol} = N(0, \phi_{ol}(s), D_R)$ equals the number of open-loop RHP poles. ∎

The multivariable Nyquist theorem (MNT) states that closed-loop stability requires the number of encirclements made by the determinant of the RDM locus to be equal to the number of unstable open-loop poles. Encirclements can be counted relative to the origin $(0, j0)$ or, as in classical Nyquist diagrams, about $(-1, j0)$.

Stability margins for multivariable systems can be derived using the MNT by assuming that the controller $K(s)$ stabilizes the nominal plant $G(s)$ and that gain and phase uncertainties are large enough to change the number of encirclements made by the determinant of the RDM locus. The assumption that the nominal plant is stabilized by the controller tells us that the RDM encircles the origin P_{ol} times in the proper sense. Gain and phase margins can be computed by inserting a gain and phase variation $ke^{i\phi}$ in between the controller $K(s)$ and plant $G(s)$ and solving for the gain k (with $\phi = 0$) and phase ϕ (with $k = 1$) that destabilizes the system. To proceed in a more general manner, consider the stability analysis model shown in Figure 15.19, where the uncertainties in the system (gain and phase uncertainties) are represented in a block matrix $\Delta(s)$ and the nominal plant and controller are represented in a matrix $M(s)$. Techniques for deriving these models will be presented in the next section.

The stability analysis question is as follows: how large can the uncertainties $\Delta(s)$ become before the system becomes unstable? The loop transfer

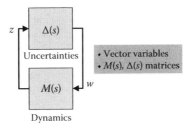

FIGURE 15.19
System block diagram with uncertainties.

function matrix $L(s)$ for this system is $L(s) = \Delta(s)M(s)$, with the RDM given by $I + L(s) = I + \Delta(s)M(s)$. Using the MNT, for the system to become unstable, the uncertainties $\Delta(s)$ must change the number of encirclements made by the $\det[I + L(s)]$ locus.

As long as the RDM $I + L(s)$ is nonsingular (for $s = j\omega$ along the D-contour), the number of encirclements made by the $\det[I + L(s)]$ locus will not change. This is best explained by examining the $\det[I + L(s)]$ locus as s transverses the D_R contour. Fundamental to this approach is the assumption that the nominal closed-loop system is stable, that is, the control design stabilizes the open-loop system.

Assuming that the nominal closed-loop system is stable, $\phi_{cl}(s)$ is a stable polynomial, in that it has no RHP zeros. Let $f(s) = \det[I + L(s)]$, and represent $f(j\omega)$ with its magnitude and phase as

$$f(j\omega) = |f(j\omega)|e^{j\phi(\omega)} \tag{15.78}$$

as s transverses the D_R contour in the s-plane.

Consider the $j\omega$ axis path A shown in Figure 15.20, where $0 \leq \omega \leq +\infty$. The section A locus of $f(j\omega)$ is shown in Figure 15.21a. At low frequencies, the

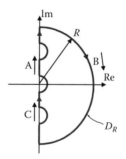

FIGURE 15.20
Nyquist D_R contour.

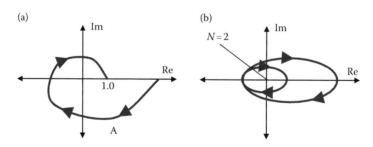

FIGURE 15.21
Counting encirclements.

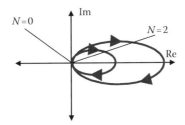

FIGURE 15.22
Counting encirclements.

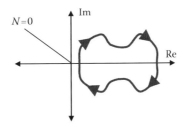

FIGURE 15.23
Counting encirclements.

magnitude of $f(j\omega)$ is large due to the magnitude of $L(j\omega)$. As $\omega \to \infty$, the loop transfer matrix (LTM) $L(j\omega) \to 0$, resulting in the $\det[I + L(j\omega)] = 1$ $(1, j0)$. Along the infinite radius path B, $s = e^{j\psi}R$, with $R \to \infty$ and $-\pi/2 \le \psi \le \pi/2$. When $R \to \infty$, $L(j\omega) \to 0$. This results in encirclements of the point $(1, j0)$. Section C will be the complex conjugate of the section A path. Figure 15.21b shows the entire locus and the number of encirclements N. Figure 15.21b shows there are two clockwise encirclements of the origin.

The number of encirclements N of the $\det[I + L(s)]$ locus must be equal to the number of open-loop unstable poles, P_{ol}, if the closed-loop system is to be stable. If the $\det[I + L(s)]$ were equal to zero, then the number of encirclements would be indeterminate, or at least not equal to P_{ol}. This is shown in Figure 15.22. In order for the number of encirclements to change, $\det[I + L(s)]$ must equal zero at some frequency.

If $\phi_{ol}(s)$ is a stable polynomial, then $P_{ol} = 0$. An example $\det[I + L(s)]$ locus for this condition is shown in Figure 15.23. In order for stable system to be destabilized by uncertainties Δ, the origin must be encircled.

15.3.2.3 Stability Margins for MIMO Systems

Uncertainty models used for stability analysis may be categorized as unstructured or structured. If the system's uncertainty is modeled as a full single-block matrix, the uncertainty is unstructured. If the uncertainty is modeled as a block diagonal matrix, the uncertainty is structured. Both unstructured and structured uncertainty analysis procedures use singular value theory to measure the size of complex valued matrices.

The following robustness theorems that are used to derive stability margins for multivariable systems are derived from an application of the MNT. Consider the control system shown in Figure 15.18. The basic problem is to determine the robustness of the design in the presence of uncertainties. This design has the state space realization using the triple (A, B, K) with the LTM given by

$$L(s) = K(sI - A)^{-1}B. \tag{15.79}$$

We wish to determine to what extent the parameters in the LTM can vary without compromising the stability of the closed-loop system. From the previous section, we know that

$$\det[I + L(s)] = \frac{\phi_{cl}(s)}{\phi_{ol}(s)} \tag{15.80}$$

where

$\phi_{ol}(s) = \det[sI - A]$: open-loop characteristic polynomial
$\phi_{cl}(s) = \det[sI - A + BK]$: closed-loop characteristic polynomial.

Using the MNT, stability for this system can be stated as follows.

The system of Figure 15.18 will be closed-loop stable in the sense that $\phi_{cl}(s)$ has no closed RHP zeros if and only if for all R sufficiently large,

$$N(0, \det[I + L(s)], D_R) = -P \tag{15.81}$$

or equivalently

$$N(-1, -1 + \det[I + L(s)], D_R) = -P \tag{15.82}$$

where D_R is the standard Nyquist D contour, which encloses all P closed RHP zeros of $\phi_{ol}(s)$. Note that $N(b_1, f(s), D)$ is indeterminate if $\phi(s_0) = b_1$ for some s_0 on the contour D.

The stability robustness of a multivariable system can be observed by the near singularity of its RDM, $I + L(s)$, at some frequency $s = j\omega_0$. If $I + L(s)$ is nearly singular, then a small change in $L(s)$ could make $I + L(s)$ singular. From a SISO viewpoint, this is the distance from the $(-1, j0)$ point in the complex plane made by the gain loci $L(j\omega)$. If the gain loci then encircle the $(-1, j0)$, point instability results. The robustness theory discussed here gives an analogous distance measure for multivariable systems.

Application of the MNT above is of little applicability as a robustness indicator because $\det[I + L(s)]$ does not indicate the near singularity of $I + L(s)$. The MNT only determines absolute stability. To determine the degree of robustness for a multivariable system, we determine how nearly singular the RDM is by computing its singular values versus frequency.

Examining the magnitude of the singular values of the RDM will indicate how close the matrix is to being singular. This measure of closeness to singularity is used in forming a multivariable gain margin, similar to the classical gain margin. However, as with many matrix norms, there is a restriction on the applicability of the singular value analysis. This restriction states that the compensated system described using the nominal $L(s)$ is closed-loop stable.

Classical gain and phase margins are used to measure the robustness of SISO systems to perturbations in the feedback loop. Singular values are used in measuring the robustness of multivariable systems. Let $L'(s)$ denote the perturbed LTM, which represents the actual system and differs from the nominal LTM $L(s)$ because of uncertainties in the open-loop plant model. Assume that $L'(s)$ has the state space realization (A', B', K') and open- and closed-loop polynomials given by

$$\phi'_{ol}(s) = \det[sI - A']\tag{15.83}$$

$$\phi'_{cl}(s) = \det[sI - A' + B'K'],$$

respectively. Define $L(s,\varepsilon)$ as a matrix of rational transfer functions with real coefficients that are continuous in ε for all ε such that $0 \le \varepsilon \le 1$ and for all $s \in D_R$, which satisfies $L(s,0) = L(s)$ and $L(s,1) = L'(s)$. Using these definitions of the perturbed model, we are ready to state the following fundamental robustness theorem.

Theorem 15.2

The polynomial $\phi'_{cl}(s)$ has no closed RHP zeros, and the perturbed feedback system is stable if the following hold:

  1. a) $\phi_{ol}(s)$ and $\phi'_{ol}(s)$ have the same number of closed RHP zeros.
     b) $\phi_{cl}(s)$ has no closed RHP zeros.
  2. $\det[I + L(s,\varepsilon)] = 0 \; \forall \; (s,\varepsilon)$ in $D_R \times [0,1]$ and $\forall \; R$ sufficiently large.

∎

This theorem states that the closed-loop perturbed system will be stable; if, by continuously deforming the Nyquist loci for the nominal system into that of the perturbed system $I + L(s,\varepsilon)$, the number of encirclements of the critical point is the same for $L'(s)$ and $L(s)$, then no closed RHP zeros were introduced into $\phi'_{cl}(s)$, resulting in a stable closed-loop system.

This theorem is used to develop simple tests that are developed for different types of model error characterizations. Just as there is not a unique representation for dynamic systems, there are many different forms for describing their modeling errors. The most common model error characterizations are additive errors and multiplicative errors (also described as relative or absolute errors). The classical gain and phase margins are associated with multiplicative error models since these margins are multiplicative in nature.

Let $\Delta(s)$ denote the modeling error under consideration. The additive model error is given by

$$\Delta_a(s) = L'(s) - L(s), \tag{15.84}$$

and the multiplicative model error is given by

$$\Delta_m(s) = [L'(s) - L(s)]L^{-1}(s). \tag{15.85}$$

The perturbed LTM can be constructed using Equations 15.84 and 15.85. For the additive error model, we have

$$L(s,\varepsilon) = L(s) + \varepsilon\Delta_a(s), \tag{15.86}$$

and for the multiplicative error model, we have

$$L(s,\varepsilon) = [I + \varepsilon\Delta_m(s)]L(s). \tag{15.87}$$

Both Equations 15.84 and 15.85 imply the same $L(s,\varepsilon)$ using different model error characterizations. In both Equations 15.84 and 15.85, $L(s,\varepsilon)$ is given by

$$L(s,\varepsilon) = (1 - \varepsilon)L(s) + \varepsilon L'(s) \tag{15.88}$$

showing that $L(s,\varepsilon)$ is continuous in ε for $\varepsilon \in [0,1]$ and for all $s \in D_R$.

We have now defined the true perturbed plant model in terms of its nominal design model and the uncertainty matrix. The fundamental robustness theorem uses the RDM $I + L(s,\varepsilon)$ to determine if the number of encirclements of the critical point will change with the uncertainties. This happens when $I + L(s,\varepsilon)$ becomes singular, in which case $\det[I + L(s,\varepsilon)] = 0$.

Using the multiplicative error characterization, the RDM is

$$I + L(s,\varepsilon) = I + L(s) + \varepsilon\Delta_m(s)L(s) \tag{15.89}$$

or

$$I + L(s,\varepsilon) = A + B \tag{15.90}$$

with $A = I + L(s)$ and $B = \varepsilon\Delta_m(s)L(s)$. For the perturbed system to be unstable, viewed through a change in the number of encirclements of $\det[I + L(s,\varepsilon)]$, the matrix $A + B$ must be singular for some $\varepsilon \in [0,1]$ and $s \in D_R$. We know that $A = I + L(s)$ is nonsingular (the RDM of the nominal design) since the nominal design is closed-loop stable. Thus, if the uncertainty is going to create instability, then the matrix $B = \varepsilon\Delta_m(s)L(s)$, when added to A, must make $A + B$ singular.

$A + B$ Argument

The minimum singular value $\underline{\sigma}(A)$ measures the near singularity of the matrix A. Assume that the matrix $A + B$ is singular. If $A + B$ is singular, then $A + B$ is rank deficient. Since $A + B$ is rank deficient, then there exists a vector $x \neq 0$ with unit magnitude ($\|x\|_2 = 1$) such that $(A + B)x = 0$ (x is in the null space of $A + B$). This leads to $Ax = -Bx$ with $\|Ax\|_2 = \|Bx\|_2$. Using the above singular value definitions in Equation 15.57 and $\|x\|_2 = 1$, we obtain the following inequality:

$$\underline{\sigma}(A) \le \|Ax\|_2 = \|Bx\|_2 \le \|B\|_2 = \bar{\sigma}(B). \tag{15.91}$$

If the matrix $A + B$ is singular, then $\underline{\sigma}(A) \le \bar{\sigma}(B)$. For $A + B$ to be nonsingular, $\underline{\sigma}(A) > \bar{\sigma}(B)$. This is precisely how the stability robustness tests are derived. ∎

Theorem 15.3: Stability Robustness Theorem— Additive Uncertainty Model

The polynomial $\phi'_{cl}(s)$ has no closed RHP zeros and the perturbed feedback system is stable if the following hold:

1. $\phi_{cl}(s)$ has no closed RHP zeros.
2. $\underline{\sigma}(I + L(s)) > \bar{\sigma}(\Delta_a(s)) \; \forall \; s \in D_R$ and for all R sufficiently large, with $\Delta_a(s)$ given by Equation 15.84.

See Wise [14] and references therein for proof of this theorem. ∎

Theorem 15.4: Stability Robustness Theorem— Multiplicative Uncertainty Model

The polynomial $\phi'_{cl}(s)$ has no closed RHP zeros and the perturbed feedback system is stable if the following hold:

1. $\phi_{cl}(s)$ has no closed RHP zeros.
2. $\underline{\sigma}(I + L^{-1}(s)) > \bar{\sigma}(\Delta_m(s)) \; \forall \; s \in D_R$ and for all R sufficiently large, with $\Delta_m(s)$ given by Equation 15.85.

∎

The proof of this theorem uses the singularity of the $A + B$ argument. Stability of the perturbed closed-loop system is guaranteed for a nonsingular $I + L(s,\varepsilon)$. Thus,

$$I + L(s,\varepsilon) = L(s)(I + L^{-1}(s) + \varepsilon\Delta_m(s)). \tag{15.92}$$

Here we assume that $L^{-1}(s)$ exists. If $I + L(s,\varepsilon)$ is to be singular, then the matrix $I + L^{-1}(s) + \varepsilon\Delta_m(s)$ must be singular. Thus, to be nonsingular,

$$\underline{\sigma}(I + L^{-1}(s)) > \bar{\sigma}(\varepsilon\Delta_m(s)) \tag{15.93}$$

or

$$\underline{\sigma}(I + L^{-1}(s)) > |\varepsilon|\,\bar{\sigma}(\Delta_m(s))$$
$$\underline{\sigma}(I + L^{-1}(s)) > \bar{\sigma}(\Delta_m(s)). \tag{15.94}$$

Depending upon the model error characterization, either additive or multiplicative, the robustness test is different. Theorems 15.3 and 15.4 are sufficient tests for stability. As long as the singular value frequency responses do not overlap, stability is guaranteed. These theorems can be used to derive multivariable stability margins, also called singular value stability margins. They are a natural extension of classical SISO gain and phase margins to multivariable systems.

Consider the SISO system shown in Figure 15.24. Gain and phase margins for this system are computed by inserting a gain and phase variation $k_i e^{i\phi_i}$ in between the controller $K(s)$ and plant $G(s)$ and solving for the gain k_i (with $\phi_i = 0$) and phase ϕ_i (with $k_i = 1$) that destabilizes the system.

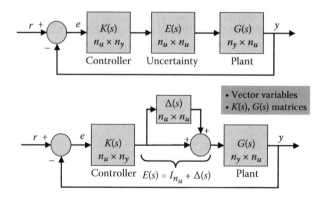

FIGURE 15.24
Uncertainty models.

First, consider the computation of a gain margin at the input to the plant. Place in each input channel a scalar gain $\varepsilon_i \in C$, with $E(s) = \text{diag}[\varepsilon_i] \in C^{n_u \times n_u}$ modeling these gains as a matrix. For the nominal condition with no uncertainty, $\varepsilon_i = 1$, and the system is stable. Positive and negative gain margins would indicate how large and small, respectively, the scalar gain ε_i needs to be to destabilize the system. Our analysis problem will focus on independent uncertainties in each channel, with the gain margin relating to the smallest gain uncertainty that can destabilize the system. Figure 15.24 indicates how this gain uncertainty enters into the block diagram and how it can be represented using $\Delta(s)$.

As shown in Figure 15.24, $\Delta(s) = E(s) - I$. For the nominal control system, let

$$\min_{\omega} \underline{\sigma}(I + L^{-1}) = \beta_\sigma.$$

From Theorem 15.4, stability is guaranteed if $\underline{\sigma}(I + L^{-1}(s)) > \bar{\sigma}(\Delta(s))$. For $\Delta(s) = E(s) - I$, $E(s) \in C^{n_u \times n_u}$, the singular values of $\Delta(s)$ are

$$\sigma_i(\Delta(s)) = \sigma_i(E(s) - I) = |\varepsilon_i - 1|.$$

If the largest $|\varepsilon_i - 1|$ is smaller than β_σ, then for $\varepsilon_i \in \mathbb{R}$

$$1 - \beta_\sigma \le \varepsilon_i \le 1 + \beta_\sigma, \tag{15.95}$$

which guarantees a gain margin of $[1 - \beta_\sigma, 1 + \beta_\sigma]$ for the system. If we consider the phase margin problem, $\varepsilon_i = \exp(j\phi_i(\omega))$, $\phi_i(\omega) \in \mathbb{R}$, $E(s) = \text{diag}[\exp(j\phi_i(\omega))]$, then

$$|\varepsilon_i - 1| = \left| e^{j\phi_i(\omega)} - 1 \right| \le \beta_\sigma$$

$$= \left| \cos(\phi_i(\omega)) - 1 + j\sin(\phi_i(\omega)) \right| \le \beta_\sigma$$

$$= \left(\cos^2(\phi_i(\omega)) - 2\cos(\phi_i(\omega)) + 1 + \sin^2(\phi_i(\omega)) \right)^{\frac{1}{2}} \le \beta_\sigma$$

$$= (2(1 - \cos(\phi_i(\omega))))^{\frac{1}{2}} \le \beta_\sigma$$

$$= \left(4\sin^2\left(\frac{\phi_i(\omega)}{2} \right) \right)^{\frac{1}{2}} \le \beta_\sigma$$

which guarantees a phase margin of $\pm 2\sin^{-1}\dfrac{\beta_\sigma}{2}$ for the system.

15.3.2.3.1 Singular Value Stability Margins

1. RDM:

   Let $\min_{\omega} \underline{\sigma}(I + L) = \alpha_\sigma$, then

   $$GM_{I+L} = \left[\frac{1}{1+\alpha_\sigma}, \frac{1}{1-\alpha_\sigma} \right]; PM_{I+L} = \pm 2 \sin^{-1} \frac{\alpha_\sigma}{2}. \tag{15.96}$$

2. Stability robustness matrix:

   Let $\min_{\omega} \underline{\sigma}(I + L^{-1}) = \beta_\sigma$, then

   $$GM_{I+L^{-1}} = \left[1 - \beta_\sigma, 1 + \beta_\sigma \right]; PM_{I+L^{-1}} = \pm 2 \sin^{-1} \frac{\beta_\sigma}{2} \tag{15.97}$$

   $$GM = GM_{I+L} \cup GM_{I+L^{-1}}, PM = PM_{I+L} \cup PM_{I+L^{-1}}. \tag{15.98}$$

Note that the best minimum singular value from the RDM is $\min_{\omega} \underline{\sigma}(I + L) = \alpha_\sigma = 1$ (at high frequencies $L \to 0$). Substituting this into Equation 15.96 produces a gain margin interval of $GM_{I+L} = \left[\frac{1}{2}, +\infty \right]$. Converting to decibels produces $GM_{I+L} = [-6, +\infty]$ dB. Similarly, the best minimum singular value from the stability robustness matrix is $\min_{\omega} \underline{\sigma}(I + L^{-1}) = \beta_\sigma = 1$ (at low frequencies $L^{-1} \to 0$). Substituting this into Equation 15.97 produces a gain margin interval of $GM_{I+L^{-1}} = [0, 2]$. Converting to decibels produces $GM_{I+L^{-1}} = [-\infty, +6]$ dB.

15.3.3 Control System Robustness Analysis

15.3.3.1 Analysis Models for Uncertain Systems

Stability analysis models for multivariable systems can be formed to analyze gain and phase uncertainties, neglected and/or mismodeled dynamics, real parameter uncertainties, and combinations thereof using methods identical to forming models for SISO systems. These models can be easily formed using block diagram algebra, signal flow graph methods, or algebraic manipulation of loop equations. The resulting models will have a "structure" associated with them depending upon the specific problem, and the analysis will depend upon the structure.

Consider the multivariable control system shown in Figure 15.25. The block diagram shows uncertainties Δ_1 at the input to the plant and uncertainties Δ_2 at the output of the plant. The uncertainties Δ_1 and Δ_2 can be constructed

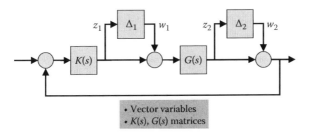

FIGURE 15.25
Expanded system block diagram with uncertainties.

to model any type of uncertainty, depending upon the analysis question at hand (gain and phase margins, uncertain actuator or sensor dynamics, etc.). The RDM for the block diagram in Figure 15.25 is $I + K(I + \Delta_2)G(I + \Delta_1)$. To ease analysis of this system, the block diagram shown in Figure 15.25 is transformed into the general analysis model shown in Figure 15.19.

Figure 15.19 illustrates a general control system analysis model in which the matrix $\Delta(s)$ models uncertainties and $M(s)$ is a transfer function matrix modeling the dynamics between the output from the uncertainties and its input. The utility of this diagram is that it isolates the uncertainties from the known dynamics (plant +controller). Using this form, it is easier to determine the "smallest" Δ that destabilizes the system. We will use this ΔM representation of the dynamics for many of our stability analysis problems. The matrix $M(s)$ models the dynamics in the system that is assumed to be known and any weighting filters used to normalize the uncertainties.

The matrix $\Delta(s)$ will be a block diagonal matrix, with each matrix or scalar uncertainty in the system, $\Delta_i(s)$, located on the diagonal of $\Delta(s)$. The matrix $M(s)$ is a block matrix where the *ij*th block is the transfer function matrix from the output of the *j*th uncertainty $\Delta_j(s)$ to the input of the *i*th uncertainty $\Delta_i(s)$.

Consider the loop equations from Figure 15.26 written as

$$z_1 = K(s)(z_2 + w_2)$$

$$z_2 = G(s)(z_1 + w_1).$$

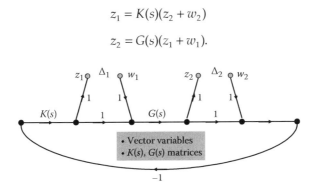

FIGURE 15.26
Signal flow graph model for writing loop equations.

Substituting the z_2 expression into the z_1 equation and manipulating yields

$$z_1 = K(s)(G(s)(z_1 + w_1) + w_2)$$

$$z_1 = K(s)G(s)(z_1 + w_1) + K(s)w_2$$

$$(I - K(s)G(s))z_1 = K(s)G(s)w_1 + K(s)w_2$$

$$z_1 = (I - K(s)G(s))^{-1} K(s)G(s)w_1 + (I - K(s)G(s))^{-1} K(s)w_2.$$

Substituting the z_1 expression into the z_2 equation and manipulating yields

$$z_2 = G(s)(K(s)(z_2 + w_2) + w_1)$$

$$z_2 = G(s)K(s)(z_2 + w_2) + G(s)w_1$$

$$(I - G(s)K(s))z_2 = G(s)K(s)w_2 + G(s)w_1$$

$$z_2 = (I - G(s)K(s))^{-1} G(s)w_1 + (I - G(s)K(s))^{-1} G(s)K(s)w_2.$$

Combining these two expressions and writing in matrix form yields

$$\begin{bmatrix} z_1 \\ z_2 \end{bmatrix} = \underbrace{\begin{bmatrix} (I - K(s)G(s))^{-1} K(s)G(s) & (I - K(s)G(s))^{-1} K(s) \\ (I - G(s)K(s))^{-1} G(s) & (I - G(s)K(s))^{-1} G(s)K(s) \end{bmatrix}}_{M(s)} \begin{bmatrix} w_1 \\ w_2 \end{bmatrix}. \quad (15.99)$$

The loop equations for the uncertainties modeled in the system can be written as

$$\begin{bmatrix} w_1 \\ w_2 \end{bmatrix} = \underbrace{\begin{bmatrix} \Delta_1 & 0 \\ 0 & \Delta_2 \end{bmatrix}}_{\Delta(s)} \begin{bmatrix} z_1 \\ z_2 \end{bmatrix}. \quad (15.100)$$

The RDM is now $I - \Delta M$. Many of the robustness tests focus on the singularity of the RDM. This form isolates the uncertainties from the known system dynamics and controller, making it easier to determine the size of the uncertainties that cause the return difference matrix to become singular.

15.3.3.2 Singular Value Robustness Tests

Singular value robustness tests are derived by examining the singularity of the RDM. If $\det[I - \Delta M] = 0$, then from the $A + B$ argument of the preceding section, we know that

$$\underline{\sigma}[I] > \bar{\sigma}[\Delta M]. \tag{15.101}$$

Using $\bar{\sigma}[\Delta M] \le \bar{\sigma}[M]\bar{\sigma}[\Delta]$, and the fact that $\underline{\sigma}[I] = 1$, we obtain what is referred to as the small gain theorem (SGT):

$$\bar{\sigma}[\Delta] < 1 / \bar{\sigma}[M]. \tag{15.102}$$

The SGT is a sufficient test for stability. If it is violated, the system may still be stable. It is also typically conservative, depending upon the structure of the uncertainty. This conservatism is introduced when the bound $\bar{\sigma}[\Delta M] \le \bar{\sigma}[M]\bar{\sigma}[\Delta]$ is used. Bounding $\bar{\sigma}[\Delta M]$ with the product $\bar{\sigma}[M]\bar{\sigma}[\Delta]$ ignores any structure that may be present in Δ and can produce a conservative analysis.

15.3.3.3 Real Stability Margin

At a fixed frequency ω, the real margin algorithm [13] maps the space D of uncertain parameters into the Nyquist plane using the MNT. This procedure is shown in Figure 15.27 for a three-dimensional parameter space. The solid cube in the parameter space represents all allowable combinations of uncertain parameters. The vertices of this cube are the extreme variations allowed for each parameter. The uncertain parameters and nominal system dynamics are modeled using the ΔM representation shown in Figure 15.18.

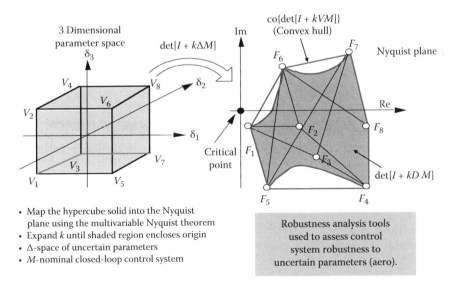

- Map the hypercube solid into the Nyquist plane using the multivariable Nyquist theorem
- Expand k until shaded region encloses origin
- Δ-space of uncertain parameters
- M-nominal closed-loop control system

Robustness analysis tools used to assess control system robustness to uncertain parameters (aero).

FIGURE 15.27
Real margin stability analysis.

If there is a combination of parameters Δ_0 that destabilizes the system, then $\det[I + \Delta_0 M] = 0$.

The bounds on the allowable parameter variations will form an n_p-dimensional polytope in the parameter space (in two dimensions, this is a rectangle). By scaling the parameter space, the polytope is transformed into a hypercube (a square in two dimensions). The scaled parameter-space hypercube solid, shown in Figure 15.27, maps into the shaded region in the Nyquist plane using the $\det[I + k\Delta M]$ function. The scalar real margin k multiplies Δ and is expanded until the shaded region encloses the origin (the nominal design matrix M is stable, so $k = 0$ is a stable point).

When mapping the hypercube solid into the Nyquist plane, it is not computationally feasible to compute the shaded region in Figure 15.27. However, the convex hull enclosing this region is easily computed. The convex hull enclosing this region is formed by mapping the hypercube vertices V_i into the points F_i in the Nyquist plane. The outer boundary enclosing all vertex points F_i is then the convex hull co{$\det[I + kVM]$} (here $\det[I + kVM]$ represents the entire set of vertex points mapped into the complex plane). The real margin k is then expanded until the origin is contained in co{$\det[I + kVM]$}. This provides a lower bound on k, denoted k_l. An upper bound, denoted k_u, is formed by splitting the parameter space into subdomains and recomputing k for each subdomain. This procedure is then repeated, improving the accuracy of k. As the parameter space is split into smaller and smaller subdomains, the unstable region in the parameter space is determined. (The general idea is that the union of infinitesimal slices in the parameter space, mapped into the Nyquist plane, approaches the true image of the hypercube.) The real margin k must be computed at each frequency along the $j\omega$ axis. The minimum k, versus frequency, is then used as the real margin, denoted k_m. This defines the real margin as

$$k_m = \min_{\Delta}\left\{k \in [0, \infty) \mid \det(I - k\Delta M) = 0\right\}. \tag{15.103}$$

15.4 Optimal Flight Control Design

Robust high-performance flight control system design requirements are generally driven by high maneuver rates needed for postlaunch agile maneuvers, off boresight target updates, and/or terminal homing. The solution will tend toward a high-bandwidth autopilot design that satisfies the actuation position and rate limitations and is robust. Robustness concerns are often related to large launch envelopes and uncertainties in plant dynamics and aerodynamics throughout the flight envelope [15]. Optimal control methods offer promising solutions to this complex control problem and have proven themselves in application.

This section presents the robust servomechanism design model. This model creates a problem formulation such that when solved using optimal control theory, it produces an autopilot that tracks commands from guidance with zero error and has excellent stability and robustness properties.

15.4.1 Robust Servomechanism Linear Quadratic Regulator

This section presents the method for constructing the robust servomechanism linear quadratic regulator (RSLQR). Consider the following finite dimensional LTI first-order state space model:

$$\dot{x} = Ax + Bu + Ew$$
$$y = Cx + Du + Fw \qquad (15.104)$$

with w an unmeasurable disturbance, and $x \in \mathbb{R}^{n_x}$, $u \in \mathbb{R}^{n_u}$, and $y \in \mathbb{R}^{n_y}$. The command input vector $r \in \mathbb{R}^{n_r}$ has dimension less than the number of outputs (i.e., $n_y < n_r$), and it is assumed that the following pth-order differential equation for $r(t)$ is known, that is,

$$\overset{(p)}{r} = \sum_{i=1}^{p} a_i \overset{(p-i)}{r} \qquad (15.105)$$

where a_i are known scalars, and the superscript denotes the order of the derivative. The polynomial formed by the Laplace transformation of Equation 15.105 is

$$a(s) = s^p + \sum_{i=1}^{p} a_i s^{p-i} \qquad (15.106)$$

and describes a known class of inputs without knowledge of their magnitudes. For the disturbance inputs, we assume the same model as used for $r(t)$:

$$\overset{(p)}{w} = \sum_{i=1}^{p} a_i \overset{(p-i)}{w} \qquad (15.107)$$

where $w(0) = w_0$ is unknown. Define the error signal as

$$e = y_c - r \qquad (15.108)$$

where y_c is a partition of the system outputs, $y = [y_c \quad y_{nc}]^T$, $y_c \in \mathbb{R}^{n_{y_c}}$. Tracking in y_c is regulation in e. The controller design objective is to make the command error approach zero $e \to 0$ ($y_c \to r$) as $t \to \infty$, in the presence of unmeasurable disturbances $w(t)$, in a robust manner with respect to the plant description. Differentiating the error expression, Equation 15.108, p times yields

$$\overset{(p)}{e} - \sum_{i=1}^{p} a_i \overset{(p-i)}{e} = \overset{(p)}{y_c} - \sum_{i=1}^{p} a_i \overset{(p-i)}{y_c} - \underbrace{\left(\overset{(p)}{r} - \sum_{i=1}^{p} a_i \overset{(p-i)}{r} \right)}_{=0}. \tag{15.109}$$

The second term of the right-hand side of Equation 15.109 is zero by definition. Using Equation 15.104, we have

$$\overset{(p-i)}{y_c} = C_c \overset{(p-i)}{x} + D_c \overset{(p-i)}{u} + F_c \overset{(p-i)}{w}. \tag{15.110}$$

Substituting this into Equation 15.109 yields

$$\overset{(p)}{e} - \sum_{i=1}^{p} \alpha_i \overset{(p-i)}{e} = C_c \left[\overset{(p)}{x} - \sum_{i=1}^{p} \alpha_i \overset{(p-i)}{x} \right] + D_c \left[\overset{(p)}{u} - \sum_{i=1}^{p} \alpha_i \overset{(p-i)}{u} \right] + F_c \left[\overset{(p)}{w} - \sum_{i=1}^{p} \alpha_i \overset{(p-i)}{w} \right]. \tag{15.111}$$

The third term of Equation 15.110 is also zero by definition. This equation defines a set of simultaneous differential equations. Let ξ and μ be defined as

$$\xi = \overset{(p)}{x} - \sum_{i=1}^{p} \alpha_i \overset{(p-i)}{x}; \quad \mu = \overset{(p)}{u} - \sum_{i=1}^{p} \alpha_i \overset{(p-i)}{u}, \tag{15.112}$$

which are linear combinations of the derivatives of the state and control satisfying the reference command model dynamics. Substituting these into the error equation yields

$$\overset{(p)}{e} - \sum_{i=1}^{p} \alpha_i \overset{(p-i)}{e} = C_c \xi + D_c \mu. \tag{15.113}$$

Differentiating Equation 15.112 yields $\dot{\xi} = \overset{(p+1)}{x} - \sum_{i=1}^{p} \alpha_i \overset{(p-i+1)}{x}$. Using Equation 15.104, we have

$$\dot{\xi} = A \underbrace{\left[\overset{(p)}{x} - \sum_{i=1}^{p} \alpha_i \overset{(p-i)}{x} \right]}_{\xi} + B \underbrace{\left[\overset{(p)}{u} - \sum_{i=1}^{p} \alpha_i \overset{(p-i)}{u} \right]}_{\mu} + E \underbrace{\left[\overset{(p)}{w} - \sum_{i=1}^{p} \alpha_i \overset{(p-i)}{w} \right]}_{=0}. \tag{15.114}$$

A new state vector z may be defined as follows:

$$z = \left[e \quad \dot{e} \quad \cdots \quad \overset{(p-1)}{e} \quad \xi \right]^{T}. \tag{15.115}$$

This new state vector z has dimension $\left(n_x + pn_{y_c} \right)$. Differentiating Equation 15.115 yields the robust servomechanism design model system defined as

$$\dot{z} = \tilde{A}z + \tilde{B}\mu \tag{15.116}$$

with $\tilde{A}$ and $\tilde{B}$ given by

$$\tilde{A} = \begin{bmatrix} 0 & I & 0 & \cdots & 0 & 0 \\ 0 & 0 & I & & 0 & 0 \\ & & & \ddots & & \\ 0 & 0 & & 0 & I & 0 \\ a_p I & a_{p-1} I & \cdots & a_2 I & a_1 I & C_c \\ 0 & \cdots & \cdots & \cdots & 0 & A \end{bmatrix} \quad \tilde{B} = \begin{bmatrix} 0 \\ 0 \\ \vdots \\ 0 \\ D_c \\ B \end{bmatrix}. \tag{15.117}$$

The RSLQR state feedback autopilot is obtained by applying linear quadratic regulator theory to Equation 15.116. By regulating z, we regulate to zero both e and ξ. In steady state, this allows the state vector x to be non-zero in which y_c tracks the command r. This control formulation adds integral control action acting on the command error and creates a controller who is "type p," which is required to track the class of signals described in Equations 15.105 and 15.106.

Consider a constant input command r. This gives $\dot{r} = 0$ ($p = 1$) with $a_1 = 0$ (Equation 15.106). The state space system using Equation 15.116 is given by

$$\dot{z} = \underbrace{\begin{bmatrix} 0 & C_c \\ 0 & A \end{bmatrix}}_{\tilde{A}} z + \underbrace{\begin{bmatrix} D_c \\ B \end{bmatrix}}_{\tilde{B}} \mu. \tag{15.118}$$

LQR control theory is applied to Equation 15.118 using the performance index (PI):

$$J = \int_0^\infty (z^T Q z + \mu^T R \mu) \, d\tau. \tag{15.119}$$

The optimal steady-state control law for μ using state feedback is formed by solving the algebraic Riccati equation (ARE) using Q and R from Equation 15.119. The resulting steady-state $n_u \times (p + n_x)$-dimensional feedback controller gain matrix K_c is partitioned as

$$K_c = [K_I \quad K_x] \tag{15.120}$$

where K_I multiplies the feedback of the integral on the command error vector, and K_x multiplies the feedback of the states. The optimal control u is obtained by integrating μ, that is,

$$u = \int \mu \, d\tau = -\int [K_I \quad K_x] z \, d\tau = -[K_I \quad K_x] \int \begin{bmatrix} e \\ \dot{x} \end{bmatrix} d\tau = -K_I \int e - K_x x \tag{15.121}$$

This controller mechanization yields integral control action on the command error to provide zero steady-state error command-following. The state vector x must be available for feedback. The implementation of this state feedback design is shown in Figure 15.18. We see from this approach that integral error control is added to the baseline plant dynamics. The "type" of controller, the number of integrators added, depends upon the class of signal to be tracked. For constant commands assumed here, $p = 1$, and a single integrator is added, producing a type 1 controller. For ramp-type commands, $p = 2$, and two integrators are added, producing a type 2 controller. In practice, type 1 controllers have been found to be acceptable in missile autopilot design problems.

The disturbance models from Equation 15.104 (Equation 15.107) satisfy the same differential equation as the command (Equation 15.105). This says that for constant commands, the controller will reject constant disturbances, with the magnitude of the disturbance unknown. For ramp commands, the controller will reject ramp disturbances, etc. The integral control action of the robust servomechanism is similar to what a "classical" integral control autopilot would provide.

15.4.2 Design Summary for RSLQR

The RSLQR incorporates integral control into an LQR state feedback design to build a type 1 controller. This will produce a controller that achieves zero steady-state error to constant commands. The autopilot design model in state space form is

$$\dot{x} = \tilde{A}x + \tilde{B}u + Fr$$

$$x = \begin{bmatrix} \int e_r \\ x_p \end{bmatrix}; \quad \tilde{A} = \begin{bmatrix} 0 & C_c \\ 0 & A \end{bmatrix}; \quad \tilde{B} = \begin{bmatrix} D_c \\ B \end{bmatrix} \quad (15.122)$$

$$\tilde{A} = \begin{bmatrix} 0 & 1 & 0 & 0 & 0 \\ 0 & -1.3046 & 1 & -0.2142 & 0 \\ 0 & 47.711 & 0 & -104.83 & 0 \\ 0 & 0 & 0 & 0 & 1 \\ 0 & 0 & 0 & -4624. & -81.6 \end{bmatrix}; \quad \tilde{B} = \begin{bmatrix} 0 \\ 0 \\ 0 \\ 0 \\ 4624. \end{bmatrix}.$$

A gain scheduled autopilot is designed by discretizing the flight envelope with α, Mach number, and altitude or dynamic pressure, linearizing the dynamics at the flight condition to form the LTI model (Equation 15.122), and solving an infinite time LQR problem with its associated ARE. From the ARE, the constant state feedback gain matrix is computed. The following is a summary of these LQR design equations:

$$\dot{x} = \tilde{A}x + \tilde{B}u \qquad \tilde{A}, \tilde{B} - \text{constant} \qquad x \in R^{n_x}, u \in R^{n_u} \qquad (15.123)$$

$$J = \int_0^\infty (x^T Q x + u^T R u)\, d\tau \qquad Q = Q^T \geq 0, \ R = R^T > 0 \qquad (15.124)$$

$$(\tilde{A}, \tilde{B}) \text{ controllable; } \left(\tilde{A}, Q^{1/2}\right) \text{ detectable} \qquad (15.125)$$

$$\text{ARE: } P\tilde{A} + \tilde{A}^T P + Q - P\tilde{B}R^{-1}\tilde{B}^T P = 0 \qquad u = -R^{-1}\tilde{B}^T P x = -K_c x. \quad (15.126)$$

Using the feedback gain matrix computed in Equation 15.126, the closed-loop system is given by

$$\dot{x} = \left(\tilde{A} - \tilde{B}K_c\right)x = \tilde{A}_{cl}x. \qquad (15.127)$$

It is the robustness properties of this system under feedback that we wish to investigate. The feedback gain matrix K_c obviously plays a central role in establishing these properties, and the selection of the numerical values is important. If the gains were selected using a pole placement method, the robustness properties would be different from that of the LQR.

There are many ways to select the LQR penalty matrices Q and R in Equation 15.126 that influence the feedback gain matrix K_c. In general, the gains are proportional to how "large" Q is and how "small" R is, with the magnitude of the gains proportional to Q/R. Methods like Bryson's rule [16] give good design guidelines for regulators, but in missile autopilot design, the goal is to achieve as high a bandwidth as possible (as fast a response as possible) subject to stability, robustness, and implementation constraints. The method used here to select Q and R is from Wise [17], and it uses LQR design charts to view performance versus bandwidth and selects numerical values by imposing constraints on stability margin or actuator usage. The LQR design charts are created by parameterizing the LQR penalty matrices as follows:

$$Q = q_{11} \begin{bmatrix} 1 & & & \\ & 0 & 0 & \\ & & 0 & \\ & 0 & 0 & \\ & & & 0 \end{bmatrix} ; \quad R = 1; \tag{15.128}$$

and numerically sweeping q_{11} [using q_{11} = logspace(1,6,100)]. At each value of q_{11}, the ARE is solved, the gains are computed, and the design is evaluated in both the frequency and time domains. In the time domain, a step response is used to compute rise time, settling time, and actuator usage. These data are plotted versus loop gain crossover frequency forming the design charts.

Figure 15.28 illustrates the rise time and settling time versus loop gain crossover frequency for the missile data in Equation 15.122. The selected design point is indicated in the figure by a circle on the curve. The LQR

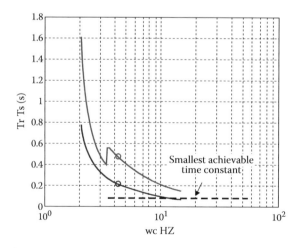

FIGURE 15.28
Rise time and settling time versus loop gain crossover frequency.

penalty is $q_{11} = 0.16681$. The gain matrix computed from the ARE, K_c, and closed-loop system matrix $\tilde{A}_{cl}$, Equation 15.127, are

$$K_c = [0.0681 \quad 0.0099 \quad -0.7494 \quad 2.9394 \quad 0.0101];$$

$$\tilde{A} - \tilde{B}K_c = \begin{bmatrix} 0 & 1.0 & 0 & 0 & 0 \\ 0 & -1.7497e+000 & -1.5516e+003 & 0 & -2.5475e+002 \\ 0 & -4.1241e-002 & 0 & -1.5110e+002 & 0 \\ 0 & 0 & 0 & 0 & 1.0 \\ -3.1503e+002 & -4.5702e+001 & 3.4651e+003 & -1.8216e+004 & -1.2837e+002 \end{bmatrix}.$$

15.4.3 Guaranteed Margins from LQR

The LQR has excellent stability guarantees. We will briefly review these guarantees, examine the closed-loop system's robustness to model uncertainties, and look at how adaptive control theory can improve them further. Start with the ARE given in Equation 15.126. Add and subtract sP from the ARE to obtain

$$-P(sI - \tilde{A}) - (-sI - \tilde{A}^T)P - P\tilde{B}R^{-1}\tilde{B}^T P + Q = 0. \qquad (15.129)$$

Let $\Phi(s) = (sI - \tilde{A})^{-1}$. Multiply by $B^T\Phi^T(-s)$ on the left and $\Phi(s)\tilde{B}$ on the right:

$$B^T P\Phi B + B^T\Phi^T(-s)PB + B^T\Phi^T(-s)PBR^{-1}B^T P\Phi B = B^T\Phi^T(-s)Q\Phi B. \qquad (15.130)$$

Add $R > 0$ from the PI to both sides. Using the state feedback control in Equation 15.126, the LTM at the input to the plant is

$$L(s) = R^{-1}\tilde{B}^T P\Phi(s)\tilde{B} = K_c\Phi(s)\tilde{B}. \qquad (15.131)$$

Using the LTM, we have

$$R \underbrace{R^{-1}B^T P\Phi B}_{L(s)} + \underbrace{B^T\Phi^T(-s)PBR^{-1}}_{L^T(-s)} R + \underbrace{B^T\Phi^T(-s)PBR^{-1}}_{L^T(-s)} R \underbrace{R^{-1}B^T P\Phi B}_{L(s)} + R$$

$$= B^T\Phi^T(-s)Q\Phi B + R \qquad (15.132)$$

which reduces to

$$R + RL(s) + L^T(-s)R + L^T(-s)RL(s) = R + B^T\Phi^T(-s)Q\Phi B$$

$$(I + L(s))^* R(I + L(s)) = R + B^T\Phi^T(-s)Q\Phi B. \qquad (15.133)$$

The matrix $B^T\Phi^T(-s)Q\Phi B$ is Hermitian positive semidefinite. Subtracting this matrix creates the inequality

$$(I + L(s))^* R(I + L(s)) \geq R. \tag{15.134}$$

Replacing $R > 0$ with $\lambda_{min}(R)I$ on the left and $\lambda_{max}(R)I$ on the right yields

$$(I + L(s))^* \lambda_{min}(R)I(I + L(s)) \geq \lambda_{max}(R)I$$

$$(I + L(s))^* (I + L(s)) \geq \frac{\lambda_{max}(R)}{\lambda_{min}(R)} I \geq I \tag{15.135}$$

$$\left\| (I + L(s)) \right\|^2 \geq 1.$$

which says that the RDM is always greater than 1 in magnitude. For a single-input system, at the plant input, this equates to the Nyquist loci never entering a unit disk centered at $(-1,j0)$ in the complex plane. These conditions are shown in Figure 15.29. This promises a $[-6\text{ dB}, +\infty]$ gain margin and $\pm60°$ phase margin. This property is guaranteed from using the optimal gains and would not necessarily be present for other gain matrices. Several questions arise as follows: How does this classical margin relate to the robustness properties relative to knowing the aerodynamic parameters, accurately knowing the actuator dynamics, and/or any unmodeled high-frequency dynamics (flexible body dynamics)? Also, how does adaptive control contribute to the robustness and overall performance?

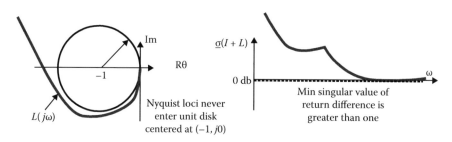

FIGURE 15.29
Loop transfer function and return difference frequency responses for state feedback LQR.

15.5 Adaptive Control Augmentation of Baseline Control

In this section, full state feedback model reference adaptive control (MRAC), u_{ad}, is used to augment the baseline robust servomechanism control in Equation 15.126, $u = u_{bl} + u_{ad}$, to improve stability and command tracking in the presence of system uncertainties. The open-loop system to be used in forming the adaptive control is

$$\dot{x} = \tilde{A}x + \tilde{B}(u + f(x)) + Fr \tag{15.136}$$

where $f(x) \in \mathbb{R}^{n_p}$ describes matched uncertainties, possibly nonlinear, in the dynamics (this will be expanded and discussed in the next section). The baseline control using the robust servomechanism from Equation 15.126 is calculated ignoring these uncertainties. The baseline control takes the form

$$u_{bl} = -K_c x. \tag{15.137}$$

Substituting the baseline control into the open loop dynamics, Equation 15.136 yields the following closed loop system:

$$\dot{x} = (\tilde{A} - \tilde{B}K_c)x + \tilde{B}f(x) + Fr. \tag{15.138}$$

Without the uncertainties, that is, when $f(x) = 0$, the dynamics in Equation 15.138 coincides with the model:

$$\dot{x} = \underbrace{(\tilde{A} - \tilde{B}K_c)}_{A_{ref}} x + \underbrace{(F)}_{B_{ref}} r. \tag{15.139}$$

These system dynamics, (A_{ref}, B_{ref}), yield the desired stability and command-following performance that is sought for the autopilot. As such, we choose the reference model for the MRAC using these dynamics:

$$\dot{x}_{ref} = A_{ref} x_{ref} + B_{ref} r. \tag{15.140}$$

As is the baseline control, these reference model dynamics are gain scheduled with flight condition.

There are different feedback/feed-forward architectures that can be used to define the MRAC increment u_{ad} that will augment the baseline control. Here, we choose a form that contains a state feedback term and a term that compensates for the matched system uncertainties $f(x)$. Using this form, the control input u becomes

$$u = u_{bl} + u_{ad} = \underbrace{-K_c x}_{u_{bl}} + \underbrace{\hat{k}_x^T x - \hat{f}(x)}_{u_{ad}} \cdot \tag{15.141}$$

In Equation 15.141, $\hat{k}_x$ is the incremental adaptive feedback gain, and $\hat{f}(x)$ is the online approximator of the matched system uncertainty $f(x)$. A feedforward neural network (NN) with N_0 fixed radial basis functions (RBFs) in its hidden inner layer is used as the function approximation mechanism:

$$f(x) = \sum_{i=1}^{N_0} \theta_i \phi_i(x) + \varepsilon_f(x) = \Theta^T \Phi(x) + \varepsilon_f(x). \tag{15.142}$$

It is well known [18] that for a sufficiently large number of RBF neurons N_0, there exists an ideal outer-layer RBF NN weights matrix Θ, which provides function approximation on a compact n_p-dimensional x domain $X_p \subset \mathbb{R}^{n_p}$, within the approximation tolerance ε_0^*:

$$\left\| f(x) - \Theta \Phi(x) \right\| \le \left\| \varepsilon_f(x) \right\| \le \varepsilon_0^*, \quad \forall x \in X_p. \tag{15.143}$$

Since the matrix Θ in Equation 15.142 is not known, an online estimate, $\hat{\Theta}$, is used. Thus, the corresponding function approximation error becomes

$$f(x) - \hat{f}(x) = \Theta^T \Phi(x) + \varepsilon_f(x) - \hat{\Theta}^T \Phi(x) = (\Theta - \hat{\Theta})^T \Phi(x) + \varepsilon_f(x). \tag{15.144}$$

Substituting the total control into Equation 15.136 yields

$$\dot{x} = \tilde{A}x + \tilde{B}\left(-K_c x + \hat{k}_x^T x - \hat{f}(x) + f(x)\right) + Fr$$

$$= \left(\tilde{A} - \tilde{B}K_c + \tilde{B}\hat{k}_x^T\right)x + Fr + \tilde{B}(\Theta^T \Phi(x) + \varepsilon_f(x) - \hat{\Theta}^T \Phi(x)) \tag{15.145}$$

$$= \left(\tilde{A} - \tilde{B}\left(K_c - \hat{k}_x^T\right)\right)x + Fr + \tilde{B}((\Theta - \hat{\Theta})^T \Phi(x) + \varepsilon_f(x)).$$

To form the adaptive control law, we derive error dynamics between the above closed-loop system and the reference model in Equation 15.140 and establish conditions on the adaptive gains to achieve bounded error tracking for all bounded reference commands r. In this process, we must assume that the matching conditions are satisfied. These conditions require that there exist ideal gains k_x^T, not necessarily known, such that

$$\left(\tilde{A} - \tilde{B}\left(K_c - k_x^T\right)\right) = A_{ref}. \tag{15.146}$$

These conditions are central to how adaptive control enters into the system dynamics and is able to improve system robustness. It is important to establish reference dynamics that can be achieved by the system.

The tracking error between the system and the reference model is defined as $e = x - x_{ref}$. Differentiating and substituting yields

$$\dot{e}(t) = \dot{x}(t) - \dot{x}_{ref}(t)$$

$$= \tilde{A}x + \tilde{B}(u + f(x)) + Fr - A_{ref}x_{ref} - B_{ref}r \pm A_{ref}x$$

$$= \tilde{A}x + \tilde{B}\left(-K_c x + \hat{k}_x^T x - \hat{f}(x) + f(x)\right) + Fr - A_{ref}x_{ref} - B_{ref}r \pm A_{ref}x$$

$$= A_{ref}(x - x_{ref}) + \left(\tilde{A} + \tilde{B}\left(-K_c + \hat{k}_x^T\right) - A_{ref}\right)x + \tilde{B}(\Theta^T \Phi(x) + \varepsilon_f(x) - \hat{\Theta}^T \Phi(x))$$

$$= A_{ref}e + \tilde{B}\left(\left(\hat{k}_x - k_x\right)^T x - (\hat{\Theta} - \Theta)^T \Phi(x) + \varepsilon_f(x)\right)$$

$$= A_{ref}e + \tilde{B}\left(\Delta K_x^T x - \Delta\Theta^T \Phi(x) + \varepsilon_f(x)\right) \qquad (15.147)$$

where $\Delta K_x = \hat{k}_x - k_x$ and $\Delta\Theta = \hat{\Theta} - \Theta$. Note that the estimation error in approximating $f(x)$ with $\hat{f}(x)$ is bounded as long as $x \in X_p$. This requires maintaining the state x within this compact region.

Using a Lyapunov-based design approach coupled with Barbalat's lemma, bounded output tracking can be achieved through an online parameter adaptation process. This requires forming a Lyapunov function and showing that its time derivative is negative within a compact subset of the extended system state space.

Define the candidate Lyapunov function as

$$V(e, \Delta K_x, \Delta\Theta) = e^T P e + \text{trace}\left(\Delta K_x^T \Gamma_x^{-1} \Delta K_x\right) + \text{trace}\left(\Delta\Theta^T \Gamma_\Theta^{-1} \Delta\Theta\right) \qquad (15.148)$$

where P, Γ_x, and Γ_Θ are symmetric positive-definite matrices, and P is the unique solution to the following algebraic Lyapunov equation:

$$PA_{ref} + A_{ref}^T P = -Q. \qquad (15.149)$$

Next, differentiate V to obtain

$$\dot{V} = \dot{e}^T P e + e^T P \dot{e} + 2\,\text{trace}\left(\Delta K_x^T \Gamma_x^{-1} \dot{\hat{k}}_x\right) + 2\,\text{trace}\left(\Delta\Theta^T \Gamma_\Theta^{-1} \dot{\hat{\Theta}}\right)$$

$$= \left(A_{ref}e + \tilde{B}\left(\Delta K_x^T x + \Delta\Theta^T \Phi(x) + \varepsilon_f(x)\right)\right)^T P e$$

$$\quad + e^T P\left(A_{ref}e + \tilde{B}\left(\Delta K_x^T x + \Delta\Theta^T \Phi(x) + \varepsilon_f(x)\right)\right)$$

$$\quad + 2\,\text{trace}\left(\Delta K_x^T \Gamma_x^{-1} \dot{\hat{K}}_x\right) + 2\,\text{trace}\left(\Delta\Theta^T \Gamma_\Theta^{-1} \dot{\hat{\Theta}}\right) \qquad (15.150)$$

$$= e^T (A_{ref} P + PA_{ref})e + 2e^T P\tilde{B}\left(\Delta K_x^T x + \Delta\Theta^T \Phi(x) + \varepsilon_f(x)\right)$$

$$\quad + 2\,\text{trace}\left(\Delta K_x^T \Gamma_x^{-1} \dot{\hat{K}}_x\right) + 2\,\text{trace}\left(\Delta\Theta^T \Gamma_\Theta^{-1} \dot{\hat{\Theta}}\right).$$

Now, using Equation 15.149,

$$\dot{V} = -e^T Q e + 2 e^T P \tilde{B} \varepsilon_f(x) + 2 e^T P \tilde{B} \Delta K_x^T x + 2 \operatorname{trace}\left(\Delta K_x^T \Gamma_x^{-1} \dot{\hat{K}}_x\right)$$
$$+ 2 e^T P \tilde{B} \Delta \Theta^T \Phi(x) + 2 \operatorname{trace}\left(\Delta \Theta^T \Gamma_\Theta^{-1} \dot{\hat{\Theta}}\right). \tag{15.151}$$

Next, use the identity $a^T b = \operatorname{trace}(ba^T)$ to rearrange terms in Equation 15.151 to obtain

$$\dot{V} = -e^T Q e + 2 e^T P B \Lambda \varepsilon_f(x) + 2 \operatorname{trace}\left(\Delta K_x^T \left\{\Gamma_x^{-1} \dot{\hat{K}}_x + x e^T P B\right\}\right)$$
$$+ 2 \operatorname{trace}\left(\Delta \Theta^T \left\{\Gamma_\Theta^{-1} \dot{\hat{\Theta}} + \Phi(x) e^T P B\right\}\right). \tag{15.152}$$

We want this expression to be negative. Suppose we choose expressions for the adaptive parameters $\hat{k}_x$ and $\hat{\Theta}$ as follows:

$$\dot{\hat{k}}_x = -\Gamma_x x e^T P B \tag{15.153}$$
$$\dot{\hat{\Theta}} = -\Gamma_\Theta \Phi(x) e^T P B.$$

Then

$$\dot{V} = -e^T Q e - 2 e^T P \tilde{B} \varepsilon_f(x) \le -\lambda_{\min}(Q)\|e\|^2 + 2\|e\|\|P\tilde{B}\|\varepsilon_0^*. \tag{15.154}$$

This will be negative outside the compact set E given by

$$E = \left\{ e : \|e\| \le \frac{2\|PB\|\varepsilon_0^*}{\lambda_{\min}(Q)} \right\}. \tag{15.155}$$

The key to the implementation of the adaptive laws is the introduction of a projection operator [19] to bound the adaptive parameters:

$$\dot{\hat{k}}_x = \operatorname{Proj}\left(\hat{k}_x, -\Gamma_x x e^T P \tilde{B}\right)$$
$$\dot{\hat{\Theta}} = \operatorname{Proj}\left(\hat{\Theta}, -\Gamma_\Theta \Phi(x) e^T P B\right). \tag{15.156}$$

The bounded adaption parameters combined with Equation 15.155 provide sufficient conditions for bounded tracking outside of the corresponding compact set. Equation 15.155 also defines the size of the tracking error and shows that its upper bound is proportional to the size of the approximation domain ε_0^*. Consequently, in order to make the tracking error smaller, one would need to increase the number of RBFs to decrease the function approximation error, $\varepsilon_{f(x)}$. In addition, a dead zone operation [20] must be added to prevent noise from causing parameter drift. The dead zone will also prevent the adaptive increment from adjusting the baseline control when the model error is small. Damping can also be added to the parameter adaption laws, called σ-modification and e-modification [8]. Lavretsky [21] provides a μ-modification that will prevent undesirable behavior in the presence of position and rate-limited actuation.

15.6 Robustness Analysis of Optimal Baseline Control

Section 15.3.1 showed that the nominal state feedback LQR autopilot provides infinite gain margin and at least a 60° phase margin. These excellent stability margins are guaranteed for complex uncertainties $ke^{j\phi}$ at the plant input but do not necessarily reflect the robustness to uncertainties in the real parameter coefficients that constitute the model of the dynamics. In this section, we introduce real parameter uncertainties and investigate the robustness of the baseline control system to these uncertainties. This analysis plays an important role in the validation and verification of the adaptive flight control system. What one will find is that each control architecture, and gains used in the architecture, possesses different sensitivity to uncertainties in the model parameters. We will also see, via nonlinear simulation, that adaptive control can extend the system's performance and robustness to uncertainties.

It is well known that classical gain and phase margins, including vector margin, do not necessarily mean that the system is robust to real parameter uncertainties. The state feedback system in Equation 15.127 has a scalar loop gain at the plant input and is described by $L(s) = K_c(sI - \tilde{A})\tilde{B}$. A Nyquist plot of $L(s)$ is shown in Figure 15.30 and displays the excellent gain margin and phase margin of the LQR control. In transfer function form

$$L(s) = \frac{(s_i + 40.8 \pm 54.4j)(s_i + 7.5 \pm 5.1j)}{s(s + 7.6)(s - 6.3)(s_i + 40.8 \pm 54.4j)} = \frac{(s_i + 7.5 \pm 5.1j)}{s(s + 7.6)(s - 6.3)} \tag{15.157}$$

note that the open-loop system is unstable ($M_\alpha > 0$), and the actuator poles are exactly cancelled in $L(s)$ (at the plant input only). The gain margin is $[-9,+\infty]$ db and the phase margin ±60. Next, consider a scalar real uncertainty δ_K at the input to the plant as shown in Figure 15.31. The closed-loop characteristic polynomial is $s(s + 7.6)(s - 6.3) + \delta_K(s_i + 7.5 \pm 5.1j)$, with Figure 15.32 displaying

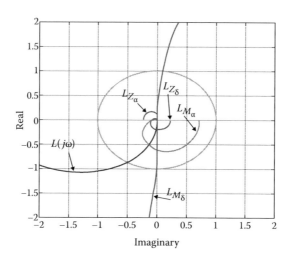

FIGURE 15.30
Nyquist plot of $L(s)$ and loop gains with loop broken at parameter uncertainties.

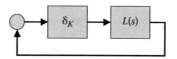

FIGURE 15.31
Scalar uncertainty inserted at plant input.

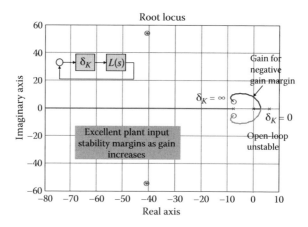

FIGURE 15.32
Root locus varying δ_K in Figure 15.31.

the root locus. The system is stable for all gain values $\delta_K > 0.28$. This equates to a gain margin of $(-11, +\infty)$ dB. This root locus displays the excellent plant input stability margins of the baseline LQR control.

15.6.1 Robustness Analysis Model

Consider real parameter uncertainties δ_i, $\delta_i \in D_i \subset \mathbb{R}$, $i = 1, ..., n_p$, and define $\Delta = \text{diag}[\delta_i]$. A model of the closed-loop system is shown in Figure 15.33. Let $D = D_1 \times D_2 \times ...D_{n_p}$ denote the set of admissible parameter uncertainties, $\Delta \in D$. The design goal is to have the baseline + adaptive controls provide robust stability and command tracking for real parameter uncertainties contained within this set. The size of D is determined by analyzing the robustness properties of the system using the baseline LQR control. We would like for the baseline control to provide stability (not necessarily performance) for all $\Delta \in D$. The adaptive increment would then be designed to augment the baseline control and recover the command tracking performance throughout $\Delta \in D$, thus providing robust stability and performance. The analysis task is to determine the size of D in which the baseline control provides stability.

To analyze the effects of uncertainties in the real aerodynamic parameters that constitute the model, the ΔM stability analysis model shown in Figure 15.18 is used. The matrix M describes the nominal stable closed-loop system, and $\Delta \in D$ describes the uncertain real parameters. The uncertainties in the aero parameters in $\tilde{A}_{cl}$ are modeled using $p_i = p_{i0}(1 + \delta_i)$. Uncertainties in the four aerodynamic stability derivatives Z_α, Z_δ, M_α, and M_δ are considered. Thus, $\delta_i \in D_i \subset \mathbb{R}$, $i = 1, ..., 4$, and $D = D_1 \times ... \times D_4$. A state space analysis model for $M(s)$ is formed by using the method of Morton and McAfoos [11] and Morton [12]. The closed-loop system is written isolating the uncertainties as follows:

$$\tilde{A}_{cl} = A_0 + \sum_{i=1}^{n_p} E_i \delta_i \tag{15.158}$$

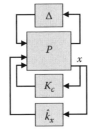

FIGURE 15.33
Closed-loop system with uncertainties.

where

$$A_0 = \begin{bmatrix} 0 & 1 & 0 & 0 & 0 \\ 0 & \dfrac{Z_\alpha}{V} & 1 & \dfrac{Z_\delta}{V} & 0 \\ 0 & M_\alpha & 0 & M_\delta & 0 \\ 0 & 0 & 0 & 0 & 1 \\ -\omega_a^2 k_1 & -\omega_a^2 k_2 & -\omega_a^2 k_3 & -\omega_a^2(1+\omega_a^2 k_4) & -2\zeta_a\omega_a - \omega_a^2 k_5 \end{bmatrix} ;$$

$$E_1 = \begin{bmatrix} 0 & 0 & 0 & 0 & 0 \\ 0 & \dfrac{Z_\alpha}{V} & 0 & 0 & 0 \\ 0 & 0 & 0 & 0 & 0 \\ 0 & 0 & 0 & 0 & 0 \\ 0 & 0 & 0 & 0 & 0 \end{bmatrix} ; E_2 = \begin{bmatrix} 0 & 0 & 0 & 0 & 0 \\ 0 & 0 & 0 & \dfrac{Z_\delta}{V} & 0 \\ 0 & 0 & 0 & 0 & 0 \\ 0 & 0 & 0 & 0 & 0 \\ 0 & 0 & 0 & 0 & 0 \end{bmatrix} ; E_3 = \begin{bmatrix} 0 & 0 & 0 & 0 & 0 \\ 0 & 0 & 0 & 0 & 0 \\ 0 & M_\alpha & 0 & 0 & 0 \\ 0 & 0 & 0 & 0 & 0 \\ 0 & 0 & 0 & 0 & 0 \end{bmatrix} ;$$

$$E_4 = \begin{bmatrix} 0 & 0 & 0 & 0 & 0 \\ 0 & 0 & 0 & 0 & 0 \\ 0 & 0 & 0 & M_\delta & 0 \\ 0 & 0 & 0 & 0 & 0 \\ 0 & 0 & 0 & 0 & 0 \end{bmatrix} ;$$

where δ_1 is the uncertainty in $\dfrac{Z_\alpha}{V}$, δ_2 is the uncertainty in $\dfrac{Z_\delta}{V}$, δ_3 is the uncertainty in M_α, and δ_4 is the uncertainty in M_δ.

When using state space models to analyze parameter uncertainties, the rank of the matrix E_i is used to describe the perturbation. The four parameters here are all rank 1 perturbations. Each multiplicative uncertainty δ_i represents a percentage variation in a parameter. A state space triple (A_M, B_M, C_M) is formed for the matrix M by decomposing the matrices E_i using an SVD, $E_i = U\Sigma V^*$. For the nonzero singular values in Σ (if rank 1, then there is only one), the columns of B_M are formed using the singular vector(s) from U, and the rows of C_M are formed from the singular vector(s) in V. Each matrix E_i is written as

$$\underbrace{U\Sigma^{\frac{1}{2}}}_{b_i} I_{k_i} \underbrace{\Sigma^{\frac{1}{2}}V^*}_{c_i} .$$
(15.159)

The closed-loop matrix becomes

$$A_{cl} = A_0 + \sum_{i=1}^{n_p} b_i \delta_i c_i.$$
(15.160)

Next, consider breaking the loop at the uncertainty. The resulting model is

$$\dot{x} = A_0 x + \sum_{i=1}^{n} b_i u_i \tag{15.161}$$

$$y_i = c_i x.$$

The loop can be closed with

$$u_i = \delta_i y_i. \tag{15.162}$$

This results in

$$\dot{x} = A_0 x + \sum_{i=1}^{n} b_i \delta_i c_i x = \left(A_0 + \sum_{i=1}^{n} b_i \delta_i c_i \right) x = A_{cl} x. \tag{15.163}$$

The state space triple for M is given by

$$A_M = A_0 \quad B_M = [b_1 \quad \cdots \quad b_n] \quad C_M = \begin{bmatrix} c_1 \\ \vdots \\ c_n \end{bmatrix} \quad M(s) = C_M (sI - A_M)^{-1} B_M. \tag{15.164}$$

First, consider each aerodynamic uncertainty δ_i individually. In this case, $M(j\omega)$ in Figure 15.18 is a scalar. Figure 15.30 shows the Nyquist plot for the loop gain, denoted $L_{Z_\alpha}, L_{Z_\delta}, L_{M_\alpha}$, and L_{M_δ}, respectively. Since each of these is a scalar, a simple root locus for each parameter can be computed to demonstrate system sensitivity to that parameter variation. Figure 15.34 illustrates the positive gain root locus plots for each of the four aero coefficients, highlighting the gain at which they cause the system to go unstable. Analyzing each uncertainty individually shows that the system is not overly sensitive. This will not be the case when all four are analyzed simultaneously.

15.6.2 Real Margin Analysis

The real margin analysis of Section 15.3.4 is useful in quantifying the controller's sensitivity to uncertainties in the aerodynamics. Using the model in Equation 15.164 that incorporates the baseline LQR control, the real margin is computed versus frequency. The minimum of k defines the real margin k_m. Figure 15.35 shows a plot of k versus frequency, with the minimum $k_m = 0.4996$ occurring at $\omega = 8.7041$ rad/s. This represents a multiplicative variation of 50% in each of the four parameters. The admissible parameter

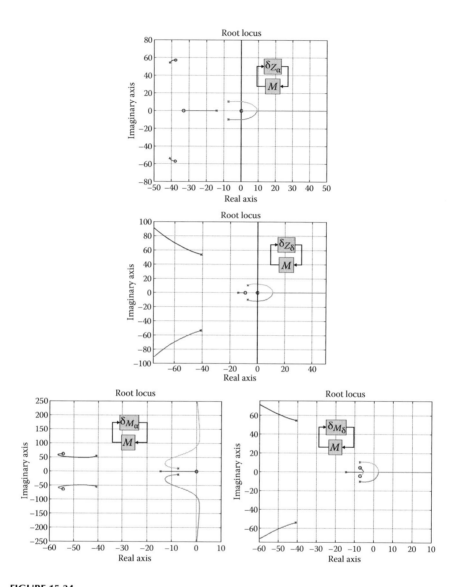

FIGURE 15.34
Root locus plots varying each aerodynamic uncertainty as scalar.

uncertainties are a hypercube defined with unit vertices scaled by $k_m =$ 0.4996. The convex hull co{det[$I + kVM$]} formed by the vertices is shown in Figure 15.36. It is evident from the figure that the hypercube vertex 14 is on the origin and represents the smallest destabilizing aero uncertainties. That vertex is defined as

$$\Delta_0 = k_m \mathrm{diag}[-1 \quad -1 \quad 1 \quad -1]. \tag{15.165}$$

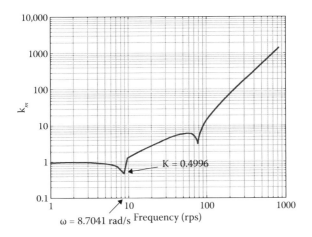

FIGURE 15.35
Real margin k_m versus frequency.

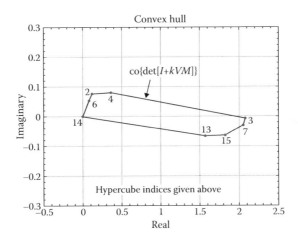

FIGURE 15.36
Convex hull formed by critical vertices.

If we perturb the aerodynamics using Equation 15.165 and substitute the uncertainties into the $\tilde{A}$ in Equation 15.127, the resulting eigenvalues become $\lambda_1 = -11.4183$, $\lambda_{2,3} = -49.1306 \pm 60.2178i$, and $\lambda_{4,5} = \pm8.7046i$, indicating loss of stability.

15.6.2.1 *Comment*

Our robust stability analysis of the LQR state feedback autopilot indicates that the excellent plant input stability margins provided by the optimal control law do not provide large robustness margins to uncertainties in the aerodynamic parameters. The reason for this can be seen and understood from Equation 15.157 and the root locus plots examining each uncertainty individually. At the plant input, Equation 15.157 and Figure 15.32 show desirable zero dynamics for L and how the root sensitivity evolves to an increase in the loop gain. In comparison, Figure 15.34 shows that the zero dynamics for a system under the aerodynamic uncertainties is significantly different. These zero dynamics are the numerator polynomials for the transfer functions from $M(s) = C_M(sI - A_M)^{-1}B_M$ and are determined by how the aerodynamic parameters enter into the dynamics. Since the closed-loop matrix A_M contains the baseline feedback gains, these zero dynamics and the resulting real margin are influenced by the robust servomechanism architecture and feedback gains. However, there is nothing in our design process that allows us to achieve the excellent command-following performance and input stability margins and then shape these zero dynamics (independently) to make the system more robust.

15.7 Robust Stability Analysis Using Nonlinear Simulation

In this section, simulation results will be used to demonstrate the robustness to matched and unmatched uncertainties. The stability analysis results from the previous section will be used to investigate the benefits that adaptive control introduces into the closed-loop system. In Section 15.4.2, the baseline control was designed to stabilize and track commands using integral control. Under no uncertainty this baseline control works very well. Even though the baseline LQR control provides infinite positive gain margin at the plant input, the robust stability analysis of Section 15.6 showed that a relatively small (50%) multiplicative perturbation on the four primary aero coefficients could destabilize the baseline control.

The MRAC in Section 15.5 was designed to compensate for matched uncertainties, $f(x)$, that enter into the model as follows:

$$\dot{x} = \tilde{A}x + \tilde{B}(u + f(x)) + Fr. \tag{15.166}$$

These matched uncertainties can be a nonlinear state-dependent function. For simulation purposes, assume that the matched uncertainty is a linear function of the state and cancels the baseline control, that is, $f(x) = K_c x$. The open-loop system is unstable, so under this matched uncertainty, the adaptive control has to provide stabilization and command tracking. Figure 15.37 shows the nonlinear simulation response to a step $3°$ AOA command varying the adaptive learning rate where $\Gamma_x = \Gamma_0$ from Equation 15.153 with $Q = \text{diag}[0 \ 1 \ 1 \ 0 \ 0]$ in Equation 15.149. We see that as the adaptive

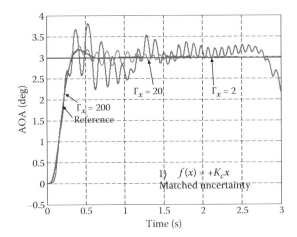

FIGURE 15.37
Nonlinear simulation testing robustness to matched uncertainties.

learning rate increases, the adaptive response significantly improves and cancels the matched uncertainty.

Simulation testing has shown that the adaptive control algorithms are very good at cancelling matched uncertainties. This is guaranteed by the MRAC theory.

Next, consider no matched uncertainty, $f(x) = 0$, but introduce nonmatched parametric uncertainties Δ_0 as analyzed in Section 15.3. These uncertainties enter into the model as

$$\dot{x} = \left(\tilde{A} + \Delta_0 \right) x + \tilde{B}(u + f(x)) + Fr \tag{15.167}$$

which we see are not matched with the control. If we introduce $\Delta_0 = 0.5 \, \text{diag}[-1 \quad -1 \quad 1 \quad -1]$, we know from Section 15.6.2 that the system under baseline control has two poles on the $j\omega$-axis. Figure 15.38 shows the system response with no matched uncertainty, $f(x) = 0$, with no adaptive control and with adaptive control varying the learning rate with $\Gamma_x = \Gamma_0$. Under no control, the system is an oscillator. We see that for small learning rates, the response is unacceptable. For a very large learning rate (probably larger than can actually be used), the system stabilizes with an offset that is slowly converging to the command. (The simulation was run for 40 s to determine that it was in fact converging to the command, albeit slowly.) This uncertainty model is not adequately compensated by the MRAC augmentation to the baseline control. We know that nonmatched uncertainties are not guaranteed to be cancelled or compensated by the MRAC theory. The uncertainties must be in the form where

$$\Delta = \tilde{B}f(x) \tag{15.168}$$

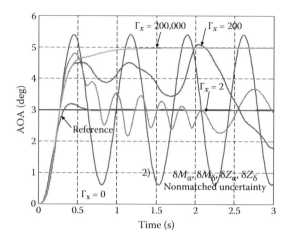

FIGURE 15.38
Nonlinear simulation testing robustness to nonmatched uncertainties.

where $f(x)$ is some function, possibly nonlinear, modeling the uncertainty. The uncertainties that can be compensated must influence the $\tilde{A}$ matrix only in the bottom row of the matrix, as shown below:

$$\tilde{A} = \begin{bmatrix} 0 & 1 & 0 & 0 & 0 \\ 0 & \dfrac{Z_\alpha}{V} & 1 & \dfrac{Z_\delta}{V} & 0 \\ 0 & M_\alpha & 0 & M_\delta & 0 \\ 0 & 0 & 0 & 0 & 1 \\ \times & \times & \times & \times & \times \end{bmatrix} ; \quad \tilde{B} = \begin{bmatrix} 0 \\ 0 \\ 0 \\ 0 \\ \omega^2 \end{bmatrix}. \qquad (15.169)$$

This is due to the fact that $\tilde{B}$ is zero except for the last element. For example, let $\tilde{A} + \Delta$ be of the form

$$A = \begin{bmatrix} 0 & 1 & 0 & 0 & 0 \\ 0 & \dfrac{Z_\alpha}{V} & 1 & \dfrac{Z_\delta}{V} & 0 \\ 0 & M_\alpha & 0 & M_\delta & 0 \\ 0 & 0 & 0 & 0 & 1 \\ 6 & 9 & 200 & 6 & 12 \end{bmatrix} \qquad (15.170)$$

where different parametric uncertainties were added to each column of $\tilde{A}$, but each was parallel to $\tilde{B}$. Figure 15.39 shows the nonlinear simulation results using $\Gamma_x = \Gamma_0 = 200$ and larger. We see that the response accurately tracks the reference command.

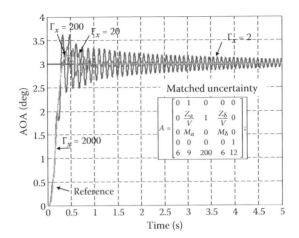

FIGURE 15.39
Nonlinear simulation testing robustness to matched uncertainties.

Our simulation analysis has shown that the MRAC augmentation provides robustness to matched uncertainties. The MRAC was able to stabilize the nonmatched parametric uncertainties, which is still a benefit, but produced a steady-state error in tracking the command. It improves the system's ability to compensate for uncertainties, but these uncertainties must be matched. The matching condition that must be satisfied if one is to achieve stability and command performance requires that the uncertainties lie within the range space of $\tilde{B}$. There are other methods, like backstepping, that allow one to compensate for nonmatched uncertainties. In addition, L1 adaptive control has also demonstrated this property.

Our results have also demonstrated that tuning is required to achieve a desired response. The MRAC has the Lyapunov matrix Q, the learning rates Γ_x and Γ_θ, and the RBFs $\Phi(x)$ as tuning knobs. Each of these must be properly set to achieve the desired response and will need to be scheduled with flight condition.

15.8 Summary and Conclusions

This chapter described a missile autopilot design that uses a baseline control, designed using an LQR and augmented with an adaptive increment to extend the system's robustness to uncertainties. The baseline control was shown to possess excellent plant input gain and phase margins but did not provide large margins to parametric uncertainties introduced in the system A matrix, as one would desire. The adaptive control was found to provide excellent robustness to matched uncertainties, as expected.

In the introduction, several questions were posed. We now address these questions within the context of missile autopilot design based upon the results presented in the paper.

What makes a control system (missile autopilot) design robust? From a linear system robust control perspective, at every loop break point in the architecture where uncertainties are to be injected, the system needs to be positive real. This property will provide the desired gain and phase margins to the uncertainties. The author is not aware of any control design method that can achieve this. Thus, stability robustness and command tracking performance must be analyzed.

Is it adequate to measure robustness using plant input gain margin and phase margin? No. This should not be the *only* analysis method used to assess robustness. As shown here, small parametric uncertainties can destabilize a system that has excellent plant input and output stability margins. Practitioners know that they must check margins at both the plant input and output (all control and sensor channels). To understand how sensitive the design is to parametric uncertainties, or unmodeled dynamics, the appropriate analysis model must be analyzed.

Do classical-type margins provide adequate robustness measures to uncertain possibly nonlinear aerodynamics and/or mismodeled/neglected high-frequency dynamics? No, small uncertainties can destabilize a system that has excellent classical margins. Using a vector margin derived from the return difference dynamics does give improved analysis information. Simulation analysis must be used when addressing nonlinear aerodynamics.

How does the choice of the feedback control law influence the robustness? Results show that the zero dynamics play a key role in the robustness properties. As the architecture is changed, these dynamics will also change.

How does one achieve high performance and still be robust using linear, model-based design approaches? Using models (A, B) to design the control law makes the design dependent upon how accurately the model is known. As gains are increased to get faster performance, the model accuracy must increase. By augmenting these approaches with adaptive increments, the system robustness and performance can be extended.

Can adaptive control theory be used in some way to maintain performance and provide robustness? Adaptive control extends the robustness and can address a bigger class of uncertainties, as compared with a linear control. Clearly, by combining the best from linear system theory and nonlinear and adaptive control theory, one can obtain a design that will achieve both performance and robustness.

References

1. Doyle, J.C., Glover, K., Khargonekar, P., and Francis, B., State-space solutions to standard H2 and H∞ control problems, *IEEE Transactions on Automatic Control*, AC-34(8), August 1989, 831–847.
2. Reichert, R.T., Robust autopilot design using μ-synthesis, *Proceedings of the American Control Conference*, San Diego, CA, May 1990, IEEE, Piscataway, NY.
3. Wise, K.A., Mears, B.C., and Poolla, K., Missile autopilot design using H∞ optimal control with μ-synthesis, *Proceedings of the American Control Conference*, San Diego, CA, May 1990, IEEE, Piscataway, NY.
4. Bibel, J.E. and Stalford, H.L., Mu-synthesis autopilot design for a flexible missile, *Proceedings of the Aerospace Sciences Mtg*, Reno, NV, AIAA-91-0586, January 1991, American Institute of Aeronautics and Astronautics, Reston, VA.
5. Kwakernak, H. and Sivan, R., *Linear Optimal Control*, Wiley, New York, 1971.
6. Williams, D.E., Madiwale, A.N., and Freidland, B., *Modern Control Theory For Autopilots*, Air Force Armament Lab., Eglin AFB, FL, AFATL-TR-85-10, June 1985.
7. Wise, K.A., A trade study on missile autopilot design using optimal control theory, *Proceedings of the AIAA Guidance, Navigation and Contr. Conf.*, Hilton Head Island, SC, AIAA-2007-6673, 2007, American Institute of Aeronautics and Astronautics, Reston, VA.
8. Narendra, K.S., and Annaswamy, A.M., *Stable Adaptive Systems*, Dover, Mineola, NY, 2005.
9. Wise, K.A., Lavretsky, E., and Hovakimyan, N., Adaptive Control of Flight: Theory, Applications, and Open Problems, 2006 ACC, Minneapolis, MN, June 2006. Also presented at the Thirteenth Yale Workshop on Adaptive and Learning Systems on May 30–June 1 at Becton Center, Yale University, New Haven, CT, 2005.
10. Wise, K.A., Lavretsky, E., Zimmerman, J., Francis-Jr., J., Dixon, D., and Whitehead, B.T., Adaptive Control of a Sensor Guided Munition, 2005 AIAA GNC Conference, San Francisco, CA, August 2005.
11. Morton, B.G. and McAfoos, R.M., A Mu-test for robustness analysis of a real parameter variation problem, *Proceedings of the American Control Conference*, 1985, IEEE, Piscataway, NY.
12. Morton, B.G., New applications of mu to real parameter variation problems, *Proceedings of the 25th IEEE Conference on Decision and Control*, 1985, IEEE, Piscataway, NY.
13. Wise, K.A., Missile autopilot robustness using the real multiloop stability margin, *Journal of Guidance, Control, and Dynamics*, 16(2), 1993, 354–362.
14. Wise, K.A., Singular value robustness tests for missile autopilot uncertainties, *Journal of Guidance, Control, and Dynamics*, 14(3), 1991, 597–606.
15. Nesline, F.W. and Zarchan, P., Why modern controllers go unstable in practice, *Journal of Guidance, Control, and Dynamics*, 7(4), 1984, 495–500.
16. Bryson, Jr., A.E. and Ho, Y., *Applied Optimal Control: Optimization, Estimation, and Control*, Blaisdell, Waltham, MA, 1969.
17. Wise, K.A., Bank-to-turn missile autopilot design using loop transfer recovery, *Journal of Guidance, Control, and Dynamics*, 13(1), 1990, 145–152.

18. Haykin, S., *Neural Networks: A Comprehensive Foundation*, 2nd edition, Prentice Hall, New York, NY, 1999.
19. Pomet, J.B. and Praly, L., Adaptive nonlinear regulation: Estimation from Lyapunov equation, *IEEE Transactions on Automatic Control*, 37(6), 1992, 729–740.
20. Ioannou, P. A. and Fidan, B., *Adaptive Control Tutorial*, SIAM, Philadelphia, PA, 2006.
21. Lavretsky, E. and Hovakimyan, N., Positive μ-modification for stable adaptation in a class of nonlinear systems with actuator constraints, *Proceedings of American Control Conference*, Boston, MA, Institute of Electronics and Electrical Engineers, Piscataway, NJ, June 2004.
22. Douglas, R., Mackler, S., and Speyer, J., Robust hover control for a short takeoff/vertical landing aircraft, *Proceedings of the AIAA GNC Conference*, pp. 150–163, Portland OR, American Institute of Aeronautics and Astronautics, Reston, VA, August 1990.

Index

Page numbers followed by *f* indicate figures.

Printed and bound by CPI Group (UK) Ltd, Croydon, CR0 4YY

23/10/2024

01778367-0001